MicroRNA and Cancer

MicroRNA and Cancer

Editor

Paola Tucci

MDPI • Basel • Beijing • Wuhan • Barcelona • Belgrade • Manchester • Tokyo • Cluj • Tianjin

Editor
Paola Tucci
Department of Pharmacy
Health and Nutritional
Sciences
University of Calabria
Rende
Italy

Editorial Office
MDPI
St. Alban-Anlage 66
4052 Basel, Switzerland

This is a reprint of articles from the Special Issue published online in the open access journal *Cancers* (ISSN 2072-6694) (available at: www.mdpi.com/journal/cancers/special_issues/MaC).

For citation purposes, cite each article independently as indicated on the article page online and as indicated below:

LastName, A.A.; LastName, B.B.; LastName, C.C. Article Title. *Journal Name* **Year**, *Volume Number*, Page Range.

ISBN 978-3-0365-4416-8 (Hbk)
ISBN 978-3-0365-4415-1 (PDF)

About the Editor

Paola Tucci

Prof. Paola Tucci is an Associate Professor of Biochemistry at the University of Calabria in Italy, Department of Pharmacy, Health and Nutritional Science.

She achieved her Degree cum Laude in 2003 at the University of Calabria, followed by a PhD in "Pharmacology and Biochemistry of Cell Death" (2007, University of Calabria), and a Medical Specialization in "Clinical Biochemistry" cum Laude (2017, University of Rome "Tor Vergata"). She worked as a Post-Doctoral Research Fellow at the Medical Research Council (MRC) Toxicology Unit, University of Cambridge (UK), and later as a Research Assistant at the University of Rome "Tor Vergata". In 2011 she returned to Italy, at the University of Calabria, first as an Assistant Professor of Biochemistry and later as an Associate Professor, position she currently holds.

She has published over 40 peer-reviewed papers in high-level international scientific journals, such as *Blood*, *Cell Death and Differentiation*, *PNAS* and *Oncogene*, and she has been invited to speak at several international meetings. In 2013, she was awarded the "Bioeconomy Rome 2013" prize at the Accademia Nazionale dei Lincei in Rome, for her research in the field of translational medicine in neurodegeneration.

Her main work focuses on the regulation of microRNAs by members of the p53 family, p63 and p73, especially in tumorigenesis, metabolism and in the neuronal development, and more recently on biological evaluations of molecules of natural origin, and of delivery systems of these molecules, on normal and tumoral cell lines.

Editorial

The Role of microRNAs in Cancer: Functions, Biomarkers and Therapeutics

Paola Tucci

Department of Pharmacy, Health and Nutritional Sciences, University of Calabria, 87036 Rende, Italy; paola.tucci@unical.it; Tel.: +39-0984493185

MicroRNAs (miRs) are small non-coding RNAs acting as post-transcriptional regulators of gene expression with important roles in almost all biological pathways, including development, differentiation, cell cycle, proliferation, and apoptosis. Deregulated miR expression has been detected in numerous cancers, where miRs act as both oncogene and tumor suppressors. Considering their important roles in tumorigenesis, miRs have been investigated as prognostic and diagnostic biomarkers and as useful targets for therapeutic intervention.

This Special Issue of *Cancers* focuses on the identification and characterization of new miR targets involved in cancer pathogenesis and on their role as potential biomarkers for early detection and diagnosis, as well as for the development of new cancer therapy. Some more recent and exciting advances in the field are collected and presented here, providing new ideas for discussion of future perspectives among researchers working on this hot topic.

The review by Katsaraki et al. [1] summarizes and highlights the current knowledge concerning the multifaceted role of miRs both in normal B-cell development and B-cell chronic lymphocytic leukemia progression, prognosis and therapy.

Recent studies revealed differences in the miR expression profiles in tissues from patients with ovarian cancer and healthy individuals. For example, the expression of miR-200a, miR-200b, and miR-200c was significantly higher than that in normal tissues, whereas mir-199a, miR-140, miR-145, and miR-125b1 displayed low expression in ovarian cancer tissues. A potential explanation for a global decrease in miR expression may be inhibition of DICER, as shown by Wilczynski et al. [2]. Down-regulation of DICER has been detected in epithelial ovarian cancer and was associated with the up-regulation of the oncogenic miRNA-103/107. Although the results of their study do not highlight any clinical or prognostic role of the miRs, the miR-103/miR-107/DICER axis may be one of the key regulators of cancer aggressiveness [2].

In their paper, Jasinski-Bergner et al. [3] describe for the first time the impact and relevance of factors involved in 2′-O-methylation and pseudouridylation of different RNA species, such as miRs, related to the processes of tumor formation and progression in a malignant melanoma model. The up-regulation of the RNA modifying proteins is, indeed, a prognostic factor in this tumor and the impact of these molecules on miRs would lead to the identification of new proteins involved in the miRs deregulation, thus suggesting that both RNA factors and miRs involved in this process, represent suitable targets for tumor therapy and putative novel prognostic markers [3].

MiRs are not only able to distinguish normal tissues from tumor, but can also distinguish and characterize the different tumoral subgroups. A few studies have dedicated their attention to the detection and monitoring of renal cancer, one of the most common cancers worldwide with a nearly non-symptomatic course until the advanced stages of the disease. Kajdasz et al. [4], through a meta-analysis study, investigate and validate the changes in miRs expression in renal cancer patients. Furthermore, deregulated miRs are

Citation: Tucci, P. The Role of microRNAs in Cancer: Functions, Biomarkers and Therapeutics. *Cancers* **2022**, *14*, 872. https://doi.org/10.3390/cancers14040872

Received: 3 February 2022
Accepted: 7 February 2022
Published: 10 February 2022

differentially expressed in the different renal cancer subtypes and can be used to distinguish the subclass [4].

In case of medulloblastoma, the most common malignant brain tumor in children, studies have mostly concentrated on the clinical entity of the single disease rather than in the four molecular subgroups [5]. Each subgroup has a different cell of origin, prognosis, and may require specific therapeutic strategies. The review by Bevacqua et al. [5] summarizes the role of miRs in the four medulloblastoma subgroups, their potential as biomarkers for early diagnosis and prognosis, highlighting the potential of these miRs in providing new opportunities to treat the different clinical and biological features between subgroups.

The role of various miRs was investigated in relation to metabolic disorders, as suggested by Sidorkiewicz et al., showing the alterations in miR profiles in endometrial cancer patients with insulin resistance [6].

On the other hand, another study by Lee et al. [7] links the expression of miRs to the effects of cigarette smoking, a major risk factor of lung cancer by inducing DNA methylation. Regarding this, the authors found that miR-584-5p expression was down-regulated by the methylation and this resulted in increased migration and invasion in smoking-related lung cancer cells by targeting the oncogenic protein YKT6 [7]. Thus, the tumor suppressor miR-584-5p might be used as molecular biomarker for this kind of cancer.

One of the factors contributing to the complexity of tumor growth, metastasis, and patient survival in breast cancer is the level of hypoxia (oxygen deficiency) [8]. To counteract hypoxia, cancerous cells secrete growth factors that facilitate angiogenesis in the tumor microenvironment to deliver the required oxygen and nutrients to tumoral cells, as well as oxidative stress, epithelial to mesenchymal transition, cell migration, and inflammation in cancer. Gervin et al. [8] identify and investigates a novel function of miR-526b and miR-655 in breast cancer, showing that hypoxia enhances oncogenic functions of these miRs in breast cancer cells and promotes the expression of tumor-associated angiogenic marker and tumoral progression.

During carcinogenesis, miRs play important roles in regulating the maintenance and acquisition of cancer stem cells. Two papers discuss on this issue [9,10]. In the first paper, Liao et al. [9] identify in colorectal stem-like cancer cells the microRNA-210-Stathmin1 axis, critical for inducing microtubule destabilization, decreasing cell elasticity and thus facilitate cell motility and metastasis. In the second, Fitriana at al. [10] summarize the latest finding on the role of miRs in regulating cancer stemness with regard to the head and neck cancers, and analyzed them as useful targets for potential clinical application.

As tumor cells can release miRs resistant to digestion by RNases through their encapsulation into microvesicles or binding to lipoproteins, the use of circulating miRs as biomarkers for different cancer types is a rapidly developing area. miRs can be detected in biological fluids, allowing non-invasive diagnosis to discriminate malignant lesions from benign lesions. Gajek et al. [11] summarize the latest findings on the utility of miRs as potential biomarkers for ovarian cancer diagnosis and prognosis as circulating miR profiles reflect the tumor profiles. Furthermore, by modulating the sensitivity of the cancer cells to chemotherapeutic agents they might serve as promising therapeutic for multidrug-resistance ovarian cancer [11].

Giussani et al. [12] analyze circulating miRs in plasma sample from patients enrolled in the clinical study and identified 5 miRs (miR-625, miR-423-5p, miR-370-3p, miR-181c, and miR-301b) that, properly combined, are able to distinguish malignant from benign breast disease in women [12].

Since recent studies have shown that exosomes promote the generation of a metastatic niche by transferring functional molecules, Eun et al. [13] investigated the role of circulating exosomal miRs in cancer metastasis, and found that the exo-miR-1307-5p was significantly overexpressed in hepatocellular carcinoma and correlated with progression and metastasis in patients with advanced-stage. For a precision treatment strategy, the identification of metastasis driver molecules in blood would help classify patients in accordance with the risk of metastasis during the initial staging process [13].

Finding biomarkers for metastasis is also important for identifying melanoma tumors. The aim of the pilot study by Bustos et al. [14] was to demonstrate the utility of circulating cell-free miRs as potential blood biomarkers for stage III and IV melanoma patients compared to serum lactate dehydrogenase which is currently an accepted biomarker for stage IV, but it has limited utility for stage III melanoma patients. Thus, they identified several miRs suitable for real-time monitoring treatment response of patients with metastatic melanoma [14].

In summary, this Special Issue of *Cancers* is a collection of articles (nine research articles and five reviews articles) discussing the role of miRs in cancer. The identification and characterization of new cancer-relevant miRs may be used to facilitate patient diagnosis and prognosis of different tumors. Moreover, miR profiles can define relevant tumoral subtypes. We can also monitor miR changes to predict therapeutic responses as a non-invasive detection method. Lastly, the importance of miRs in cancer has paved the way for new diagnostic and therapeutic opportunities. Even though substantial questions must be answered, with the advances in in vivo delivery systems, the administration of miR-based therapeutics is feasible and safe in humans, and they could represent a suitable target for the clinical treatment, able to change the medical practice in the foreseeable future.

Conflicts of Interest: The author declares no conflict of interest.

References

1. Katsaraki, K.; Karousi, P.; Artemaki, P.I.; Scorilas, A.; Pappa, V.; Kontos, C.K.; Papageorgiou, S.G. MicroRNAs: Tiny Regulators of Gene Expression with Pivotal Roles in Normal B-Cell Development and B-Cell Chronic Lymphocytic Leukemia. *Cancers* **2021**, *13*, 593. [CrossRef] [PubMed]
2. Wilczynski, M.; Kielbik, M.; Senderowska, D.; Krawczyk, T.; Szymanska, B.; Klink, M.; Malinowski, A. MiRNA-103/107 in Primary High-Grade Serous Ovarian Cancer and Its Clinical Significance. *Cancers* **2020**, *12*, 2680. [CrossRef] [PubMed]
3. Jasinski-Bergner, S.; Blümke, J.; Wickenhauser, C.; Seliger, B. Relevance of 2′-O-Methylation and Pseudouridylation for the Malignant Melanoma. *Cancers* **2021**, *13*, 1167. [CrossRef] [PubMed]
4. Kajdasz, A.; Majer, W.; Kluzek, K.; Sobkowiak, J.; Milecki, T.; Derebecka, N.; Kwias, Z.; Bluyssen, H.A.; Wesoly, J. Identification of RCC Subtype-Specific microRNAs–Meta-Analysis of High-Throughput RCC Tumor microRNA Expression Data. *Cancers* **2021**, *13*, 548. [CrossRef] [PubMed]
5. Bevacqua, E.; Farshchi, J.; Niklison-Chirou, M.V.; Tucci, P. Role of MicroRNAs in the Development and Progression of the Four Medulloblastoma Subgroups. *Cancers* **2021**, *13*, 6323. [CrossRef] [PubMed]
6. Sidorkiewicz, I.; Jóźwik, M.; Niemira, M.; Krętowski, A. Insulin Resistance and Endometrial Cancer: Emerging Role for microRNA. *Cancers* **2020**, *12*, 2559. [CrossRef] [PubMed]
7. Lee, S.B.; Park, Y.S.; Sung, J.S.; Lee, J.W.; Kim, B.; Kim, Y.H. Tumor Suppressor miR-584-5p Inhibits Migration and Invasion in Smoking Related Non-Small Cell Lung Cancer Cells by Targeting YKT6. *Cancers* **2021**, *13*, 1159. [CrossRef] [PubMed]
8. Gervin, E.; Shin, B.; Opperman, R.; Cullen, M.; Feser, R.; Maiti, S.; Majumder, M. Chemically Induced Hypoxia Enhances miRNA Functions in Breast Cancer. *Cancers* **2020**, *12*, 2008. [CrossRef] [PubMed]
9. Liao, T.-T.; Cheng, W.-C.; Yang, C.-Y.; Chen, Y.-Q.; Su, S.-H.; Yeh, T.-Y.; Lan, H.-Y.; Lee, C.-C.; Lin, H.-H.; Lin, C.-C.; et al. The microRNA-210-Stathmin1 Axis Decreases Cell Stiffness to Facilitate the Invasiveness of Colorectal Cancer Stem Cells. *Cancers* **2021**, *13*, 1833. [CrossRef]
10. Fitriana, M.; Hwang, W.-L.; Chan, P.-Y.; Hsueh, T.-Y.; Liao, T.-T. Roles of microRNAs in Regulating Cancer Stemness in Head and Neck Cancers. *Cancers* **2021**, *13*, 1742. [CrossRef] [PubMed]
11. Gajek, A.; Gralewska, P.; Marczak, A.; Rogalska, A. Current Implications of microRNAs in Genome Stability and Stress Responses of Ovarian Cancer. *Cancers* **2021**, *13*, 2690. [CrossRef] [PubMed]
12. Giussani, M.; Ciniselli, C.M.; De Cecco, L.; Lecchi, M.; Dugo, M.; Gargiuli, C.; Mariancini, A.; Mancinelli, E.; Cosentino, G.; Veneroni, S.; et al. Circulating miRNAs as Novel Non-Invasive Biomarkers to Aid the Early Diagnosis of Suspicious Breast Lesions for Which Biopsy Is Recommended. *Cancers* **2021**, *13*, 4028. [CrossRef]
13. Eun, J.W.; Seo, C.W.; Baek, G.O.; Yoon, M.G.; Ahn, H.R.; Son, J.A.; Sung, S.; Kim, D.W.; Kim, S.S.; Cho, H.J.; et al. Circulating Exosomal MicroRNA-1307-5p as a Predictor for Metastasis in Patients with Hepatocellular Carcinoma. *Cancers* **2020**, *12*, 3819. [CrossRef] [PubMed]
14. Bustos, M.A.; Gross, R.; Rahimzadeh, N.; Cole, H.; Tran, L.T.; Tran, K.D.; Takeshima, L.; Stacey LStern, S.L.; O'Day, S.; Hoon, D.S.B. A Pilot Study Comparing the Efficacy of Lactate Dehydrogenase Levels Versus Circulating Cell-Free microRNAs in Monitoring Responses to Checkpoint Inhibitor Immunotherapy in Metastatic Melanoma Patients. *Cancer* **2020**, *12*, 3361. [CrossRef] [PubMed]

Review

MicroRNAs: Tiny Regulators of Gene Expression with Pivotal Roles in Normal B-Cell Development and B-Cell Chronic Lymphocytic Leukemia

Katerina Katsaraki [1], Paraskevi Karousi [1], Pinelopi I. Artemaki [1], Andreas Scorilas [1], Vasiliki Pappa [2], Christos K. Kontos [1,*] and Sotirios G. Papageorgiou [2,*]

1 Department of Biochemistry and Molecular Biology, Faculty of Biology, National and Kapodistrian University of Athens, 15701 Athens, Greece; kkatsaraki@biol.uoa.gr (K.K.); pkarousi@biol.uoa.gr (P.K.); partemaki@biol.uoa.gr (P.I.A.); ascorilas@biol.uoa.gr (A.S.)

2 Second Department of Internal Medicine and Research Unit, University General Hospital "Attikon", 12462 Athens, Greece; vaspappa@med.uoa.gr

* Correspondence: chkontos@biol.uoa.gr (C.K.K.); sotirispapageorgiou@hotmail.com (S.G.P.); Tel.: +30-210-727-4616 (C.K.K.); +30-210-583-2519 (S.G.P.)

Simple Summary: The involvement of miRNAs in physiological cellular processes has been well documented. The development of B cells, which is dictated by a miRNA-transcription factor regulatory network, suggests a typical process partly orchestrated by miRNAs. Besides their contribution in normal hematopoiesis, miRNAs have been severally reported to be implicated in hematological malignancies, a typical example of which is B-cell chronic lymphocytic leukemia (B-CLL). Numerous studies have attempted to highlight the regulatory role of miRNAs in B-CLL or establish some of them as molecular biomarkers or therapeutic targets. Thus, a critical review summarizing the current knowledge concerning the multifaceted role of miRNAs in normal B-cell development and B-CLL progression, prognosis, and therapy, is urgent. Moreover, this review aims to highlight important miRNAs in both normal B-cell development and B-CLL and discuss future perspectives concerning their regulatory potential and establishment in clinical practice.

Citation: Katsaraki, K.; Karousi, P.; Artemaki, P.I.; Scorilas, A.; Pappa, V.; Kontos, C.K.; Papageorgiou, S.G. MicroRNAs: Tiny Regulators of Gene Expression with Pivotal Roles in Normal B-Cell Development and B-Cell Chronic Lymphocytic Leukemia. *Cancers* **2021**, *13*, 593. https://doi.org/10.3390/cancers13040593

Academic Editor: Paola Tucci

Received: 28 December 2020

Accepted: 1 February 2021

Published: 3 February 2021

Abstract: MicroRNAs (miRNAs) represent a class of small non-coding RNAs bearing regulatory potency. The implication of miRNAs in physiological cellular processes has been well documented so far. A typical process orchestrated by miRNAs is the normal B-cell development. A stage-specific expression pattern of miRNAs has been reported in the developmental procedure, as well as interactions with transcription factors that dictate B-cell development. Besides their involvement in normal hematopoiesis, miRNAs are severally implicated in hematological malignancies, a typical paradigm of which is B-cell chronic lymphocytic leukemia (B-CLL). B-CLL is a highly heterogeneous disease characterized by the accumulation of abnormal B cells in blood, bone marrow, lymph nodes, and spleen. Therefore, timely, specific, and sensitive assessment of the malignancy is vital. Several studies have attempted to highlight the remarkable significance of miRNAs as regulators of gene expression, biomarkers for diagnosis, prognosis, progression, and therapy response prediction, as well as molecules with potential therapeutic utility. This review seeks to outline the linkage between miRNA function in normal and malignant hematopoiesis by demonstrating the main benchmarks of the implication of miRNAs in the regulation of normal B-cell development, and to summarize the key findings about their value as regulators, biomarkers, or therapeutic targets in B-CLL.

Keywords: miRNAs; normal B-cell development; B-CLL; miRNA-transcription factor network; regulation; biomarker; therapy; prognosis; diagnosis; progression; prediction

1. Introduction

B-cell chronic lymphocytic leukemia (B-CLL) is the most common type of leukemia in the Western World affecting mainly elders. It accounts for 7% of non-Hodgkin lymphomas and it is classified as a low-grade B-cell non-Hodgkin lymphoma. Biologically, it is characterized by the accumulation of abnormal B cells in the blood, bone marrow, lymph nodes and the spleen. Approximately 80–90% of B-CLL patients have chromosomal abnormalities, with the most common one being in the chromosomal region 13q14.3 [1]. Furthermore, numerous mutated genes have been characterized in the genome of B-CLL patients which occur mainly in the *NOTCH1*, *MYD88*, *TP53*, *ATM*, *SF3B1* and *BIRC3* genes. B-CLL is a highly heterogeneous disease and therefore, timely, specific, and sensitive assessment of this malignancy and its progression is vital. Therefore, in the last decades, several molecular and clinical prognostic markers have been proposed for B-CLL assessment. Age, Binet and Rai staging systems, deletions in chromosomes 11q, 13q, 17p, serum markers such as β2-microglobulin, cell surface markers such as CD38, *IGHV* mutational status, z-associated protein-70 (ZAP70) and lipoprotein lipase (LPL) represent some of these markers [2].

MicroRNAs (miRNAs) are small non-coding RNAs (sncRNAs), with an average length of 22 nucleotides. These sncRNAs are transcribed and processed by Drosha and Dicer enzymes and target mRNAs post-transcriptionally via recruitment of Argonaute (AGO) proteins and therefore, recruitment of the relative RISC complexes to a complementary mRNA target sequence. This recruitment leads to mRNA degradation and/or repression of translation. Furthermore, miRNAs appear as a complex regulatory network as multiple miRNAs target the same mRNA and multiple mRNA sequences are targeted by one specific miRNA. During the last few years, the role of miRNAs has been evaluated in various diseases including cancer. These studies highlighted the remarkable significance of miRNAs as regulators, biomarkers for diagnosis, prognosis, progression and prediction, as well as molecules with therapeutic utility [3]. Unique miRNAs and other factors contributing to B-CLL pathogenesis are summarized in Figure 1.

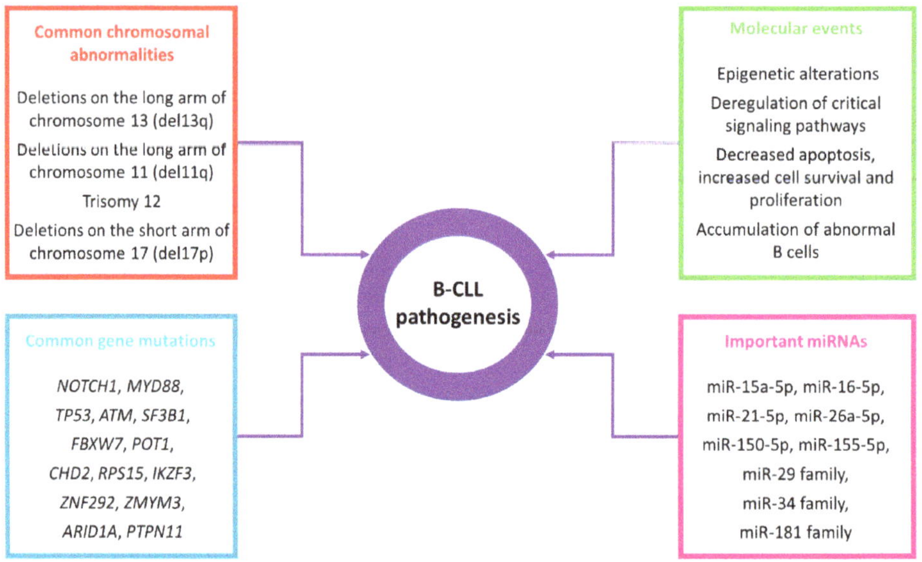

Figure 1. Common genetic alterations, molecular events, and miRNAs contributing to B-CLL pathogenesis.

Until today, various studies delineate the regulatory role of miRNAs in hematological malignancies, as they are critically implicated in B-cell development, proliferation and migration, by affecting numerous pathways including the BCR signaling, the mitogen-activated protein kinase (MAPK/ERK), the phosphatidylinositol 3–kinase (PI3K)/serine/threonine kinase AKT, and the nuclear factor κB (NFkB). Since 2002, when Calin et al. reported a downregulation in the expression levels of miR-15a-5p and miR-16-5p as a result of the deletion located at the chromosome 13q14 [4], which is the most common among B-CLL patients, great progress has been made in order to reveal the role of miRNAs in B-CLL. At present, miRNAs have been characterized as valuable regulators and biomarkers for the assessment of the occurrence and progression of B-CLL as well as evaluators of therapy and promising molecules for therapeutic strategies.

Even though the implication of miRNAs in normal B-cell hematopoiesis, lymphomas and leukemias has been extensively studied [5], scientific knowledge in this context remains limited with only specific miRNAs being studied broadly. Moreover, a review summarizing and highlighting their significant involvement in normal B-cell development and B-CLL is still missing. Therefore, in this review, we aim to highlight the importance of miRNAs in normal B-cell development both in the bone marrow and the periphery and delineate their roles in B-CLL regulation, therapy and disease assessment. Moreover, we propose promising topics for future, targeted research concerning the regulatory potential of miRNAs and their establishment in clinical practice.

2. miRNAs: How Are They Involved in Normal B-Cell Development?

The development of B cells is a multi-step and tightly regulated process. In brief, the developmental procedure consists of the following stages: the developmental phases, that take place in the bone marrow, include the transformation of hematopoietic stem cells (HSCs) to common lymphoid progenitors, pro-B cells, pre-B cells, and immature B cells. These developmental stages are characterized by rearrangements in the Ig genomic loci, called VDJ recombination. This process leads to the formation of Ig heavy chains at the pro-B stage and Ig light chains at the pre-B-cell stage, which compose a transmembrane protein called BCR. Next, the BCR-expressing immature B cells undergo the central tolerance checkpoint, during which those expressing a self-reactive antibody are eliminated. Following, B cells migrate in the spleen where naïve B cells bind to an antigen and differentiate into follicular or marginal zone B cells. This differentiation strongly depends on BCR signals. Next, marginal zone B cells populate the marginal zone, while follicular B cells enter the germinal center, forming three distinct zones: the dark, light, and mantle zone. Subsequently, germinal center B cells differentiate into memory or plasma cells [6].

2.1. miRNAs in Bone Marrow B-Cell Development

miRNAs have been multifariously implicated in the regulation of B-cell development, affecting most of the stages composing this process. The global impact of miRNAs is demonstrated by the fact that the lack of DGCR8, a key molecule for proper miRNA biogenesis, led to elevated apoptosis rate of early B cells [7,8]. Moreover, the expression of each miRNA during B-cell development seems to be stage-specific, as it was uncovered by the study of Spierings et al. [9]. Other studies support this finding as well; miR-181a-5p, miR-150-5p, miR-132-3p, and miR-126-3p were differentially expressed among developmental stages [10,11]; moreover, the ectopic overexpression of miR-181a-5p in common lymphoid progenitors led to an increase in the total number of B cells, indicating its involvement in B-cell development [12]. In contrast, overexpression of a member of the miR-23a cluster, miR-23a-5p, in HSCs led to a decrease in total B-cell number [13].

Besides their involvement in the general context of B-cell development, several miRNAs have been demonstrated to affect specific early stages of the developmental process by interacting with transcription factors crucial for normal B-cell development. The main transcription factors that orchestrate the development of B cells are TCF3, EBF1, and PAX5 [14,15]. TCF3 is required for the initiation and maintenance of the developmental procedure, as well as for the recruitment of EBF1 and PAX5 [16]; both these transcriptional factors are essential for early B-cell differentiation, as they participate in the formation of a functional BCR [15,17]. Interestingly, in absence of EBF1, the developmental process is not abolished, since miR-126-3p was shown to partly rescue B-cell development, by inducing the expression of genes that are required for the process, including *RAG1* and *RAG2*, two recombinases responsible for VDJ recombination [11,18]; this fact comes in line with its overexpression in early developmental stages. Other transcription factors are implicated too, including FOXP1, which controls RAG1 and RAG2 expression [19], and EGR1, a transcriptional regulator essential for B-cell differentiation [20]. In this context, miR-191-5p was shown to target *Foxp1, Tcf3,* and *Egr1* in mice, exerting a potential controversial role in the developmental procedure since both its deletion and overexpression led to inefficient development of the B cells [21]. *Egr1* is also decreased upon miR-146a-5p overexpression in mice, leading to B-cell malignancies [20].

Such a regulatory network, consisting of miRNAs and transcription factors, has been described to arrest the developmental procedure at the pro- to pre-B-cell transition as well; this transition is an important checkpoint in the B-cell development process. Overexpression of miR-150-5p in common lymphoid progenitors, blocked pro- to pre-B-cell transition, due to its binding to *MYB*, a transcription factor whose deletion completely abolishes B-cell development; this fact explains the above-mentioned stage-specific expression of miR-150-5p, as normally it is not detected in early developmental stages [10,22,23]. Similar to miR-150-5p, miR-132-3p overexpression in mouse HSCs blocked the developmental procedure at the pro-B-cell stage by targeting another transcription factor, namely *Sox4*, which is required for the survival of pro-B cells and is involved in Rag1 expression; the expression of miR-132-3p is stage-specific as well and is likely to be induced by BCR signaling, which means that it is normally expressed after the transition stage [24]. These typical examples indicate that the stage-specific expression pattern is not stochastic but strictly linked to each miRNA utility and function in the developmental process.

Additionally, miR-24-3p, which belongs to the previously mentioned miR-23a cluster, had similar results by targeting the transcription factor *MYC* [25]; this information can explain the B-cell development blocking, caused by the miR-23a cluster [10]. Additionally, MYC controls the expression of the miR-17/92 cluster; its overexpression arrests B cells at the pro-B-cell stage, by inhibition of the proapoptotic protein BCL2L11, and of PTEN, which is implicated in the vital PI3K signaling pathway [26,27]. This axis is also implicated in immature B cells; the targeting of *PTEN* and *BCL2L11* by this cluster and by miR-148a-3p promotes the survival of immature B cells, as well as the production of self-reactive antibodies, and thus leads to their elimination at the central tolerance checkpoint [28–30]. All this information highlights the great and multifaceted impact of miRNAs in early-stage B-cell development.

2.2. *miRNAs in Peripheral B-Cell Development*

Concerning the development in the periphery, several miRNAs have been reported to play a role, too, with most of them affecting germinal center B to plasma cell transition. miR-155-5p is essential for plasma cell production, as it inhibits SPI1 proto-oncogene, leading to downregulation of PAX5 [31]; PAX5 downregulation is essential for terminal B-cell differentiation. Moreover, miR-148a-3p promotes germinal center to plasma cell transition, by inhibiting *BACH2* and *MITF*; these transcription factors repress the transcription factors *PRDM1* and *IRF4*, both involved in premature plasma cell differentiation, by initiating cascades of gene expression changes and inducing class-switch recombination [32], a process of further recombination of the Ig genes, essential for terminal B-cell differentiation [6]. On the contrary, miR-125b-5p was shown to negatively regulate the B-cell differentiation in germinal centers, through targeting the transcription factors *PRDM1* and *IRF4* [33], thus its physiological silencing is required for normal B-cell development [34]. This fact is also supported by a study that demonstrated that transgenic mice overexpressing miR-125b-5p developed lethal B-cell malignancies [35]. *PRDM1* was also shown to be targeted by two members of the miR-30 family, namely miR-30b-5p and miR-30d-5p, and miR-9-5p, leading to similar results [36].

Besides germinal center B cells, miRNAs were shown to affect marginal zone B cells, too, which are involved in the early rapid response to infection. miR-146a-5p overexpression was demonstrated to reduce the total number of marginal zone B cells, by targeting *NUMB* which prevents TP53 degradation and upregulates the Notch signaling pathway;

this pathway promotes the development of marginal zone lymphocytes [37–39]. Aberrant development of marginal zone B cells is also a consequence of miR-142-5p. Moreover, its deficiency in nude mice led to a deregulation of gene expression in mature B cells, due to higher levels of B cell-activating factor receptor (BAFFR; also known as TNFRSF13C), which is a direct target of miR-142-5p and enhances B-cell survival, leading to robust B-cell proliferation [40].

All this information, as well as additional miRNAs participating in B-cell development, are summarized in Table 1; those miRNAs that were shown to have a great impact in the developmental process are presented in Figure 2.

Table 1. miRNAs implicated in normal B-cell development.

	miRNA	**Target**	**Effect**	**References**
Bone marrow	miR-181a-5p	-	Promotes B-cell development	[12]
	miR-23a-5p	-	Inhibits B-cell development	[13]
	miR-191-5p	Tcf3, Foxp1, and Egr1	Acts as a rheostat for early B-cell development in mice	[21]
	miR-126-3p	RAG1 and RAG2	Rescues B-cell development in absence of EBF1	[11]
	miR-146a-5p	Egr1	Downregulates Egr1, leading to B-cell malignancies	[20]
	miR-150-5p	MYB	Inhibits pro- to pre-B-cell transition	[10,22]
	miR-132-3p	Sox4		[24]
	miR-34a-5p	FOXP1		[41]
	miR-24-3p	MYC		[13,25]
	miR-17/92 cluster	BCL2L11 and PTEN	Inhibits pro- to pre-B-cell transition; promotes the survival of immature B cells	[26]
	miR-21-5p	-	Inhibits pre-B cell apoptosis	[42]
	miR-148a-3p	BCL2L11, PTEN, and GADD45A	Promotes the survival of immature B cells	[28]
	miR-210-3p	-	Inhibits autoantibody production in mice	[43]
Periphery	miR-29a-3p	-	Is essential for the formation of germinal center B cells in mice	[44]
	miR-125b-5p	PRDM1 and IRF4	Inhibits the differentiation of germinal center B cells	[33–35]
	miR-9-5p miR-30b-5p/miR-30d-5p	PRDM1		[36]
	miR-223-3p	LMO2		[36]
	miR-155-5p	SPI1	Promotes the differentiation of germinal center B cells	[31]
	miR-148a-3p	BACH2 and MITF		[32]
	miR-146a-5p	NUMB	Inhibits the formation of marginal zone B cells	[38]
	miR-142-5p	BAFFR	Inhibits the formation of marginal zone B cells; regulates gene expression in mature B cells	[40]

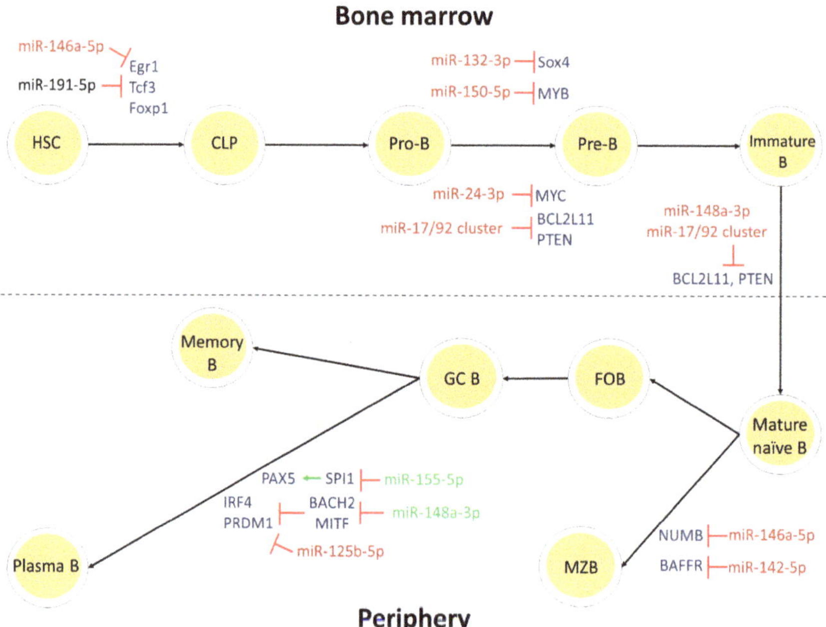

Figure 2. The main steps of B-cell development, and some of the miRNAs implicated in this process. Red color in lines and miRNA red font color indicate inhibition of expression, while green color in arrows and miRNA font indicates induction of expression. Black font color indicates a miRNA acting as a rheostat for the developmental process. CLP, common lymphoid progenitor; FOB, follicular B cell; MZB marginal zone B cell; GC B, germinal center B cell.

3. miRNAs: Regulators, Biomarkers and Potential Therapeutic Entities in B-CLL

3.1. miRNAs as Regulators in B-CLL

miRNAs appear as important molecules in the regulation of B-CLL either by directly regulating key factors that are involved in B-CLL pathogenesis or after the alteration of their levels by epigenetic modifications. They can act as oncogenic molecules or as tumor suppressors with a remarkable involvement in several B-CLL signaling pathways. They appear as crucial regulators of cellular procedures such as early B-cell development [45], cell metabolism and autophagy [46,47], of signaling pathways including the NFkB pathway [20,48], the Hedgehog (Hh) signaling pathway [49], as well as of key molecules in B-CLL such as BTK [50], TCL1A [51–53], BCL2 [54], TERT [55], and the heat shock proteins (HSPs) [56]. Moreover, they have an exceptional involvement in distinct parts of the BCR signaling pathway, which is abnormally activated in B-CLL, as well as, in its downstream pathways [57]. This interaction leads to the regulation of apoptosis, survival, proliferation and migration of leukemic B cells.

Initially, a differential expression of specific miRNAs was found between BCR stimulated and unstimulated cells [58]. Interestingly, in leukemic B cells, high expression of miR-155-5p mimics led to an increased BCR stimulation compared to controls. On the contrary, miR-155-5p inhibition resulted in a reduced calcium flux which is a result of BCR stimulation reduction [59]. In another study, miR-150-5p was found to regulate BCR signaling in B-CLL by regulating the expression levels of *GAB1* and *FOXP1*. GAB1 is an adaptor molecule that recruits numerous factors, including PI3K, enhancing the BCR signaling. FOXP1 is a transcription factor that is strongly expressed after B-cell activation. Characteristically, transfection of cells with miR-150-5p mimics decreased the levels of *GAB1* and *FOXP1*. Moreover, sensitive B-CLL samples to BCR stimulation had higher levels of *GAB1* and *FOXP1* mRNAs. Additionally, insensitive to BCR stimulation samples had high levels of miR-150-5p, reinforcing the idea of the connection between high miR-150-5p levels with low *GAB1* and *FOXP1* expression which results in a reduced sensitivity [60]. Additionally, FOXP1 was also found to be controlled by miR-34a-5p with its levels being reduced during DNA damage response leading to a limitation of BCR signaling [61]. Furthermore, a recent study revealed the involvement of miR-29 family in CD40 signaling by targeting *TRAF4* and proposed a novel regulatory axis in B-CLL, modulated by the BCR activity [62].

Another study revealed the involvement of miR-21-5p in BCR-mediated MAPK/ERK signaling. miR-21-5p was found to downregulate *SPRY2* expression, leading to a decrease of SPRY2 levels in MEC-1 cells, a human B-CLL cell line. SPRY2 is an inhibitory protein that interacts with RAF1, BRAF, and SYK for the downregulation of the MAPK/ERK signaling in leukemic B cells. Therefore, high expression of miR-21-5p leads to an upregulation of MAPK/ERK signaling and a relative result to survival and proliferation of the leukemic B cells [63]. However, tumor-suppressive miRNAs which contribute negatively to the MAPK/ERK signaling pathway have also been identified in B-CLL [64,65]. As described previously, miRNAs are also involved in the regulation of the PI3K/AKT pathway, either by directly influencing the pathway or by downregulating the expression of *PTEN*, a tumor suppressor of this pathway [66,67]. This pathway regulates cellular growth, metabolism and survival.

Another important example of miRNAs regulation in B-CLL is the connection of miR-15a-5p and miR-16-5p with TP53 and both miR-34b-3p and miR-34c-5p–produced by genes located in chromosomes 13q, 17p, and 11q, respectively–and their relation to B-CLL pathogenesis and patients' outcome. The deletion of 13q14 in B-CLL patients results to downregulation of miR-15a-5p and miR-16-5p levels and high levels of its targets, BCL2 and MCL1 anti-apoptotic proteins, leading to a reduction of apoptosis. Furthermore, low miR-15a-5p and miR-16-5p levels, caused by 13q14 deletion, results in an upregulation of their target, TP53, which leads to increased levels of miR-34b-3p and miR-34c-5p, thus leading to reduction of ZAP-70 levels and its downstream pathways, as well as to an indolent B-CLL phenotype [68].

Epigenetic modifications of miRNAs may lead to significant alterations in their expression and function. Baer et al. found a negative correlation between the DNA methylation status of miRNA promoters and the expression of their corresponding mRNA targets [69], proposing a strong epigenetic effect in the miRNA expression after a methylation change in the DNA promoter region. Two other studies highlighted the significance of methylation in the *MIR34B/MIR34C* promoter region. The methylation status of the *MIR34B/MIR34C* promoter region differed between normal and B-CLL cohorts [70,71], proposing a correlation between *MIR34B/MIR34C* promoter methylation, the downregulation of mir-34b/mir-34c and the respective mature miRNAs, and subsequently the reduction of their tumor-suppressive activity in B-CLL patients. Furthermore, acetylation alteration may differentiate the levels of specific miRNAs as it was found that histone deacetylases may mediate the silencing of miR-15a-5p, miR-16-5p, and miR-29b-3p in B-CLL [72]. This alteration is of high importance as these miRNAs are significant regulators in B-CLL. Furthermore, it is important to mention that epigenetic alterations in specific miRNAs, such as RNA editing, which were changes of adenosine to inosine and cytosine to uracil, have been found altered between leukemic and normal B cells and may differentiate the targets of these miRNAs [73].

Recent studies have also examined the interaction of miRNAs with a relatively new RNA type, circular RNAs (circRNAs), in the context of B-CLL pathogenesis. CircRNAs, which are mainly produced by backsplicing events in pre-mRNAs, bear a circular structure and can regulate cell function, multifariously, including sponging of miRNAs, which consequently limits the effect of miRNAs. A relative example is circ-CBFB, which was predicted to sponge miR-607, leading to an upregulated expression of the miRNA target, *FZD3*, which is a receptor for WNT proteins. This upregulation leads to the activation of the WNT/β-catenin pathway and B-CLL progression [74]. Additionally, the downregulation of circ_0132266, which acts as a sponge for miR-337-3p, resulted in a downregulation of *PML*, which is the target molecule of the latter and a tumor suppressor. The downregulation of PML leads to increased cell viability [75].

All the aforementioned information delineates a great involvement of miRNAs in B-CLL regulation as they can modulate oncosuppressors and oncogenes, possess a key role in a plethora of signaling pathways and can also be epigenetically regulated having different functions. Specific miRNAs with a regulatory role in B-CLL are summarized in Table 2.

Table 2. miRNAs with a regulatory effect in B-CLL.

miRNA	Target	Effect	References
miR-29 family	TRAF4	Suppresses CD40 signaling	[62]
miR-202-3p	SUFU	Regulates Hedgehog (Hh) signaling	[49]
miR-607	FZD3	Enhances WNT/β-catenin signaling	[74]
miR-708-5p	IKBKB	Suppresses NFkB signaling	[48]
miR-22-3p	PTEN, CDKN1B, and BIRC5	Enhances PI3K/AKT signaling	[66]
miR-3151-5p	MADD and PIK3R2	Suppresses MAPK/ERK and PI3K/AKT signaling	[64]
miR-126-3p	PIK3R2	Suppresses MAPK/ERK signaling	[65]
miR-21-5p	SPRY2	Enhances BCR and MAPK/ERK signaling	[63]
miR-150-5p	GAB1 and FOXP1	Suppresses BCR signaling	[58,60]
miR-34a-5p	FOXP1		[61]
miR-155-5p	SHIP1	Enhances BCR signaling	[59]
	-	Regulates cell survival	[76]
miR-221-3p; miR-222-3p	CDKN1B	Regulates cell proliferation	[77]
miR-15a-5p; miR-16-5p	BCL2	Regulates cell survival	[54]
	TP53	Regulates cell proliferation	[55,68]
	ROR1	Regulates cell survival and proliferation	[78]
miR-26a-5p	EZH2	Regulates cell survival and proliferation	[79]
	PTEN		[67]
miR-214-3p	PTEN		[67]
miR-337-3p	PML		[75]
miR-106b-5p	ITCH		[80]
miR-28-5p	NDRG2		[81]
miR-650			
miR-181a-5p/miR-181b-5p	-	Suppresses cell growth	[58]
miR-210-3p; miR-425-5p; miR-1253; miR-4269; miR-4667-3p	BTK	Promotes apoptosis	[50]
miR-130a-3p	ATG2B and DICER1	Inhibits autophagy and regulates cell survival	[46]
miR-125b-5p	PCTP, LIPA, GSS, HK2, IKZF4, and TP53	Regulates metabolic adaptation to cancer transformation	[47]
miR-29b-3p; miR-34b-5p; miR-181b-5p; miR-484	TCL1A	Regulates multiple signaling pathways and cell survival	[51–53]

3.2. miRNAs as Diagnostic, Prognostic and Predictive Biomarkers in B-CLL

In 2004, Calin et al. found deregulation in the expression levels of numerous miRNAs between leukemic B cells and normal CD5$^+$ B cells, as well as, among distinct molecular B-CLL subtypes [82]. This study indicated for the first time a potential value of miRNAs as diagnostic and prognostic biomarkers for B-CLL. One year later, the authors proposed the first miRNA-signature for prognosis and progression of this malignancy which could be used in order to distinguish between *IGHV*-mutated/ZAP70-positive and *IGHV*-unmutated/ZAP70-negative patients [83].

Today, many scientists propose miRNAs as ideal biomarkers as they are highly stable and can be easily detected and quantified in blood, other body fluids and fresh or paraffin-embedded tissues. Over the years, numerous other miRNAs have been proposed in B-CLL as biomarkers for risk assessment, prognosis, prediction and progression. In the last few years, research has focused on assessing the value of specific miRNAs for the disease rather than a panel of numerous miRNAs. A typical example is miR-181b-5p, which has been identified as a biomarker of progression from indolent to aggressive B-CLL [84]. Moreover, miR-155-5p expression levels were elevated in individuals with monoclonal B-cell lymphocytosis compared to normal blood donors and in patients with B-CLL compared to patients with monoclonal B-cell lymphocytosis [85]. Furthermore, miR-155-5p levels were characterized as a valuable prognostic biomarker and biomarker for the risk assessment of B-CLL development [86]. Moreover, it is worth mentioning that other miRNAs appear as promising biomarkers in distinct subgroups of B-CLL patients with different cytogenetic abnormalities [87–89].

Based on the high potential of miRNAs as biomarkers, there are studies, which incorporate miRNAs in prognostic models, which provide B-CLL–specific prognostic scores. In this context, Stamatopoulos et al. proposed a molecular prognostic score for the disease which includes the expression levels of miR-29c-3p, ZAP-70 and LPL for the stratification of B-CLL patients into three distinct groups concerning treatment-free and overall survival (OS) prognosis [90]. Another system, the 21FK score, has been proposed for prognosis assessment of B-CLL patients. This score examines the expression levels of miR-21-5p by qRT-PCR, chromosomal abnormalities with fluorescence in situ hybridization, and karyotype in peripheral blood mononuclear cells (PBMCs) of B-CLL patients, in order to stratify patients for OS prognosis assessment [91]. Patients with a high 21FK score had a shorter OS time and therefore worse prognosis. Even though these scoring systems could provide a satisfactory prognosis, they have not been adopted in clinical practice.

Several studies revealed the predictive biomarker utility of miRNAs in B-CLL. These studies assess whether the treatment strategy will be effective in patients with distinct characteristics. For instance, particular miRNAs have emerged as predictive biomarkers in order to distinguish fludarabine resistant or rituximab resistant patients from responsive individuals for each therapeutic strategy [92,93]. Moreover, it is worth mentioning a study that found that miR-34a-5p low expression was associated with TP53 inactivation, which is a tumor suppressor with a key role in the induction of apoptosis of leukemic B cells, regardless of 17p deletion or *TP53* mutation. This inactivation of TP53 was observed independently of the mutational status of the *TP53* gene or deletions in 17p chromosomal region, where the *TP53* gene is located. Consequently, low levels of miR-34a-5p denote apoptosis resistance and fludarabine refractory disease [94]. Additionally, the expression levels of miR-21-5p, miR-148a-3p and miR-222-3p could also serve for the discrimination of fludarabine-refractory B-CLL patients from fludarabine-sensitive ones. Inhibition of miR-21-5p and miR-222-3p was found to increase caspase activity in fludarabine-treated *TP53*-mutant MEG-01 chronic myelogenous leukemia cells, suggesting these two miRNAs as key factors of acquisition of resistance to fludarabine [95]. Another noteworthy study noticed an inversive correlation between circulating miR-125b-5p and miR-532-3p expression levels, rituximab-induced lymphodepletion and CD20 expression on CD19+ T cells in patients with B-CLL [93]. All the aforementioned information highlights the high and complicated importance of miRNAs for the risk assessment of B-CLL development, progression and treatment prediction. Unique miRNAs, miRNA signatures and scoring systems, which have been proposed for the assessment of B-CLL are summarized in Table 3.

Table 3. miRNAs as candidate biomarkers in B-CLL.

miRNA	Localization	miRNA Expression	Biomarker Utility	References
miR-20b-5p	PBMCs [1]	Lower levels in patients with poor prognosis	Prognosis	[96]
miR-21-5p; miR-125b-5p; miR-148a-3p; miR-181a-5p; miR-221-3p; miR-222-3p; miR-532-3p	PBMCs [1]	Lower levels in responsive patients	Prediction of response	[92,93,95]
miR-29a-3p; miR-34a-5p		Higher levels in responsive patients		[92,94]
miR-181b-5p	PBMCs [1]	Higher levels in indolent vs. aggressive disease	Prediction of progression	[84]
miR-744-5p		Lower levels in patients with shorter time to first treatment		[97]
miR-4524a-5p		High levels in patients with shorter time to first treatment		
miR-92a-3p	PBMCs [1]	Lower levels in B-CLL patients vs. non-leukemic controls	Diagnosis	[98]
		Lower levels in patients with poor prognosis	Prognosis	
miR-155-5p	Plasma	Higher levels in B-CLL patients vs. non-leukemic controls	Diagnosis	[86]
	PBMCs [1]	Higher levels in patients with poor prognosis	Prognosis	
	Purified B cells	Lower levels in responsive patients	Prediction of response	[85]
miRNA signature	Serum	-	Diagnosis	[99]
miRNA signature	-			[100]
miRNA signature	-		Prediction of progression	[89]
miRNA signature	PBMCs [1]		Diagnosis; prognosis	[82]
miRNA signature	Purified B cells			[101]
miRNA signature	PBMCs [1]		Prognosis; prediction of progression	[83]
Scoring system, including miR-21-5p	PBMCs [1]	Higher levels in patients with poor prognosis	Prognosis	[91]
Scoring system, including miR-29c-3p		Lower levels in patients with poor prognosis		[90]

[1] Peripheral blood mononuclear cells.

3.3. miRNAs in B-CLL Therapy

The treatment of B-CLL consists of numerous therapeutic strategies and involves alkylating agents, glucocorticoids, purine analogs, monoclonal antibodies and bone marrow transplantation. Moreover, target-specific therapies have also emerged, targeting BCR receptor, BTK, PI3K and apoptosis-related proteins. Taking into consideration the multifaceted roles of miRNAs in B-CLL regulation, the fact that they are naturally produced molecules by organisms and that their levels may be easily regulated with miRNA-mimics or miRNA-antagomiRs, miRNAs appear as promising therapeutic molecules and therapeutic targets for this disease.

Specific studies emphasized the therapeutic potential of miRNAs. Interestingly, Salerno et al. described an increase of drug sensitivity in the New Zealand Black (NZB) mouse cell line, LNC, after the correction of the miR-15a-5p and miR-16-5p defect. NZB mouse model has a genetically determined age-associated increase in malignant B-1 clones and decreased expression of miR-15a-5p and miR-16-5p in B-1 cells. A cell cycle arrest in the G_1 phase was observed after the exogenous addition of miR-16-5p mimics. This observation is in correlation with the decrease of *CCND1* levels–an overexpressed gene in some human B-CLL cases–as miR-15a-5p and miR-16-5p target the 3′ UTR of *CCND1*. Moreover, a synergetic effect of miR-16-5p and chemotherapeutic agents was observed in the induction of apoptosis [102]. In another study where lentiviral vectors were used for the in vivo restoration of miR-15a-5p and miR-16-5p in NZB mouse model, mice appeared with a moderate B-CLL phenotype. The lentivirus delivering system assisted in low systemic toxicity levels and limited off-target effects. These results came in line with those of another study, which demonstrated that the effect of the restoration of these two miRNAs were the increased expression of miR-15a-5p and miR-16-5p both in transduced cells and serum and the decreased viability of B-1 cells [103].

Numerous other miRNAs have also been associated as regulators of B-CLL therapy [104,105], with miR-181a-5p and miR-181b-5p constituting promising examples. Leukemic B cells from *TP53*^{wt} patients were transfected with miR-181a-5p and miR-181b-5p mimics resulting in a significant increase in apoptosis compared to controls, with no effect being observed in B-CLL patients with a decreased expression of *TP53* [106]. Moreover, miR-181b-5p was found to affect the levels of TCL1A, AKT, and both phosphorylated ERK1 and ERK2, to reduce leukemic cell expansion and to increase survival of a treated transgenic mouse model [107].

miRNAs have also been characterized as oncogenic for numerous malignancies leading to a required downregulation of these miRNAs for the improvement of the cancer patients' outcome. In B-CLL, ibrutinib suppresses the expression of oncogenic miRNAs, leading to a downregulation of malignant B-cell proliferation [108]. Moreover, in another study, the downregulation of miR-17-5p expression levels was proposed as a potential therapeutic strategy for B-CLL. An in-vitro administration of antagomiR-17-5p, which is a miRNA inhibitory oligonucleotide molecule, in MEC-1 cells significantly reduced miR-17-5p expression levels and cell proliferation. Moreover, tumors generated by MEC-1 cells injected into severe combined immunodeficiency mice, which were treated with antagomiR-17-5p, presented an inhibition of growth and complete remission in 20% of the cases. Furthermore, antagomiR-17-5p treated mice possessed a longer median OS in comparison to the controls, while no signs of toxicity were observed [109].

Besides in vivo lentiviral delivery of miRNA mimics as a therapeutic strategy for B-CLL, antibody-based strategies have also been proposed for the delivery of miRNA mimics and antagomiRs. These strategies include the construction of particles which include the selected miRNAs and are conjugated with antigen-specific antibodies for characteristic markers of leukemic B cells, such as CD38 and ROR1. Therefore, these "vehicles" are attached to leukemic B cells which are expressing these markers [110,111]. The findings of studies proposing miRNAs as molecules for therapeutic strategies are summarized in Table 4.

Table 4. miRNAs with a therapeutic interest with regard to B-CLL.

miRNA	Experimental Approach	Effect	References
miR-15a-5p; miR-16-5p	Human cells	Restoration of cell cycle control	[102]
	Mouse model	Drug sensitization and induction of apoptosis in mice upon its upregulation	[102,103,112]
miR-155-3p	Human cells	Regulation of chemoresistance	[104]
miR-222-3p	Human cells	Reduced cell viability and proliferation upon its downregulation	[105]
miR-181a-5p/miR-181b-5p	Human cells	Increased apoptosis of cells	[106,107]
	Mouse model	Reduced leukemic cell expansion and increase of survival in mice upon its upregulation	[107]
miR-34a-5p; miR-146b-5p	Human cells	Inhibition of cell proliferation upon its downregulation	[108]
miR-17-5p	Human cells	Reduced cell proliferation	[109]
	Mouse model	Reduced tumor growth and increased survival in mice upon its downregulation	
miR-26a-5p	Human cells	Induction of apoptosis with CD38-targeted delivery	[110]
	Mouse model		
miR-29b-3p	Mouse model	Induction of cell cycle arrest with ROR1-targeted delivery	[111]

The use of miRNA mimics or antagomiRs has entered clinical trials with promising results for targeted therapy of various diseases. Therefore, these molecules appear as promising therapeutic agents for the future as they could expand even more the field of personalized medicine. However, for miRNA mimics or antagomiRs to succeed, off-target phenomena and severe effects should be eliminated, and delivery strategies should be improved. However, prior to all these ameliorations, it is important to clarify the miRNA regulatory disease-specific network of miRNAs with deregulated levels in B-CLL, in order to distinguish the most promising miRNAs for treatment, and therefore achieve a miRNA-based therapy.

3.4. *Viral miRNAs in B-CLL*

Epstein-Barr virus (EBV) is a ubiquitous oncogenic human herpesvirus implicated in lymphomas, such as Burkitt's lymphoma, while recent studies have associated EBV infection with B-CLL progression [113]. More specifically, it was shown that B-CLL patients with elevated levels of EBV DNA load had significantly shorter OS time intervals and were characterized by therapy resistance, compared to the ones with lower levels of EBV DNA load [114,115]. However, further investigation is required regarding the potential mechanisms of EBV-driven oncogenesis in B-CLL patients.

A proposed mode of action, via which EBV sustains viral infection, evades host immunity and potentially leads to oncogenesis, is based on viral miRNAs [116]. EBV was the first virus shown to encode viral miRNAs. To date, 44 mature miRNAs that could be classified into two clusters (*BHRF1* cluster and *BART* cluster) have been identified in EBV. Particularly, the miRNAs from the *BHRF1* cluster are expressed during lytic infection, inhibit apoptosis, and favor proliferation of infected cells to enable the early phase of viral propagation. Considering the role of these viral miRNAs, miR-BHRF1-1 has been further investigated in the context of B-CLL. Precisely, the expression levels of miR-BHRF1-1 were significantly higher in the plasma of B-CLL patients compared to the plasma of healthy individuals, while elevated levels of this miRNA were positively associated with advanced Rai stages. Furthermore, B-CLL patients with elevated expression levels of miR-BHRF1-1 were characterized by shorter OS time intervals, compared to the ones with lower expression levels [117]. These findings designate the potential value of this miRNA as biomarker. Additionally, high expression levels of miR-BHRF1-1 were positively correlated with high miR-155-5p expression levels. This observation is quite significant, since miR-155-5p plays a decisive role in both normal B-cell development and B-CLL progression, as it has been thoroughly analyzed above. The association between these two miRNAs has been also observed in another study, which showed that cellular miR-155-5p was induced by the viral miRNA and that miR-155-5p played a key role in B-cell immortalization [118]. Finally, infection of leukemic B cells with miR-BHRF1-1 reduces the levels of the key tumor suppressor, TP53 [117]. This interaction is, also, supported by an independent study, which suggested that this miRNA exerts its role in B-CLL via downregulation of TP53 and uncovered the therapeutic potential of miR-BHRF1-1 [119].

All these findings advocate the implication of EBV miRNAs in B-CLL onset and progression. Even though this research field is still in its infancy, the elucidation of the role of viral miRNAs in B-CLL progression is quite promising since it could further assist the discovery of novel biomarkers and therapeutic strategies.

4. Future Perspectives

Although extensive research has been conducted, only a few miRNAs appear as important regulatory molecules both in normal B-cell development and B-CLL, as depicted in Figure 3. These include members of the miR-17/92 cluster, miR-21-5p, miR-29 family, miR-34 family, miR-125b-5p, miR-150-5p, miR-155-5p, and miR-181 family. In particular, miR-34a-5p was found to act similarly in both situations targeting *FOXP1*, either by blocking the development of B cells in normal development or by limiting proliferation and survival of leukemic B cells. miR-150-5p appears to function in a similar way, as well. On the contrary, miR-181 isomiRs possess an opposite function, as they promote normal B-cell development, but reduce leukemic cell expansion in B-CLL.

The implication of the aforementioned miRNAs in normal B-cell development indicates that fully unraveling the regulatory network that orchestrates the typical developmental pathway of B cells can gain new insights in B-CLL understanding. Even though a great number of studies delineate the roles of miRNAs in normal B-cell development and B-CLL, there is no clear evidence of a large regulatory network of miRNAs in these two different conditions. Moreover, no safe conclusions can be drawn, as scientific knowledge seems to focus either on specific miRNAs that could regulate for instance a specific pathway or on miRNA signatures with significant differences in their expression levels between different stages of the developmental process or distinct B-CLL conditions. Additionally, some findings seem contradictory such as the paradigm of cellular and serum circulating miR-150-5p that was found to possess an opposite prognostic significance in patients with B-CLL [120]. Moreover, specific miRNAs such as miR-125a-5p and miR-34a-5p which were proposed for the prediction of Richter syndrome, a lethal complication in B-CLL patients [121], may appear as ideal molecules for further research in order to distinguish molecular pathways that contribute to distinct subgroups of B-CLL patients. Taking into consideration the significant presence of miRNAs both in normal B-cell development and B-CLL, further research is required to shed light on the involvement of miRNAs in normal B-cell development and the pathogenetic events that lead to B-CLL.

Another important aspect is that polymorphisms in pre-miRNAs and genes which are involved in miRNAs biogenesis pathway were found to contribute to the risk of B-CLL [122]. miRNAs can be epigenetically regulated by multiple processes and therefore possess a different regulatory effect. These epigenetic regulations may remarkably enlarge the regulatory potential of miRNAs and appear as another promising area for research.

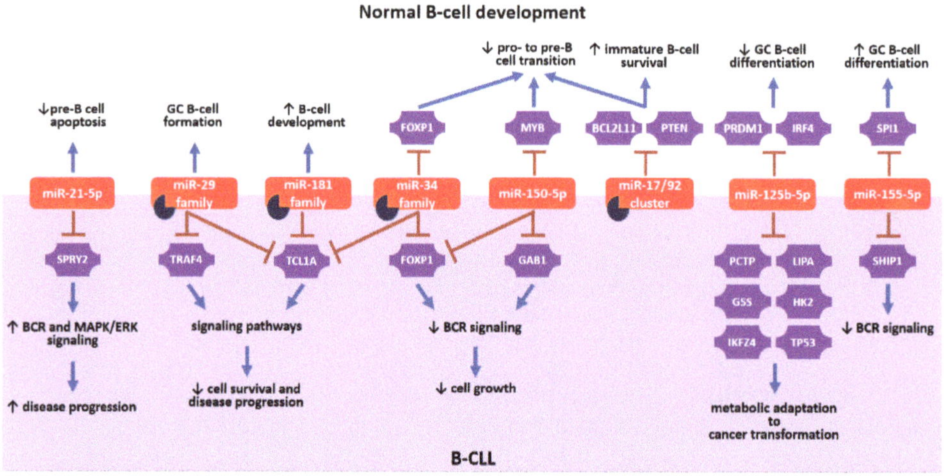

Figure 3. The regulatory roles and specific targets of miRNAs, which are involved in normal B-cell development and B-CLL. Red inhibitory signs indicate the downregulation of specific targets by miRNAs and blue arrows indicate the regulatory impact of the miRNAs. Partial circles indicate the potential therapeutic utility of the miRNAs. miRNAs can promote (↑) or suppress (↓) numerous biological processes. With regard to the miR-17/92 cluster, only miR-17-5p has been proposed as a therapeutic target for B-CLL. GC B, germinal center B cell.

Increasing evidence shows specific miRNAs with deregulated levels in B-CLL, in comparison with normal individuals, suggesting miRNAs as biomarkers. Moreover, the alteration of their levels is a tightly regulated and finely tuned process. Therefore, elucidating the regulatory effect of miRNAs with deregulated levels may reveal other miRNAs with therapeutic potency. Furthermore, specific miRNAs have already been identified as molecules with significant therapeutic utility. Their impact may be extremely targeted as they regulate the expression of their specific targets and therefore, miRNAs may appear as ideal agents for combinational therapy. Lastly, therapeutic miRNA-based strategies have entered clinical trials for numerous diseases. Therefore, more targeted research is required in order to clarify the specific therapeutic potential of miRNAs in B-CLL.

5. Conclusions

miRNAs are undoubtedly implicated in numerous stages of the normal B-cell development either by blocking the development or by facilitating it. In B-CLL, they act as oncogenes or as oncosuppressors with key involvement in signaling pathway regulation. Moreover, they can be epigenetically regulated which can potentially lead to other regulatory effects. In addition, they are valuable biomarkers for diagnosis, prognosis and prediction for B-CLL patients and appear as promising molecules for therapeutic strategies. Even though the multifaceted role of miRNAs in B-CLL has been extensively studied in the last few years, important information is still missing, while no molecule has emerged as a validated regulator in numerous pathogenetic pathways of this malignancy. As a result, their regulatory potency accompanied by their therapeutic ability are two topics that require further targeted research.

Funding: This research received no external funding.

Institutional Review Board Statement: Not applicable.

Informed Consent Statement: Not applicable.

Data Availability Statement: Not applicable.

Acknowledgments: This work was supported by the Special Account for Research Grants of the National and Kapodistrian University of Athens.

Conflicts of Interest: The authors declare no conflict of interest.

References

1. Dohner, H.; Stilgenbauer, S.; Benner, A.; Leupolt, E.; Krober, A.; Bullinger, L.; Dohner, K.; Bentz, M.; Lichter, P. Genomic aberrations and survival in chronic lymphocytic leukemia. *N. Engl. J. Med.* **2000**, *343*, 1910–1916. [CrossRef]
2. International CLL-IPI working group. An international prognostic index for patients with chronic lymphocytic leukaemia (CLL-IPI): A meta-analysis of individual patient data. *Lancet Oncol.* **2016**, *17*, 779–790. [CrossRef]
3. Hayes, J.; Peruzzi, P.P.; Lawler, S. MicroRNAs in cancer: Biomarkers, functions and therapy. *Trends Mol. Med.* **2014**, *20*, 460–469. [CrossRef]
4. Calin, G.A.; Dumitru, C.D.; Shimizu, M.; Bichi, R.; Zupo, S.; Noch, E.; Aldler, H.; Rattan, S.; Keating, M.; Rai, K.; et al. Frequent deletions and down-regulation of micro- RNA genes miR15 and miR16 at 13q14 in chronic lymphocytic leukemia. *Proc. Natl. Acad. Sci. USA* **2002**, *99*, 15524–15529. [CrossRef]
5. Vasilatou, D.; Papageorgiou, S.; Pappa, V.; Papageorgiou, E.; Dervenoulas, J. The role of microRNAs in normal and malignant hematopoiesis. *Eur. J. Haematol.* **2010**, *84*, 1–16. [CrossRef]
6. Melchers, F. Checkpoints that control B cell development. *J. Clin. Investig.* **2015**, *125*, 2203–2210. [CrossRef] [PubMed]
7. Koralov, S.B.; Muljo, S.A.; Galler, G.R.; Krek, A.; Chakraborty, T.; Kanellopoulou, C.; Jensen, K.; Cobb, B.S.; Merkenschlager, M.; Rajewsky, N.; et al. Dicer ablation affects antibody diversity and cell survival in the B lymphocyte lineage. *Cell* **2008**, *132*, 860–874. [CrossRef] [PubMed]
8. Brandl, A.; Daum, P.; Brenner, S.; Schulz, S.R.; Yap, D.Y.; Bosl, M.R.; Wittmann, J.; Schuh, W.; Jack, H.M. The microprocessor component, DGCR8, is essential for early B-cell development in mice. *Eur. J. Immunol.* **2016**, *46*, 2710–2718. [CrossRef] [PubMed]
9. Spierings, D.C.; McGoldrick, D.; Hamilton-Easton, A.M.; Neale, G.; Murchison, E.P.; Hannon, G.J.; Green, D.R.; Withoff, S. Ordered progression of stage-specific miRNA profiles in the mouse B2 B-cell lineage. *Blood* **2011**, *117*, 5340–5349. [CrossRef] [PubMed]
10. Xiao, C.; Calado, D.P.; Galler, G.; Thai, T.H.; Patterson, H.C.; Wang, J.; Rajewsky, N.; Bender, T.P.; Rajewsky, K. MiR-150 controls B cell differentiation by targeting the transcription factor c-Myb. *Cell* **2007**, *131*, 146–159. [CrossRef] [PubMed]
11. Okuyama, K.; Ikawa, T.; Gentner, B.; Hozumi, K.; Harnprasopwat, R.; Lu, J.; Yamashita, R.; Ha, D.; Toyoshima, T.; Chanda, B.; et al. MicroRNA-126-mediated control of cell fate in B-cell myeloid progenitors as a potential alternative to transcriptional factors. *Proc. Natl. Acad. Sci. USA* **2013**, *110*, 13410–13415. [CrossRef] [PubMed]
12. Chen, C.Z.; Li, L.; Lodish, H.F.; Bartel, D.P. MicroRNAs modulate hematopoietic lineage differentiation. *Science* **2004**, *303*, 83–86. [CrossRef] [PubMed]
13. Kurkewich, J.L.; Bikorimana, E.; Nguyen, T.; Klopfenstein, N.; Zhang, H.; Hallas, W.M.; Stayback, G.; McDowell, M.A.; Dahl, R. The mirn23a microRNA cluster antagonizes B cell development. *J. Leukoc. Biol.* **2016**, *100*, 665–677. [CrossRef] [PubMed]
14. Hagman, J.; Lukin, K. Transcription factors drive B cell development. *Curr. Opin. Immunol.* **2006**, *18*, 127–134. [CrossRef] [PubMed]
15. Medvedovic, J.; Ebert, A.; Tagoh, H.; Busslinger, M. Pax5: A master regulator of B cell development and leukemogenesis. *Adv. Immunol.* **2011**, *111*, 179–206. [CrossRef] [PubMed]
16. Kwon, K.; Hutter, C.; Sun, Q.; Bilic, I.; Cobaleda, C.; Malin, S.; Busslinger, M. Instructive role of the transcription factor E2A in early B lymphopoiesis and germinal center B cell development. *Immunity* **2008**, *28*, 751–762. [CrossRef]
17. Hagman, J.; Ramirez, J.; Lukin, K. B lymphocyte lineage specification, commitment and epigenetic control of transcription by early B cell factor 1. *Curr. Top. Microbiol. Immunol.* **2012**, *356*, 17–38. [CrossRef]
18. Nechanitzky, R.; Akbas, D.; Scherer, S.; Gyory, I.; Hoyler, T.; Ramamoorthy, S.; Diefenbach, A.; Grosschedl, R. Transcription factor EBF1 is essential for the maintenance of B cell identity and prevention of alternative fates in committed cells. *Nat. Immunol.* **2013**, *14*, 867–875. [CrossRef]
19. Hu, H.; Wang, B.; Borde, M.; Nardone, J.; Maika, S.; Allred, L.; Tucker, P.W.; Rao, A. Foxp1 is an essential transcriptional regulator of B cell development. *Nat. Immunol.* **2006**, *7*, 819–826. [CrossRef]
20. Contreras, J.R.; Palanichamy, J.K.; Tran, T.M.; Fernando, T.R.; Rodriguez-Malave, N.I.; Goswami, N.; Arboleda, V.A.; Casero, D.; Rao, D.S. MicroRNA-146a modulates B-cell oncogenesis by regulating Egr1. *Oncotarget* **2015**, *6*, 11023–11037. [CrossRef]
21. Blume, J.; Zietara, N.; Witzlau, K.; Liu, Y.; Sanchez, O.O.; Puchalka, J.; Winter, S.J.; Kunze-Schumacher, H.; Saran, N.; Duber, S.; et al. miR-191 modulates B-cell development and targets transcription factors E2A, Foxp1, and Egr1. *Eur. J. Immunol.* **2019**, *49*, 121–132. [CrossRef] [PubMed]
22. Zhou, B.; Wang, S.; Mayr, C.; Bartel, D.P.; Lodish, H.F. miR-150, a microRNA expressed in mature B and T cells, blocks early B cell development when expressed prematurely. *Proc. Natl. Acad. Sci. USA* **2007**, *104*, 7080–7085. [CrossRef] [PubMed]
23. Greig, K.T.; de Graaf, C.A.; Murphy, J.M.; Carpinelli, M.R.; Pang, S.H.; Frampton, J.; Kile, B.T.; Hilton, D.J.; Nutt, S.L. Critical roles for c-Myb in lymphoid priming and early B-cell development. *Blood* **2010**, *115*, 2796–2805. [CrossRef]
24. Mehta, A.; Mann, M.; Zhao, J.L.; Marinov, G.K.; Majumdar, D.; Garcia-Flores, Y.; Du, X.; Erikci, E.; Chowdhury, K.; Baltimore, D. The microRNA-212/132 cluster regulates B cell development by targeting Sox4. *J. Exp. Med.* **2015**, *212*, 1679–1692. [CrossRef] [PubMed]

25. Kong, K.Y.; Owens, K.S.; Rogers, J.H.; Mullenix, J.; Velu, C.S.; Grimes, H.L.; Dahl, R. MIR-23A microRNA cluster inhibits B-cell development. *Exp. Hematol.* **2010**, *38*, 629–640.e1. [CrossRef] [PubMed]
26. Lai, M.; Gonzalez-Martin, A.; Cooper, A.B.; Oda, H.; Jin, H.Y.; Shepherd, J.; He, L.; Zhu, J.; Nemazee, D.; Xiao, C. Regulation of B-cell development and tolerance by different members of the miR-17 approximately 92 family microRNAs. *Nat. Commun.* **2016**, *7*, 12207. [CrossRef] [PubMed]
27. Benhamou, D.; Labi, V.; Getahun, A.; Benchetrit, E.; Dowery, R.; Rajewsky, K.; Cambier, J.C.; Melamed, D. The c-Myc/miR17-92/PTEN Axis Tunes PI3K Activity to Control Expression of Recombination Activating Genes in Early B Cell Development. *Front. Immunol.* **2018**, *9*, 2715. [CrossRef]
28. Gonzalez-Martin, A.; Adams, B.D.; Lai, M.; Shepherd, J.; Salvador-Bernaldez, M.; Salvador, J.M.; Lu, J.; Nemazee, D.; Xiao, C. The microRNA miR-148a functions as a critical regulator of B cell tolerance and autoimmunity. *Nat. Immunol.* **2016**, *17*, 433–440. [CrossRef]
29. Xiao, C.; Srinivasan, L.; Calado, D.P.; Patterson, H.C.; Zhang, B.; Wang, J.; Henderson, J.M.; Kutok, J.L.; Rajewsky, K. Lymphoproliferative disease and autoimmunity in mice with increased miR-17-92 expression in lymphocytes. *Nat. Immunol.* **2008**, *9*, 405–414. [CrossRef]
30. Benhamou, D.; Labi, V.; Novak, R.; Dai, I.; Shafir-Alon, S.; Weiss, A.; Gaujoux, R.; Arnold, R.; Shen-Orr, S.S.; Rajewsky, K.; et al. A c-Myc/miR17-92/Pten Axis Controls PI3K-Mediated Positive and Negative Selection in B Cell Development and Reconstitutes CD19 Deficiency. *Cell Rep.* **2016**, *16*, 419–431. [CrossRef]
31. Lu, D.; Nakagawa, R.; Lazzaro, S.; Staudacher, P.; Abreu-Goodger, C.; Henley, T.; Boiani, S.; Leyland, R.; Galloway, A.; Andrews, S.; et al. The miR-155-PU.1 axis acts on Pax5 to enable efficient terminal B cell differentiation. *J. Exp. Med.* **2014**, *211*, 2183–2198. [CrossRef] [PubMed]
32. Porstner, M.; Winkelmann, R.; Daum, P.; Schmid, J.; Pracht, K.; Corte-Real, J.; Schreiber, S.; Haftmann, C.; Brandl, A.; Mashreghi, M.F.; et al. miR-148a promotes plasma cell differentiation and targets the germinal center transcription factors Mitf and Bach2. *Eur. J. Immunol.* **2015**, *45*, 1206–1215. [CrossRef] [PubMed]
33. Gururajan, M.; Haga, C.L.; Das, S.; Leu, C.M.; Hodson, D.; Josson, S.; Turner, M.; Cooper, M.D. MicroRNA 125b inhibition of B cell differentiation in germinal centers. *Int. Immunol.* **2010**, *22*, 583–592. [CrossRef] [PubMed]
34. Li, G.; So, A.Y.; Sookram, R.; Wong, S.; Wang, J.K.; Ouyang, Y.; He, P.; Su, Y.; Casellas, R.; Baltimore, D. Epigenetic silencing of miR-125b is required for normal B-cell development. *Blood* **2018**, *131*, 1920–1930. [CrossRef] [PubMed]
35. Enomoto, Y.; Kitaura, J.; Hatakeyama, K.; Watanuki, J.; Akasaka, T.; Kato, N.; Shimanuki, M.; Nishimura, K.; Takahashi, M.; Taniwaki, M.; et al. Emu/miR-125b transgenic mice develop lethal B-cell malignancies. *Leukemia* **2011**, *25*, 1849–1856. [CrossRef]
36. Zhang, J.; Jima, D.D.; Jacobs, C.; Fischer, R.; Gottwein, E.; Huang, G.; Lugar, P.L.; Lagoo, A.S.; Rizzieri, D.A.; Friedman, D.R.; et al. Patterns of microRNA expression characterize stages of human B-cell differentiation. *Blood* **2009**, *113*, 4586–4594. [CrossRef]
37. He, Y.; Pear, W.S. Notch signalling in B cells. *Semin. Cell Dev. Biol.* **2003**, *14*, 135–142. [CrossRef]
38. King, J.K.; Ung, N.M.; Paing, M.H.; Contreras, J.R.; Alberti, M.O.; Fernando, T.R.; Zhang, K.; Pellegrini, M.; Rao, D.S. Regulation of Marginal Zone B-Cell Differentiation by MicroRNA-146a. *Front. Immunol.* **2016**, *7*, 670. [CrossRef]
39. Luo, Z.; Mu, L.; Zheng, Y.; Shen, W.; Li, J.; Xu, L.; Zhong, B.; Liu, Y.; Zhou, Y. NUMB enhances Notch signaling by repressing ubiquitination of NOTCH1 intracellular domain. *J. Mol. Cell Biol.* **2020**, *12*, 345–358. [CrossRef]
40. Kramer, N.J.; Wang, W.L.; Reyes, E.Y.; Kumar, B.; Chen, C.C.; Ramakrishna, C.; Cantin, E.M.; Vonderfecht, S.L.; Taganov, K.D.; Chau, N.; et al. Altered lymphopoiesis and immunodeficiency in miR-142 null mice. *Blood* **2015**, *125*, 3720–3730. [CrossRef]
41. Rao, D.S.; O'Connell, R.M.; Chaudhuri, A.A.; Garcia-Flores, Y.; Geiger, T.L.; Baltimore, D. MicroRNA-34a perturbs B lymphocyte development by repressing the forkhead box transcription factor Foxp1. *Immunity* **2010**, *33*, 48–59. [CrossRef] [PubMed]
42. Medina, P.P.; Nolde, M.; Slack, F.J. OncomiR addiction in an in vivo model of microRNA-21-induced pre-B-cell lymphoma. *Nature* **2010**, *467*, 86–90. [CrossRef] [PubMed]
43. Mok, Y.; Schwierzeck, V.; Thomas, D.C.; Vigorito, E.; Rayner, T.F.; Jarvis, L.B.; Prosser, H.M.; Bradley, A.; Withers, D.R.; Martensson, I.L.; et al. MiR-210 is induced by Oct-2, regulates B cells, and inhibits autoantibody production. *J. Immunol.* **2013**, *191*, 3037–3048. [CrossRef]
44. van Nieuwenhuijze, A.; Dooley, J.; Humblet-Baron, S.; Sreenivasan, J.; Koenders, M.; Schlenner, S.M.; Linterman, M.; Liston, A. Defective germinal center B-cell response and reduced arthritic pathology in microRNA-29a-deficient mice. *Cell. Mol. Life Sci. CMLS* **2017**, *74*, 2095–2106. [CrossRef] [PubMed]
45. Underbayev, C.; Kasar, S.; Ruezinsky, W.; Degheidy, H.; Schneider, J.S.; Marti, G.; Bauer, S.R.; Fraidenraich, D.; Lightfoote, M.M.; Parashar, V.; et al. Role of mir-15a/16-1 in early B cell development in a mouse model of chronic lymphocytic leukemia. *Oncotarget* **2016**, *7*, 60986–60999. [CrossRef]
46. Kovaleva, V.; Mora, R.; Park, Y.J.; Plass, C.; Chiramel, A.I.; Bartenschlager, R.; Dohner, H.; Stilgenbauer, S.; Pscherer, A.; Lichter, P.; et al. miRNA-130a targets ATG2B and DICER1 to inhibit autophagy and trigger killing of chronic lymphocytic leukemia cells. *Cancer Res.* **2012**, *72*, 1763–1772. [CrossRef]
47. Tili, E.; Michaille, J.J.; Luo, Z.; Volinia, S.; Rassenti, L.Z.; Kipps, T.J.; Croce, C.M. The down-regulation of miR-125b in chronic lymphocytic leukemias leads to metabolic adaptation of cells to a transformed state. *Blood* **2012**, *120*, 2631–2638. [CrossRef]
48. Baer, C.; Oakes, C.C.; Ruppert, A.S.; Claus, R.; Kim-Wanner, S.Z.; Mertens, D.; Zenz, T.; Stilgenbauer, S.; Byrd, J.C.; Plass, C. Epigenetic silencing of miR-708 enhances NF-kappaB signaling in chronic lymphocytic leukemia. *Int. J. Cancer* **2015**, *137*, 1352–1361. [CrossRef]

49. Farahani, M.; Rubbi, C.; Liu, L.; Slupsky, J.R.; Kalakonda, N. CLL Exosomes Modulate the Transcriptome and Behaviour of Recipient Stromal Cells and Are Selectively Enriched in miR-202-3p. *PLoS ONE* **2015**, *10*, e0141429. [CrossRef]
50. Bottoni, A.; Rizzotto, L.; Lai, T.H.; Liu, C.; Smith, L.L.; Mantel, R.; Reiff, S.; El-Gamal, D.; Larkin, K.; Johnson, A.J.; et al. Targeting BTK through microRNA in chronic lymphocytic leukemia. *Blood* **2016**, *128*, 3101–3112. [CrossRef]
51. Pekarsky, Y.; Santanam, U.; Cimmino, A.; Palamarchuk, A.; Efanov, A.; Maximov, V.; Volinia, S.; Alder, H.; Liu, C.G.; Rassenti, L.; et al. Tcl1 expression in chronic lymphocytic leukemia is regulated by miR-29 and miR-181. *Cancer Res.* **2006**, *66*, 11590–11593. [CrossRef] [PubMed]
52. Sivina, M.; Hartmann, E.; Vasyutina, E.; Boucas, J.M.; Breuer, A.; Keating, M.J.; Wierda, W.G.; Rosenwald, A.; Herling, M.; Burger, J.A. Stromal cells modulate TCL1 expression, interacting AP-1 components and TCL1-targeting micro-RNAs in chronic lymphocytic leukemia. *Leukemia* **2012**, *26*, 1812–1820. [CrossRef] [PubMed]
53. Vasyutina, E.; Boucas, J.M.; Bloehdorn, J.; Aszyk, C.; Crispatzu, G.; Stiefelhagen, M.; Breuer, A.; Mayer, P.; Lengerke, C.; Dohner, H.; et al. The regulatory interaction of EVI1 with the TCL1A oncogene impacts cell survival and clinical outcome in CLL. *Leukemia* **2015**, *29*, 2003–2014. [CrossRef] [PubMed]
54. Cimmino, A.; Calin, G.A.; Fabbri, M.; Iorio, M.V.; Ferracin, M.; Shimizu, M.; Wojcik, S.E.; Aqeilan, R.I.; Zupo, S.; Dono, M.; et al. miR-15 and miR-16 induce apoptosis by targeting BCL2. *Proc. Natl. Acad. Sci. USA* **2005**, *102*, 13944–13949. [CrossRef] [PubMed]
55. Rampazzo, E.; Bojnik, E.; Trentin, L.; Bonaldi, L.; Del Bianco, P.; Frezzato, F.; Visentin, A.; Facco, M.; Semenzato, G.; De Rossi, A. Role of miR-15a/miR-16-1 and the TP53 axis in regulating telomerase expression in chronic lymphocytic leukemia. *Haematologica* **2017**, *102*, e253–e256. [CrossRef] [PubMed]
56. Rodriguez-Vicente, A.E.; Quwaider, D.; Benito, R.; Misiewicz-Krzeminska, I.; Hernandez-Sanchez, M.; de Coca, A.G.; Fisac, R.; Alonso, J.M.; Zato, C.; de Paz, J.F.; et al. MicroRNA-223 is a novel negative regulator of HSP90B1 in CLL. *BMC Cancer* **2015**, *15*, 238. [CrossRef] [PubMed]
57. Rozovski, U.; Calin, G.A.; Setoyama, T.; D'Abundo, L.; Harris, D.M.; Li, P.; Liu, Z.; Grgurevic, S.; Ferrajoli, A.; Faderl, S.; et al. Signal transducer and activator of transcription (STAT)-3 regulates microRNA gene expression in chronic lymphocytic leukemia cells. *Mol. Cancer* **2013**, *12*, 50. [CrossRef] [PubMed]
58. Kluiver, J.L.; Chen, C.Z. MicroRNAs regulate B-cell receptor signaling-induced apoptosis. *Genes Immun.* **2012**, *13*, 239–244. [CrossRef]
59. Cui, B.; Chen, L.; Zhang, S.; Mraz, M.; Fecteau, J.F.; Yu, J.; Ghia, E.M.; Zhang, L.; Bao, L.; Rassenti, L.Z.; et al. MicroRNA-155 influences B-cell receptor signaling and associates with aggressive disease in chronic lymphocytic leukemia. *Blood* **2014**, *124*, 546–554. [CrossRef]
60. Mraz, M.; Chen, L.; Rassenti, L.Z.; Ghia, E.M.; Li, H.; Jepsen, K.; Smith, E.N.; Messer, K.; Frazer, K.A.; Kipps, T.J. miR-150 influences B-cell receptor signaling in chronic lymphocytic leukemia by regulating expression of GAB1 and FOXP1. *Blood* **2014**, *124*, 84–95. [CrossRef]
61. Cerna, K.; Oppelt, J.; Chochola, V.; Musilova, K.; Seda, V.; Pavlasova, G.; Radova, L.; Arigoni, M.; Calogero, R.A.; Benes, V.; et al. MicroRNA miR-34a downregulates FOXP1 during DNA damage response to limit BCR signalling in chronic lymphocytic leukaemia B cells. *Leukemia* **2019**, *33*, 403–414. [CrossRef] [PubMed]
62. Sharma, S.; Pavlasova, G.M.; Seda, V.; Cerna, K.A.; Vojackova, E.; Filip, D.; Ondrisova, L.; Sandova, V.; Kostalova, L.; Zeni, P.F.; et al. miR-29 Modulates CD40 Signaling in Chronic Lymphocytic Leukemia by Targeting TRAF4: An Axis Affected by BCR inhibitors. *Blood* **2020**. [CrossRef]
63. Shukla, A.; Rai, K.; Shukla, V.; Chaturvedi, N.K.; Bociek, R.G.; Pirruccello, S.J.; Band, H.; Lu, R.; Joshi, S.S. Sprouty 2: A novel attenuator of B-cell receptor and MAPK-Erk signaling in CLL. *Blood* **2016**, *127*, 2310–2321. [CrossRef] [PubMed]
64. Wang, L.Q.; Wong, K.Y.; Rosen, A.; Chim, C.S. Epigenetic silencing of tumor suppressor miR-3151 contributes to Chinese chronic lymphocytic leukemia by constitutive activation of MADD/ERK and PIK3R2/AKT signaling pathways. *Oncotarget* **2015**, *6*, 44422–44436. [CrossRef]
65. Guinn, D.; Lehman, A.; Fabian, C.; Yu, L.; Maddocks, K.; Andritsos, L.A.; Jones, J.A.; Flynn, J.M.; Jaglowski, S.M.; Woyach, J.A.; et al. The regulation of tumor-suppressive microRNA, miR-126, in chronic lymphocytic leukemia. *Cancer Med.* **2017**, *6*, 778–787. [CrossRef]
66. Palacios, F.; Abreu, C.; Prieto, D.; Morande, P.; Ruiz, S.; Fernandez-Calero, T.; Naya, H.; Libisch, G.; Robello, C.; Landoni, A.I.; et al. Activation of the PI3K/AKT pathway by microRNA-22 results in CLL B-cell proliferation. *Leukemia* **2015**, *29*, 115–125. [CrossRef]
67. Zou, Z.J.; Fan, L.; Wang, L.; Xu, J.; Zhang, R.; Tian, T.; Li, J.Y.; Xu, W. miR-26a and miR-214 down-regulate expression of the PTEN gene in chronic lymphocytic leukemia, but not PTEN mutation or promoter methylation. *Oncotarget* **2015**, *6*, 1276–1285. [CrossRef]
68. Fabbri, M.; Bottoni, A.; Shimizu, M.; Spizzo, R.; Nicoloso, M.S.; Rossi, S.; Barbarotto, E.; Cimmino, A.; Adair, B.; Wojcik, S.E.; et al. Association of a microRNA/TP53 feedback circuitry with pathogenesis and outcome of B-cell chronic lymphocytic leukemia. *JAMA* **2011**, *305*, 59–67. [CrossRef]
69. Baer, C.; Claus, R.; Frenzel, L.P.; Zucknick, M.; Park, Y.J.; Gu, L.; Weichenhan, D.; Fischer, M.; Pallasch, C.P.; Herpel, E.; et al. Extensive promoter DNA hypermethylation and hypomethylation is associated with aberrant microRNA expression in chronic lymphocytic leukemia. *Cancer Res.* **2012**, *72*, 3775–3785. [CrossRef]

70. Deneberg, S.; Kanduri, M.; Ali, D.; Bengtzen, S.; Karimi, M.; Qu, Y.; Kimby, E.; Mansouri, L.; Rosenquist, R.; Lennartsson, A.; et al. microRNA-34b/c on chromosome 11q23 is aberrantly methylated in chronic lymphocytic leukemia. *Epigenetics* **2014**, *9*, 910–917. [CrossRef]
71. Wang, L.Q.; Kwong, Y.L.; Wong, K.F.; Kho, C.S.; Jin, D.Y.; Tse, E.; Rosen, A.; Chim, C.S. Epigenetic inactivation of mir-34b/c in addition to mir-34a and DAPK1 in chronic lymphocytic leukemia. *J. Transl. Med.* **2014**, *12*, 52. [CrossRef]
72. Sampath, D.; Liu, C.; Vasan, K.; Sulda, M.; Puduvalli, V.K.; Wierda, W.G.; Keating, M.J. Histone deacetylases mediate the silencing of miR-15a, miR-16, and miR-29b in chronic lymphocytic leukemia. *Blood* **2012**, *119*, 1162–1172. [CrossRef] [PubMed]
73. Gassner, F.J.; Zaborsky, N.; Feldbacher, D.; Greil, R.; Geisberger, R. RNA Editing Alters miRNA Function in Chronic Lymphocytic Leukemia. *Cancers* **2020**, *12*, 1159. [CrossRef] [PubMed]
74. Xia, L.; Wu, L.; Bao, J.; Li, Q.; Chen, X.; Xia, H.; Xia, R. Circular RNA circ-CBFB promotes proliferation and inhibits apoptosis in chronic lymphocytic leukemia through regulating miR-607/FZD3/Wnt/beta-catenin pathway. *Biochem. Biophys. Res. Commun.* **2018**, *503*, 385–390. [CrossRef]
75. Wu, W.; Wu, Z.; Xia, Y.; Qin, S.; Li, Y.; Wu, J.; Liang, J.; Wang, L.; Zhu, H.; Fan, L.; et al. Downregulation of circ_0132266 in chronic lymphocytic leukemia promoted cell viability through miR-337-3p/PML axis. *Aging* **2019**, *11*, 3561–3573. [CrossRef] [PubMed]
76. Chen, N.; Feng, L.; Lu, K.; Li, P.; Lv, X.; Wang, X. STAT6 phosphorylation upregulates microRNA-155 expression and subsequently enhances the pathogenesis of chronic lymphocytic leukemia. *Oncology Lett.* **2019**, *18*, 95–100. [CrossRef]
77. Frenquelli, M.; Muzio, M.; Scielzo, C.; Fazi, C.; Scarfo, L.; Rossi, C.; Ferrari, G.; Ghia, P.; Caligaris-Cappio, F. MicroRNA and proliferation control in chronic lymphocytic leukemia: Functional relationship between miR-221/222 cluster and p27. *Blood* **2010**, *115*, 3949–3959. [CrossRef]
78. Rassenti, L.Z.; Balatti, V.; Ghia, E.M.; Palamarchuk, A.; Tomasello, L.; Fadda, P.; Pekarsky, Y.; Widhopf, G.F., 2nd; Kipps, T.J.; Croce, C.M. MicroRNA dysregulation to identify therapeutic target combinations for chronic lymphocytic leukemia. *Proc. Natl. Acad. Sci. USA* **2017**, *114*, 10731–10736. [CrossRef]
79. Kopparapu, P.K.; Bhoi, S.; Mansouri, L.; Arabanian, L.S.; Plevova, K.; Pospisilova, S.; Wasik, A.M.; Croci, G.A.; Sander, B.; Paulli, M.; et al. Epigenetic silencing of miR-26A1 in chronic lymphocytic leukemia and mantle cell lymphoma: Impact on EZH2 expression. *Epigenetics* **2016**, *11*, 335–343. [CrossRef]
80. Sampath, D.; Calin, G.A.; Puduvalli, V.K.; Gopisetty, G.; Taccioli, C.; Liu, C.G.; Ewald, B.; Liu, C.; Keating, M.J.; Plunkett, W. Specific activation of microRNA106b enables the p73 apoptotic response in chronic lymphocytic leukemia by targeting the ubiquitin ligase Itch for degradation. *Blood* **2009**, *113*, 3744–3753. [CrossRef]
81. Yang, Y.Q.; Tian, T.; Zhu, H.Y.; Liang, J.H.; Wu, W.; Wu, J.Z.; Xia, Y.; Wang, L.; Fan, L.; Li, J.Y.; et al. NDRG2 mRNA levels and miR-28-5p and miR-650 activity in chronic lymphocytic leukemia. *BMC Cancer* **2018**, *18*, 1009. [CrossRef] [PubMed]
82. Calin, G.A.; Liu, C.G.; Sevignani, C.; Ferracin, M.; Felli, N.; Dumitru, C.D.; Shimizu, M.; Cimmino, A.; Zupo, S.; Dono, M.; et al. MicroRNA profiling reveals distinct signatures in B cell chronic lymphocytic leukemias. *Proc. Natl. Acad. Sci. USA* **2004**, *101*, 11755–11760. [CrossRef] [PubMed]
83. Calin, G.A.; Ferracin, M.; Cimmino, A.; Di Leva, G.; Shimizu, M.; Wojcik, S.E.; Iorio, M.V.; Visone, R.; Sever, N.I.; Fabbri, M.; et al. A MicroRNA signature associated with prognosis and progression in chronic lymphocytic leukemia. *N. Engl. J. Med.* **2005**, *353*, 1793–1801. [CrossRef] [PubMed]
84. Visone, R.; Veronese, A.; Rassenti, L.Z.; Balatti, V.; Pearl, D.K.; Acunzo, M.; Volinia, S.; Taccioli, C.; Kipps, T.J.; Croce, C.M. miR-181b is a biomarker of disease progression in chronic lymphocytic leukemia. *Blood* **2011**, *118*, 3072–3079. [CrossRef]
85. Ferrajoli, A.; Shanafelt, T.D.; Ivan, C.; Shimizu, M.; Rabe, K.G.; Nouraee, N.; Ikuo, M.; Ghosh, A.K.; Lerner, S.; Rassenti, L.Z.; et al. Prognostic value of miR-155 in individuals with monoclonal B-cell lymphocytosis and patients with B chronic lymphocytic leukemia. *Blood* **2013**, *122*, 1891–1899. [CrossRef]
86. Papageorgiou, S.G.; Kontos, C.K.; Diamantopoulos, M.A.; Bouchla, A.; Glezou, E.; Bazani, E.; Pappa, V.; Scorilas, A. MicroRNA-155-5p Overexpression in Peripheral Blood Mononuclear Cells of Chronic Lymphocytic Leukemia Patients Is a Novel, Independent Molecular Biomarker of Poor Prognosis. *Dis. Markers* **2017**, *2017*, 2046545. [CrossRef]
87. Mraz, M.; Malinova, K.; Kotaskova, J.; Pavlova, S.; Tichy, B.; Malcikova, J.; Stano Kozubik, K.; Smardova, J.; Brychtova, Y.; Doubek, M.; et al. miR-34a, miR-29c and miR-17-5p are downregulated in CLL patients with TP53 abnormalities. *Leukemia* **2009**, *23*, 1159–1163. [CrossRef]
88. Mertens, D.; Philippen, A.; Ruppel, M.; Allegra, D.; Bhattacharya, N.; Tschuch, C.; Wolf, S.; Idler, I.; Zenz, T.; Stilgenbauer, S. Chronic lymphocytic leukemia and 13q14: miRs and more. *Leuk. Lymphoma* **2009**, *50*, 502–505. [CrossRef]
89. Visone, R.; Rassenti, L.Z.; Veronese, A.; Taccioli, C.; Costinean, S.; Aguda, B.D.; Volinia, S.; Ferracin, M.; Palatini, J.; Balatti, V.; et al. Karyotype-specific microRNA signature in chronic lymphocytic leukemia. *Blood* **2009**, *114*, 3872–3879. [CrossRef]
90. Stamatopoulos, B.; Meuleman, N.; De Bruyn, C.; Pieters, K.; Anthoine, G.; Mineur, P.; Bron, D.; Lagneaux, L. A molecular score by quantitative PCR as a new prognostic tool at diagnosis for chronic lymphocytic leukemia patients. *PLoS ONE* **2010**, *5*. [CrossRef]
91. Rossi, S.; Shimizu, M.; Barbarotto, E.; Nicoloso, M.S.; Dimitri, F.; Sampath, D.; Fabbri, M.; Lerner, S.; Barron, L.L.; Rassenti, L.Z.; et al. microRNA fingerprinting of CLL patients with chromosome 17p deletion identify a miR-21 score that stratifies early survival. *Blood* **2010**, *116*, 945–952. [CrossRef] [PubMed]
92. Moussay, E.; Palissot, V.; Vallar, L.; Poirel, H.A.; Wenner, T.; El Khoury, V.; Aouali, N.; Van Moer, K.; Leners, B.; Bernardin, F.; et al. Determination of genes and microRNAs involved in the resistance to fludarabine in vivo in chronic lymphocytic leukemia. *Mol. Cancer* **2010**, *9*, 115. [CrossRef] [PubMed]

93. Gagez, A.L.; Duroux-Richard, I.; Lepretre, S.; Orsini-Piocelle, F.; Letestu, R.; De Guibert, S.; Tuaillon, E.; Leblond, V.; Khalifa, O.; Gouilleux-Gruart, V.; et al. miR-125b and miR-532-3p predict the efficiency of rituximab-mediated lymphodepletion in chronic lymphocytic leukemia patients. A French Innovative Leukemia Organization study. *Haematologica* **2017**, *102*, 746–754. [CrossRef]
94. Zenz, T.; Mohr, J.; Eldering, E.; Kater, A.P.; Buhler, A.; Kienle, D.; Winkler, D.; Durig, J.; van Oers, M.H.; Mertens, D.; et al. miR-34a as part of the resistance network in chronic lymphocytic leukemia. *Blood* **2009**, *113*, 3801–3808. [CrossRef] [PubMed]
95. Ferracin, M.; Zagatti, B.; Rizzotto, L.; Cavazzini, F.; Veronese, A.; Ciccone, M.; Saccenti, E.; Lupini, L.; Grilli, A.; De Angeli, C.; et al. MicroRNAs involvement in fludarabine refractory chronic lymphocytic leukemia. *Mol. Cancer* **2010**, *9*, 123. [CrossRef]
96. Papageorgiou, S.G.; Kontos, C.K.; Tsiakanikas, P.; Stavroulaki, G.; Bouchla, A.; Vasilatou, D.; Bazani, E.; Lazarakou, A.; Scorilas, A.; Pappa, V. Elevated miR-20b-5p expression in peripheral blood mononuclear cells: A novel, independent molecular biomarker of favorable prognosis in chronic lymphocytic leukemia. *Leuk. Res.* **2018**, *70*, 1–7. [CrossRef] [PubMed]
97. Kaur, G.; Ruhela, V.; Rani, L.; Gupta, A.; Sriram, K.; Gogia, A.; Sharma, A.; Kumar, L.; Gupta, R. RNA-Seq profiling of deregulated miRs in CLL and their impact on clinical outcome. *Blood Cancer J.* **2020**, *10*, 6. [CrossRef]
98. Papageorgiou, S.G.; Diamantopoulos, M.A.; Kontos, C.K.; Bouchla, A.; Vasilatou, D.; Bazani, E.; Scorilas, A.; Pappa, V. MicroRNA-92a-3p overexpression in peripheral blood mononuclear cells is an independent predictor of prolonged overall survival of patients with chronic lymphocytic leukemia. *Leuk. Lymphoma* **2019**, *60*, 658–667. [CrossRef]
99. Casabonne, D.; Benavente, Y.; Seifert, J.; Costas, L.; Armesto, M.; Arestin, M.; Besson, C.; Hosnijeh, F.S.; Duell, E.J.; Weiderpass, E.; et al. Serum levels of hsa-miR-16-5p, hsa-miR-29a-3p, hsa-miR-150-5p, hsa-miR-155-5p and hsa-miR-223-3p and subsequent risk of chronic lymphocytic leukemia in the EPIC study. *Int. J. Cancer* **2020**, *147*, 1315–1324. [CrossRef]
100. Moussay, E.; Wang, K.; Cho, J.H.; van Moer, K.; Pierson, S.; Paggetti, J.; Nazarov, P.V.; Palissot, V.; Hood, L.E.; Berchem, G.; et al. MicroRNA as biomarkers and regulators in B-cell chronic lymphocytic leukemia. *Proc. Natl. Acad. Sci. USA* **2011**, *108*, 6573–6578. [CrossRef]
101. Negrini, M.; Cutrona, G.; Bassi, C.; Fabris, S.; Zagatti, B.; Colombo, M.; Ferracin, M.; D'Abundo, L.; Saccenti, E.; Matis, S.; et al. microRNAome expression in chronic lymphocytic leukemia: Comparison with normal B-cell subsets and correlations with prognostic and clinical parameters. *Clin. Cancer Res. Off. J. Am. Assoc. Cancer Res.* **2014**, *20*, 4141–4153. [CrossRef] [PubMed]
102. Salerno, E.; Scaglione, B.J.; Coffman, F.D.; Brown, B.D.; Baccarini, A.; Fernandes, H.; Marti, G.; Raveche, E.S. Correcting miR-15a/16 genetic defect in New Zealand Black mouse model of CLL enhances drug sensitivity. *Mol. Cancer Ther.* **2009**, *8*, 2684–2692. [CrossRef] [PubMed]
103. Kasar, S.; Salerno, E.; Yuan, Y.; Underbayev, C.; Vollenweider, D.; Laurindo, M.F.; Fernandes, H.; Bonci, D.; Addario, A.; Mazzella, F.; et al. Systemic in vivo lentiviral delivery of miR-15a/16 reduces malignancy in the NZB de novo mouse model of chronic lymphocytic leukemia. *Genes Immun.* **2012**, *13*, 109–119. [CrossRef] [PubMed]
104. Fonte, E.; Apollonio, B.; Scarfo, L.; Ranghetti, P.; Fazi, C.; Ghia, P.; Caligaris-Cappio, F.; Muzio, M. In vitro sensitivity of CLL cells to fludarabine may be modulated by the stimulation of Toll-like receptors. *Clin. Cancer Res. Off. J. Am. Assoc. Cancer Res.* **2013**, *19*, 367–379. [CrossRef] [PubMed]
105. Dehkordi, K.A.; Chaleshtori, M.H.; Sharifi, M.; Jalili, A.; Fathi, F.; Roshani, D.; Nikkhoo, B.; Hakhamaneshi, M.S.; Sani, M.R.M.; Ganji-Arjenaki, M. Inhibition of MicroRNA miR-222 with LNA Inhibitor Can Reduce Cell Proliferation in B Chronic Lymphoblastic Leukemia. *Indian J. Hematol. Blood Transfus. Off. J. Indian Soc. Hematol. Blood Transfus.* **2017**, *33*, 327–332. [CrossRef]
106. Zhu, D.X.; Zhu, W.; Fang, C.; Fan, L.; Zou, Z.J.; Wang, Y.H.; Liu, P.; Hong, M.; Miao, K.R.; Liu, P.; et al. miR-181a/b significantly enhances drug sensitivity in chronic lymphocytic leukemia cells via targeting multiple anti-apoptosis genes. *Carcinogenesis* **2012**, *33*, 1294–1301. [CrossRef]
107. Bresin, A.; Callegari, E.; D'Abundo, L.; Cattani, C.; Bassi, C.; Zagatti, B.; Narducci, M.G.; Caprini, E.; Pekarsky, Y.; Croce, C.M.; et al. miR-181b as a therapeutic agent for chronic lymphocytic leukemia in the Emicro-TCL1 mouse model. *Oncotarget* **2015**, *6*, 19807–19818. [CrossRef]
108. Saleh, L.M.; Wang, W.; Herman, S.E.; Saba, N.S.; Anastas, V.; Barber, E.; Corrigan-Cummins, M.; Farooqui, M.; Sun, C.; Sarasua, S.M.; et al. Ibrutinib downregulates a subset of miRNA leading to upregulation of tumor suppressors and inhibition of cell proliferation in chronic lymphocytic leukemia. *Leukemia* **2017**, *31*, 340–349. [CrossRef]
109. Dereani, S.; Macor, P.; D'Agaro, T.; Mezzaroba, N.; Dal-Bo, M.; Capolla, S.; Zucchetto, A.; Tissino, E.; Del Poeta, G.; Zorzet, S.; et al. Potential therapeutic role of antagomiR17 for the treatment of chronic lymphocytic leukemia. *J. Hematol. Oncol.* **2014**, *7*, 79. [CrossRef]
110. D'Abundo, L.; Callegari, E.; Bresin, A.; Chillemi, A.; Elamin, B.K.; Guerriero, P.; Huang, X.; Saccenti, E.; Hussein, E.; Casciano, F.; et al. Anti-leukemic activity of microRNA-26a in a chronic lymphocytic leukemia mouse model. *Oncogene* **2017**, *36*, 6617–6626. [CrossRef]
111. Chiang, C.L.; Goswami, S.; Frissora, F.W.; Xie, Z.; Yan, P.S.; Bundschuh, R.; Walker, L.A.; Huang, X.; Mani, R.; Mo, X.M.; et al. ROR1-targeted delivery of miR-29b induces cell cycle arrest and therapeutic benefit in vivo in a CLL mouse model. *Blood* **2019**, *134*, 432–444. [CrossRef] [PubMed]
112. Cutrona, G.; Matis, S.; Colombo, M.; Massucco, C.; Baio, G.; Valdora, F.; Emionite, L.; Fabris, S.; Recchia, A.G.; Gentile, M.; et al. Effects of miRNA-15 and miRNA-16 expression replacement in chronic lymphocytic leukemia: Implication for therapy. *Leukemia* **2017**, *31*, 1894–1904. [CrossRef] [PubMed]
113. Dolcetti, R.; Carbone, A. Epstein-Barr virus infection and chronic lymphocytic leukemia: A possible progression factor? *Infect. Agents Cancer* **2010**, *5*, 22. [CrossRef] [PubMed]

114. Visco, C.; Falisi, E.; Young, K.H.; Pascarella, M.; Perbellini, O.; Carli, G.; Novella, E.; Rossi, D.; Giaretta, I.; Cavallini, C.; et al. Epstein-Barr virus DNA load in chronic lymphocytic leukemia is an independent predictor of clinical course and survival. *Oncotarget* **2015**, *6*, 18653–18663. [CrossRef] [PubMed]
115. Liang, J.H.; Gao, R.; Xia, Y.; Gale, R.P.; Chen, R.Z.; Yang, Y.Q.; Wang, L.; Qu, X.Y.; Qiu, H.R.; Cao, L.; et al. Prognostic impact of Epstein-Barr virus (EBV)-DNA copy number at diagnosis in chronic lymphocytic leukemia. *Oncotarget* **2016**, *7*, 2135–2142. [CrossRef] [PubMed]
116. Iizasa, H.; Kim, H.; Kartika, A.V.; Kanehiro, Y.; Yoshiyama, H. Role of Viral and Host microRNAs in Immune Regulation of Epstein-Barr Virus-Associated Diseases. *Front. Immunol.* **2020**, *11*, 367. [CrossRef]
117. Ferrajoli, A.; Ivan, C.; Ciccone, M.; Shimizu, M.; Kita, Y.; Ohtsuka, M.; D'Abundo, L.; Qiang, J.; Lerner, S.; Nouraee, N.; et al. Epstein-Barr Virus MicroRNAs are Expressed in Patients with Chronic Lymphocytic Leukemia and Correlate with Overall Survival. *EBioMedicine* **2015**, *2*, 572–582. [CrossRef]
118. Linnstaedt, S.D.; Gottwein, E.; Skalsky, R.L.; Luftig, M.A.; Cullen, B.R. Virally induced cellular microRNA miR-155 plays a key role in B-cell immortalization by Epstein-Barr virus. *J. Virol.* **2010**, *84*, 11670–11678. [CrossRef]
119. Xu, D.M.; Kong, Y.L.; Wang, L.; Zhu, H.Y.; Wu, J.Z.; Xia, Y.; Li, Y.; Qin, S.C.; Fan, L.; Li, J.Y.; et al. EBV-miR-BHRF1-1 Targets p53 Gene: Potential Role in Epstein-Barr Virus Associated Chronic Lymphocytic Leukemia. *Cancer Res. Treat. Off. J. Korean Cancer Assoc.* **2020**, *52*, 492–504. [CrossRef]
120. Stamatopoulos, B.; Van Damme, M.; Crompot, E.; Dessars, B.; Housni, H.E.; Mineur, P.; Meuleman, N.; Bron, D.; Lagneaux, L. Opposite Prognostic Significance of Cellular and Serum Circulating MicroRNA-150 in Patients with Chronic Lymphocytic Leukemia. *Mol. Med.* **2015**, *21*, 123–133. [CrossRef]
121. Balatti, V.; Tomasello, L.; Rassenti, L.Z.; Veneziano, D.; Nigita, G.; Wang, H.Y.; Thorson, J.A.; Kipps, T.J.; Pekarsky, Y.; Croce, C.M. miR-125a and miR-34a expression predicts Richter syndrome in chronic lymphocytic leukemia patients. *Blood* **2018**, *132*, 2179–2182. [CrossRef] [PubMed]
122. Martin-Guerrero, I.; Gutierrez-Camino, A.; Lopez-Lopez, E.; Bilbao-Aldaiturriaga, N.; Pombar-Gomez, M.; Ardanaz, M.; Garcia-Orad, A. Genetic variants in miRNA processing genes and pre-miRNAs are associated with the risk of chronic lymphocytic leukemia. *PLoS ONE* **2015**, *10*, e0118905. [CrossRef] [PubMed]

20, 12, 2680

Article

MiRNA-103/107 in Primary High-Grade Serous Ovarian Cancer and Its Clinical Significance

Milosz Wilczynski [1,*], Michal Kielbik [2], Daria Senderowska [3], Tomasz Krawczyk [4], Bozena Szymanska [5], Magdalena Klink [2], Jan Bieńkiewicz [1], Hanna Romanowicz [4], Filip Frühauf [6] and Andrzej Malinowski [7]

1 Department of Operative Gynecology, Endoscopy and Gynecologic Oncology, Polish Mother's Memorial Hospital Research Institute, 281 Rzgowska Str., 93-338 Lodz, Poland; jan.bienkiewicz@umed.lodz.pl

2 Institute of Medical Biology, Polish Academy of Sciences, 106 Lodowa Str., 93-232 Lodz, Poland; mkielbik@cbm.pan.pl (M.K.); mklink@cbm.pan.pl (M.K.)

3 Department of Molecular Medicine, Medical University of Lodz, Al. Kościuszki 4, 90-419 Lodz, Poland; daria.domanska@umed.lodz.pl

4 Department of Clinical Pathology, Polish Mothers' Memorial Hospital-Research Institute, 281 Rzgowska Str., 93-338 Lodz, Poland; tk.pracownia@gmail.com (T.K.); hanna-romanowicz@wp.pl (H.R.)

5 The Central Laboratory of Medical University in Lodz, 6/8 Mazowiecka Str., 92-215 Lodz, Poland; bozena.szymanska@umed.lodz.pl

6 Gynecologic Oncology Center, Department of Obstetrics and Gynecology, First Faculty of Medicine, Charles University in Prague and General University Hospital in Prague, 128 00 Prague, Czech Republic; Filip.Fruehauf@vfn.cz

7 Department of Surgical and Endoscopic Gynecology, Medical University in Lodz, Al. Kościuszki 4, 90-419 Lodz, Poland; amalinowski@kki.pl

* Correspondence: jrwil@wp.pl; Tel.: +48-42-2711131

Received: 21 August 2020; Accepted: 17 September 2020; Published: 19 September 2020

Simple Summary: MiRNA-103/107-DICER axis may be one of the key regulators of cancer aggressiveness. Data on miRNA-103/107 in high grade serous ovarian cancer is scarce. We aimed to assess miRNA-103/107 expression levels in high grade serous ovarian cancer tissues and relate them to patients' clinicopathological data. MiRNA-103/107, DICER expression levels were also evaluated in selected ovarian cancer cell lines. Clinical and prognostic significance of miRNA-103/107 was not confirmed in our study. However, the results of our study support the possible existence of miRNA-103/107- DICER axis in ovarian cancer.

Abstract: High levels of miRNA-103/107 are associated with poor outcomes in the case of breast cancer patients. MiRNA-103/107-DICER axis may be one of the key regulators of cancer aggressiveness. MiRNA-103/107 expression levels have never been related to patients' clinicopathological data in epithelial ovarian cancer. We aimed to assess miRNA-103/107 expression levels in high grade serous ovarian cancer tissues. Expression levels of both miRNAs were related to the clinicopathological features and survival. We also evaluated expression levels of miRNA-103/107 and DICER in selected ovarian cancer cell lines (A2780, A2780cis, SK-OV-3, OVCAR3). We assessed the relative expression of miRNA-103/107 (quantitative reverse transcription-polymerase chain reaction) in fifty archival formalin-fixed paraffin-embedded tissue samples of primary high grade serous ovarian cancer. Then, miRNA-103/107 and DICER expression levels were evaluated in selected ovarian cancer cell lines. Additionally, DICER, N-/E-cadherin protein levels were assessed with the use of western blot. We identified miRNA-107 up-regulation in ovarian cancer in comparison to healthy tissues ($p = 0.0005$). In the case of miRNA-103, we did not observe statistically significant differences between cancerous and healthy tissues ($p = 0.07$). We did not find any correlations between miRNA-103/107 expression levels and clinicopathological features. Kaplan–Meier survival (disease-free and overall survival) analysis revealed that both miRNAs could not be considered as prognostic factors. SK-OV-3 cancer cell lines were characterized by high expression of miRNA-103/107, relatively low expression of DICER (western-blot), and relatively high N-cadherin levels in comparison to other ovarian cancer cell lines. Clinical and prognostic significance of miRNA-103/107 was not confirmed in our study.

Keywords: miRNA; ovarian cancer; survival; prognostic factor

---

1. Introduction

Ovarian cancer is characterized by a high fatality rate and is responsible for approximately 2–3% of all cancer deaths. The early-stage disease has a 5-year survival of 93%. Unfortunately, the majority of patients are diagnosed at the FIGO III or IV stage of the disease, for which the 5-year survival is much lower [1]. Epithelial ovarian cancer (EOC) is the most common type of ovarian malignancy, as only 10% of tumors are of non- epithelial origin. Serous high-grade carcinomas (HGSOC) are the most prevalent type of EOC, which are characterized by TP53 mutations and relatively poor outcome [2].

MicroRNAs (miRNAs) are small (approximately 18–25 nt), single-stranded non-coding RNAs that are evolutionarily conserved among species [3]. MiRNA genes are transcribed by RNA polymerase II to pri-miRNAs, which are cleaved by the microprocessor complex (nuclease Drosha together with DGCR8 protein) to create hairpin-like pre-miRNAs. Pre-miRNAs bind to the Exportin- 5 (RanGTP-dependent transporter) and are exported to the cytoplasm. RNase III-type nuclease enzyme DICER splits pre-miRNAs into two single-stranded forms of miRNA (miRNA-3p and miRNA-5p). One strand creates the mature miRNA, and the second passenger strand is usually destroyed. However, it is possible that the passenger strand can also be selected as a mature form of miRNA. Mature forms of miRNAs are incorporated into the RNA-induced silencing complex (RISC) and bind to the 3′ untranslated regions (UTRs) of their mRNA targets, which causes posttranscriptional suppression/activation of translation or mRNA cleavage [3,4]. MiRNAs possess unique abilities to affect the expression of genes and take part in such cellular processes as proliferation, differentiation, invasion, migration, epithelial-to-mesenchymal transition (EMT), or apoptosis [3,4]. Numerous authors showed miRNAs to be dysregulated in multiple cancers. Aberrant expression of miRNAs has been reported in multiple neoplasms and related to the stage of the disease or clinical outcome [5,6]. A large body of evidence suggests that miRNAs play a crucial role in carcinogenesis, tumor progression, and metastasis. Several miRNAs may be up-regulated in specific neoplasms; however, a global miRNA reduction in human cancers seems to be a common phenomenon [7]. A potential explanation for a global decrease in miRNA expression may be inhibition of DICER. Down-regulation of DICER has been detected in epithelial ovarian cancer (EOC). Meritt et al. reported that low DICER expression was associated with advanced-stage disease and reduced median survival [8]. Martello et al. identified a miRNA family (miRNA-103/107) that inhibited miRNA biosynthesis by targeting DICER [9]. They reported that high levels of miR-103/107 are associated with metastasis and poor outcome in breast cancer. Inhibition of miRNA-103/107 in malignant cells resulted in the attenuation of migratory and metastatic properties. Martello et al. concluded that the up-regulation of miRNA-103/107 is responsible for the induction of EMT, attained by down-regulating miRNA-200 levels. However, the role of miRNA-103/107 in EOC has not been elucidated yet, as the evidence is scarce. The Cancer Genome Atlas (TCGA) provided data on miRNA-103/107 expression levels in EOC [10]. Yang et al. identified miRNA-103 as an oncogene in serous ovarian cancer, which promotes invasion and metastasis via the down-regulation of DICER1 [11]. Apart from those two studies, there is no evidence on the clinical significance of both miRNAs. Several authors reported aberrant miRNA103/107 expression levels in other tumors. Yu et al. identified miRNA-103/107 up-regulation in bladder cancer specimens and revealed its oncogenic role in cell proliferation and PI3K/AKT signaling partially through PTEN dependent mechanism [12]. However, some authors showed that miRNA-107 might also be considered as a tumor suppressor.

Tang et al. described that ectopic expression of miRNA-107 suppressed cell proliferation and was associated with the down-regulation of cyclin E1 (CCNE1) expression [13].

Although miRNA-103/107 may be one of the key-regulators of EOC carcinogenesis through the down-regulation of DICER, its prognostic and clinical significance has not been thoroughly evaluated. Thus, we decided to investigate miRNA-103/107 expression levels in primary HGSOC tissues and relate it to clinicopathological characteristics, with particular attention to overall survival (OS) and progression-free survival (PFS).

2. Results

2.1. miRNA-103/107 Evaluation in FFPE Tissues

We included fifty patients who were diagnosed with serous high-grade ovarian cancer between 2010 and 2018 (clinical data are presented in Table 1).

Table 1. Pathological characteristics and survival data.

Serous, High-Grade Ovarian Cancer Patients	
Cases number (N)	50
Age (range, years)	60.1 (30–83)
FIGO Stage	
I	6
II	1
III(N)	30
IV(N)	13
Recurrence/Progression	
No (N)	9
Yes (N)	41
Platinum-Resistant	
No (N)	38
Yes (N)	12
PFS (Range, months)	29 (2–103)
OS (Range, months)	42.3 (14–112)
Overall Survival	
No (N)	28
Yes (N)	22

PFS—progression-free survival (until the first recurrence/progression); OS—overall survival.

Quantitative Real-time PCR was performed in fifty formalin-fixed paraffin-embedded (FFPE) samples of high grade serous ovarian cancer and ten FFPE samples of normal Fallopian tubes' fimbriae containing normal surface epithelium. Fallopian tubes' fimbriae were chosen as controls, since the majority of serous carcinomas appear to arise from lesions in the distal fallopian tube [14]. An experienced pathologist assessed all samples. Careful microdissection of representative tissue areas was performed. MiRNA-103/107 expression was assessed in all fifty ovarian cancer samples and controls. We were not able to evaluate DICER expression, since mRNA derived from FFPE tissues is sensitive to chemical modifications and degradation. MiRNAs are known to be stable and less affected by the embedding process and degradation over time [15]. We performed immunohistochemical analysis of DICER in our set of FFPE tissue samples (Figure 1). We discovered that majority of ovarian cancer samples showed a reduced level of DICER protein. We found only six samples which were characterized by strong cytoplasmic positivity in cancer cells. What is worth mentioning is the fact that among six samples with strong positivity, there were five samples with low miRNA-103/107 levels.

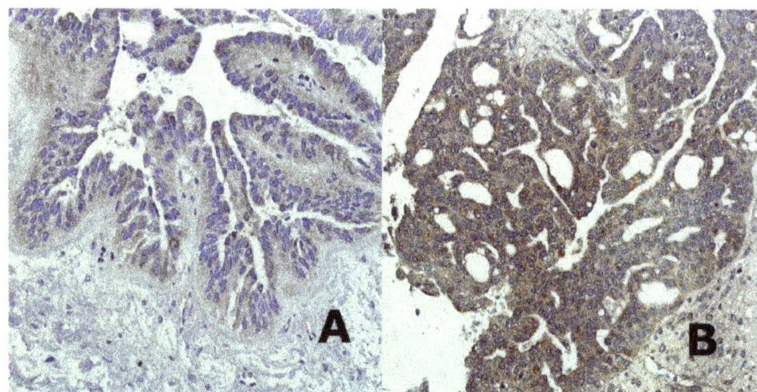

Figure 1. IHC (original magnification 40×) staining of ovarian cancer tissues: (**A**) Weak cytoplasmic positivity for DICER, (**B**) strong cytoplasmic positivity for DICER.

MiRNA-107 expression levels, which were determined in primary high-grade ovarian cancer tissues, showed up-regulation in comparison to Fallopian tubes' fimbriae containing normal surface epithelium (Figure 2, mean RQ for cancer—1.69, lower quartile—0.99, upper quartile—2.99; mean RQ for controls—0.6, lower quartile—0.55, upper quartile—0.89; $p = 0.0005$). In the case of miRNA-103, we did not observe statistically significant differences between cancerous and healthy tissues. However, miRNA-103 showed a trend towards up-regulation in high-grade ovarian cancer samples (mean RQ for cancer—1.24, lower quartile—0.67, upper quartile—1.66; mean RQ for controls—0.77, lower quartile—0.63, upper quartile—0.94; $p = 0.07$). Furthermore, we found a positive correlation between miRNA-103 and miRNA-107 expression values (Spearman's rank correlation coefficient, $p = 0.00006$). We evaluated miRNA-103/107 expression levels regarding such clinicopathological data as overall survival, progression-free survival, FIGO stages, platinum sensitivity, Ca125/HE4 levels, age, and BMI (body mass index). Statistical analysis revealed that there was no correlation between FIGO stages and expression levels of selected miRNAs (ANOVA, miRNA-103—$p = 0.45$; miRNA-107—$p = 0.25$). Similarly, we did not find any significant association between miRNA-103/107 expression levels and BMI (Spearman's rank correlation coefficient, miRNA-103—$p = 0.25$; miRNA-107—$p = 0.83$) or age (Spearman's rank correlation coefficient, miRNA-103—$p = 0.19$; miRNA-107—$p = 0.75$). Spearman's rank correlation coefficient indicated that neither Ca125 nor HE4 showed any relationship with expression levels of selected miRNAs (Spearman's rank correlation coefficient, miRNA-103: Ca125—$p = 0.54$/HE4—$p = 0.35$; miRNA-107: Ca125–$p = 0.26$/HE4—$p = 0.17$). We also evaluated miRNA-103/107 expression levels in the early stage versus the advanced stage of cancer, however, we did not find any correlations (FIGO stage I vs. II, III, IV, Mann–Whitney U test: miRNA-103—$p = 0.8$, miRNA-107—$p = 0.3$). Expression levels (low vs. high) of both miRNAs did not differ between platinum-sensitive and platinum-resistant patients (chi-square test, $p = 0.11$). MiRNA103/107 expression levels in regard to all clinicopathological data are presented in Table 2.

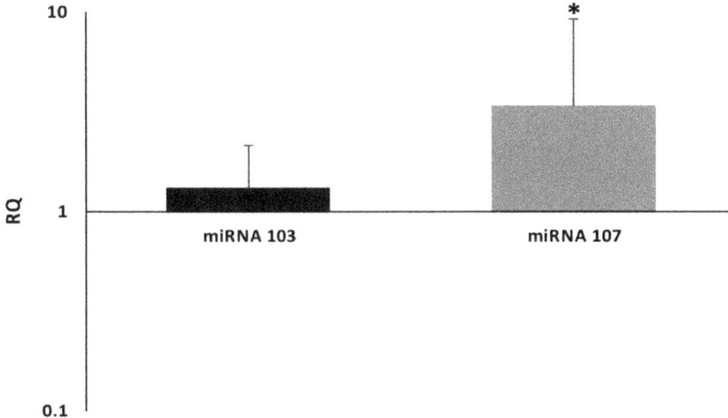

Figure 2. MiRNA-103 and -107 expression in ovarian cancer and control tissue samples. Statistically significant higher expression of miRNA-107: * Mann–Whitney U test, $p = 0.0005$. RQ = 1 for control tissue samples.

Table 2. Basic results of miRNA-103/107 expression regarding clinicopathological data.

	MiRNA-103	MiRNA-107
Mean RQ	1.69	1.24
miRNA expression vs. control tissue, *p* values	A trend towards up-regulation, $p = 0.07$	Up-regulation, $p = 0.0005$
Clinicopathological data, *p* values		
BMI	$p = 0.25$	$p = 0.84$
Age	$p = 0.19$	$p = 0.75$
FIGO stage	$p = 0.45$	$p = 0.25$
FIGO stage I vs. II,III,IV	$p = 0.8$	$p = 0.3$
Ca125	$p = 0.54$	$p = 0.26$
HE4	$p = 0.35$	$p = 0.17$
Deceased vs. Alive	$p = 0.46$	$p = 0.39$
Progression (yes vs. no)	$p = 0.45$	$p = 0.06$
Platinum-resistance (yes vs. no)	$p = 0.37$	$p = 0.21$

2.2. *Survival Analysis*

We divided patients into two groups based on miRNA expression levels (low and high miRNA-103/107) in order to perform Kaplan–Maier survival analysis. Both DFS (disease-free survival) and OS (overall survival) were assessed in regard to miRNA-103/107 expression levels (Figures 3 and 4). Statistical analysis revealed that there were not any significant differences in survival between patients in low and high miRNA-103/107 subgroups.

We also decided to perform survival analysis (OS) based on the data derived from TCGA. OncoLnc is a commonly available tool for exploring survival correlations, and for downloading clinical data coupled to expression data for miRNAs [16]. OncoLnc contains data from studies performed by TCGA. We used data from TCGA cohort of 470 ovarian cancer patients and performed Kaplan–Meier analysis. Due to the high number of cases, we decided to compare the bottom third to the top third according to expression values (cut-off values: RQ for the bottom third 43.8, RQ for the upper third 77.1). Kaplan–Meier survival analysis of TCGA data revealed that patients characterized by high miRNA-107 levels had more favorable overall survival. The survival difference between patients with low and high miRNA-107 levels was significant. However, we did not find any significant differences in survival between patients in low and high miRNA-103 subgroups (Figure 5).

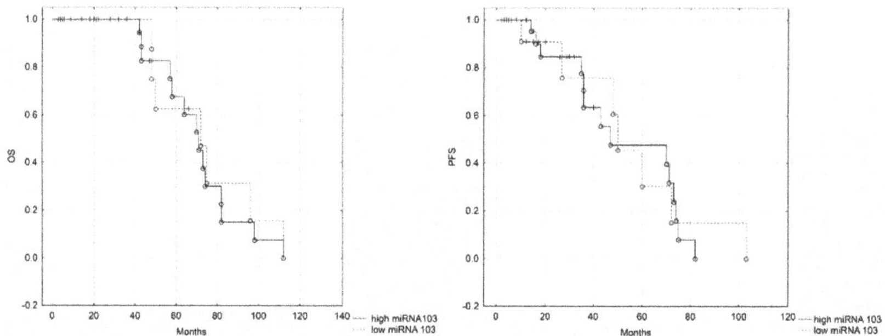

Figure 3. Kaplan–Meier analysis of OS (overall survival) and PFS (progression-free survival)—miRNA-103 (long rank test, $p = 0.68$, $p = 0.78$, respectively).

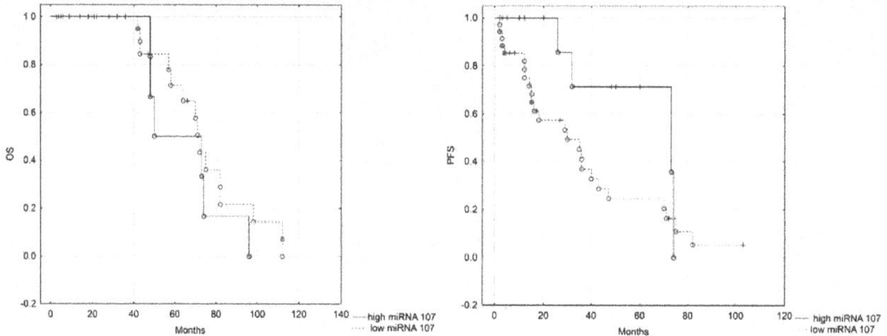

Figure 4. Kaplan–Meier analysis of OS (overall survival) and PFS (progression-free survival)—miRNA-107 (long rank test, $p = 0.34$, $p = 0.09$, respectively).

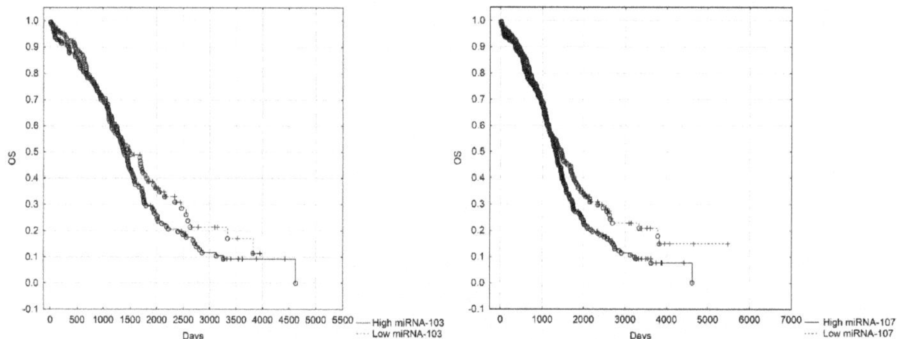

Figure 5. Kaplan–Meier analysis of OS (overall survival) in the Cancer Genome Atlas (TCGA) cohort—miRNA-103 and miRNA-107 (long rank test, $p = 0.16$, $p = 0.01$, respectively).

2.3. *Evaluation of miRNA-103/107, DICER, and N-/E-Cadherin in Selected Ovarian Cancer Cell Lines*

After miRNA-103/107 evaluation in FFPE tissues, we also decided to assess miRNA-103/107 expression levels in the most commonly used, commercially available ovarian cancer cell lines. We hypothesized that potential up-regulation of miRNA-103/107 might be associated with lower DICER expression levels and partial loss of E-cadherin. Such findings would support the possible existence of the miRNA-103/107-DICER axis in ovarian cancer. According to Martello et al., the miRNA-103/107-DICER axis plays a crucial role in epithelial plasticity and fosters the invasive behavior of cancer cells [9]. Since we failed to show any significant clinical correlations in regard to miRNA-103/107 in high grade serous ovarian cancer, we decided to perform a preliminary and simple assessment of various cancer cell lines in order to verify the potential existence of the miRNA-103/107-DICER axis in the case of ovarian cancer. We decided to evaluate DICER, N-/E-cadherin, and miRNA-103/107 expression levels in selected ovarian cancer cell lines. Loss of E-cadherin, together with the presence of N-cadherin are hallmarks of epithelial to mesenchymal transition. We decided to choose cell lines that are commonly used as a template for ovarian cancer: A2780 (cisplatin-sensitive, EC50 = 6 μM) and A2780cis (cisplatin-resistant, EC50 = 30 μM) cell lines; group 2—resistant to cisplatin, but differing in the degree of aggressiveness SK-OV-3 (aggressive, EC50 = 38μM) and OVCAR-3 (non-aggressive, EC50 = 20 μM). SK-OV-3, OVCAR -3, A2780, and A2780cis cancer cell lines present different properties in regard to aggressiveness, chemo-sensitivity, epithelial character, or metastatic potential. SK-OV-3 is considered to be an aggressive and chemoresistant ovarian cancer cell line, probably derived from clear cell carcinoma. OVCAR-3 is a typical epithelial serous-like ovarian cancer cell line that seems to be modestly resistant to cisplatin and less aggressive than SK-OV-3. A2780 (cisplatin not resistant) and A2780cis (cisplatin-resistant) cancer cell lines are probably derived from undifferentiated or endometrioid ovarian carcinoma [17].

Such diversification among ovarian cancer cell lines allowed us to evaluate DICER, N-/E-cadherin, and miRNA-103/107 expression in molecularly different types of cells and increased our chances of detecting any correlations among them. We hypothesized that the cell lines with the highest miRNA-103/107 expression levels would be associated with low DICER, which would contribute to a partial loss of E-cadherin and higher expression of N-cadherin at the same time. The highest miRNA-103/107 expression levels were detected in SK-OV-3 and A2780cis ovarian cancer cells (Figure 6). We performed mRNA assessment by RT-PCR and did not detect low DICER expression in SK-OV-3 in comparison to other ovarian cancer cells (Figure 6). The discrepancy between mRNA levels and protein expression is a common phenomenon [18]. Thus, we decided to use the western blotting method in order to re-evaluate DICER expression in selected ovarian cancer tissues (Figures 7 and 8).

Interestingly, there was a difference between the level of DICER protein assessed by western-blotting and the expression of its mRNA in the case of A2780 and SK-OV-3 cell lines. A low DICER protein level characterized SK-OV-3 cells as opposed to its mRNA level. A2780 cells were characterized by high DICER protein level as opposed to its mRNA level. We compared A2780 (cisplatin-sensitive) and A2780cis (cisplatin-resistant) ovarian cancer cells in regard to miRNA-103/107 and DICER expression levels. We found that level of DICER protein was significantly higher in A2780 cells in comparison to other tested cell lines (Mann–Whitney U test $p \leq 0.007$). Furthermore, we observed higher expression levels of miRNA-103/107 in A2780cis cells (Figure 6; Mann–Whitney U test miRNA-103—$p \leq 0.02$; miRNA-107—$p \leq 0.02$). Similarly, SK-OV-3 (cisplatin-resistant) ovarian cancer cells showed significantly higher expression levels of miRNA-103/107 in comparison to A2780 cells (cisplatin-sensitive) (Figure 6; Mann–Whitney U test miRNA-103—$p \leq 0.02$; miRNA-107—$p \leq 0.02$).

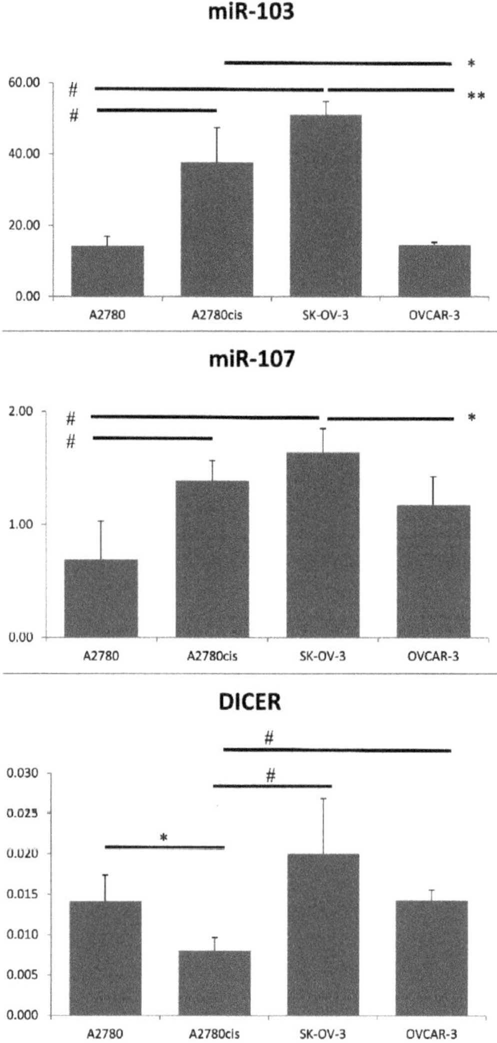

Figure 6. mRNA level of DICER, miRNA-103, and miRNA-107 in A2780, A2780cis, SK-OV-3, and OVCAR-3 cell lines. A2780, A2780cis, SK-OV-3, OVCAR3 ovarian cancer cell lines were cultured, harvested, and total RNA or miRNA was isolated from cells. DICER, miRNA-103, and miRNA-107 expression was measured using the qRT-PCR assay. Data are presented as means $2^{-\Delta CT}$ ± SD from four independent experiments ($n = 4$). $2^{-\Delta CT}$ is an absolute value representing the expression level of each gene in a particular cell line. Statistically significant increase of miRNA-103: #A2780 vs. A2780cis or SK-OV-3, $p \leq 0.02$ (Mann–Whitney U test). Statistically significant decrease of miRNA-103: * A2780cis vs. OVCAR-3; ** SK-OV-3 vs. OVCAR-3, $p \leq 0.02$ (Mann–Whitney U test). Statistically significant increase of miRNA-107: # A2780 vs. A2780cis or SK-OV-3, $p \leq 0.02$ (Mann–Whitney U test). Statistically significant decrease of miRNA-107: * SK-OV-3 vs. OVCAR-3, $p \leq 0.02$ (Mann–Whitney U test). Statistically significant decrease in the expression of DICER: * A2780 vs. A2780cis, $p \leq 0.007$ (Mann–Whitney U test). Statistically significant increase in the expression of DICER: # A2780cis vs. SK-OV-3 or OVCAR-3, $p \leq 0.007$ (Mann–Whitney U test).

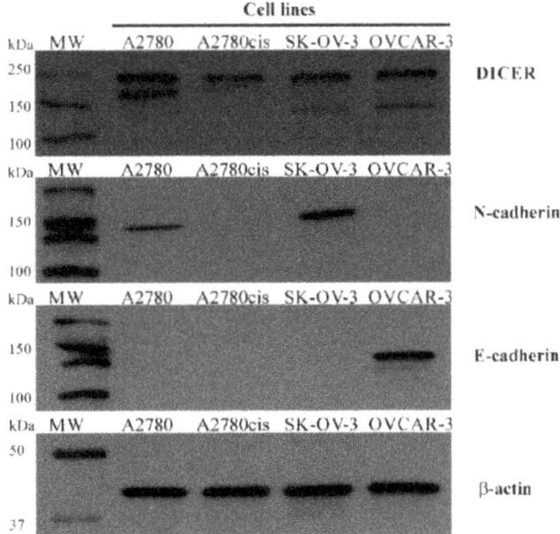

Figure 7. The total level of DICER, N-cadherin and E-cadherin proteins in A2780, A2780cis, SK-OV-3, and OVCAR-3 cell lines-Immunoblots. A2780, A2780cis, SK-OV-3, OVCAR3 ovarian cancer cell lines were cultured, harvested, and lysed with RIPA buffer. This Figure contains representative immunoblots of DICER, N-cadherin, E-cadherin, and β-actin proteins level, which were obtained with the Immunoblotting-ECL method.

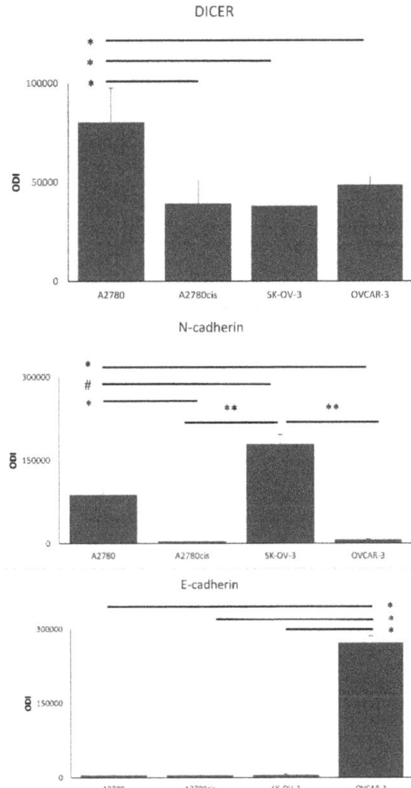

Figure 8. The total level of DICER, N-cadherin and E-cadherin proteins in A2780, A2780cis, SK-OV-3, and OVCAR-3 cell lines—densitometric analysis. A2780, A2780cis, SK-OV-3, OVCAR3 ovarian cancer cell lines were cultured, harvested, and lysed with RIPA buffer. The level of DICER, N-cadherin, E-cadherin, and β-actin proteins was evaluated with the Immunoblotting-ECL method. Bands were

quantified by densitometric analysis, and data are presented as the optical density intensity (ODI) of the area under each band's peak ± SD from five independent experiments ($n = 5$). Statistically significant decrease in the level of DICER: * A2780 vs. A2780cis or SK-OV-3 or OVCAR-3, $p \leq 0.007$ (Mann–Whitney U test). Statistically significant decrease in the level of N-cadherin: * A2780 vs. A2780cis or OVCAR-3; ** SK-OV-3 vs. A2780cis or OVCAR-3, $p \leq 0.0006$ (Mann–Whitney U test). Statistically significant increase in the level of N-cadherin: # A2780 vs. SK-OV-3, $p \leq 0.0006$ (Mann–Whitney U test). Statistically significant decrease in the level of E-cadherin: * OVCAR-3 vs. A2780 or A2780cis or SK-OV-3, $p \leq 0.00005$ (Mann–Whitney U test).

3. Discussion

Our results showed that miRNA-107 is up-regulated in primary high-grade serous ovarian cancer tissues in comparison to normal epithelium derived from the Fallopian tube's fimbriae. Nevertheless, we did not find any clinical correlations with miRNA-107 expression levels. We also did not confirm that miRNA-103 is up-regulated in cancerous tissues. Such results may be caused by a small number of samples (i.e., 50 patients). However, we believe that our investigation is not without merit. To the best of our knowledge, there have not been any studies on ovarian cancer that evaluated miRNA-103/107 expression levels in regard to clinicopathological data. The only data on miRNA-103/107 survival comes from TCGA research. We assessed the OS in TCGA cohort of patients. Kaplan–Meier analysis indicated that low miRNA-107 expression levels were associated with improved survival.

Expression levels of both miRNAs were also determined in other types of tumors, such as gastric or bladder cancers. MiRNA-103/107 expression was assessed by Yu et al. in bladder cancer samples. Both miRNAs were up-regulated in the bladder cancer specimens and positively correlated with the tumor stage [12]. Li et al. demonstrated that overexpression of miRNA-107 in gastric cancer might be associated with gastric cancer metastasis. Their results suggested that miRNA-107 promotes cancer metastasis through the down-regulation of DICER1 [19].

MiRNA-103/107- DICER axis has been described in breast cancer cells by Martello et al. According to them, invasive and metastatic properties of cancer cells are empowered by the up-regulation of miRNA-103/107 [9]. High levels of miRNA-103/107 affect DICER expression and cause its attenuation. Martello et al. associated the miRNA-103/107-DICER axis with epithelial to mesenchymal transition (EMT), identifying miRNA-200 family members as downstream mediators of the axis. Thus, high levels of both miRNAs may lead to more invasive and metastatic abilities of cancer cells. ZEB1 and ZEB2 are miRNA-200 targets and crucial genes for mesenchymality. According to Martello et al., both of them are down-regulated in antagomir-103/107-treated cells to about 50%. Additionally, the authors performed Kaplan–Meier survival analysis, which showed that high levels of miR-103/107 are associated with metastasis and poor outcome.

Identification of miRNAs that may control DICER expression levels is crucial in full understanding of miRNA-dependent carcinogenesis, tumor growth, and metastasis. Only a few authors have assessed MiRNA-103 expression levels in ovarian cancer. Wilczynski et al. assessed miRNA-103 by qRT-PCR in 48 samples derived from advanced serous ovarian cancer patients and found no differences between primary tumor and healthy ovarian tissue [20]. Yang et al. performed qRT-PCR to compare miR-103 expression levels in ovarian cancer and healthy ovarian tissues, however, they used only five fresh tissue samples of serous ovarian cancer [11]. Yang et al. showed that miRNA-103 was significantly up-regulated in ovarian cancer samples in comparison to healthy ovarian tissues. Furthermore, they reported that overexpression of miRNA-103 in cancer cell lines led to the enhancement of migration or invasion and a significant reduction of DICER1 levels.

DICER seems to be one of the key regulators of miRNA's expression and action in cancerous cells. Several authors reported that DICER expression might be associated with patients' prognosis. Meritt et al. performed qRT-PCR with validation by immunohistochemistry in 111 samples of EOC (2 endometrioid, 109 serous) and reported that low DICER expression was significantly associated with advanced tumor stage [9]. Moreover, they found that high DICER expression was associated with increased survival among EOC patients. Flavin et al. reported similar results and showed that high DICER expression might be associated with low metastatic lesions. However, survival analysis revealed that DICER expression did not affect the survival rates [21]. Such results stay in corcondance with our hypothesis. The results of survival analysis in TCGA cohort of patients showed that low miRNA-107 levels were associated with improved survival. The analysis of TCGA patients failed to show a possible impact of miRNA-103 levels of survival, but we believe it might have been changed if more patients had been included. Low miRNA-103/107 expression levels should be associated with high DICER levels. The up-regulation of DICER has been identified by Meritt et al. as a favorable factor in EOC patients. It seems that low levels of miRNA-107 and high levels of DICER should be both considered as a mark of improved prognosis among EOC patients.

Different DICER expression levels characterize ovarian cancer cell lines. Wang et al. reported low DICER levels in cisplatin-resistant A2780 cells in comparison to A2780 cisplatin-sensitive line. Furthermore, the down-regulation of DICER decreased the sensitivity of A2780 cancer cells and inhibited cisplatin-induced apoptosis [22]. In the study by Kuang et al., DICER down-regulation promoted cell proliferation and was significantly decreased in cisplatin-resistant A2780 cells compared with parental A2780 cells [23]. Our results showed that A2780cis is characterized by lower DICER expression levels than parental A2780 ovarian cancer cells. Moreover, we demonstrated that A2780cis cells presented a significantly higher expression of miRNA-103 and miRNA-107 values in comparison to A2780 cells. Such results indicate the possible existence of the miRNA-103/107-DICER axis in the case of ovarian cancer. High levels of both miRNAs were coexistent with DICER down-regulation in A2780cis cells. Furthermore, we found that SK-OV-3 cancer cells were characterized by the highest miRNA-103/107 expression levels and relatively low level of DICER protein among four selected ovarian cancer cell lines. It is worth noticing that SK-OV-3 cells also presented high levels of N-cadherin, which may be evidence for the possible shift of these cells towards the mesenchymal phenotype.

The possible existence of miRNA-103/107-DICER axis, that had previously been described in the case of breast cancer, convinced us to perform a study dedicated to the clinical usefulness of miRNA-103/107 expression levels in high-grade serous ovarian cancer patients. We did not aim to define the exact molecular ways of action of both miRNAs. MiRNA-103/107 expression levels have never been related to prognostic clinicopathological tumor characteristics. The results of the study indicated that miRNA-103/107 did not have any clinical or prognostic significance in our sample of patients. Furthermore, evaluation of DICER in FFPE tissues was not possible, due to the high amount of degraded mRNA. We found that majority of our serous high-grade ovarian cancer samples were characterized by reduced levels of DICER protein in immunohistochemistry, and it is what we expected in regard to the potential existence of the miRNA-103/107-DICER axis. Therefore, we aimed to perform a preliminary evaluation of SK-OV-3, OVCAR-3, A2780, and A2780cis cancer cell lines in order to verify a possible existence of the miRNA-103/107-DICER axis in the case of ovarian cancer. We selected various ovarian cancer cell lines that are derived from different histotypes of EOC. We expected that selected cells should vary between each other in regard to DICER and miRNA-103/103 expression levels. The cells with the highest miRNA-103/107 expression levels should be characterized by the lowest DICER expression at the same time. Our results confirmed that the miRNA-103/107-DICER axis might exist in the case of ovarian cancer, however, we are far from drawing any final conclusions.

Lack of clinical significance of both miRNAs might be related to the small set of patients or the fact that up-regulation of miRNA-103/107 may happen only in few ovarian cancer cells that gain migratory potential and form metastases.

4. Materials and Methods

Ethical approval was obtained from the Polish Mother's Memorial Hospital Research Institute Ethics Committee (21 May 2019, approval number 71/2019). All individuals who participated in the study provided their consent. All procedures performed in studies involving human participants were in accordance with the ethical standards of the institutional and/or national research committee and with the 1964 Helsinki declaration and its later amendments or comparable ethical standards. All patients were operated in the Department of Operative Gynecology, Endoscopy and Gynecologic Oncology, Polish Mother's Memorial Hospital Research Institute, Lodz, Poland. Total hysterectomy with bilateral salpingo-oophorectomy, omentectomy, appendectomy was performed in all cases. The extent of the surgery was individually modified in order to obtain optimal cytoreduction. Standard platinum-taxane chemotherapy was introduced in all cases as a first-line treatment. We defined platinum-resistant tumors when there was a relapse/progression within six months after completion of the chemotherapy. Serum levels of CA125 (cancer antigen 125), serum human epididymis antigen-4 (HE4), and ROMA index (Risk of Malignancy Algorithm) were obtained from patients.

4.1. *Assessment of Formalin-Fixed, Paraffin-Embedded Tissues*

RNA isolation was performed with the use of a miRNeasy FFPE Kit (Qiagen, Hilden, Germany), according to the manufacturer's protocol. Spectrophotometry (PicoDrop, 260/280 nm) was used in order to assess the quality of the samples. Reverse transcription was performed with the use of the TaqMan® MicroRNA Reverse Transcription Kit (Applied Biosystems, Waltham, MA, USA) and miRNA-specific primers (RT primer), according to the manufacturer's protocol.

Quantitative Real-time PCR was performed with the use of standard TaqMan® MicroRNA Assays (Applied Biosystems): hsa-miR-103 (Assay ID: 000439), hsa-miR-107 (Assay ID: 000443) and RNU6B (endogenous control, Assay ID: 001093). The 10 µL qPCR reaction mixture included 0.7 µL RT product, 5 µL TaqMan Fast PCR Master Mix (Applied Biosystems) and 1 µL TaqMan miRNA Assay (20×). The reactions were incubated in a 96-well plate at 95 °C for 10 min, followed by 40 cycles of 95 °C for 5 s and 60 °C for 20 s. All reactions were run in duplicate (Applied Biosystems 7900HT Fast Real-Time PCR System). We used Detection System 2.3 Software (Applied Biosystems, Waltham, MA, USA) for the quantification of miRNA. Relative expression was calculated according to the Ct method $2^{-\Delta\Delta Ct}$.

4.2. *Quantitative Real-Time PCR Method for Selected Ovarian Cancer Cell Lines*

A2780 and A2780cis were purchased from ECACC General Cell Collection (UK), while SK-OV-3, and OVCAR-3 were purchased from ATCC (USA). All cell lines are of epithelial origin, have adherent growth as a monolayer, and defined resistance or sensitivity to cisplatin. The growth medium consists of RPMI 1640 with 10% FBS and the addition of penicillin and streptomycin (100 U/mL/100 µg/mL). Moreover, cisplatin in the concentration of 1µM was added to the A2780cis cell line every 2–3 passages in order to maintain its resistance to cisplatin. A2780, A2780cis, SK-OV-3, OVCAR3 ovarian cancer cell lines were cultured on culture flasks until they reached 90% of confluence. All cell lines were harvested, centrifuged, and total RNA or miRNA was isolated from cells using TRIzol® Reagent (Life Technologies, Waltham, MA, USA) or GeneMatrix Universal RNA/miRNA Purification Kit (EURX, Gdansk, Poland), respectively, according to the manufacturer procedure. Next, the quality control of isolated RNA and miRNA with the use of Nanodrop with ND 1000 Software (ThermoFisher Scientific, Waltham, MA, USA) was performed. RNA (5 µg) was processed directly to cDNA synthesis using Maxima First Strand cDNA Synthesis Kits for RT-qPCR (Life Technologies), according to the manufacturers manual. Similarly, the mature RNU6B, miR-103, and miR-107 were processed directly to cDNA with the use of the Taqman MicroRNA Reverse transcription kit. The human β-actin, DICER, RNU6B, miR-103, miR-107 expressions were quantified using TaqMan Gene Expression Assays (Applied Biosystems, CA, USA) by real-time PCR using ABI 7900-HT detection system (Applied Biosystems, Life Technologies), according to the manufacturer's protocol. Controls with no template cDNA were performed with each assay. Relative quantitation of gene expression was calculated using the comparative CT (ΔΔCT) method. The obtained data were analyzed with ABI 7900-HT (RQ manager software v1.2) and DataAssist software v3.01 and are presented as the RQ value- representing fold change in gene expression normalized to the reference genes (β-actin or RNU6B) and relative to the control. Additionally, data are presented as $2^{-\Delta CT}$, which represents the absolute value of the mRNA level of each evaluated gene in a particular cell line.

4.3. Immunoblotting-ECL (Western Blotting Method)

A2780, A2780cis, SK-OV-3, OVCAR3 ovarian cancer cell lines were cultured on culture flasks until they reached 90% of confluence. All cell lines were harvested, centrifuged, and lysed with RIPA lysing buffer (ThermoFisher Scientific) with the addition of 1 mM PMSF and 1% of halt protease and phosphatase inhibition cocktail for 30 min on ice. The lysates were stored at −70°C until further analysis. The amount of protein in each sample was measured using a DC Protein Assay kit. The cell lysates containing equal levels of proteins were run on a 10% SDS-PAGE mini-protean precast TGX gel, and proteins were transferred to PVDF membranes using a Trans-Blot Turbo Transfer System (Bio-Rad, Hercules, CA, USA) at 2.5A for 10 min. Afterward, the membranes were blocked with SuperBlock Blocking Buffer for 30 min and blotted with mouse monoclonal anti-DICER (1:500, Thermo Fisher Scientific, catalog # MA5-31353,Waltham, MA, USA), rabbit monoclonal anti-N-cadherin (1:1000, Cell Signaling Technology, N-Cadherin (D4R1H) XP® Rabbit mAb #13116), rabbit monoclonal anti-E-cadherin (1:1000, Cell Signaling Technology, E-Cadherin (24E10) Rabbit mAb #3195, Danvers, MA, USA) or mouse IgG anti-βactin antibody (1:4000, Thermo Fisher Scientific, catalog # MA1-744) (1 h, room temperature). After washing the membranes (5 times in 2× TBS-Tween 20), they were incubated with secondary antibodies, HRP-conjugated goat anti-rabbit IgG (1:4000) or HRP-conjugated goat anti-mouse IgG (1:4000) (1 h, room temperature), and again washed five times. Proteins were detected by the incubation of membranes with ECL Western Blotting Substrate. Proteins in blots underwent densitometric analysis using a FluoroChem MultiImage FC Cabinet (Alpha Innotech Corporation, San Leandro, CA, USA) and Alpha Ease FC software 3.1.2 (Alpha Innotech Corporation, San Leandro, CA, USA). The results are presented as the optical density intensity (ODI) of the area under each band's peak.

4.4. Immunohistochemistry

Immunohistochemical staining was performed in formalin-fixed, paraffin-embedded tissues, which were cut into 4 μm slices on a microtome. Deparaffinization and rehydration were performed in xylene and ethanol (standard protocol). Slices were boiled in 0.01 M citrate buffer (pH 6.0) for 20 min, and then incubated with 3% hydrogen peroxide for 5 min. After washings with PBS, sections were incubated for 0.5 h with mouse DICER monoclonal antibody (1:400, Thermo Fisher Scientific, CL0378, Catalog # MA5-31353). After washes, a horseradish peroxidase kit was used for antibody detection and diaminobenzidine for chromagen visualization (Dako EnVision Detection Systems).

4.5. Statistical Analysis

Statistical analysis was performed with Excel 2013 (Microsoft, Redmond, Washington) and STATISTICA 13.1 software (Statsoft, Tulsa, OK, USA). Statistical significance was defined by p value lower than 0.05. Kruskal–Wallis test, median test (Pearson's chi-squared test), unpaired t-test, and Mann–Whitney U-test were used. Relative expression levels (RQ values) were determined with the use of the delta-delta CT method, adjusted to the expression of selected endogenous control.

5. Conclusions

Further investigation may clarify the role of the miRNA-103/107-DICER axis in ovarian cancer and its potential clinical/prognostic significance. The results of our study show that miRNA-103/107 did not have any clinical or prognostic significance. Nevertheless, we believe that our study is not without merit, as the evidence on the miRNA-103/107-DICER axis in ovarian cancer is extremely scarce.

Author Contributions: Conceptualization, M.W.; Methodology, M.W., M.K. (Michal Kielbik); Validation, M.W., B.S., M.K. (Michal Kielbik); Formal Analysis, D.S., M.W., M.K. (Michal Kielbik); Investigation, M.W., M.K. (Michal Kielbik), B.S., T.K., H.R., J.B.; Resources, M.W., A.M.; Data Curation, M.W.; Writing—Original Draft Preparation, M.W., F.F.; Writing—Review & Editing, F.F., J.B., M.K. (Michal Kielbik); Visualization, M.W., M.K. (Michal Kielbik); Supervision, A.M., M.K. (Magdalena Klink); Project Administration, M.W.; Funding Acquisition, M.W., M.W.—corresponding author who designed the study, acquired funding, collected clinical data and tissue samples, performed molecular experiments, wrote the paper and analysed data, M.K. (Michal Kielbik)—molecular analysis of cancer cell lines, visualization, methodology, review. D.S.—statistical analysis, T.K.—histopathological assessment of samples, B.S.—PCR quantification, M.K. (Magdalena Klink)—molecular analysis of cancer cell lines, validation, J.B.—clinical data collection, review, H.R.—histopathological assessment of samples, F.F.—review, writing the manuscript (language corrections), A.M.—final approval of the manuscript, clinical assessment of patients. All authors have read and agreed to the published version of the manuscript.

Funding: The study was financed by National Science Centre, Poland, grant no. 2018/02/X/NZ5/02278 (MINIATURA).

Conflicts of Interest: Authors declare that there is no conflict of interest.

References

1. Howlader, N.; Noone, A.M.; Krapcho, M.; Miller, D.; Brest, A.; Yu, M.; Ruhl, J.; Tatalovich, Z.; Mariotto, A.; Lewis, D.R.; et al. *SEER Cancer Statistics Review, 1975–2016*; National Cancer Institute: Bethesda, MD, USA, 2018.
2. Vang, R.; Shih, I.-M.; Kurman, R.J. Ovarian Low-grade and High-grade Serous Carcinoma: Pathogenesis, Clinicopathologic and Molecular Biologic Features, and Diagnostic Problems. *Adv. Anat. Pathol.* **2009**, *16*, 267–282. [CrossRef] [PubMed]
3. Bartel, B. MicroRNAs: Target Recognition and Regulatory Functions. *Cell* **2009**, *136*, 215–233. [CrossRef] [PubMed]
4. Ambros, V. The functions of animal microRNAs. *Nature* **2004**, *431*, 350–355. [CrossRef] [PubMed]
5. Ventura, A.; Jacks, T. MicroRNAs and Cancer: Short RNAs Go a Long Way. *Cell* **2009**, *136*, 586–591. [CrossRef] [PubMed]
6. Chen, C.-Z. MicroRNAs as oncogenes and tumor suppressors. *N. Engl. J. Med.* **2005**, *353*, 1768–1771. [CrossRef]
7. Kumar, M.S.; Lu, J.; Mercer, K.L.; Golub, T.R.; Jacks, T. Impaired microRNA processing enhances cellular transformation and tumorigenesis. *Nat. Genet.* **2007**, *39*, 673–677. [CrossRef]
8. Merritt, W.M.; Lin, Y.G.; Han, L.Y.; Kamat, A.A.; Spannuth, W.A.; Schmandt, R.; Urbauer, D.; Pennacchio, L.A.; Cheng, J.-F.; Nick, A.M.; et al. Dicer, Drosha, and Outcomes in Patients with Ovarian Cancer. *N. Engl. J. Med.* **2008**, *359*, 2641–2650. [CrossRef]
9. Martello, G.; Rosato, A.; Ferrari, F.; Manfrin, A.; Cordenonsi, M.; Dupont, S.; Enzo, E.; Guzzardo, V.; Rondina, M.; Spruce, T.; et al. A MicroRNA Targeting Dicer for Metastasis Control. *Cell* **2010**, *141*, 1195–1207. [CrossRef]
10. TCGA Research Network. Integrated genomic analyses of ovarian carcinoma. *Nature* **2011**, *474*, 609–615. [CrossRef]
11. Yang, Z.Y.; Wei, X.; Zhou, X.S.; Liu, Y.; Gong, C.; Gao, Q.L. MicroRNA-103 promote ovarian cancer metastasis through targeting Dicer1. *Chin. J. Cancer Prev. Treat.* **2014**, *21*, 1757–1763.
12. Yu, Q.F.; Liu, P.; Li, Z.Y.; Zhang, C.F.; Chen, S.Q.; Li, Z.H.; Zhang, G.Y.; Li, J.C. MiR-103/107 induces tumorigenicity in bladder cancer cell by suppressing PTEN. *Eur. Rev. Med. Pharmacol. Sci.* **2018**, *22*, 8616–8623. [PubMed]
13. Tang, Z.; Fang, Y.; Du, R. MicroRNA-107 induces cell cycle arrests by directly targeting cyclin E1 in ovarian cancer. *Biochem. Biophys. Res. Commun.* **2019**, *512*, 331–337. [CrossRef] [PubMed]
14. Erickson, B.K.; Conner, M.G.; Landen, C. The role of the fallopian tube in the origin of ovarian cancer. *Am. J. Obstet. Gynecol.* **2013**, *209*, 409–414. [CrossRef]
15. Liu, A.; Xu, X. MicroRNA isolation from formalin-fixed, paraffin-embedded tissues. *Breast Cancer* **2011**, *724*, 259–267. [CrossRef]
16. Anaya, J. OncoLnc: Linking TCGA survival data to mRNAs, miRNAs, and lncRNAs. *PeerJ Comput. Sci.* **2016**, *2*, 67. [CrossRef]
17. Shaw, T.J.; Senterman, M.K.; Dawson, K.; Crane, C.A.; Vanderhyden, B.C. Characterization of intraperitoneal, orthotopic, and metastatic xenograft models of human ovarian cancer. *Mol. Ther.* **2004**, *10*, 1032–1042. [CrossRef]
18. Vogel, C.; Marcotte, E. Insights into the regulation of protein abundance from proteomic and transcriptomic analyses. *Nat. Rev. Genet.* **2012**, *13*, 227–232. [CrossRef]

19. Li, X.; Zhang, Y.; Shi, Y.; Dong, G.; Liang, J.; Han, Y.; Wang, X.; Zhao, Q.; Ding, J.; Wu, K.; et al. MicroRNA-107, an oncogene microRNA that regulates tumour invasion and metastasis by targeting DICER1 in gastric cancer. *J. Cell. Mol. Med.* **2011**, *15*, 1887–1895. [CrossRef]
20. Wilczynski, M.; Żytko, E.; Danielska, J.; Szymańska, B.; Dzieniecka, M.; Nowak, M.; Malinowski, J.; Owczarek, D.; Wilczyński, J.R. Clinical significance of miRNA-21, -103, -129, -150 in serous ovarian cancer. *Arch. Gynecol. Obstet.* **2018**, *297*, 741–748. [CrossRef]
21. Flavin, R.; Smyth, P.C.; Finn, S.P.; Laios, A.; O'Toole, S.A.; Barrett, C.; Ring, M.; Denning, K.M.; Li, J.; Aherne, S.T.; et al. Altered eIF6 and Dicer expression is associated with clinicopathological features in ovarian serous carcinoma patients. *Mod. Pathol.* **2008**, *21*, 676–684. [CrossRef]
22. Wang, X.; Chen, H.; Wen, Y.; Yang, X.; Han, Q.; Jiang, P.; Huang, Z.; Cai, J.; Wang, Z. Dicer affects cisplatin-mediated apoptosis in epithelial ovarian cancer cells. *Mol. Med. Rep.* **2018**, *18*, 4381–4387. [CrossRef] [PubMed]
23. Kuang, Y.; Cai, J.; Li, D.; Han, Q.; Cao, J.; Wang, Z. Repression of Dicer is associated with invasive phenotype and chemoresistance in ovarian cancer. *Oncol. Lett.* **2013**, *5*, 1149–1154. [CrossRef] [PubMed]

 cancers

Article

Relevance of 2′-O-Methylation and Pseudouridylation for the Malignant Melanoma

Simon Jasinski-Bergner [1,*], Juliane Blümke [1], Claudia Wickenhauser [2] and Barbara Seliger [1,3]

1 Institute for Medical Immunology, Faculty of Medicine, Martin Luther University Halle-Wittenberg, Magdeburger Straße 2, 06108 Halle (Saale), Germany; juliane.bluemke@gmail.com (J.B.); Barbara.seliger@uk-halle.de (B.S.)
2 Institute for Pathology, Faculty of Medicine, Martin Luther University Halle-Wittenberg, 06108 Halle (Saale), Germany; claudia.wickenhauser@uk-halle.de
3 Fraunhofer Institute for Cell Therapy and Immunology, 04103 Leipzig, Germany
* Correspondence: simon.jasinski@uk-halle.de; Tel.: +49-(345)-557-4736; Fax: +49-(345)-557-4055

Simple Summary: This study investigates the expression, the histological localization, and the influence of the factors involved in 2′-O-methylation and pseudouridylation on prognostic relevant markers, proliferation markers, overall survival, molecular immune surveillance and evasion mechanisms within the malignant melanoma. Statistically significant positive correlations to the expression of markers involved in cell proliferation were observed. The upregulation of the RNA modifying factors was of prognostic relevance in this tumor disease with a negative impact on the overall survival of melanoma patients. Furthermore, the factors involved in 2′-O-methylation and pseudouridylation were statistically significant negative correlated to the expression of human leukocyte antigen class I genes as well as of components of the antigen processing machinery.

Abstract: The two RNA modifications 2′-O-methylation and pseudouridylation occur on several RNA species including ribosomal RNAs leading to an increased translation as well as cell proliferation associated with distinct functions. Using malignant melanoma (MM) as a model system the proteins mediating these RNA modifications were for the first time analyzed by different bioinformatics tools and public available databases regarding their expression and histological localization. Next to this, the impact of these RNA-modifying factors on prognostic relevant processes and marker genes of malignant melanoma was investigated and correlated to immune surveillance and evasion strategies. The RNA modifying factors exerted statistically significant positive correlations to the expression of genes involved in cell proliferation and were statistically significant negative correlated to the expression of human leukocyte antigen class I genes as well as of components of the antigen processing machinery in malignant melanoma. Upregulation of the RNA modifying proteins was of prognostic relevance in this tumor disease with a negative impact on the overall survival of melanoma patients. Furthermore, the expression of known oncogenic miRs, which are induced in malignant melanoma, directly correlated to the expression of factors involved in these two RNA modifications.

Keywords: snoRNA; 2′-O-methylation; pseudouridylation; microRNA; malignant melanoma

Citation: Jasinski-Bergner, S.; Blümke, J.; Wickenhauser, C.; Seliger, B. Relevance of 2′-O-Methylation and Pseudouridylation for the Malignant Melanoma. *Cancers* **2021**, *13*, 1167. https://doi.org/10.3390/cancers13051167

Academic Editor: Paola Tucci

Received: 12 February 2021
Accepted: 4 March 2021
Published: 9 March 2021

1. Introduction

Malignant melanoma (MM) refers to a neoplasm of melanocytes comprising of neural crest-derived cells located in particular in the stratum basal of the skin's epidermis and in the uvea of the eye. In contrast to keratinocytes, melanocytes are not linked to the basal lamina by hemidesmosomes and do not have desmosomes to neighboring keratinocytes. Instead, melanocytes as well as adjacent keratinocytes express, e.g., E-cadherin on their cell surface [1], which is an important factor contributing to the cell migration and invasion of melanoma cells after malignant transformation of melanocytes. Due to this migratory ability, growth characteristics as well as other properties including resistance to radiation and certain chemotherapeutics [2], MM represents the deadliest type of skin cancer with increasing incidences [3]. UV light exposition leading to DNA damages and oxidative stress in melanocytes [4], nevus number, pigmentation characteristics as well as genetic mutations are known predispositions for the processes leading to malignant transformation [5]. Concerning the hereditary predisposition, it is known that between 5–10% of melanoma cases are familial [6] and mainly induced by germ line mutations in tumor suppressor genes involved in proliferation control and DNA repair of cells.

Benign nevi defined as growth arrested, clonal neoplasms of melanocytes initiated by well-defined oncogenic mutations in the mitogen-activated protein kinase pathway, such as the BRAFV600E-activating mutation [7], may develop dysplastic features and are classified as MM precursor lesions.

An essential impact on the cell proliferation per se and exemplarily referred to MM is confined by ribosomes. An average mammalian cell has between 5 and 10 million ribosomes, which are large multi-subunit ribonucleoprotein complexes and are required for protein synthesis. Therefore, the formation of ribosomes and their linked translational efficiency (TE) directly influence cell proliferation, metabolism and vitality—parameters that are modulated upon malignant transformation [8] and indicative for neoplastic progression [9]. Thus, an enhanced TE is directly proportional to the ribosomal density, since the number of ribosomes on a coding transcript is linked to the efficacy of translation [10–12]. Indeed, an increase of ribosomes has been described in various cancers suggesting that the number and modifications of ribosomes drive tumorigenesis [13].

The size of the nucleoli as major location of the ribosomal assembly represents one important cytomorphologic parameter for the determination of malignancy in melanocytic lesions and has prognostic relevance [14]. Interestingly, expression of MYC as well as mutations in the tumor suppressor genes TP53 and RB1 enhanced ribosomal biogenesis [14,15].

Next to processes of transcription, the splicing and assembly of the eukaryotic 80 S ribosomes RNA modifications including 2′-O-methylation as well as the pseudouridylation occur in nucleoli, which involve a class of non-coding small RNAs termed sno-RNAs. Both processes represent the most common modifications of rRNAs and are important steps for their maturation as well as stabilization. In mammalian cells, the 5.8S rRNA, the 18S and the 28S rRNA have in total > 100 2′-O-methylations and 95 pseudouridinylations [16,17]. In addition to the rRNAs, the class of spliceosomal snoRNAs contain both modifications underlining the influence of snoRNAs for the splicing process [18]. The altered splicing pattern after malignant transformation involve mRNAs of many cancer related and tumor biological important genes [19]. In addition, transfer RNA (tRNA) molecules, microRNAs (miRs) and messenger RNAs (mRNAs) can be modified by 2′-O-methylation [20].

For the 2′-O-methylation, a methyl residue is added to the ribose backbone, whereby fibrillarin (FBL) is the methyltransferase using S-adenosyl methionine (SAM) as a methyl donor (Figure 1A). FBL acts in a complex with nucleolar protein (NOP) 56, NOP58, 15.5K (SNU13) and the guide RNA [21]. The introduction of the methyl residue leads to steric alterations and increases the hydrophobicity thereby protecting RNA molecules from nucleolytic attacks [22]. The sum of all 2′-O-methylations within a RNA molecule can also affect the secondary structure and therefore possible interactions including RNA/RNA interactions, RNA-protein interactions [23,24] as well as mRNA splicing, stability and translation [25].

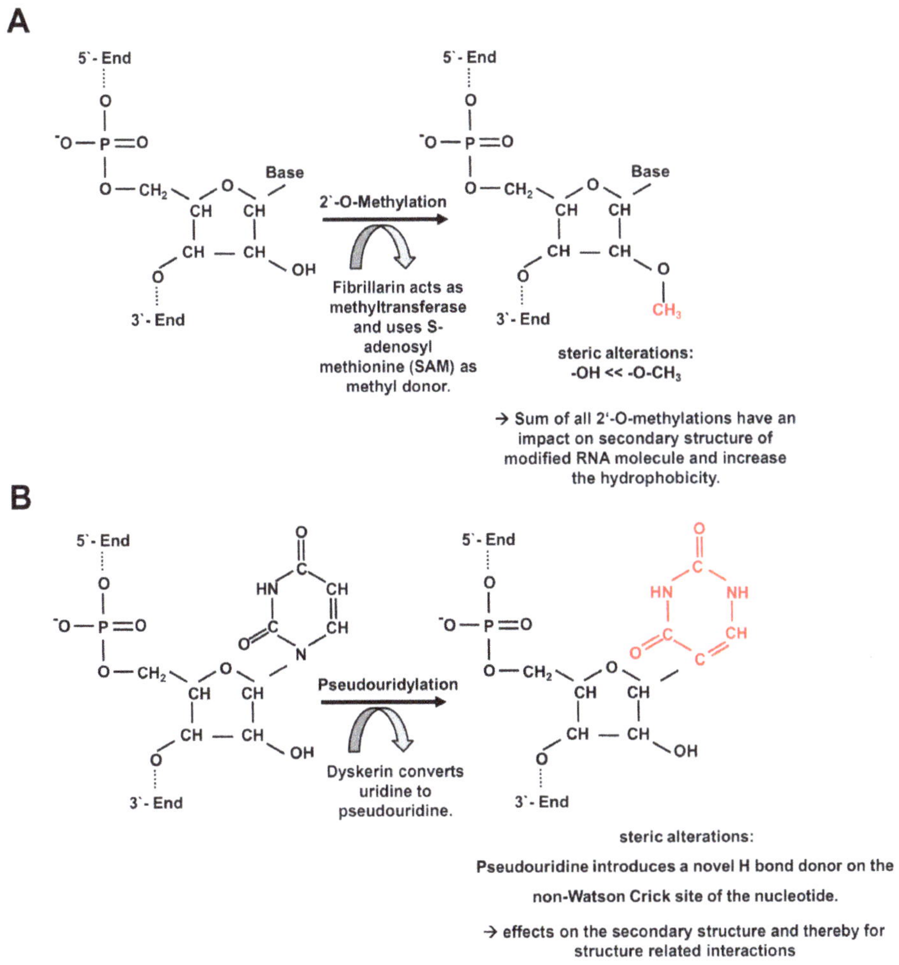

Figure 1. Molecular RNA modifications on RNAs involving snoRNAs. (**A**) 2′-O-methylation and (**B**) pseudouridylation.

Many sites of the pseudouridylation are evolutionarily conserved [26,27]. In humans, the uridine is converted to pseudouridine in rRNA, sno/scaRNA and snRNA by dyskerin (DKC1), which is a component of a complex consisting of one H/ACA snoRNA and four core proteins, namely GAR1 ribonucleoprotein (GAR1), NHP2 ribonucleoprotein (NHP2), NOP10 ribonucleoprotein (NOP10) and DKC1 itself [28]. The substitution of uridine with pseudouridine introduces a novel H bond donor on the non-Watson Crick site of the nucleotide affecting the secondary structure and consequently structure related interactions [29] (Figure 1B). Pseudouridylation is involved in many important processes of gene expression including spliceosomal small nuclear ribonucleoprotein (snRNP) biogenesis, efficiency of pre-mRNA splicing and translation fidelity [30]. Another human pseudouridine synthase is PUS10 involved in the miR maturation, which results after its depletion in a reduced expression of mature miRs [31]. Furthermore, the 2′-O-methylation and the pseudouridylation have also an impact on the binding of RNA-binding proteins (RBPs) and are required for an appropriate mRNA splicing [25,30], which is altered in various tumor entities including MM [32–34]. In this study, the expression of key molecules involved in 2′-O-methylation and pseudouridylation is addressed in more detail using MM as a model.

2. Results

2.1. Expression Pattern and Localization of RNA-Modifying Proteins in the Skin

According to the information provided by The Human Protein Atlas, the protein expression involved in the pseudouridylation including DKC1, GAR1, NHP2 and NOP10 were only localized in the nucleus, while the staining of the proteins involved in 2′-O-methylation, namely FBL, NOP56, NOP58 as well as 15.5K (SNU13) revealed a more heterogenic localization summarized in Table 1. FBL, NOP56 and NOP58 were located in the nucleus and with the exception of NOP56 strongly expressed in the epidermis and in the highly proliferation active cells of the stratum germinativum. In contrast, the staining of 15.5K revealed a cytoplasmic and membranous expression almost exclusively in cells of the epidermis layers of the skin (Table 1, Figure 2).

The correlations of factors involved in 2′-O-methylation and pseudouridylation with melanoma relevant proliferation markers, prognostic markers, and with genes responsible for immune surveillance as well as for immune evasion and positive as well as negative regulations were seen as indicated and visualized in the summarizing heatmap in Figure 3. The respective R and p values are listed within Tables 2–4. The results will be addressed in more details in the following paragraphs.

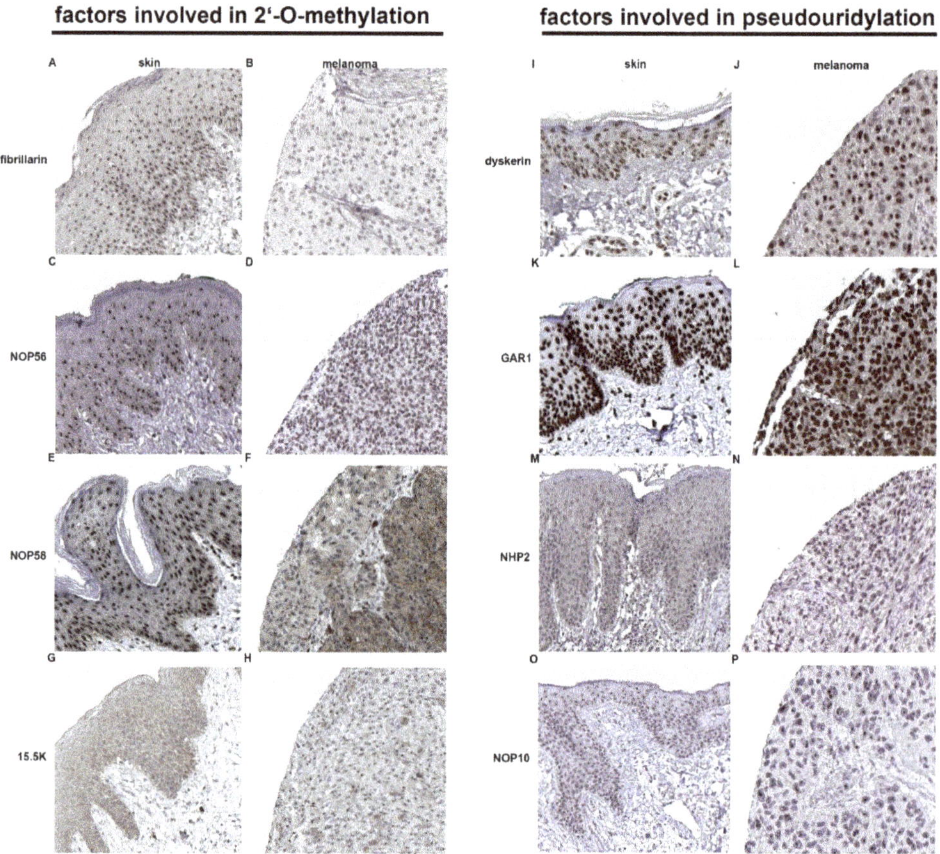

Figure 2. Representative immunohistochemical stainings of selected factors involved in 2′-O-methylation and pseudouridylation in human skin and melanoma specimen.

Table 1. Localization of the proteins involved into 2′-O-methylation and pseudouridylation in skin and MM specimen based upon the information provided by The Human Protein Atlas.

RNA Modification	Factor	Localization within Skin	Localization within MM
factors involved into 2′-O-methylation	FBL	nuclear, extra strong within the cells of the epidermis especially in the stratum germinativum	nuclear
	NOP56	nuclear	nuclear
	NOP58	nuclear strongest expression within the cells of the epidermal layers, especially within the stratum germinativum	cytoplasmic and nuclear
	15.5K (SNU13)	cytoplasmic and membranous restricted to the cells of the epidermis	nuclear
factors involved into pseudouridylation	DKC1	nuclear	nuclear
	GAR1	nuclear	nuclear
	NHP2	nuclear	nuclear
	NHP10	nuclear	nuclear

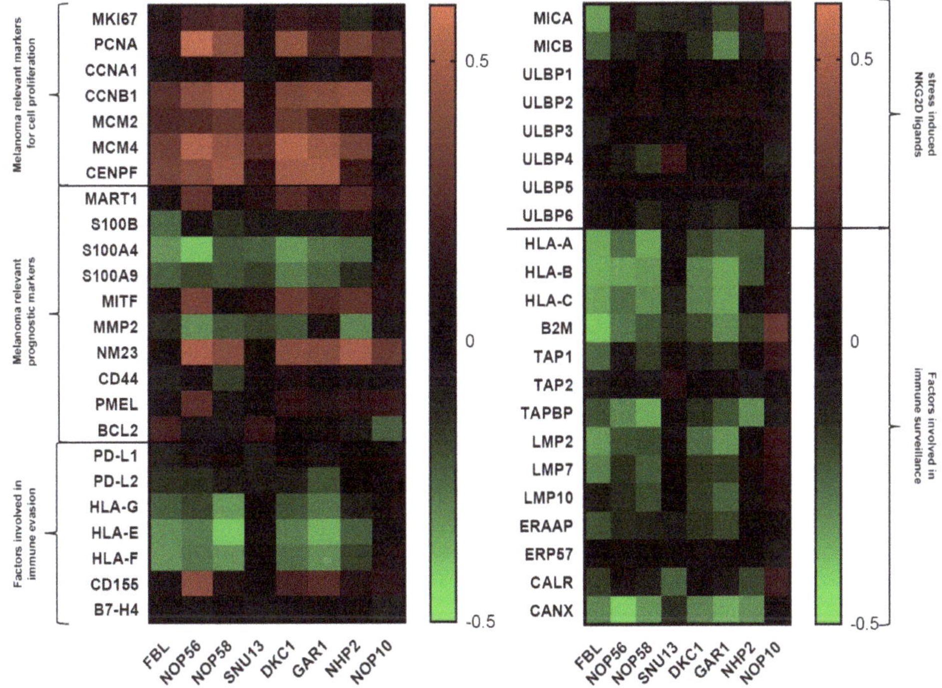

Figure 3. Summarizing visualization as a heatmap of the positive and negative correlations of factors involved in 2′-O-methylation and pseudouridylation with melanoma relevant markers of cell proliferation, prognostic markers, and genes involved in immune surveillance as well as immune evasion.

Table 2. Correlation of the factors involved in 2′-O-methylation and pseudouridylation with known proliferation markers in MM.

Correlated Expression	FBL	NOP56	NOP58	SNU13	DKC1	GAR1	NHP2	NOP10
MKI67	R = 0.081 p = 0.241	R = 0.216 $p = 1.47 \times 10^{-3}$	R = 0.235 $p = 5.22 \times 10^{-4}$	R = 0.050 p = 0.468	R = 0.185 $p = 6.50 \times 10^{-3}$	R = 0.154 p = 0.024	R = −0.133 p = 0.053	R = 0.011 p = 0.874
PCNA	R = 0.152 p = 0.026	R = 0.581 $p = 9.59 \times 10^{-21}$	R = 0.448 $p = 5.58 \times 10^{-12}$	R = 0.106 p = 0.122	R = 0.463 $p = 9.06 \times 10^{-13}$	R = 0.238 $p = 4.55 \times 10^{-4}$	R = 0.372 $p = 1.92 \times 10^{-8}$	R = 0.258 $p = 1.36 \times 10^{-4}$
CCNA1	R = −0.046 p = 0.499	R = 0.065 p = 0.341	R = 0.131 p = 0.056	R = −0.079 p = 0.250	R = 0.060 p = 0.383	R = −0.063 p = 0.356	R = 0.066 p = 0.339	R = 0.172 p = 0.011
CCNB1	R = 0.265 $p = 8.62 \times 10^{-5}$	R = 0.439 $p = 1.78 \times 10^{-11}$	R = 0.492 $p = 1.88 \times 10^{-14}$	R = 0.136 p = 0.047	R = 0.398 $p = 1.50 \times 10^{-9}$	R = 0.423 $p = 1.10 \times 10^{-10}$	R = 0.463 $p = 9.44 \times 10^{-13}$	R = 0.165 p = 0.015
MCM2	R = 0.236 $p = 4.94 \times 10^{-4}$	R = 0.251 $p = 2.07 \times 10^{-4}$	R = 0.302 $p = 6.94 \times 10^{-6}$	R = 0.166 p = 0.015	R = 0.325 $p = 1.16 \times 10^{-6}$	R = 0.233 $p = 5.80 \times 10^{-4}$	R = 0.123 p = 0.072	R = 0.031 p = 0.656
MCM4	R = 0.334 $p = 5.60 \times 10^{-7}$	R = 0.546 $p = 4.78 \times 10^{-18}$	R = 0.466 $p = 6.15 \times 10^{-13}$	R = 0.262 $p = 1.04 \times 10^{-4}$	R = 0.554 $p = 1.34 \times 10^{-18}$	R = 0.482 $p = 7.93 \times 10^{-14}$	R = 0.377 $p = 1.29 \times 10^{-8}$	R = 0.057 p = 0.403
CENPF	R = 0.363 $p = 4.69 \times 10^{-8}$	R = 0.428 $p = 6.01 \times 10^{-11}$	R = 0.495 $p = 1.33 \times 10^{-14}$	R = 0.162 p = 0.018	R = 0.509 $p = 1.62 \times 10^{-15}$	R = 0.515 $p = 6.59 \times 10^{-16}$	R = 0.198 $p = 3.61 \times 10^{-3}$	R = 0.040 p = 0.556

Based on TCGA data of 214 human MM samples provided by the R2 database (https://hgserver1.amc.nl/ (accessed on 12 February 2021)), statistically significant positive correlations ($p < 0.05$) of proteins in 2′-O-methylation and pseudouridylation to proliferation markers of MM were found and highlighted in red. Background color: One group of factors belong to the brighter grey. The 2nd group to the darker grey.

Table 3. Correlation of the factors involved in 2′-O-methylation and pseudouridylation with known prognostic markers in MM.

Correlated Expression	FBL	NOP56	NOP58	SNU13	DKC1	GAR1	NHP2	NOP10
MART1	R = 0.009 p = 0.893	R = 0.304 p= 6.12 × 10^{-6}	R = −0.024 p = 0.728	R = 0.134 p= 0.050	R = 0.265 p= 8.59 × 10^{-5}	R = 0.203 p= 2.92 × 10^{-3}	R = 0.259 p= 1.25 × 10^{-4}	R = 0.115 p = 0.094
S100B	R = −0.248 p= 2.53 × 10^{-4}	R = −0.038 p = 0.577	R = −0.152 p= 0.026	R = −0.085 p = 0.218	R = −0.090 p = 0.191	R = −0.067 p = 0.327	R = 0.133 p = 0.053	R = 0.021 p = 0.757
S100A4	R = −0.325 p= 1.20 × 10^{-6}	R = −0.414 p= 2.79 × 10^{-10}	R = −0.223 p= 1.01 × 10^{-3}	R = −0.240 p= 3.96 × 10^{-4}	R = −0.342 p= 2.98 × 10^{-7}	R = −0.274 p= 4.75 × 10^{-5}	R = −0.258 p= 1.33 × 10^{-4}	R = 0.042 p = 0.539
S100A9	R = −0.234 p= 5.57 × 10^{-4}	R = −0.192 p= 4.92 × 10^{-3}	R = −0.212 p= 1.81 × 10^{-3}	R = −0.185 p= 6.72 × 10^{-3}	R = −0.295 p= 1.13 × 10^{-5}	R = −0.234 p= 5.58 × 10^{-4}	R = −0.147 p= 0.031	R = 0.104 p = 0.130
MITF	R = 0.076 p = 0.271	R = 0.388 p= 4.05 × 10^{-9}	R = 0.050 p = 0.468	R = 0.124 p = 0.070	R = 0.345 p= 2.24 × 10^{-7}	R = 0.256 p= 1.51 × 10^{-4}	R = 0.304 p= 5.90 × 10^{-6}	R = 0.124 p = 0.070
MMP2	R = −0.129 p = 0.059	R = −0.321 p= 1.67 × 10^{-6}	R = −0.271 p= 5.96 × 10^{-5}	R = −0.190 p= 5.20 × 10^{-3}	R = −0.226 p= 8.54 × 10^{-4}	R = −0.040 p = 0.561	R = −0.309 p= 4.10 × 10^{-6}	R = −0.050 p = 0.465
NM23	R = −0.005 p = 0.944	R = 0.534 p= 3.56 × 10^{-17}	R = 0.424 p= 9.77 × 10^{-11}	R = 0.053 p = 0.445	R = 0.442 p= 1.21 × 10^{-11}	R = 0.425 p= 8.80 × 10^{-11}	R = 0.531 p= 5.60 × 10^{-17}	R = 0.321 p= 1.66 × 10^{-6}
CD44	R = −0.091 p = 0.184	R = −0.026 p = 0.705	R = −0.174 p= 0.011	R = 0.034 p = 0.618	R = 0.101 p = 0.139	R = 0.012 p = 0.863	R = −0.040 p = 0.557	R = −0.040 p = 0.556
PMEL	R = −0.033 p = 0.630	R = 0.281 p= 2.96 × 10^{-5}	R = −0.061 p = 0.371	R = 0.029 p = 0.670	R = 0.171 p= 0.012	R = 0.126 p = 0.066	R = 0.160 p= 0.019	R = 0.179 p= 8.58 × 10^{-3}
BCL2	R = 0.217 p= 1.40 × 10^{-3}	R = −0.059 p = 0.392	R = −0.072 p = 0.292	R = 0.183 p= 7.21 × 10^{-3}	R = 0.008 p = 0.907	R = 0.031 p = 0.652	R = −0.101 p = 0.140	R = −0.235 p= 5.15 × 10^{-4}

Based on TCGA data of 214 MM samples provided by the R2 database (https://hgserver1.amc.nl/ (accessed on 12 February 2021)) statistically significant positive (red) or negative (green) correlations ($p < 0.05$) of proteins in 2′-O-methylation and pseudouridylation to prognostic markers of MM are highlighted. Background color: One group of factors belong to the brighter grey. The 2nd group to the darker grey.

Table 4. Correlation of the expression of factors involved in 2′-O-methylation and pseudouridylation with preselected molecules involved into immune recognition/evasion.

	Correlated Expression	FBL	NOP56	NOP58	SNU13 (NHP2L1)	DKC1	GAR1	NHP2	NOP10
molecules contributing to recognition by immune effector cells	MICA	R = −0.341 p = 3.10 × 10^{-7}	R = 0.100 p = 0.146	R = −0.152 p = 0.026	R = −0.128 p = 0.061	R = −0.112 p = 0.103	R = −0.189 p = 5.59 × 10^{-3}	R = 0.121 p = 0.079	R = 0.245 p = 2.88 × 10^{-4}
	MICB	R = −0.241 p = 3.72 × 10^{-4}	R = −0.119 p = 0.082	R = −0.069 p = 0.314	R = −0.014 p = 0.834	R = −0.149 p = 0.029	R = −0.304 p = 5.75 × 10^{-6}	R = −0.106 p = 0.121	R = 0.175 p = 0.010
	ULBP1	R = 0.109 p = 0.112	R = 0.020 p = 0.767	R = 0.109 p = 0.112	R = 0.011 p = 0.877	R = 0.067 p = 0.328	R = 0.056 p = 0.412	R = 0.077 p = 0.264	R = 0.029 p = 0.677
	ULBP2	R = 0.047 p = 0.496	R = 0.078 p = 0.255	R = 0.107 p = 0.117	R = −0.020 p = 0.766	R = 0.096 p = 0.162	R = 0.120 p = 0.080	R = 0.093 p = 0.176	R = −0.023 p = 0.742
	ULBP3	R = −0.091 p = 0.185	R = 0.028 p = 0.688	R = 0.005 p = 0.947	R = −0.008 p = 0.912	R = 0.036 p = 0.603	R = 0.058 p = 0.398	R = 0.060 p = 0.380	R = 0.098 p = 0.154
	ULBP4	R = 0.048 p = 0.488	R = −0.104 p = 0.129	R = −0.171 p = 0.012	R = 0.202 p = 2.93 × 10^{-3}	R = −0.015 p = 0.825	R = −0.008 p = 0.906	R = −0.024 p = 0.729	R = −0.107 p = 0.117
	ULBP5	R = −0.011 p = 0.867	R = −0.007 p = 0.915	R = 0.018 p = 0.791	R = 0.022 p = 0.744	R = −0.001 p = 1.000	R = 0.069 p = 0.313	R = −0.022 p = 0.752	R = 0.024 p = 0.730
	ULBP6	R = −0.005 p = 0.940	R = 0.020 p = 0.772	R = −0.132 p = 0.055	R = −0.029 p = 0.674	R = −0.093 p = 0.175	R = −0.138 p = 0.044	R = 0.025 p = 0.715	R = −0.004 p = 0.951
	HLA-A	R = −0.364 p = 4.16 × 10^{-8}	R = −0.261 p = 1.13 × 10^{-4}	R = −0.370 p = 2.47 × 10^{-8}	R = −0.054 p = 0.431	R = −0.195 p = 4.12 × 10^{-3}	R = −0.238 p = 4.53 × 10^{-4}	R = −0.229 p = 7.55 × 10^{-4}	R = 0.110 p = 0.109
	HLA-B	R = −0.369 p = 2.66 × 10^{-8}	R = −0.353 p = 1.16 × 10^{-7}	R = −0.342 p = 2.81 × 10^{-7}	R = −0.024 p = 0.723	R = −0.305 p = 5.60 × 10^{-6}	R = −0.365 p = 3.92 × 10^{-8}	R = −0.218 p = 1.30 × 10^{-3}	R = 0.114 p = 0.096
	HLA-C	R = −0.351 p = 1.29 × 10^{-7}	R = −0.264 p = 9.52 × 10^{-5}	R = −0.323 p = 1.35 × 10^{-6}	R = −0.126 p = 0.066	R = −0.298 p = 9.08 × 10^{-6}	R = −0.360 p = 5.96 × 10^{-8}	R = −0.048 p = 0.481	R = 0.145 p = 0.034
	B2M	R = −0.407 p = 6.26 × 10^{-10}	R = −0.281 p = 2.97 × 10^{-5}	R = −0.224 p = 9.60 × 10^{-4}	R = −0.150 p = 0.028	R = −0.198 p = 3.67 × 10^{-3}	R = −0.318 p = 2.04 × 10^{-6}	R = −0.178 p = 9.12 × 10^{-3}	R = 0.314 p = 2.73 × 10^{-6}
	TAP1	R = −0.251 p = 2.05 × 10^{-4}	R = −0.074 p = 0.278	R = −0.176 p = 9.93 × 10^{-3}	R = 0.067 p = 0.331	R = −0.114 p = 0.097	R = −0.154 p = 0.024	R = 0.018 p = 0.788	R = 0.189 p = 5.58 × 10^{-3}
	TAP2	R = −0.102 p = 0.137	R = −0.042 p = 0.538	R = −0.064 p = 0.355	R = 0.144 p = 0.035	R = 0.027 p = 0.695	R = −0.031 p = 0.655	R = −0.081 p = 0.241	R = 0.034 p = 0.624
	TAPBP	R = −0.210 p = 1.99 × 10^{-3}	R = −0.309 p = 3.94 × 10^{-6}	R = −0.374 p = 1.70 × 10^{-8}	R = 0.070 p = 0.311	R = −0.156 p = 0.022	R = −0.219 p = 1.23 × 10^{-3}	R = −0.328 p = 9.18 × 10^{-7}	R = −0.088 p = 0.19
	LMP2	R = −0.342 p = 2.96 × 10^{-7}	R = −0.214 p = 1.60 × 10^{-3}	R = −0.209 p = 2.07 × 10^{-3}	R = −0.038 p = 0.577	R = −0.262 p = 1.03 × 10^{-4}	R = −0.338 p = 4.19 × 10^{-7}	R = −0.014 p = 0.833	R = 0.190 p = 5.41 × 10^{-3}

Table 4. *Cont.*

	Correlated Expression	FBL	NOP56	NOP58	SNU13 (NHP2L1)	DKC1	GAR1	NHP2	NOP10
	LMP7	R = −0.299 p =8.42 × 10^{-6}	R = −0.159 p= 0.020	R = −0.233 p= 5.83 × 10^{-4}	R = 0.055 p = 0.420	R = −0.164 p= 0.016	R = −0.180 p= 8.24 × 10^{-3}	R = 0.092 p = 0.179	R = 0.165 p= 0.015
	LMP10	R = −0.092 p = 0.180	R = −0.148 p= 0.030	R = −0.205 p= 2.55 × 10^{-3}	R = 0.008 p = 0.909	R = −0.180 p= 8.33 × 10^{-3}	R = −0.245 p= 2.99 × 10^{-4}	R = −0.057 p = 0.406	R = 0.138 p= 0.044
	ERAAP	R = −0.208 p =2.24 × 10^{-3}	R = −0.141 p= 0.040	R = −0.122 p = 0.075	R = −0.104 p = 0.131	R = −0.161 p= 0.019	R = −0.197 p= 3.73 × 10^{-3}	R = −0.006 p = 0.932	R = 0.036 p = 0.602
	ERP57	R = 0.004 p = 0.951	R = 0.077 p = 0.262	R = 0.014 p = 0.841	R = −0.031 p = 0.648	R = 0.081 p = 0.239	R = −0.017 p = 0.804	R = 0.012 p = 0.859	R = −0.073 p = 0.285
	CALR	R = −0.169 p= 0.013	R = 0.073 p = 0.288	R = −0.102 p = 0.138	R = −0.241 p= 3.64 × 10^{-4}	R = −0.010 p = 0.888	R = −0.072 p = 0.296	R = −0.191 p = 5.11 × 10^{-3}	R = 0.204 p = 2.74 × 10^{-3}
	CANX	R = −0.303 p =6.34 × 10^{-6}	R = −0.430 p= 4.64 × 10^{-11}	R = −0.332 p= 6.62 × 10^{-7}	R = −0.147 p= 0.031	R = −0.300 p= 8.14 × 10^{-6}	R = −0.370 p= 2.48 × 10^{-8}	R = −0.320 p = 1.79 × 10^{-6}	R = −0.022 p = 0.744
molecules contributing to immune evasion	PD-L1 (B7-H1)	R = −0.107 p = 0.117	R = −0.051 p = 0.460	R = −0.004 p = 0.954	R = −0.103 p = 0.132	R = 0.010 p = 0.888	R = −0.062 p = 0.363	R = 0.003 p = 0.966	R = 0.041 p = 0.548
	PD-L2 (PDCD1LG2)	R = −0.151 p = 0.027	R = −0.150 p= 0.028	R = −0.092 p = 0.182	R = −0.096 p = 0.163	R = −0.114 p = 0.097	R = −0.215 p= 1.59 × 10^{-3}	R = 0.014 p = 0.838	R = 0.068 p = 0.319
	HLA-G	R = −0.230 p =6.90 × 10^{-4}	R = −0.190 p= 5.32 × 10^{-3}	R = −0.345 p= 2.20 × 10^{-7}	R = −0.008 p = 0.905	R = −0.217 p= 1.39 × 10^{-3}	R = −0.280 p= 3.19 × 10^{-5}	R = −0.111 p = 0.104	R = 0.081 p = 0.236
	HLA-E	R = −0.321 p =1.59 × 10^{-6}	R = −0.290 p= 1.65 × 10^{-5}	R = −0.418 p= 1.95 × 10^{-10}	R = −0.045 p = 0.509	R = −0.296 p= 1.06 × 10^{-5}	R = −0.380 p= 9.50 × 10^{-9}	R = −0.236 p = 4.92 × 10^{-4}	R = −0.013 p = 0.845
	HLA-F	R = −0.323 p =1.41 × 10^{-6}	R = −0.274 p= 4.71 × 10^{-5}	R = −0.329 p= 8.52 × 10^{-7}	R = −0.005 p = 0.944	R = −0.267 p= 7.79 × 10^{-5}	R = −0.318 p= 1.98 × 10^{-6}	R = −0.182 p = 7.47 × 10^{-3}	R = 0.098 p = 0.154
	CD155 (PVR)	R = 0.029 p = 0.678	R = 0.436 p= 2.41 × 10^{-11}	R = 0.098 p = 0.151	R = 0.011 p = 0.874	R = 0.227 p= 8.28 × 10^{-4}	R = 0.263 p= 1.00 × 10^{-4}	R = 0.125 p= 0.067	R = 0.148 p= 0.030
	B7-H4 (VTCN1)	R = −0.003 p = 0.963	R = −0.059 p = 0.388	R = −0.016 p = 0.814	R = −0.048 p = 0.482	R = −0.031 p = 0.657	R = 0.034 p = 0.624	R = 0.062 p = 0.369	R = −0.086 p = 0.210

Bioinformatics analysis of TCGA data of 214 melanoma specimens provided by the R2 database (https://hgserver1.amc.nl/ (accessed on 12 February 2021)) was performed. Statistically significant positive (red) or negative (green) correlations ($p < 0.05$) are highlighted. Background color: One group of factors belong to the brighter grey. The 2nd group to the darker grey.

2.2. Correlation of RNA-Modifying Proteins with Tumor Cell Proliferation

To address whether the expression of factors involved in 2′-O-methylation and pseudouridylation correlates with the pathological increased proliferation rates in MM, the expression data of these RNA-modifying factors were analyzed in The Cancer Genome Atlas (TCGA) data sets consisting of 214 samples from MM patients provided by the R2 data base. The following proliferation markers known to play a role in MM were investigated: Ki67 (MKI67), PCNA, cyclin A (CCNA1), cyclin B (CCNB1), MCM2, MCM4, and mitosin (CENPF) [35–38].

As summarized in Table 2, positive correlations between both RNA modifying factors and the proliferation markers analyzed were found. The effect was the strongest for MCM4, which correlated statistically significant to all factors involved into 2′-O-methylation and pseudouridylation with the exception of NOP10. In addition, PCNA, cyclin B and mitosin showed strong correlations, while the expression of CCNA1 did not correlate statistically significant with any of these factors.

Based on the strong positive correlation between the factors involved in 2′-O-methylation and pseudouridylation with most of the clinical relevant proliferation markers, a possible correlation between these factors and prognostic marker genes relevant for MM was evaluated such as melanoma antigen recognized by T cells (MART)1, S100 calcium binding protein B (S100B), S100A4, S100A9, melanocyte inducing transcription factor (MITF), matrix metallopeptidase 2 (MMP2), nucleoside diphosphate kinase 1 (NM23), cluster of differentiation (CD) 44, premelanosome protein (PMEL), and BCL2 apoptosis regulator (BCL2) using the same TCGA data sets [39–43] (Table 3). Next to these marker genes the invasion depth termed Breslow's depth is of prognostic value. The average Breslow score of the 214 MM patients analyzed was 2.5 mm. All other clinical parameters of this MM patient cohort ($n = 214$) of the analyzed microarray data can be obtained from the original literature Jönsson et al., 2015 [44].

The two markers MART1 and MITF, but also NM23 showed statistically significant positive correlations to the expression of RNA modifying proteins. In contrast, the S100 family members S100B, S100A4, S100A9 and MMP2 exhibiting a prognostic potential in MM exerted a statistically significant negative correlation, whereas CD44 did not show any correlation and PMEL as well as BCL2 were only weakly positivey correlated.

To address whether the factors involved in 2′-O-methylation and pseudouridylation are of central importance for prognosis in MM, the same TCGA data set was used for the generation of Kaplan–Meier plots. As shown in Figure 4A–H, a direct correlation of the expression level of these factors with the overall survival (OS) of MM patients was detected and statistically significant for FBL, NOP58, and GAR1. In contrast, low NOP10 expression levels predicted a statistically non-significant correlation with a worse patients' outcome (Figure 4). However, it is noteworthy that the use of TCGA data cannot replace extensive protein based analyses of ex vivo MM specimens.

2.3. Correlation of the Expression of RNA Modifying Factors with Immune Modulatory Genes

Due to the increased implementation of immunotherapies for the treatment of hematopoietic and solid tumors, it was analyzed whether the factors involved in 2′-O-methylation and pseudouridylation directly or indirectly affect transcripts participating in the immune surveillance of tumor cells. Therefore, their expression patterns were correlated to molecules involved in the immune recognition/evasion of malignant and/or virus transformed cells from immune effector cells. Using the same melanoma data set, an impressive statistically significant negative correlation between immunological relevant molecules, in particular the human leucocyte antigen (HLA) class Ia and Ib and components of the HLA class I antigen processing machinery (APM), with the factors involved in 2′-O-methylation and pseudouridylation was detected (Table 4). This might also explain the negative correlation of these factors with the OS of MM patients, since their impaired expression was associated with a reduced anti-tumoral immune cell response [45] (Figure 4). The divergent statistically significant positive correlation pattern of NOP10 to molecules involved in immune surveillance was also associated with a statistically significant positive correlation to the OS in MM patients (Figure 4H). However, further studies are needed to investigate whether the statistically significant negative correlated immunological relevant genes are directly negatively regulated upon a reduced activity/expression of the factors involved 2′-O-methylation and pseudouridylation.

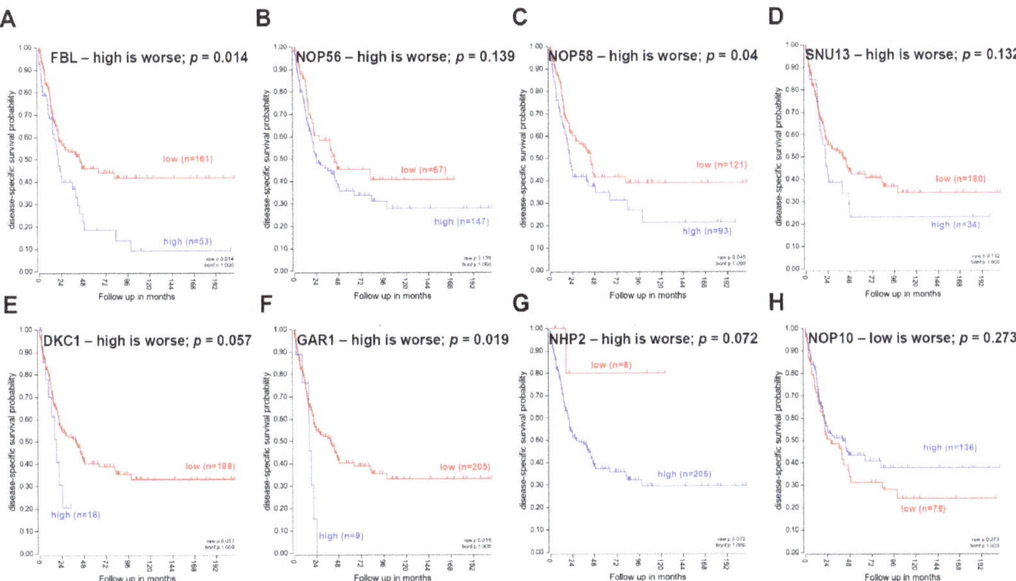

Figure 4. Impact of factors involved in 2′-O-methylation and pseudouridylation on the overall survival in MM. (**A**–**H**): Kaplan–Meier Plots were generated with TCGA data of 214 melanoma specimens provided by the R2 database (https://hgserver1.amc.nl/ (accessed on 12 February 2021)).

2.4. Correlation of miR Expression with RNA-Modifying Factors

The processes of 2′-O-methylation and pseudouridylation also occur in other RNA species with a strong impact to diverse molecular processes of malignant transformation and thus are of clinical relevance. These other RNA species include microRNAs (miRs), which are non-coding single stranded RNAs with an approximately length of ~22 nt [46], binding specifically and preferentially, but not exclusively, within the 3′- untranslated region (UTR) to their target mRNAs sequence [40,47] thereby causing a translational repression and mRNA decay [48].

Some miRs can be classified into oncogenic, tumor suppressive and/or immune modulatory miRs [48]. In addition, human viruses including herpes viruses encode for viral miRs, which also affect cancer related cellular processes [49].

Several oncogenic miRs have been reported to be frequently overexpressed in MM. Fattore and co-authors (2017) even grouped the most representative deregulated miRs in melanoma with regard to the different steps of tumor progression. These potentially oncogenic miRs with putative diagnostic and/or prognostic values include miR-9, miR-10a, miR-10b, miR-17-5p, miR-18a, miR-21, miR-26b, miR-92a, miR-221, miR-222, miR-126, miR-145, miR146, miR-182, miR-514, miR-520d and miR-527 [50].

The pseudouridylation of different human RNA species is mediated by different pseudouridine synthases, such as pseudouridine synthase (PUS) 1, TruB pseudouridine synthase 2 (TRUB2), TRUB1, PUS3, PUS4, RNA pseudouridine synthase D3 (RPUSD3), RPUSD4, PUS7, pseudouridine synthase 7 like (PUS7L), PUS10 and DKC1 [28]. From these candidates, TRUB1 has been recently reported to modulate the stability of hsa-let-7 [51]. However, it is noteworthy that the actual list of enzymes involved in 2′-O-methylation and pseudouridylation of miRs might be incomplete.

Thus far, only little is known, which of these enzymes are able to modify miRs and which alterations in the respective miR-mediated functions and/or miR stability are caused by such modifications.

To determine whether the expression of known oncogenic miRs in MM might be correlated with the expression of enzymes involved into 2′-O-methylation and pseudouridylation, the TCGA data of 214 human MM samples were further analyzed for miR expression. As shown in Table 5, there existed no evidence for a global impact of the induction of oncogenic miRs, while miR-21 was statistically significant negatively correlated to the expression of the investigated enzymes in MM.

Table 5. Correlation of miR expression with enzymes known to be involved in 2′-O-methylation or pseudouridylation of miRs.

Induced miRs in MM with Diagnostic/Prognostic Relevance	So Far in Literature Mentioned Enzymes with Putative Role for 2′-O-Methylation of miRs		So Far in Literature Mentioned Enzymes with Putative Role for Pseudouridylation of miRs	
	FBL	**HENMT1**	**DKC1**	**TRUB1**
miR-9-5p	R = 0.007 p = 0.916	R = 0.050 p = 0.469	R = −0.026 p = 0.708	R = 0.132 p = 0.053
miR-10a	R = 0.020 p = 0.767	R = −0.036 p = 0.603	R = −0.004 p = 0.957	R = −0.047 p = 0.492
miR-10b	R = −0.012 p = 0.857	R = 0.009 p = 0.896	R = −0.064 p = 0.351	R = −0.200 $p = 3.33 \times 10^{-3}$
miR-17-5p	n.d.	n.d.	n.d.	n.d.
miR-18a	n.d.	n.d.	n.d.	n.d.
miR-21	R = −0.168 p = 0.014	R = 0.017 p = 0.800	R = −0.304 $p = 5.88 \times 10^{-6}$	R = −0.139 p = 0.042
miR-26b	R = 0.191 $p = 4.96 \times 10^{-3}$	R = −0.008 p = 0.906	R = 0.112 p = 0.101	R = −0.091 p = 0.184
miR-92a	R = 0.106 p = 0.122	R = −0.055 p = 0.424	R = 0.155 p = 0.023	R = 0.059 p = 0.390
miR-221	R = −0.029 p = 0.676	R = 0.100 p = 0.144	R = 0.016 p = 0.815	R = 0.064 p = 0.352
miR-222	R = −0.015 p = 0.825	R = 0.083 p = 0.227	R = 0.083 p = 0.224	R = −0.073 p = 0.285
miR-126	R = 0.088 p = 0.198	R = −0.025 p = 0.711	R = −0.034 p = 0.620	R = −0.091 p = 0.187
miR-145	n.d.	n.d.	n.d.	n.d.
miR146a	R = −0.015 p = 0.822	R = −0.081 p = 0.235	R = 0.003 p = 0.968	R = 0.098 p = 0.154
miR-182	R = 0.008 p = 0.905	R = −0.068 p = 0.322	R = −0.045 p = 0.510	R = 0.071 p = 0.299
miR-514	R = −0.110 p = 0.109	R = 0.065 p = 0.340	R = 0.011 p = 0.874	R = 0.052 p = 0.446
miR-520d	n.d.	n.d.	n.d.	n.d.
miR-527	n.d.	n.d.	n.d.	n.d.

Statistically significant positive or negative correlations ($p < 0.05$) are highlighted in red or green. Background color: One group of factors belong to the brighter grey. The 2nd group to the darker grey.

Analysis of the TCGA data from 214 human melanoma tumors provided by the R2 database (https://hgserver1.amc.nl/ (accessed on 12 February 2021)) for correlation with enzymes known to be involved in miR 2′-O-methylation and pseudouridylation with the amount of increased and/or stabilized oncogenic miRs reported to be induced in MM.

3. Discussion

Recently, the aberrant expression pattern of DKC1, NHP2, and NOP10 in several cancer entities has been reviewed [52]. Due to their biological functions, the localization of these factors was almost completely limited to the nucleus, since 2′-O-methylation and pseudouridylation occur within the nucleoli. Furthermore, an increase within the highly proliferational active cells of the epidermal stratum germinativum was demonstrated, which reflects the involvement and increased number of ribosomes during proliferation. This was further underlined by statistically significant positive correlations to proliferation markers suggesting that some factors involved in 2′-O-methylation and pseudouridylation might also be suitable markers for proliferation themselves.

However, this study did not answer whether in the case of positive correlations between the factors involved in 2′-O-methylation and pseudouridylation and, e.g., proliferation markers these positive correlations are an indirect result due to an enhanced proliferation per se mediated by increased translational efficacy caused by modifications within ribosomal RNAs, or whether the mRNAs of the positively correlated proliferation markers are directly modified by these factors or by both mechanisms in parallel.

The positive correlations described in this study are based upon TCGA data, not taken posttranscriptional mechanisms of gene regulation into account. Therefore, in depth protein-based studies applying human melanoma tissue specimens are required to proof this hypothesis.

In regard to the observed statistically significant negative and coordinated correlations to molecules involved in immune surveillance/evasion it is noteworthy that (i) HLA Ia and Ib genes, HLA class related MICs and major APM components are located within the major histocompatibility complex (MHC) locus on chromosome 6p21 and that (ii) the adaptive immunity phylogenetically occurs with the jawed vertebrates [53] and, e.g., HLA class Ib genes are evolutionary even younger, which might have an impact on the observed negative correlations.

Due to the limitation of transcriptome based microarray data sets it is necessary to underline that also indirect effects may cause such down regulations. The increased proliferation itself might decrease the antigen processing and presentation efficacy in melanoma cells. In turn, the impaired antigen processing and presentation is a frequent immune evasion strategy of tumors [54], which might be a prerequisite for an increased proliferation rate of tumor cells. Thus far, the sequence of these events in MM has not been studied.

RNA modifications on miRs affecting their stability or decay might also indirectly contribute to the regulation of cancer related processes, even independently of the translational processes linked to the number of ribosomes. This opens a new regulatory dimension for RNA modifying enzymes (Figure 5), which has to be explored in more detail.

Thus far, it is controversially discussed whether mature mammalian miRs contain 2′-O-methylations as reported for plants and *Drosophila* [55]. In *Drosophila*, 2′-O-methylation of miRs occurred age dependent and its inhibition resulted in an accelerated neurodegeneration and shorter life span [56]. Despite discrepancies regarding the missing 2′-O-methylation in mammalian miRs, the human miR-21-5p has been shown to contain a 3′-terminal 2′-O-methylation, which enhances the stability of this oncogenic miR in lung cancer patients. Interestingly, HENMT1 was identified to act as methyltransferase [57].

Indirect effects concerning the positive or negative correlations of factors involved in 2′-O-methylation and pseudouridylation with miRs might occur. Several miRs are processed from introns after the splicing, while for the process of splicing the snoRNAs as huge group of the snRNAs in involved. These snoRNAs harbor themselves 2′-O-methylations and pseudouridylations contributing to the stability and functionality of the snoRNAs.

In this study, the impact and relevance of factors involved in 2′-O-methylation and pseudouridylation of different RNA species related to the processes of tumor formation and progression were summarized in MM as a tumor model. Using different bioinformatics tools statistically significant positive correlations between proliferation and prognostic marker genes relevant for the MM with the factors involved into 2′-O-methylation and pseudouridylation were described for the first time.

The impact of these molecules on other RNA species including miRs has recently been investigated, but needs to be further studied in detail. This will lead to the identification of proteins involved in such miR modifications, which might have an impact on the function of such modified miRs. However, the existence of miR 2′-O-methylation and pseudouridylation in general must not necessarily involve tumor relevant miRs and processes.

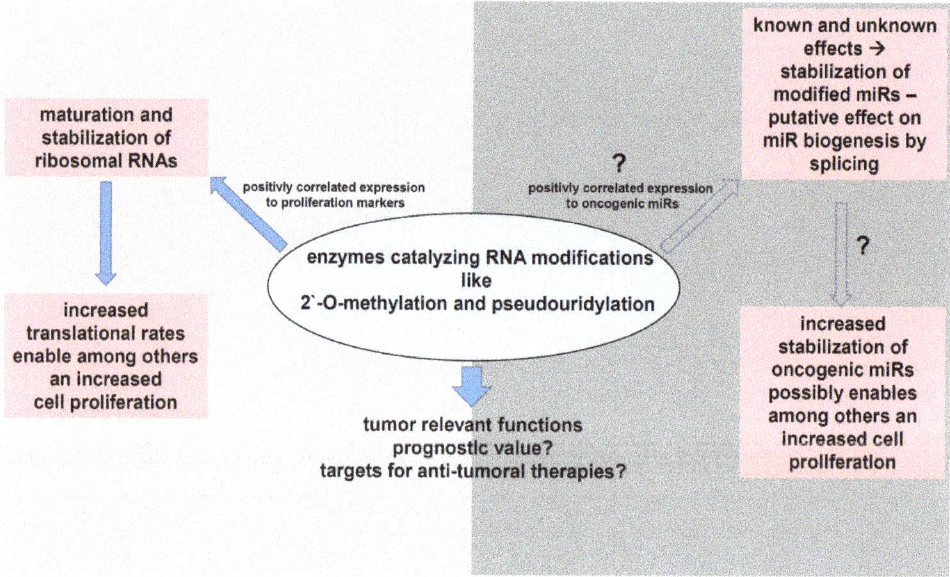

Figure 5. Impact of the RNA modifications analyzed on tumor relevant processes.

4. Materials and Methods

Immunohistochemistry

The frequency and localization of protein levels involved in the two RNA modification processes, namely FBL, NOP56, NOP58, 15.5K (SNU13), DKC1, GAR1, NHP2, and NOP10, were analyzed by evaluation of immunohistochemistry data of healthy normal skin sections and melanoma specimen provided by the free available online data base. The Human Protein Atlas [58–60].

Bioinformatic analyses of gene expression data and correlation with patients' overall survival.

The factors involved in RNA 2′-O-methylation and pseudouridylation were correlated with genes encoding for above mentioned ribosomal RNAs carrying multiple of such modifications. Unfortunately, the applied Microarrays (Illumina Human HT-12V4.0 Chips) of the 214 MM patients published by Jönsson and co-authors 2015 [44] contained only probes against the human RN5S9 transcript encoding the 5S ribosomal RNA. The factors showed a positive correlation to the RN5S9, which was statistically significant for NOP56, NOP58 and NHP2.

The correlation coefficients reflecting R values and the respective p values were calculated by the R2 database (https://hgserver1.amc.nl/ (accessed on 12 February 2021)). For visualization these R values were presented in a heatmap generated by GraphPad Prism 8 (GraphPad Software, San Diego, CA, USA).

Correlation of the expression pattern of genes involved into 2′-O-methylation and pseudouridylation with preselected genes involved in cell proliferation, prognosis and immune recognition/evasion in MM is based upon TCGA data of 214 melanoma specimens provided by the R2 database (https://hgserver1.amc.nl/ (accessed on 12 February 2021)).

Statistically significant positive or negative correlations ($p < 0.05$) are highlighted in red or green. The same TCGA data set was applied for the generation of the respective Kaplan–Meier Plots.

5. Conclusions

Thus, a link between 2′-O-methylation and pseudouridylation to cell proliferation, host immunity and oncogenic miRs exists in MM suggesting that both RNA modifications and factors involved in this process represent suitable targets for tumor therapy and putative novel prognostic markers.

Author Contributions: S.J.-B. designed the study. J.B. and S.J.-B. performed the bioinformatics analyses. B.S., C.W., and S.J.-B. wrote the manuscript. All authors have read and agreed to the published version of the manuscript.

Funding: This research was funded by the German-Israeli Foundation (GIF; I-37-414.11-2016) and Werner Jackstädt Foundation. We acknowledge the financial support of the Open Access Publication Fund of the Martin-Luther-University Halle-Wittenberg.

Institutional Review Board Statement: Not applicable.

Informed Consent Statement: Not applicable.

Data Availability Statement: The data presented in this study are available on request from the corresponding author.

Conflicts of Interest: The authors declare no conflict of interest.

Abbreviations

APM, antigen processing machinery; BCL2, BCL2 apoptosis regulator; CCNA1, cyclin A1; CD44, cluster of differentiation 44; CENPF, centromere protein F; EMT, epithelial to mesenchymal transition; GAR1, GAR1 ribonucleoprotein; HLA, human leukocyte antigen; lncRNA, long non-coding RNA; MAPK, mitogen-activated protein kinase; MART-1, Melanoma Antigen Recognized by T-cells; MCM2, minichromosome maintenance complex component 2; MCM4, minichromosome maintenance complex component 4; MHC, major histocompatibility complex; MKI67, marker of proliferation Ki-67; MITF, melanocyte inducing transcription factor; MM, malignant melanoma; MMP2, matrix metallopeptidase 2; miR, microRNA; mRNA, messenger RNA; n.d., not determined; NHP2, NHP2 ribonucleoprotein; NM23, NME/NM23 nucleoside diphosphate kinase 1; NOP, nucleolar protein; PCNA, proliferating cell nuclear antigen; PMEL; premelanosome protein; PUS1, pseudouridine synthase 1; PUS7L, pseudouridine synthase 7 like; RBP, RNA-binding protein; RPUSD3, RNA pseudouridine synthase D3; rRNA, ribosomal RNA; S, Svedberg; S100B, S100 calcium binding protein B; scaRNA, small cajal body RNA; sncRNA, small non-coding RNA; snRNA, small nuclear RNA; SNU13, small nuclear ribonucleoprotein 13; TCGA, The Cancer Genome Atlas; TE, translational efficiency; tRNA, transfer RNA; UV, ultraviolet.

References

1. Haass, N.K.; Herlyn, M. Normal human melanocyte homeostasis as a paradigm for understanding melanoma. *J. Investig. Dermatol. Symp. Proc.* **2005**, *10*, 153–163. [CrossRef]
2. Laikova, K.V.; Oberemok, V.V.; Krasnodubets, A.M.; Gal'chinsky, N.V.; Useinov, R.Z.; Novikov, I.A.; Temirova, Z.Z.; Gorlov, M.V.; Shved, N.A.; Kumeiko, V.V.; et al. Advances in the Understanding of Skin Cancer: Ultraviolet Radiation, Mutations, and Antisense Oligonucleotides as Anticancer Drugs. *Molecules* **2019**, *24*, 1516. [CrossRef]
3. Bray, F.; Ren, J.S.; Masuyer, E.; Ferlay, J. Global estimates of cancer prevalence for 27 sites in the adult population in 2008. *Int. J. Cancer* **2013**, *132*, 1133–1145. [CrossRef] [PubMed]
4. Sample, A.; He, Y.Y. Mechanisms and prevention of UV-induced melanoma. *Photodermatol. Photoimmunol. Photomed.* **2018**, *34*, 13–24. [CrossRef] [PubMed]
5. Hawkes, J.E.; Truong, A.; Meyer, L.J. Genetic predisposition to melanoma. *Semin. Oncol.* **2016**, *43*, 591–597. [CrossRef]
6. Debniak, T. Familial malignant melanoma—Overview. *Hered. Cancer Clin. Pract.* **2004**, *2*, 123–129. [CrossRef] [PubMed]
7. Damsky, W.E.; Bosenberg, M. Melanocytic nevi and melanoma: Unraveling a complex relationship. *Oncogene* **2017**, *36*, 5771–5792. [CrossRef]

8. Stepinski, D. The nucleolus, an ally, and an enemy of cancer cells. *Histochem. Cell Biol.* **2018**, *150*, 607–629. [CrossRef]
9. Pierard, G.E. Cell proliferation in cutaneous malignant melanoma: Relationship with neoplastic progression. *ISRN Dermatol.* **2012**, *2012*, 828146. [CrossRef] [PubMed]
10. Szavits-Nossan, J.; Ciandrini, L. Inferring efficiency of translation initiation and elongation from ribosome profiling. *Nucleic Acids Res.* **2020**, *48*, 9478–9490. [CrossRef]
11. von der Haar, T. Mathematical and Computational Modelling of Ribosomal Movement and Protein Synthesis: An overview. *Comput. Struct. Biotechnol. J.* **2012**, *1*, e201204002. [CrossRef]
12. Szavits-Nossan, J.; Ciandrini, L. Accurate measures of translational efficiency and traffic using ribosome profiling. *bioRxiv* **2019**. [CrossRef]
13. Pelletier, J.; Thomas, G.; Volarevic, S. Ribosome biogenesis in cancer: New players and therapeutic avenues. *Nat. Rev. Cancer* **2018**, *18*, 51–63. [CrossRef] [PubMed]
14. Penzo, M.; Montanaro, L.; Trere, D.; Derenzini, M. The Ribosome Biogenesis-Cancer Connection. *Cells* **2019**, *8*, 55. [CrossRef] [PubMed]
15. Brighenti, E.; Calabrese, C.; Liguori, G.; Giannone, F.A.; Trere, D.; Montanaro, L.; Derenzini, M. Interleukin 6 downregulates p53 expression and activity by stimulating ribosome biogenesis: A new pathway connecting inflammation to cancer. *Oncogene* **2014**, *33*, 4396–4406. [CrossRef]
16. Maden, B.E. The numerous modified nucleotides in eukaryotic ribosomal RNA. *Prog. Nucleic Acid Res. Mol. Biol.* **1990**, *39*, 241–303. [CrossRef]
17. Ofengand, J.; Bakin, A.; Wrzesinski, J.; Nurse, K.; Lane, B.G. The pseudouridine residues of ribosomal RNA. *Biochem. Cell Biol.* **1995**, *73*, 915–924. [CrossRef]
18. Jady, B.E.; Kiss, T. A small nucleolar guide RNA functions both in 2′-O-ribose methylation and pseudouridylation of the U5 spliceosomal RNA. *EMBO J.* **2001**, *20*, 541–551. [CrossRef] [PubMed]
19. Silipo, M.; Gautrey, H.; Tyson-Capper, A. Deregulation of splicing factors and breast cancer development. *J. Mol. Cell Biol.* **2015**, *7*, 388–401. [CrossRef]
20. Somme, J.; Van Laer, B.; Roovers, M.; Steyaert, J.; Versees, W.; Droogmans, L. Characterization of two homologous 2′-O-methyltransferases showing different specificities for their tRNA substrates. *RNA* **2014**, *20*, 1257–1271. [CrossRef]
21. Peng, Y.; Yu, G.; Tian, S.; Li, H. Co-expression and co-purification of archaeal and eukaryal box C/D RNPs. *PLoS ONE* **2014**, *9*, e103096. [CrossRef]
22. Sproat, B.S.; Lamond, A.I.; Beijer, B.; Neuner, P.; Ryder, U. Highly efficient chemical synthesis of 2′-O-methyloligoribonucleotides and tetrabiotinylated derivatives; novel probes that are resistant to degradation by RNA or DNA specific nucleases. *Nucleic Acids Res.* **1989**, *17*, 3373–3386. [CrossRef] [PubMed]
23. Inoue, H.; Hayase, Y.; Imura, A.; Iwai, S.; Miura, K.; Ohtsuka, E. Synthesis and hybridization studies on two complementary nona(2′-O-methyl)ribonucleotides. *Nucleic Acids Res.* **1987**, *15*, 6131–6148. [CrossRef]
24. Lacoux, C.; Di Marino, D.; Boyl, P.P.; Zalfa, F.; Yan, B.; Ciotti, M.T.; Falconi, M.; Urlaub, H.; Achsel, T.; Mougin, A.; et al. BC1-FMRP interaction is modulated by 2′-O-methylation: RNA-binding activity of the tudor domain and translational regulation at synapses. *Nucleic Acids Res.* **2012**, *40*, 4086–4096. [CrossRef] [PubMed]
25. Elliott, B.A.; Ho, H.T.; Ranganathan, S.V.; Vangaveti, S.; Ilkayeva, O.; Abou Assi, H.; Choi, A.K.; Agris, P.F.; Holley, C.L. Modification of messenger RNA by 2′-O-methylation regulates gene expression in vivo. *Nat. Commun.* **2019**, *10*, 3401. [CrossRef] [PubMed]
26. Li, X.; Ma, S.; Yi, C. Pseudouridine: The fifth RNA nucleotide with renewed interests. *Curr. Opin. Chem. Biol.* **2016**, *33*, 108–116. [CrossRef] [PubMed]
27. van der Feltz, C.; DeHaven, A.C.; Hoskins, A.A. Stress-induced Pseudouridylation Alters the Structural Equilibrium of Yeast U2 snRNA Stem II. *J. Mol. Biol.* **2018**, *430*, 524–536. [CrossRef]
28. Penzo, M.; Guerrieri, A.N.; Zacchini, F.; Trere, D.; Montanaro, L. RNA Pseudouridylation in Physiology and Medicine: For Better and for Worse. *Genes* **2017**, *8*, 301. [CrossRef]
29. Ge, J.; Yu, Y.T. RNA pseudouridylation: New insights into an old modification. *Trends Biochem. Sci.* **2013**, *38*, 210–218. [CrossRef]
30. Zhao, Y.; Dunker, W.; Yu, Y.T.; Karijolich, J. The Role of Noncoding RNA Pseudouridylation in Nuclear Gene Expression Events. *Front. Bioeng. Biotechnol.* **2018**, *6*, 8. [CrossRef]
31. Song, J.; Zhuang, Y.; Zhu, C.; Meng, H.; Lu, B.; Xie, B.; Peng, J.; Li, M.; Yi, C. Differential roles of human PUS10 in miRNA processing and tRNA pseudouridylation. *Nat. Chem. Biol.* **2020**, *16*, 160–169. [CrossRef] [PubMed]
32. Chen, J.; Weiss, W.A. Alternative splicing in cancer: Implications for biology and therapy. *Oncogene* **2015**, *34*, 1–14. [CrossRef] [PubMed]
33. Ma, F.C.; He, R.Q.; Lin, P.; Zhong, J.C.; Ma, J.; Yang, H.; Hu, X.H.; Chen, G. Profiling of prognostic alternative splicing in melanoma. *Oncol. Lett.* **2019**, *18*, 1081–1088. [CrossRef]
34. Singh, B.; Eyras, E. The role of alternative splicing in cancer. *Transcription* **2017**, *8*, 91–98. [CrossRef]
35. Guzinska-Ustymowicz, K.; Pryczynicz, A.; Kemona, A.; Czyzewska, J. Correlation between proliferation markers: PCNA, Ki-67, MCM-2 and antiapoptotic protein Bcl-2 in colorectal cancer. *Anticancer Res.* **2009**, *29*, 3049–3052.

36. Ladstein, R.G.; Bachmann, I.M.; Straume, O.; Akslen, L.A. Ki-67 expression is superior to mitotic count and novel proliferation markers PHH3, MCM4 and mitosin as a prognostic factor in thick cutaneous melanoma. *BMC Cancer* **2010**, *10*, 140. [CrossRef] [PubMed]
37. Ohsie, S.J.; Sarantopoulos, G.P.; Cochran, A.J.; Binder, S.W. Immunohistochemical characteristics of melanoma. *J. Cutan. Pathol.* **2008**, *35*, 433–444. [CrossRef]
38. Reddy, V.B.; Gattuso, P.; Aranha, G.; Carson, H.J. Cell proliferation markers in predicting metastases in malignant melanoma. *J. Cutan. Pathol.* **1995**, *22*, 248–251. [CrossRef]
39. Bresnick, A.R.; Weber, D.J.; Zimmer, D.B. S100 proteins in cancer. *Nat. Rev. Cancer* **2015**, *15*, 96–109. [CrossRef]
40. Fander, J.; Kielstein, H.; Buttner, M.; Koelblinger, P.; Dummer, R.; Bauer, M.; Handke, D.; Wickenhauser, C.; Seliger, B.; Jasinski-Bergner, S. Characterizing CD44 regulatory microRNAs as putative therapeutic agents in human melanoma. *Oncotarget* **2019**, *10*, 6509–6525. [CrossRef]
41. Ilmonen, S.; Hernberg, M.; Pyrhonen, S.; Tarkkanen, J.; Asko-Seljavaara, S. Ki-67, Bcl-2 and p53 expression in primary and metastatic melanoma. *Melanoma Res.* **2005**, *15*, 375–381. [CrossRef] [PubMed]
42. Weinstein, D.; Leininger, J.; Hamby, C.; Safai, B. Diagnostic and prognostic biomarkers in melanoma. *J. Clin. Aesthet. Dermatol.* **2014**, *7*, 13–24. [PubMed]
43. Kycler, W.; Grodecka-Gazdecka, S.; Breborowicz, J.; Filas, V.; Teresiak, M. Prognostic factors in melanoma. *Rep. Pract. Oncol. Radiother.* **2006**, *11*, 39–48. [CrossRef]
44. Cirenajwis, H.; Ekedahl, H.; Lauss, M.; Harbst, K.; Carneiro, A.; Enoksson, J.; Rosengren, F.; Werner-Hartman, L.; Torngren, T.; Kvist, A.; et al. Molecular stratification of metastatic melanoma using gene expression profiling: Prediction of survival outcome and benefit from molecular targeted therapy. *Oncotarget* **2015**, *6*, 12297–12309. [CrossRef]
45. Quandt, D.; Fiedler, E.; Boettcher, D.; Marsch, W.; Seliger, B. B7-h4 expression in human melanoma: Its association with patients' survival and antitumor immune response. *Clin. Cancer Res.* **2011**, *17*, 3100–3111. [CrossRef] [PubMed]
46. Starega-Roslan, J.; Krol, J.; Koscianska, E.; Kozlowski, P.; Szlachcic, W.J.; Sobczak, K.; Krzyzosiak, W.J. Structural basis of microRNA length variety. *Nucleic Acids Res.* **2011**, *39*, 257–268. [CrossRef]
47. Friedrich, M.; Vaxevanis, C.K.; Biehl, K.; Mueller, A.; Seliger, B. Targeting the coding sequence: Opposing roles in regulating classical and non-classical MHC class I molecules by miR-16 and miR-744. *J. Immunother. Cancer* **2020**, *8*. [CrossRef] [PubMed]
48. Jasinski-Bergner, S.; Mandelboim, O.; Seliger, B. The role of microRNAs in the control of innate immune response in cancer. *J. Natl. Cancer Inst.* **2014**, *106*. [CrossRef] [PubMed]
49. Jasinski-Bergner, S.; Mandelboim, O.; Seliger, B. Molecular mechanisms of human herpes viruses inferring with host immune surveillance. *J. Immunother. Cancer* **2020**, *8*. [CrossRef]
50. Fattore, L.; Costantini, S.; Malpicci, D.; Ruggiero, C.F.; Ascierto, P.A.; Croce, C.M.; Mancini, R.; Ciliberto, G. MicroRNAs in melanoma development and resistance to target therapy. *Oncotarget* **2017**, *8*, 22262–22278. [CrossRef]
51. Kurimoto, R.; Chiba, T.; Ito, Y.; Matsushima, T.; Yano, Y.; Miyata, K.; Yashiro, Y.; Suzuki, T.; Tomita, K.; Asahara, H. The tRNA pseudouridine synthase TruB1 regulates the maturation of let-7 miRNA. *EMBO J.* **2020**, *39*, e104708. [CrossRef]
52. Penzo, M.; Montanaro, L. Turning Uridines around: Role of rRNA Pseudouridylation in Ribosome Biogenesis and Ribosomal Function. *Biomolecules* **2018**, *8*, 38. [CrossRef] [PubMed]
53. Hirano, M.; Das, S.; Guo, P.; Cooper, M.D. The evolution of adaptive immunity in vertebrates. *Adv. Immunol.* **2011**, *109*, 125–157. [CrossRef]
54. Bukur, J.; Jasinski, S.; Seliger, B. The role of classical and non-classical HLA class I antigens in human tumors. *Semin. Cancer Biol.* **2012**, *22*, 350–358. [CrossRef] [PubMed]
55. Motorin, Y.; Marchand, V. Detection and Analysis of RNA Ribose 2′-O-Methylations: Challenges and Solutions. *Genes* **2018**, *9*, 642. [CrossRef] [PubMed]
56. Abe, M.; Naqvi, A.; Hendriks, G.J.; Feltzin, V.; Zhu, Y.; Grigoriev, A.; Bonini, N.M. Impact of age-associated increase in 2′-O-methylation of miRNAs on aging and neurodegeneration in Drosophila. *Genes Dev.* **2014**, *28*, 44–57. [CrossRef]
57. Liang, H.; Jiao, Z.; Rong, W.; Qu, S.; Liao, Z.; Sun, X.; Wei, Y.; Zhao, Q.; Wang, J.; Liu, Y.; et al. 3′-Terminal 2′-O-methylation of lung cancer miR-21-5p enhances its stability and association with Argonaute 2. *Nucleic Acids Res.* **2020**, *48*, 7027–7040. [CrossRef]
58. Ponten, F.; Jirstrom, K.; Uhlen, M. The Human Protein Atlas–a tool for pathology. *J. Pathol.* **2008**, *216*, 387–393. [CrossRef]
59. Thul, P.J.; Akesson, L.; Wiking, M.; Mahdessian, D.; Geladaki, A.; Ait Blal, H.; Alm, T.; Asplund, A.; Bjork, L.; Breckels, L.M.; et al. A subcellular map of the human proteome. *Science* **2017**, *356*. [CrossRef]
60. Uhlen, M.; Fagerberg, L.; Hallstrom, B.M.; Lindskog, C.; Oksvold, P.; Mardinoglu, A.; Sivertsson, A.; Kampf, C.; Sjostedt, E.; Asplund, A.; et al. Proteomics. Tissue-based map of the human proteome. *Science* **2015**, *347*, 1260419. [CrossRef] [PubMed]

 MDPI

Article

Identification of RCC Subtype-Specific microRNAs–Meta-Analysis of High-Throughput RCC Tumor microRNA Expression Data

Arkadiusz Kajdasz [1,*,†], Weronika Majer [2,†], Katarzyna Kluzek [1], Jacek Sobkowiak [3], Tomasz Milecki [3], Natalia Derebecka [2], Zbigniew Kwias [3], Hans A. R. Bluyssen [1] and Joanna Wesoly [2,*]

1 Laboratory of Human Molecular Genetics, Faculty of Biology, Institute of Molecular Biology and Biotechnology, Adam Mickiewicz University Poznan, Uniwersytetu Poznanskiego 6, 61-614 Poznan, Poland; k.kluzek@amu.edu.pl (K.K.); h.bluyss@amu.edu.pl (H.A.R.B.)
2 Laboratory of High Throughput Technologies, Faculty of Biology, Adam Mickiewicz University Poznan, Uniwersytetu Poznanskiego 6, 61-614 Poznan, Poland; weronika.majer@amu.edu.pl (W.M.); natalia.derebecka@amu.edu.pl (N.D.)
3 Department of Urology, Poznan University of Medical Sciences, Szwajcarska 3, 61-285 Poznan, Poland; jksobkowiak@gmail.com (J.S.); mileckito@gmail.com (T.M.); zbigniew.kwias@poczta.onet.pl (Z.K.)
* Correspondence: akajdasz@amu.edu.pl (A.K.); j.wesoly@amu.edu.pl (J.W.); Tel.: +48-61-829-58-33 (A.K.); +48-61-829-58-32 (J.W.)
† Both authors contributed equally to this work.

Simple Summary: In the majority of renal cancer cases, the disease course is non-symptomatic which frequently leads to late diagnosis of disease. Currently, there are no molecular tools dedicated to the detection and monitoring of renal cancer. Our study aimed to investigate changes in microRNA (miRNA) expression in tissue samples of renal cancer patients. We performed meta-analysis using results of 14 high-throughput studies (both, NGS and microarrays) and as a result, selected a group of miRNAs deregulated in renal cancer and its subtypes. Later, the expression changes of selected miRNA were validated in an independent sample set. We confirmed that the investigation of miRNA expression might be potentially applicable in the detection and monitoring of renal cancer and its subtypes.

Abstract: Renal cell carcinoma (RCC) is one of the most common cancers worldwide with a nearly non-symptomatic course until the advanced stages of the disease. RCC can be distinguished into three subtypes: papillary (pRCC), chromophobe (chRCC) and clear cell renal cell carcinoma (ccRCC) representing up to 75% of all RCC cases. Detection and RCC monitoring tools are limited to standard imaging techniques, in combination with non-RCC specific morphological and biochemical read-outs. RCC subtype identification relays mainly on results of pathological examination of tumor slides. Molecular, clinically applicable and ideally non-invasive tools aiding RCC management are still non-existent, although molecular characterization of RCC is relatively advanced. Hence, many research efforts concentrate on the identification of molecular markers that will assist with RCC sub-classification and monitoring. Due to stability and tissue-specificity miRNAs are promising candidates for such biomarkers. Here, we performed a meta-analysis study, utilized seven NGS and seven microarray RCC studies in order to identify subtype-specific expression of miRNAs. We concentrated on potentially oncocytoma-specific miRNAs (miRNA-424-5p, miRNA-146b-5p, miRNA-183-5p, miRNA-218-5p), pRCC-specific (miRNA-127-3p, miRNA-139-5p) and ccRCC-specific miRNAs (miRNA-200c-3p, miRNA-362-5p, miRNA-363-3p and miRNA-204-5p, 21-5p, miRNA-224-5p, miRNA-155-5p, miRNA-210-3p) and validated their expression in an independent sample set. Additionally, we found ccRCC-specific miRNAs to be differentially expressed in ccRCC tumor according to Fuhrman grades and identified alterations in their isoform composition in tumor tissue. Our results revealed that changes in the expression of selected miRNA might be potentially utilized as a tool aiding ccRCC subclass discrimination and we propose a miRNA panel aiding RCC subtype distinction.

Citation: Kajdasz, A.; Majer, W.; Kluzek, K.; Sobkowiak, J.; Milecki, T.; Derebecka, N.; Kwias, Z.; Bluyssen, H.A.R.; Wesoly, J. Identification of RCC Subtype-Specific microRNAs–Meta-Analysis of High-Throughput RCC Tumor microRNA Expression Data. *Cancers* **2021**, *13*, 548. https://doi.org/10.3390/cancers13030548

Academic Editor: Paola Tucci
Received: 14 December 2020
Accepted: 20 January 2021
Published: 1 February 2021

Keywords: microRNA; renal cancer; RCC; ccRCC; meta-analysis

1. Introduction

Renal cell carcinoma (RCC) is one of ten the most commonly occurring cancer types worldwide [1]. The occurrence of RCC is population dependent, although the general incidence is estimated to be 10 per 100,000 individuals [2]. The 5-year recovery rate of metastatic RCC patients is 12.3% [3] and is frequently a consequence of a late diagnosis. Nearly non-symptomatic disease course and lack of characteristic symptoms except flank pain, hematuria and hypertension accompanied by general fatigue, recurrently lead to the identification of RCC in advanced and/or metastatic stage, with 18% of patients displaying peripheral metastases in distal organs [4]. However, first mutations leading to the tumor development occur in childhood or adolescence, years or even decades before diagnosis [5]. Since RCC is chemo- and radiotherapy resistant the main RCC treatment is partial or complete nephrectomy [6].

Four main RCC subtypes have been identified: clear cell renal cell carcinoma (ccRCC) (75–85% of all RCC cases), papillary (pRCC) (12–14%), chromophobe (chRCC) (4–6%), clear cell papillary (ccpRCC) (~4%) also accompanied by renal oncocytoma, often benign, that comprises approximately 1% of kidney tumors [6,7]. Additionally, over the last few years a relatively rare subtype of renal tumor tubulocystic RCC (tcRCC) was described [8]. The diagnosis relays mainly on the results of imaging techniques, such as computer tomography or magnetic resonance, rarely followed by a tumor biopsy. RCC subtype differentiation is confirmed after tumor resection by histological examination of the tumor slides. Frequently, RCC tumors are difficult to distinguish due to the limitations of the imaging techniques and histological classification might be incorrect due to tumor heterogeneity. In the majority of oncocytoma cases, surgical intervention is not required but incorrect tumor classification may lead to unnecessary surgery [9]. Additionally, frequent and repetitive use of imaging techniques or biopsy could be potentially harmful to patients (excess of radiation or post-procedure complications) [10]. Therefore, new methods that could aid cancer detection and RCC classification such as noninvasive molecular biomarkers are a promising alternative.

Non-invasive biomarkers have been utilized in many cancer types and include Human Epidermal Growth Factor Receptor 2 (HER2) (in breast tumors), BRAF V600E (in metastatic melanoma), Prostate Specific Antigen (PSA) (in prostate cancer) and Carcinoembryonic Antigen (CEA) (in colorectal cancer) [11,12]. On the other hand, there are no specific non-invasive biomarkers aiding RCC diagnosis. However currently, clinical trials (e.g., RECORD-3) are focused on finding non-invasive biomarkers to monitor treatment outcome [13].

In recent years, there is an increasing interest in employing microRNA (miRNA)—small noncoding approximately 22 nucleotides long RNA—as cancer biomarkers [14]. miRNA originates during a multistep process in which long miRNA transcript, called primary-miRNA (pri-miRNA) is cleaved to ~70 nt length pre-miRNA by Drosha Ribonuclease III (DROSHA) in complex with DiGeorge Syndrome Critical Region Gene 8 (DGCR8). Pre-miRNA is further cleaved by DICER1 Ribonuclease III to ~22 nt double-stranded miRNA molecule. One of these strands is loaded into Argonaut 2 protein (AGO2) creating RNA Inducing Silencing Complex (RISC), the second strand is degraded [15]. RISC takes part in posttranscriptional gene expression by blocking mRNA translation or initiating mRNA cleavage [16] due to the presence of "seed" sequence in miRNA, complementary to mRNA usually in 3′ untranslated region (3′UTR) of targeted mRNA. Mature miRNA can occur in isoforms (iso-miRNA) processed from the same pri-miRNA and different at 5′ and 3′ ends as a result of inaccurate cleavage by DROSHA and DICER1. Those modifications can influence miRNA activity and function. Additionally, 3′ end of miRNA may be adenylated or uridylated which affects its stability. Deregulations of miRNA expression have been previously correlated with changes in protein levels engaged in proliferation, motility or cell invasiveness and in consequence promotion of tumor development and growth [17,18].

Changes in miRNA expression are well known in RCC tumors and these abnormalities can be potentially useful to distinguish RCC subtypes, although certain discrepancies between the studies can be noted [19]. The most commonly identified as downregulated in ccRCC tumor samples were miR-141, miRNA-200c [19]. On the other hand, many studies identified miRNA-210, miRNA-224 and miRNA-155 as upregulated in ccRCC tumors [20,21]. It has been also shown ccRCC and pRCC display significant changes in miRNA-424 expression, which could be helpful in RCC tumor subtype classification [22].

Here, we performed a meta-analysis of miRNA expression in ccRCC, pRCC and chRCC tumors, analyzed the expression of miRNA isoforms and examined potential causes of miRNA deregulation in ccRCC using a bioinformatics approach. After validation of miRNA expression in RCC tumors kidney tissue, we postulate that a miRNA panel could be potentially a powerful RCC classification tool and miRNA profile may be indicative of disease grades.

2. Materials and Methods

2.1. Sample Preparation

One hundred and fifteen samples of cancer tissue and 36 adjacent noncancerous kidney samples were obtained from renal cancer patients after partial or complete nephrectomy. Samples were collected the Department of Urology and Urological Oncology, Poznan University of Medical Sciences, Poland with the signed consent of patients (bioethical consent of Local Bioethical Committee at Poznan University of Medical Sciences, no. 1124/12). Tumors were classified as RCC subtypes and according to Fuhrman grade. ccRCC tumors (n = 97) were divided into Fuhrman grade 1, n = 12; grade 2, n = 35; grade 3, n = 33; grade 4, n = 17. 10 pRCC, 10 oncocytoma and 36 adjacent, histopathologically unchanged kidney tissue samples were also included in the analysis (Table S1). Samples were randomly divided into groups used in different experiments (details in Table S1). The average age of patients was 65. For the preservation of RNA, tissue fragments were collected into tubes containing RNAlater (Sigma Aldrich, St. Louis, MO, USA) and stored at −80 °C for further processing. Next, fragments of tissue were transferred into a sterile mortar and grinded in liquid nitrogen. RNA was extracted with Trizol and quantified using NanoDrop ND-1000.

2.2. Library Preparation and Sequencing

For library preparation, 1 µg of total RNA with RNA Integrity Number (RIN) equal to or above 7 was used. RNA-Seq libraries (controls, n = 17; ccRCC, n = 58 samples) were prepared with TruSeq RNA Library Preparation Kit v2 (Illumina, San Diego, CA, USA) and small RNA-Seq libraries (controls, n = 6; ccRCC, n = 26) with TruSeq Small RNA Library Preparation Kit (Illumina, San Diego, CA, USA). Quality and concentration of the libraries were tested using Agilent DNA 1000 Kit and Agilent High Sensitivity DNA Kit (Agilent Technologies Inc., Santa Clara, CA, USA). The libraries were sequenced on HiScan SQ (Illumina) with TruSeq PE Cluster Kit v3-cBot-HS (Illumina, San Diego, CA, USA, cat. no. PE-401-3001) in PE100 and SR50 modes, respectively.

The data discussed in this publication have been deposited in NCBI's Gene Expression Omnibus [23] and are accessible through GEO Series accession number GSE151428 (https://www.ncbi.nlm.nih.gov/geo/query/acc.cgi?acc=GSE151428).

2.3. Small RNA-Seq Data Processing

In small RNA-Seq analysis workflow, raw reads were trimmed with cutadapt (https://cutadapt.readthedocs.io/en/stable/), untrimmed reads and reads shorter than 10 nt were discarded. Reads were aligned to Ensembl GRCh38 human genome with bowtie2 [24]. Raw read counts were generated with featureCounts v1.6.3 [25] and differentially expressed miRNA in ccRCC tumors were identified with edgeR (v3.28) package (R v3.6). miRNA isoforms were calculated with MIRALIGNER protocol [26] and isomiRs (v1.14) (https://bioconductor.org/packages/isomiRs/) package (R v3.6) based on miRBase v22.1.

2.4. RNA-Seq Data Processing

RNA-Seq raw paired-end reads were trimmed with cutadapt. Trimmed reads were aligned to Ensembl GRCh38 human genome with STAR (v2.7) [27] and counts were obtained using featureCounts v1.6.3 [25]. Differentially expressed mRNA in ccRCC tumors were identified with edgeR (v3.28) package (R v3.6).

2.5. Meta-Analysis of miRNA Expression in RCC Tumors

Sequencing data from Exp1 were generated in our laboratory. Exp2 [28], Exp3 [29] and Exp4 [30] were downloaded from the SRA database [31] as raw reads. All small RNA-Seq experiments used in the meta-analysis were performed on fresh tissues, available Formalin-Fixed Paraffin-Embedded experiments were excluded. Exp5–Exp14 results were derived from dbDEMC2.0 database [32] (Figure S1). Significantly deregulated miRNAs in ccRCC obtained from all above experiments were compared using the dplyr package (v0.8.3; R v3.6). Venn graphs were created with the VennDiagram package (v1.6; R v3.6).

2.6. Poly(A)-RT

Synthesis of cDNA by polyadenylation reverse transcription reaction (Poly(A)-RT) was described previously [33]. 12.4 μL of RNA sample (1 μg of RNA) were added to reverse transcription mix containing: 12 μL of RT buffer (25 μL 1 M Tris-HCl, pH = 8.0; 93.75 μL 2 M KCl; 250 μL 100 mM DTT; 175 μL 1 M $MgCl_2$); 20 μL 100 μM anchor RT primer, containing universal adapter sequence; 436.25 μL H_2O; 6 μL deoxynucleotide mix (100 mM of each); 25 μL 10 mM rATP; 25 μL 40 U/μL RiboLock (ThermoFisher, Waltham, MA, USA); 0.6 μL *E. coli* poly(A) polymerase (New England Biolabs, Rowley, MA, USA) and 0.6 μL reverse transcriptase (ThermoFisher). The reaction was performed at 37 °C for 1 h followed by inactivation at 85 °C for 10 min.

2.7. qPCR

In the analysis of mRNA and miRNA expression, 2 μL of the 5 times diluted cDNA template were used per qPCR reaction. miRNA amplification was performed with miRNA specific forward primer (Table S4) and universal reverse primer complement to the adapter sequence. In all reactions Maxima SYBR Green/ROX qPCR Master Mix (2X) (ThermoFisher) was used. miRNA results were normalized to U6. Specific "iso-miRNA primers," which discriminate miRNA-363-3p or miRNA-224-5p shorter isoforms and amplify longer isoforms, were used in qPCR. Iso-miRNA results were normalized to U6 or specific miRNA. mRNA results were normalized to *GAPDH*. The expression level was determined by the $2^{-\Delta\Delta Ct}$ method [34].

2.8. Gene Ontology (GO) Analysis of miRNA Targets

Gene Ontology analysis for miRNA targets was performed with GeneMANIA (v3.5.1) [35] in Cytoscape (v3.7.1) [36]. The most probable miRNA targets clusters in networks were detected with MCODE (v1.5.1) [37] in Cytoscape (v3.7.1).

2.9. Statistical Analysis

Pearson correlation coefficient was calculated and the regression graph was created with base functions of R (v3.6). The remaining statistical analyses were performed using a two-tailed *t*-test in Microsoft Excel (NS, non-significant; * $p < 0.05$; ** $p < 0.01$ and *** $p < 0.001$), ± standard error of the mean (SEM). The number of biological replicates (n) is shown in the figure descriptions. Receiver operating characteristic (ROC) analysis was performed with the MetaboAnalyst tool [38]. Kaplan-Meier survival plots were created based on The Cancer Genome Atlas (TCGA) data with Kaplan-Meier Plotter [39].

3. Results

3.1. Small RNA-Seq and Meta-Analysis

In order to identify specifically deregulated microRNAs in RCC first we conducted small RNA-Seq experiment on ccRCC tumor tissue derived from Polish patients (Exp1, ccRCC: $n = 26$, controls: $n = 6$). The data was extended with additional publicly available data sets, derived from both NGS and microarray experiments, collected from Sequence Read Archive (SRA) and dbDMEMC 2.0 databases. The details of the experiments and analytic workflow were listed in Figure S1.

The NGS data was collected in form of raw reads, subjected to the identical data processing and included four small RNA-Seq experiments performed on ccRCC tumors (Exp1–Exp4) [28–30]. The final number of differentially expressed miRNAs (FDR < 0.05) differed per data set (Figure S1a).

Additionally, in order to retrieve the information derived from NGS experiments collected by The Cancer Genome Atlas (pRCC, chRCC) and microarray experiments (pRCC, chRCC and oncocytoma), we utilized datasets from the dbDEMC 2.0 in a form of lists of differentially expressed (DE) miRNAs. The data included: four ccRCC (Exp5–Exp8) [40–42], three pRCC (Exp9–Exp11) [40,43], two chRCC (Exp12, Exp13) [40] and one oncocytoma (Exp14) [40] experiments (Figure S1a).

After data processing, we performed a meta-analysis of deregulated miRNAs in ccRCC and compared significantly deregulated miRNAs in NGS experiments (Figure 1a) and microarrays (Figure 1b) identified 22 and 25 commonly deregulated miRNAs, respectively. Due to technical and analytical differences between compared experiments, we implemented stringent exclusion criteria: only miRNAs reported in all data sets to be deregulated, with significant FDR values were taken into account. chRCC was characterized by deregulation of 18 miRNAs (Figure 1e) and 10 miRNAs were found deregulated in pRCC (Figure 1f). Limited information on oncocytoma listed 34 deregulated miRNAs (single microarray experiment). Detailed lists of DE miRNAs, including FC and FDR parameters, are included in Supplementary File S1.

After comparison of all available data sets eight commonly deregulated miRNAs in ccRCC were selected (Figure 1c) and those included: miR-200c-3p, miR-362-5p, miR-363-3p and miR-204-5p as downregulated and miR-21-5p, miR-224-5p, miR-155-5p and miR-210-3p as upregulated (Figure 1d).

Interestingly, as depicted in Figure 1g, our analysis suggests that there are no commonly deregulated miRNAs for all RCC tumors, although we cannot exclude the possibility of missing a number of miRNA candidates due to rigorous cut-off criteria. Additionally, this analysis suggests that deregulation of miR-21-5p, miR-155-5p and miR-210-3p could be ccRCC-specific.

3.2. Validation of RCC-Specific miRNA Candidates

Next, we set out to validate the expression of potentially RCC subtype-specific miRNAs in an independent sample set using quantitative PCR (qPCR). Due to lack of availability of chRCC samples the 18 commonly deregulated miRNAs were not verified. We included 39 ccRCC, 10 pRCC and 8 oncocytoma tumor tissues, with 15 adjacent, histopathologically unchanged kidney tissue samples.

From potentially pRCC- or oncocytoma-specific miRNAs distinguished in meta-analysis (Figure 1 and Supplementary File S1) those with the highest fold change were selected to further validation.

For oncocytoma-specific miRNA validation we selected four miRNAs: downregulated miRNA-424-5p, miRNA-146b-5p and upregulated miRNA-183-5p and miRNA-218-5p. In independent sample-set miRNA-183-5p is upregulated in pRCC (22-fold change, $p < 0.01$) and oncocytoma (27-fold change, $p < 0.001$) samples, while level of miRNA-218-5p is significantly decreased in ccRCC tumors (0.17-fold change, $p = 0.048$) (Figure 2a). miRNA-424-5p, miRNA-146b-5p display similar expression levels in all samples. Our findings suggest that miRNAs selected in the meta-analysis are not oncocytoma-specific.

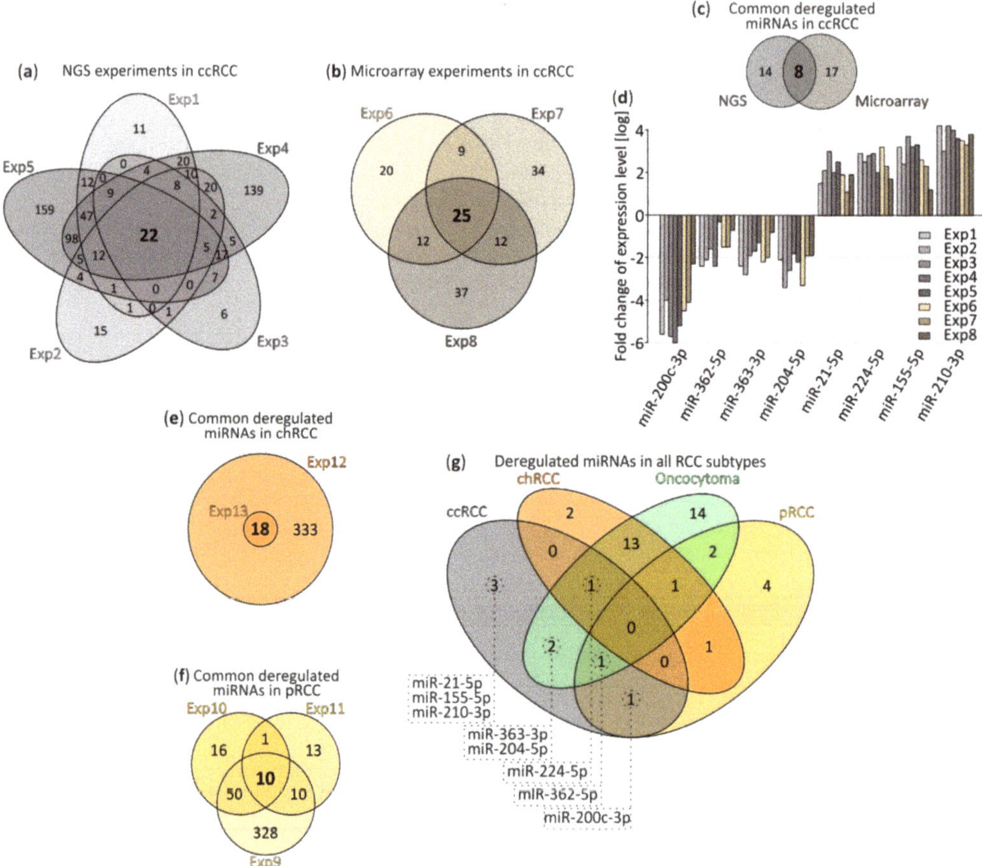

Figure 1. The meta-analysis of deregulated miRNA in renal cell carcinoma (RCC) tumor tissues. (**a**,**b**) Venn diagrams depicting commonly deregulated miRNAs in clear cell renal cell carcinoma (ccRCC) tumors. (**a**) small RNA-seq and (**b**) microarray experiments; (**c**,**d**) Comparison of commonly deregulated miRNA in chromophobe renal cell carcinoma (chRCC) (**c**) and papillary (pRCC) (**d**); (**e**) miRNA identified in both next-generation sequencing (NGS) and microarray experiments performed in ccRCC; (**f**) Comparison of eight miRNA expression levels commonly disrupted in ccRCC reported in original NGS and microarray experiments; (**g**) Comparison of commonly deregulated miRNA in ccRCC (8 miRNA), chRCC (18 miRNA), oncocytoma (34 miRNA) and pRCC (10 miRNA). ccRCC, clear renal cell carcinoma; chRCC, chromophobe renal cell carcinoma; pRCC, papillary renal cell carcinoma; Exp1–8, experiments 1–8.

According to high-throughput data miRNA-127-3p and miRNA-139-5p should be downregulated in pRCC tumors. Although, in contrast to previous reports, in our sample set miRNA-127-3p appears to be significantly upregulated in pRCC (29-fold change, $p < 0.01$). Increase in expression of miRNA-139-5p does not significantly differ in pRCC and oncocytoma (Figure 2b), however miRNA-139-5p appears to be downregulated in ccRCC (0.2-fold change, $p = 0.04$). Inconsistency of these data suggest rejection of these miRNA as pRCC-specific.

As shown in Figure 2c,d, expression profiles of eight potentially ccRCC-specific microRNAs were further investigated. Remarkably, downregulation of miR-200c-3p is observed in pRCC (0.04-fold change, $p < 0.014$) and oncocytoma (0.03-fold change, $p = 0.03$) with no change in ccRCC (for explanation see below). The remaining miRNAs do not appear to be ccRCC-specific. Although, read out of the miR-362-5p (ccRCC, 0.1-fold change,

$p < 0.01$; pRCC, 0.007-fold change, $p < 0.001$) and miR-363-3p (ccRCC, 0.15-fold change, $p < 0.01$; pRCC, 0.2-fold change, $p = 0.03$) did not reach the statistical significance in oncocytoma. Only miR-204-5p, was significantly downregulated in all RCC tumor types as compared to controls (ccRCC, 0.09-fold change, $p = 0.011$; pRCC, 0.006-fold change, $p < 0.01$; oncocytoma, 0.03-fold change, $p < 0.001$).

miRNA-21-5p was overexpressed solely in pRCC tumors (14-fold change, $p < 0.01$). Levels of both miRNA-224-5p and miRNA-210-3p were found significantly increased in ccRCC tumors (5.5-fold change, $p = 0.02$; 12-fold change, $p < 0.001$, respectively), although they were also upregulated in oncocytoma (18-fold change, $p < 0.01$) and pRCC (10-fold change, $p = 0.013$), respectively. There was no significant change in the expression of miR-21-5p, miR-224-5p and miR-210-3p among the subtypes (Figure 2d). Interestingly, significant miR-155-5p upregulation appeared to be distinctive of ccRCC (15-fold change, $p < 0.001$) (Figure 2d).

Next, we analyzed expression profiles of the eight miRNAs in ccRCC tumors with different grading, represented by nine tumors in Fuhrman grade one (G1), eleven in grade two (G2), ten in grade three (G3) and nine in grade four (G4). Figure 2e shows no significant changes in different tumor grades in case of miRNA-200c-3p. miRNA-200c-3p is one of most commonly identified as ccRCC downregulated miRNA [44]. Since we could not validate its downregulation, although also suggested by NGS data implemented in the meta-analysis, we investigated the potential reasons for such discrepancy. Firstly, we compare sequences of miR-200 family members including miR-200a, miR-200b, miR-200c, miR-141 and miR-429 (Figure S2a). Due to a significant sequence homology between the miRNA-200 family members, we hypothesized that simultaneous amplification of more than one miRNA might be the reason for the inconsistency between the studies. To further support this hypothesis, we investigated the contribution of miR-200c-3p in the complete family which is relatively low in both control and ccRCC tissues based on NGS data (Figure S2b). Additionally, the comparison of miR-200 family members expression in all available data sets shows that they could be deregulated in all RCC tumor types (Figure S2c). These observations confirming the potential cumulative readout might lead to misinterpretation of the data provided by qPCR.

Consistent downregulation throughout all tumor grades were observed in case of miR-362-5p (G1, 0.09-fold change, $p < 0.01$; G3, 0.03-fold change, $p < 0.001$; G4, 0.09-fold change, $p = 0.013$), miR-363-3p (G1, 0.3-fold change, $p = 0.019$; G3, 0.08-fold change, $p < 0.001$; G4, 0.07-fold change, $p < 0.01$) and miR-204-5p (G1, 0.3-fold change, $p = 0.042$; G3, 0.01-fold change, $p < 0.001$; G4, 0.05-fold change, $p = 0.02$) (Figure 2e). Significant upregulation of miRNA-21-5p expression is present only in the G4 (6-fold change, $p = 0.047$), although, increasing expression of this miRNA is noticeable across all Fuhrman grades (Figure 2f). The following miRNAs displayed elevated expression in all samples: miR-210-3p (G1, 21-fold change, $p < 0.001$; G2, 7.4-fold change, $p = 0.013$; G3, 12-fold change, $p < 0.01$; G4, 11-fold change, $p < 0.01$) and miR-155-5p (G1, 22-fold change, $p < 0.01$; G2, 7.4-fold change, $p = 0.02$; G3, 8-fold change, $p = 0.03$; G4, 47-fold change, $p < 0.001$). Interestingly, significant rise of miR-155-5p was clearly observed in G4 (Figure 2f). Interestingly, significant miR-224-5p overexpression was characteristic only for G1 ccRCC tumors (28-fold change, $p < 0.001$).

Validation of expression levels of eight selected miRNA revealed grade-dependent variations in ccRCC tumor tissues, whereas, a comparison of expression levels of the 14 miRNAs between ccRCC, pRCC and oncocytoma suggests differences in miRNA expression dependent on the RCC subtype.

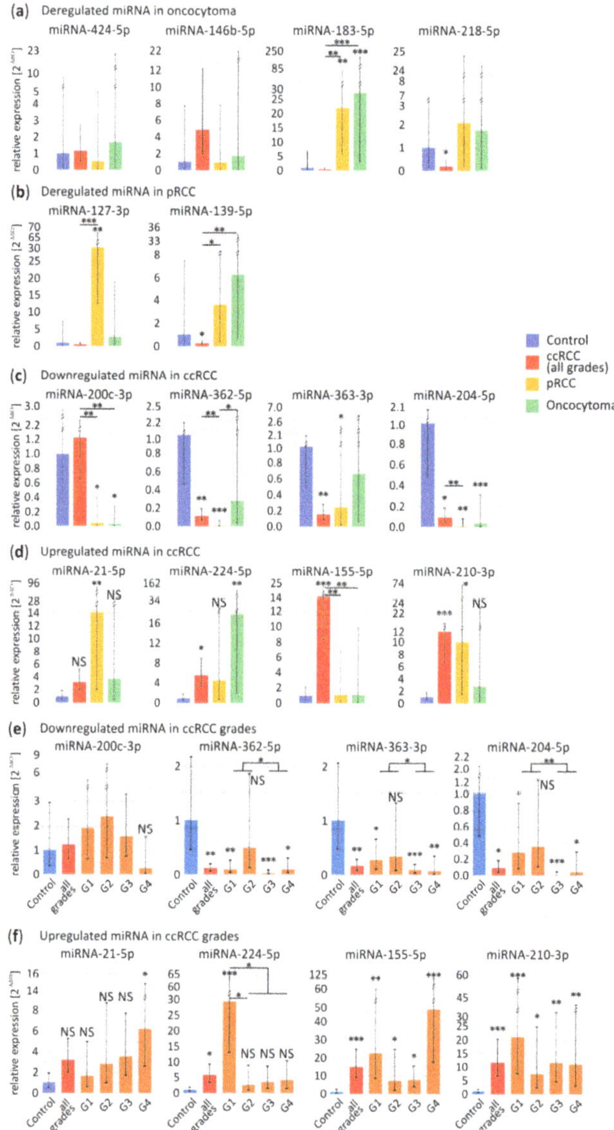

Figure 2. Validation of commonly deregulated miRNA in the independent sample set of ccRCC, pRCC and oncocytoma tumors by quantitative PCR (qPCR). (**a**,**b**) Specifically deregulated miRNA in oncocytoma (**a**) and pRCC (**b**) tumors based on a meta-analysis compared to other subtypes. Analyzed specimens included: control (n = 9), ccRCC (n = 11), pRCC (n = 10), oncocytoma (n = 8) samples; (**c**–**f**) Validation of expression disruptions of commonly deregulated miRNA in RCC tumors. Downregulated (**c**) and upregulated (**d**) miRNA in ccRCC tumors based on a meta-analysis compared to pRCC and oncocytoma. Analyzed specimens included: control (n = 15), ccRCC (n = 39), pRCC (n = 11), oncocytoma (n = 8) samples; (**e**,**f**) miRNA expression in ccRCC tumors grouped according to Fuhrman grade: downregulated (**e**) and upregulated (**f**) miRNA in ccRCC tumors based on a meta-analysis. Analyzed specimens included: control, n = 15; ccRCC (n = 39), ccRCC G1 (n = 9), ccRCC G2 (n = 11), ccRCC G3 (n = 10), ccRCC G4 (n = 9) samples. Blue bars, control tissue; red bars, ccRCC tumors; orange bars, ccRCC Fuhrman grades; yellow bars, pRCC; green bars, oncocytoma; ccRCC, clear renal cell carcinoma; pRCC, papillary renal cell carcinoma; G1-G4, ccRCC Fuhrman grades 1–4. NS, non-significant; * $p < 0.05$; ** $p < 0.01$ and *** $p < 0.001$.

3.3. ROC Analysis

Receiver operating characteristic (ROC) analysis was performed to assess the prognostic accuracy of the miRNA signatures based on real-time PCR results. The area under the curve (AUC) was calculated for each comparison. miRNA expression levels obtained from all RCC subtypes were compared with control tissue or between each other to calculate their predictive potential (Figure 3). The data suggest that downregulated miRNA-362-5p (AUC = 0.79, $p < 0.01$) and miRNA-363-3p (AUC = 0.8, $p < 0.01$) jointly with upregulated miRNA-155-5p (AUC = 0.83, $p < 0.001$) and miRNA-210-3p (AUC = 0.85, $p < 0.001$) significantly differentiates ccRCC tumors from healthy tissue. pRCC could be classified using miRNA-362-5p (AUC = 0.87, $p < 0.001$), miRNA-363-3p (AUC = 0.75, $p = 0.03$), miRNA-204-5p (AUC = 0.9, $p < 0.001$), miRNA-21-5p (AUC = 0.79, $p = 0.02$) and miRNA-210-3p (AUC = 0.86, $p = 0.01$) while oncocytoma with miRNA-204-5p (AUC = 0.9, $p < 0.001$) and miRNA-224-5p (AUC = 0.79, $p < 0.01$). Furthermore, ccRCC could be significantly distinguished from pRCC with miRNA-362-5p (AUC = 0.76, $p < 0.01$) and miRNA-155-5p (AUC = 0.79, $p < 0.01$) or from oncocytoma with miRNA-155-5p (AUC = 0.81, $p < 0.01$). Whereas, miRNA-362-5p (AUC = 0.8, $p = 0.02$) differentiates pRCC and oncocytoma. Additionally, expression changes of miRNA-224-5p shows potential to discriminate ccRCC Fuhrman grades, G1 versus G2 with AUC = 0.83 ($p = 0.02$) or G1 versus G2-4 with AUC = 0.79 ($p = 0.02$) (Figure 3a).

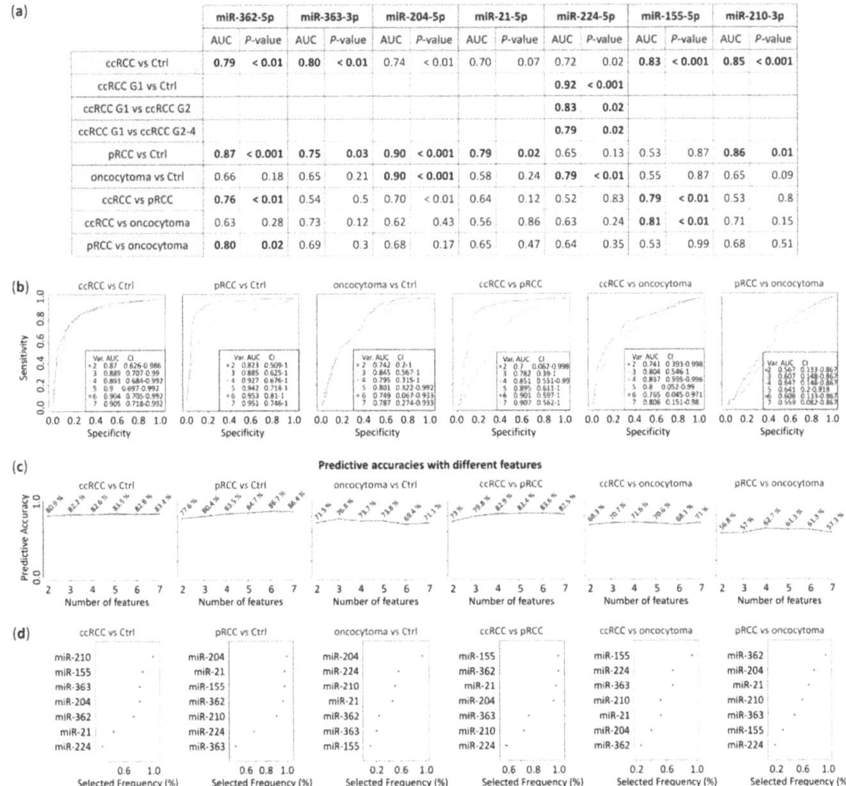

(a)

	miR-362-5p		miR-363-3p		miR-204-5p		miR-21-5p		miR-224-5p		miR-155-5p		miR-210-3p	
	AUC	*P*-value	AUC	*P*-value	AUC	*P*-value	AUC	*P*-value	AUC	*P*-value	AUC	*P*-value	AUC	*P*-value
ccRCC vs Ctrl	**0.79**	**< 0.01**	**0.80**	**< 0.01**	0.74	< 0.01	0.70	0.07	0.72	0.02	**0.83**	**< 0.001**	**0.85**	**< 0.001**
ccRCC G1 vs Ctrl									**0.92**	**< 0.001**				
ccRCC G1 vs ccRCC G2									**0.83**	**0.02**				
ccRCC G1 vs ccRCC G2-4									**0.79**	**0.02**				
pRCC vs Ctrl	**0.87**	**< 0.001**	**0.75**	**0.03**	**0.90**	**< 0.001**	**0.79**	**0.02**	0.65	0.13	0.53	0.87	**0.86**	**0.01**
oncocytoma vs Ctrl	0.66	0.18	0.65	0.21	**0.90**	**< 0.001**	0.58	0.24	**0.79**	**< 0.01**	0.55	0.87	0.65	0.09
ccRCC vs pRCC	**0.76**	**< 0.01**	0.54	0.5	0.70	< 0.01	0.64	0.12	0.52	0.83	**0.79**	**< 0.01**	0.53	0.8
ccRCC vs oncocytoma	0.63	0.28	0.73	0.12	0.62	0.43	0.56	0.86	0.63	0.24	**0.81**	**< 0.01**	0.71	0.15
pRCC vs oncocytoma	**0.80**	**0.02**	0.69	0.3	0.68	0.17	0.65	0.47	0.64	0.35	0.53	0.99	0.68	0.51

Figure 3. Receiver Operating Characteristic (ROC) analysis predictive potential of selected miRNA. (**a**) Area under the curve (AUC) with *p*-value calculated from miRNA expression level obtained from qPCR. Bold, comparisons with AUC > 0.75 and *p*-value < 0.05; (**b**) ROC multivariate analysis with seven proposed miRNAs; (**c**) Predictive accuracies with different miRNAs (red dot indicates a number of features with the highest predictive accuracy) and (**d**) frequency of selection in the test.

To propose potential panels of miRNA to distinguish RCC subtypes we performed ROC multivariate analysis (Figure 3b). Predictive accuracies with different miRNAs (Figure 3c) suggest that 5, 6 and 3 miRNAs distinguish ccRCC, pRCC and oncocytoma from healthy tissue with 83.5%, 86.7% and 76.8% of accuracy, respectively. RCC subtypes could be differentiated by 6 miRNAs (ccRCC vs. pRCC, 83.6% of accuracy), 4 miRNAs (ccRCC vs. oncocytoma, 71.6% of accuracy) or 4 miRNAs (pRCC vs. oncocytoma, 62.7% of accuracy). Selection frequency of each miRNA in the ROC test is shown on Figure 3d.

Our data suggest that proposed miRNA could have diagnostic potential and could efficiently distinguish RCC subtypes or between ccRCC tumors with Fuhrman grades.

3.4. *Iso-miRNA Analysis*

Mature miRNA could occur in isoforms that vary in length or presence of poly(A) or poly(U) tails on 3′ end. According to literature, iso-miRNA expression levels could successfully differentiate cancer types [45]. Hence, we decided to investigate iso-miRNA signatures in ccRCC using data sets obtained from Exp1 and Exp4 and validate them in ccRCC (controls, $n = 4$; ccRCC, $n = 17$).

We analyzed the percentage contribution of iso-miRNA of eight miRNAs, which according to our meta-analysis were commonly deregulated in ccRCC (Table S2). As shown in Figure 4a (upper panel), among four downregulated miRNAs, two: miR-363-3p and miR-204-5p exhibit significantly different isoform expression pattern in ccRCC tumors. In case of miR-363-3p, shortening of 3′ end is more frequent with simultaneous reduction of elongation, with 67% and 14% contribution as compared to control: 52% and 20%, respectively ($p < 0.01$ and $p = 0.047$, respectively). Similarly, poly(U) addition is more common in non-ccRCC tissue (control 15%, ccRCC 7.3%; $p < 0.01$). We also observed more reduced 3′ end lengthening of miRNA-204-5p in ccRCC (control 49%, ccRCC 28.5%; $p < 0.001$).

In the group of potentially ccRCC-specific, upregulated miRNAs (Figure 4a lower panel) miRNAs miR-21-5p displayed slight, less frequent, though statistically significant modifications, with addition of poly(A) (control 4%, ccRCC 3%; $p < 0.001$) and poly(U) (control 0.3%, ccRCC 0.2%). In case of miR-224-5p 3′ lengthening has elevated level in ccRCC (control 30%, ccRCC 50%; $p < 0.01$). Reference miRNA-210-3p has significantly lower percentage by 7% in ccRCC ($p = 0.01$). Interestingly, in case of miRNA-210-3p, we found differences in sequencing Exp1 and Exp4 concerning a tendency toward 5′ lengthening and less frequent addition of U nucleotide (Exp1, $p = 0.02$) and more frequent addition of mixed nucleotides in Exp4 ($p = 0.049$), which inclines us to carefully interpret the sequencing data and stresses the necessity for in depth, a multi-dataset study on miRNA isoforms (Figure 4a and Figure S3a).

Specific primers used in qPCR ("iso-miRNA primer") (Figure 4b,e) discriminate miRNA-363-3p and miRNA-224-5p shorter isoforms and amplify longer isoforms. As shown on an independent sample set (Figure 4c), iso-miRNA-363-3p expression was relatively similar in all tumors, regardless their grading status (G1, 0.004-fold change, $p < 0.001$; G2, 0.004-fold change, $p < 0.01$; G3, 0.001-fold change, $p < 0.01$; G4, 0.003-fold change, $p < 0.001$). Interestingly, iso-miRNA-224-5p was the most significant downregulated in G4 (0.15-fold change, $p = 0.02$) (Figure 4f), in contrast to miRNA-224-5p upregulation in G1 (Figure 3d). In relation to miR-363-3p (G1, 0.02-fold change, $p = 0.01$; G2, 0.02-fold change, $p = 0.01$; G3, 0.02-fold change, $p = 0.01$; G4, 0.01-fold change, $p < 0.01$) and miR-224-5p (G2, 0.02-fold change, $p < 0.01$; G4, 0.07-fold change, $p = 0.02$) both iso-miRNAs are significantly reduced in ccRCC (Figure 4d,g) which confirms decreased level of longer isoforms (Figure 4a).

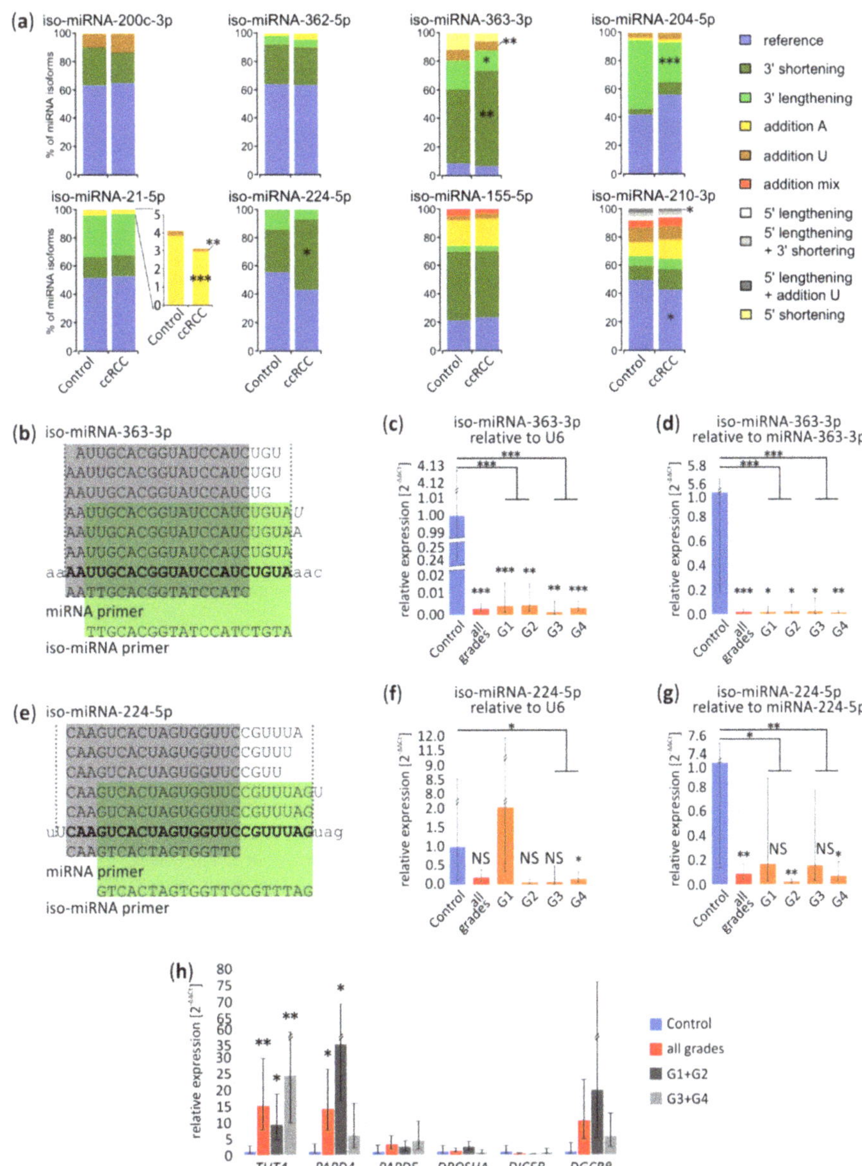

Figure 4. miRNA isoform deregulation in ccRCC tumors. (**a**) The percentage per miRNA isoform (iso-miRNA) in control and tumor samples based on Exp1 results. Part of the chart represents the percentage of iso-miRNA-21-5p is magnified to show more accurately 3′ tailing of miRNA-21-5p; (**b**,**e**) Sequences of miRNA-363-3p and miRNA-224-5p isoforms identified in kidney, respectively. "miRNA primer" (in grey) amplifies all iso-miRNA and "iso-miRNA primer" (in green) amplifies reference miRNA and longer iso-miRNA. Bold, reference miRNA sequence; small letters, pre-miRNA sequence; italic, additional U or A nucleotides; (**c**,**d**) Validation of iso-miRNA-363-3p in ccRCC tumors and controls by qPCR relative to U6 (**c**) and miRNA-363-3p (**d**); (**f**,**g**) Validation of iso-miRNA-224-5p in ccRCC tumors and controls by qPCR relative to U6 (**f**) and miRNA-224-5p (**g**); (**h**) qPCR analysis of expression of genes involved in miRNA processing and after maturation modifications in ccRCC tumors. Control (n = 4), ccRCC (n = 17) containing: G1 (n = 4), G2 (n = 3), G3 (n = 5), G4 (n = 5). NS, non-significant; * $p < 0.05$; ** $p < 0.01$ and *** $p < 0.001$.

Additionally, we analyzed expression profiles of genes involved in miRNA maturation (*DROSHA, DGCR8, DICER1*) or post-maturation processing (Terminal Uridylyl Transferase 4, *TUT4*; PAP Associated Domain Containing 4, *PAPD4;* PAP Associated Domain Containing 5, *PAPD5*) (Figure 4h). qPCR results suggest that *TUT4* and *PAPD4* are significantly upregulated in ccRCC tumors (15-fold change, $p < 0.01$ and 14-fold change, $p < 0.05$, respectively), however, expression of *PAPD4* is the highest in tumors with lower grades (G1 + G2) (34-fold change, $p = 0.03$).

Our data suggest that iso-miRNA contribution in ccRCC tumors may differ from the control tissue and observed expression shifts if validated, could aid ccRCC classification. Additionally, disruptions of miRNA expression in ccRCC could be partially explained by differences in miRNA isoform stability.

3.5. Basis of Deregulation of Selected miRNA in ccRCC

In order to investigate the character of deregulation of the eight commonly disrupted miRNAs in ccRCC we explored the possibility of co-transcriptional deregulation of miRNAs with their host genes. We utilized the data from RNA-Seq, performed on ccRCC tumors (ccRCC: $n = 60$, controls: $n = 17$) matching small RNA-Seq samples (this work). As shown in Figure S3b miR-200c, miR-204 and miR-362 host genes (*MIR200CHG, TRPM3, CLCN5*, respectively) display a statistically significant decrease in their expression. Similarly, miR-224, miR-21, miR-155 and miR-210 host genes (*GABRE, VMP1, MIR155HG* and *MIR210HG*, respectively), are elevated, supporting the mechanism of co-transcriptional regulation being the basis of seven out of eight analyzed miRNAs. The expression of miRNAs and their host genes was highly correlated, with $R = 0.97$ and statistically significant ($p < 0.001$) (Figure S3c). Since miR-363 is encoded by an intergenic locus, no data was obtained from RNA-Seq.

These data suggest that miRNAs are deregulated co-transcriptionally in ccRCC tumors. Although, more studies are necessary to investigate factors responsible for this disruption.

3.6. miRNA Functions

Survival analysis performed on the TCGA data with Kaplan-Meier Plotter on-line tool revealed that ccRCC patients with a high level of miRNA-224 (which is overexpressed in ccRCC G1) significantly classified patients into higher risk for death (hazard ratio (HR) = 1.49) (Figure 5a). Although, higher grades patients show the opposite pattern. In contrary, patients with the high level of miRNA-210, which is upregulated in ccRCC, have a lower risk for death (HR = 0.71) which is independent of the ccRCC Fuhrman's grade (Figure 5b).

One miRNA could regulate the expression of many genes and simultaneously one mRNA could be targeted by a few miRNAs. Using data from the miRTarBase database [46], we obtained a list of validated gene targets of 8 commonly deregulated miRNA in ccRCC. Subsequently, we selected genes expressed in the kidney, examined their expression using RNA-seq data from tumors derived from Polish ccRCC patients (Supplementary File S2) and followed with gene ontology enrichment analysis (GO) using GeneMANIA application in Cytoscape.

The majority of identified pathways for example, cellular response to hypoxia, response to TGF-beta, serine/threonine kinase signaling pathway are well known to be de-regulated in ccRCC, although our analysis points to other, also interesting pathways that may contribute to ccRCC etiology, involving components of the homotypic fusion and protein sorting (HOPS)-tethering complex and mRNA poly(A) tail shortening (Table 1 and Table S3).

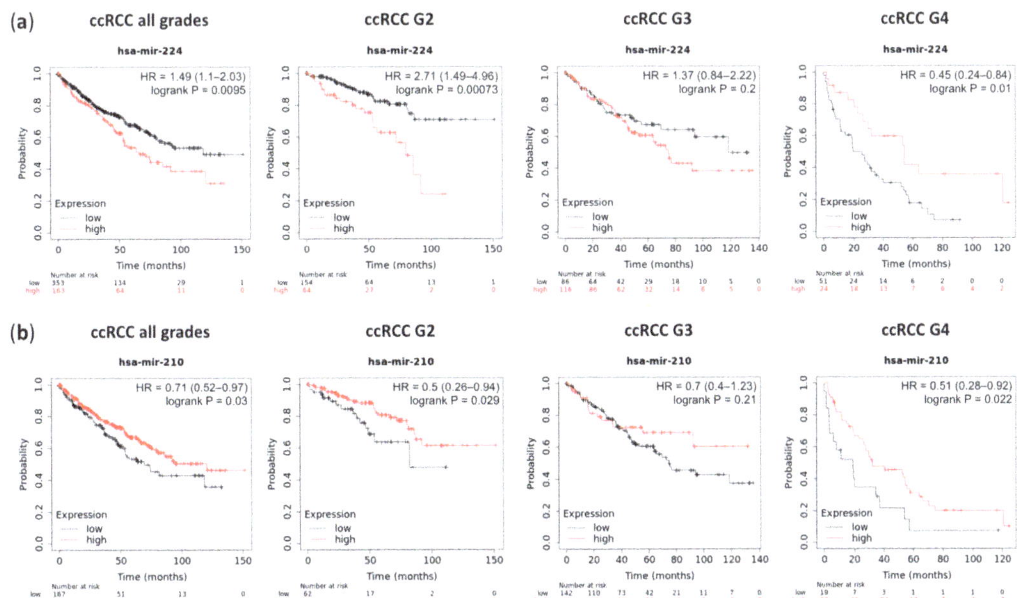

Figure 5. The survival rate analysis of ccRCC patients with deregulation expression of miRNA-224 and miRNA-210. (**a**) Patients with high expression (red line) of miRNA-224 have worst hazard ratio (HR) in ccRCC G2 than in G4; (**b**) Patients with high expression (red line) of miRNA-210 have higher survival rate although its massive upregulation in ccRCC tumors.

Table 1. Gene ontology (GO) terms for targets of commonly deregulated miRNA in ccRCC.

GO Id	Description	miRNA Targets
GO:0038093	Fc receptor signaling pathway	miR-200c-3p, miR-224-5p, miR-155-5p
GO:0002768	immune response-regulating cell surface receptor signaling pathway	miR-200c-3p, miR-204-5p, miR-224-5p
GO:0038179	neurotrophin signaling pathway	miR-200c-3p, miR-224-5p, miR-155-5p
GO:0071774	response to fibroblast growth factor	miR-200c-3p, miR-224-5p
GO:0030897	HOPS complex	miR-362-5p
GO:0000289	nuclear-transcribed mRNA poly(A) tail shortening	miR-363-3p
GO:0007178	transmembrane receptor protein serine/threonine kinase signaling pathway	miR-204-5p
GO:0071559	response to transforming growth factor beta	miR-204-5p, miR-155-5p
GO:0071214	cellular response to abiotic stimulus	miR-21-5p
GO:0034142	toll-like receptor 4 signaling pathway	miR-21-5p
GO:0019787	small conjugating protein ligase activity	miR-21-5p
GO:0019901	protein kinase binding	miR-155-5p
GO:0051169	nuclear transport	miR-155-5p
GO:0010608	posttranscriptional regulation of gene expression	miR-155-5p
GO:0071456	cellular response to hypoxia	miR-210-3p
GO:1901989	positive regulation of cell cycle phase transition	miR-210-3p
GO:0010639	negative regulation of organelle organization	miR-210-3p
GO:0007059	chromosome segregation	miR-210-3p

From networks created with GeneMANIA, we extracted the most crucial clusters (the most connected regions) of genes and networks by MCODE application. On these networks, we overlaid fold change and significance of differently expressed genes in ccRCC tumors based on RNA-Seq results as shown in Figure S4. These data suggest that targets of the selected miRNAs are involved in crucial pathways for ccRCC development such as cellular response to hypoxia, chromosome segregation or response to signaling factors (Table 1).

4. Discussion

Taking into consideration the asymptomatic course of RCC, its frequent diagnosis at advance stage and as consequence its relatively low 5-years survival rate, effective clinical and molecular markers aiding RCC classification, detection and monitoring could significantly improve disease management. Molecular alterations in RCC tumors have been extensively studied in the last decade, especially in ccRCC but neither identified mutations (e.g.,: *VHL, BAP1, PBRM1, SETD2, KDM5C, MTOR*) nor transcriptome-based ccRCC tumor sub-classification have straightforward clinical relevance [47,48]. Chromosomal rearrangements which are ccRCC drivers occur decades before diagnosis in childhood or adolescence which makes early detection of the diseases difficult [5]. The tools suitable for detection of disease initiation, early diagnosis and progression are currently not available, hence the need for identification of novel, molecular and ideally, noninvasive biomarkers.

As miRNA are short molecules, relatively stable in tissues and body fluids and have been previously shown to be deregulated in all RCC subtypes. miRNA panel could be utilized as a subtype classification, an indicator of a disease stage or treatment monitoring tool. Furthermore, miRNA regulate the expression of thousands of genes, therefore their deregulation likely plays a role in ccRCC pathogenesis.

We set off to perform a meta-analysis of miRNA expression in ccRCC, chRCC, pRCC and oncocytoma, using publicly available data sets derived from small RNA-Seq and microarray experiments. As a result, we obtained a list of miRNAs commonly deregulated in ccRCC, chRCC and pRCC. In the case of oncocytoma, only one study was available (Figure 1). Based on the meta-analysis ccRCC could be potentially classified by the comparison of expression levels of eight miRNAs: miRNA-200c-3p, miRNA-362-5p, miRNA-363-3p, miRNA-204-5p, miRNA-21-5p, miRNA-224-5p, miRNA-155-5p and miRNA-210-3p. miRNAs identified as specifically expressed in ccRCC, pRCC and oncocytoma subtypes were validated in an independent sample set (Figure 2).

All eight commonly deregulated miRNAs in ccRCC were previously described in the literature as onco-suppressors or oncogenes in various cancer types [49–51]. Several studies showed that miRNA-21-5p is upregulated in solid cancers, mainly in advanced tumors and has been linked to uncontrolled cell growth and necrosis [52]. Recently miRNA-21-5p was suggested as a potential therapeutic target, likely involved in processes of drug resistance in breast cancer and leukemia [53]. miRNA-224-5p was reported to be upregulated for example in colorectal cancer and hepatocellular carcinoma. However, its downregulation also was observed in prostate cancer [54,55]. This miRNA was shown to regulate cell signaling, proliferation and response to fibroblast and epidermal growth factors. miRNA-155-5p, which is upregulated in the majority of solid tumors [56,57], target genes are involved in tumorigenesis, DNA damage repair and inflammation. Elevated expression of miRNA-155-5p induces the formation of new blood vessels and tumor growth [58,59]. Furthermore, miR-155-5p influences hypoxia by targeting *VHL* mRNA [59] and its overexpression is additionally connected to diminished drug response and chemo- and radio-resistance of breast and colon cancer cells [57,59]. Overexpression of miRNA-210-3p correlates with a negative disease outcome in several cancers [15]. Many miRNA-210-3p targets are engaged in angiogenesis, cell survival and differentiation [15] miRNA-200c-3p is one of the most significantly downregulated miRNAs in ccRCC tumors [44]. miRNA-200c is member of miRNA-200 family (miRNA-200a, miRNA-200b, miRNA-200c, miRNA-141 and miRNA-429), commonly deregulated in other cancer types. miRNA-200c-3p targets are engaged in cell signaling, proliferation, cell invasion [60,61], cancer initiation and metastasis [62]. miRNA-362-5p has been classified as oncogenic in solid tumors [63] and could be a potential therapeutic target or prognostic factor for human cancers [63]. In gastric cancer, it is upregulated, displaying its oncogenic function by inhibiting tumor suppression cylindromatosis [63]. Its downregulation was reported in cervical cancer promotes vascular invasion and metastasis [64] miRNA-363-3p is well known as miRNA with an anti-tumor role in many human cancers such as hepatocellular carcinoma and lung cancer [65,66]. It blocks cell proliferation, migration and invasion [65]. miRNA-363-3p downregulation correlates with metastasis in colorectal and hepatocellular cancers [67]. miRNA-204-5p is an example of oncogenic miRNA with dual function [68]. In solid tumors it mainly acts as a tumor suppressor (e.g., breast, prostate cancers and metastatic lung cancer [69] and in colorectal cancer was described as an inhibitor of proliferation and promotes cancer sensitivity to chemotherapy [70]. In contrast, miRNA-204-5p has been found upregulated in leukemia, although its role in disease development is unknown [71].

Targets of selected miRNAs identified in ccRCC tumors are involved in similar pathways like in other cancer types. It is worth to mention that miRNA-210-3p, which targets are involved in oxygen metabolism (Table 1, Table S3 and Figure S4b), is upregulated in both ccRCC and pRCC tumors but not in oncocytoma (Figure 2b), where lack of hypoxia and HIF1 stabilization are documented [72,73]. Additionally, patients with high level of miRNA-210 have lower risk for death (Figure 5b). Other interesting examples are potential targets of miRNA-362-3p belong to homotypic fusion and vacuole protein sorting (HOPS) complex (Table 1, Table S3 and Figure S4a), controlling cell homeostasis, which dysfunctions are associated with various cancer types including renal cancers [74]. miRNA-363-3p targets stand out from other miRNA target genes since they are involved in post-transcriptional control of RNA metabolism (Table 1, Table S3 and Figure S4a).

Deregulation of analyzed miRNA in ccRCC could be caused by different processes. For example, upregulation of miR-210-3p in ccRCC is caused by the elevated level of HIF [75]. In general, the expression of host genes correlates with miRNA expression, which suggests that deregulation is linked to transcription (Figure S3b,c). Additionally, miRNA expression could be regulated during processing exemplified by the decreasing percentage of longer isoforms (e.g., miRNA-363-3p) (Figure 4a–d), suggestive of effective degradation after maturation. Furthermore, adenylation is a well-known mechanism of miRNA-21-5p destabilization [76]. Reduction of adenylated miRNA-21-5p isoforms in ccRCC suggests its stabilization which could act together with high transcription efficiency and explain the miRNA-21-5p increased level in ccRCC (Figure 4a). Changes in iso-miRNA tailing are potentially related to upregulation of *TUT4* and *PAPD4*, factors involved in post-maturation miRNA modifications (Figure 4h). Further investigation of the reasons of miRNA deregulation in ccRCC is needed, especially when miRNAs are considered as therapeutic targets. Potential therapeutics could inhibit or mimic mature deregulated miRNAs although regulation of their expression by targeting of the mentioned above process is also possible [77].

As a result of our meta-analysis, we would like to propose a miRNA panel that could potentially aid RCC classification, with miRNA-362-5p, miRNA-363-3p, miRNA-224-5p, miRNA-155-5p and miRNA-210-3p as classifiers of ccRCC, miRNA-362-5p, miRNA-363-3p, miRNA-21-5p, miRNA-204-5p as characteristic for pRCC and miRNA-204-5p and miRNA-224-5p for oncocytoma (Figures 1–3). Interestingly, our data suggest that miRNA-155-5p is a promising ccRCC-specific miRNA because it is unchanged in pRCC and oncocytoma (Figures 2b and 3) and is highly upregulated in all ccRCC tumor regardless Fuhrman grade (Figures 2f and 3). It was shown previously that the elevated level of this miRNA correlates with poor ccRCC outcome [78]. Additionally, our results suggest that analysis of specific miRNA isoforms (Figure 4) could increase the number of tested molecules in potential miRNA panel. Combination of the expression pattern of the seven mentioned above miRNAs is likely to be an indicative of the RCC tumor subtype. The most downregulated miRNA-200c in ccRCC seems to be a poor biomarker candidate. Regrettably, we did not validate the alteration of miRNA-200c-3p expression in ccRCC tumors likely due to low specificity of the qPCR primer. According to all NGS experiments, miRNA-200 family comprises a low percentage of miRNA-200c-3p in the kidney, therefore our results can be explained by amplification of the remaining family members (Figure S2). Unfortunately, data from high-throughput analysis of miRNA in ccpRCC and tcRCC with adjacent control tissue are not available and we did not include these RCC subtypes into the meta-analysis. However, ccpRCC and tcRCC could be mistaken with ccRCC and pRCC, respectively, during the diagnosis [7,8]. Both RCC subtypes show unique miRNA expression patterns distinguish them from the other RCCs [79,80], hence including these subtypes in meta-analysis should be considerate in the future.

A number of above-mentioned miRNAs might be useful to differentiate ccRCC tumor grading and such approach is poorly represented in the literature. For example, miRNA-224-5p is significantly upregulated in ccRCC first grade and oncocytoma (Figure 2b,f) however, its longer isoforms are significantly downregulated in ccRCC G4 (Figure 4f). Simultaneously, miRNA-21-5p is upregulated in ccRCC G4 (Figure 2f) and in pRCC (Figure 2b). Likewise, miRNA-210-3p is upregulated in ccRCC and pRCC, with no change in oncocytoma (Figure 2b). If status of these three miRNAs and its isoforms (Figure 4e–g) would be reflected in liquid biopsies or core needle biopsies [81] (non-invasive or less invasive biopsy methods than nephrectomy, respectively) it is worth to considerate their combination as ccRCC tumor grade indicator or panel distinguishing benign and malignant renal tumors (Figure 3). This is important for oncocytoma cases where surgical intervention is usually not required [82].

Although promising, suggested miRNA panel (Figure 3) requires additional validation in an independent sample set, in analyses additionally taking under consideration the heterogeneity of the tumor. Moreover, it would be interesting to test if proposed miRNA panel would be useful as a non-invasive classification test utilizing plasma and urine from RCC patients, as was shown previously that miRNA-210-3p was found circulating in the serum of mice tumors with hypoxia [83].

In conclusion, based on the meta-analysis and qPCR confirmation we propose panel of six miRNAs, with potential to distinguish ccRCC tumor grades (if extended with isoform analysis) and between RCC subtypes, which if validated further, may aid RCC classification in the future.

5. Conclusions

In conclusion, we found eight miRNAs to be commonly deregulated in ccRCC tumors; additionally, their levels can be used to distinguish RCC subtypes. Functions of these miRNAs have a potential impact on ccRCC etiology and/or development. Our results revealed that changes in the expression of selected miRNA might be potentially utilized as a tool aiding ccRCC subclass discrimination and we propose a miRNA panel that could be potentially utilized as a tool for RCC subtype distinction.

Supplementary Materials: The following are available online at https://www.mdpi.com/2072-6694/13/3/548/s1, Figure S1. The meta-analysis workflow. Figure S2. miRNA-200-3p family. Figure S3. Potential reasons of miRNA deregulation in ccRCC tumors. Figure S4. miRNA function in ccRCC tumor. Table S1. List of patients. Table S2. Identified iso-miRNAs in healthy and ccRCC kidney tissues in Exp1 and Exp4 data sets. The average percentage of miRNA isoforms. Table S3. Top 10 gene ontology (GO) terms for targets of commonly deregulated miRNA in ccRCC. Table S4. Primers used in the experimental procedures.

Author Contributions: Conceptualization, A.K., W.M.; formal analysis, A.K.; funding acquisition, J.W.; investigation, A.K., W.M., K.K., N.D.; methodology, A.K; project administration, A.K.; visualization, A.K., W.M.; writing-original draft, A.K., W.M., K.K.; writing-review & editing, H.A.R.B., J.W.; resources, J.S., T.M., Z.K.; supervision, J.W. All authors have read and agreed to the published version of the manuscript.

Funding: This work was funded by the National Science Centre Poland 2014/15/B/NZ2/00589 (J.W.).

Institutional Review Board Statement: The study was conducted according to the guidelines of the Declaration of Helsinki and approved by the Local Bioethical Committee at Poznan University of Medical Sciences (protocol code 1124/12; date of approval 2012.12.06).

Informed Consent Statement: Informed consent was obtained from all subjects involved in the study.

Data Availability Statement: Publicly available datasets were analyzed in this study. This data can be found here: https://www.ncbi.nlm.nih.gov/geo/query/acc.cgi?acc=GSE151428.

Acknowledgments: We would like to thank Tomasz Wrzesinski for technical assistance.

Conflicts of Interest: The authors declare no potential conflicts of interest.

References

1. Hsieh, J.J.; Purdue, M.P.; Signoretti, S.; Swanton, C.; Albiges, L.; Schmidinger, M.; Heng, D.Y.; Larkin, J.; Ficarra, V. Renal cell carcinoma. *Nat. Rev. Dis. Prim.* **2017**, *3*, 1–19. [CrossRef] [PubMed]
2. Medina-Rico, M.; Ramos, H.L.; Lobo, M.; Romo, J.; Prada, J.G. Epidemiology of renal cancer in developing countries: Review of the literature. *J. Can. Urol. Assoc.* **2018**, *12*, E154–E162. [CrossRef] [PubMed]
3. Ridge, C.A.; Pua, B.B.; Madoff, D.C. Epidemiology and staging of renal cell carcinoma. *Semin. Intervent. Radiol.* **2014**, *31*, 3–8. [CrossRef] [PubMed]
4. Paño Brufau, B.; Sebastià Cerqueda, C.; Buñesch Villalba, L.; Salvador Izquierdo, R.; Mellado González, B.; Nicolau Molina, C. Metastatic renal cell carcinoma: Radiologic findings and assessment of response to targeted antiangiogenic therapy by using multidetector CT. *Radiographics* **2013**, *33*, 1691–1716. [CrossRef]
5. Mitchell, T.J.; Turajlic, S.; Rowan, A.; Nicol, D.; Farmery, J.H.R.; O'Brien, T.; Martincorena, I.; Tarpey, P.; Angelopoulos, N.; Yates, L.R.; et al. Timing the Landmark Events in the Evolution of Clear Cell Renal Cell Cancer: TRACERx Renal. *Cell* **2018**, *173*, 611–623.e17. [CrossRef]
6. Grimm, M.O.; Wolff, I.; Zastrow, S.; Fröhner, M.; Wirth, M. Advances in renal cell carcinoma treatment. *Ther. Adv. Urol.* **2010**, *2*, 11–17. [CrossRef]
7. Zhou, H.; Zheng, S.; Truong, L.D.; Ro, J.Y.; Ayala, A.G.; Shen, S.S. Clear cell papillary renal cell carcinoma is the fourth most common histologic type of renal cell carcinoma in 290 consecutive nephrectomies for renal cell carcinoma. *Hum. Pathol.* **2014**, *45*, 59–64. [CrossRef]
8. Sarungbam, J.; Mehra, R.; Tomlins, S.A.; Smith, S.C.; Jayakumaran, G.; Al-Ahmadie, H.; Gopalan, A.; Sirintrapun, S.J.; Fine, S.W.; Zhang, Y.; et al. Tubulocystic renal cell carcinoma: A distinct clinicopathologic entity with a characteristic genomic profile. *Mod. Pathol.* **2019**, *32*, 701–709. [CrossRef]
9. Benatiya, M.A.; Rais, G.; Tahri, M.; Barki, A.; Sayegh, H.E.; Iken, A.; Nouini, Y.; Lachkar, A.; Benslimane, L.; Errihani, H.; et al. Renal oncocytoma: Experience of clinical urology a, urology department, chu ibn sina, rabat, morocco and literature review. *Pan Afr. Med. J.* **2012**, *12*, 2–7.
10. Dainiak, N. Inferences, Risk Modeling, and Prediction of Health Effects of Ionizing Radiation. *Health Phys.* **2016**, *110*, 271–273. [CrossRef]
11. Nilsson, J.; Skog, J.; Nordstrand, A.; Baranov, V.; Mincheva-Nilsson, L.; Breakefield, X.O.; Widmark, A. Prostate cancer-derived urine exosomes: A novel approach to biomarkers for prostate cancer. *Br. J. Cancer* **2009**, *100*, 1603–1607. [CrossRef] [PubMed]
12. Duffy, M.J. Carcinoembryonic antigen as a marker for colorectal cancer: Is it clinically useful? *Clin. Chem.* **2001**, *47*, 624–630. [CrossRef] [PubMed]
13. Knox, J.J.; Barrios, C.H.; Kim, T.M.; Cosgriff, T.; Srimuninnimit, V.; Pittman, K.; Sabbatini, R.; Rha, S.Y.; Flaig, T.W.; Page, R.D.; et al. Final overall survival analysis for the phase II RECORD-3 study of first-line everolimus followed by sunitinib versus first-line sunitinib followed by everolimus in metastatic RCC. *Ann. Oncol.* **2017**, *28*, 1339–1345. [CrossRef] [PubMed]
14. Gao, Y.; Zhao, H.; Lu, Y.; Li, H.; Yan, G. MicroRNAs as potential diagnostic biomarkers in renal cell carcinoma. *Tumor Biol.* **2014**, *35*, 11041–11050. [CrossRef]
15. Bavelloni, A.; Ramazzotti, G.; Poli, A.; Piazzi, M.; Focaccia, E.; Blalock, W.; Faenza, I. Mirna-210: A current overview. *Anticancer Res.* **2017**, *37*, 6511–6521. [CrossRef]
16. Catalanotto, C.; Cogoni, C.; Zardo, G. MicroRNA in control of gene expression: An overview of nuclear functions. *Int. J. Mol. Sci.* **2016**, *17*, 1712. [CrossRef]
17. Vychytilova-Faltejskova, P.; Kovarikova, A.S.; Grolich, T.; Prochazka, V.; Slaba, K.; Machackova, T.; Halamkova, J.; Svoboda, M.; Kala, Z.; Kiss, I.; et al. MicroRNA biogenesis pathway genes are deregulated in colorectal cancer. *Int. J. Mol. Sci.* **2019**, *20*, 4460. [CrossRef]
18. Parpart, S.; Wang, X.W. microRNA Regulation and Its Consequences in Cancer. *Curr. Pathobiol. Rep.* **2013**, *1*, 71–79. [CrossRef]
19. Rydzanicz, M.; Wrzesiński, T.; Bluyssen, H.A.R.; Wesoły, J. Genomics and epigenomics of clear cell renal cell carcinoma: Recent developments and potential applications. *Cancer Lett.* **2013**, *341*, 111–126. [CrossRef]
20. Ying, G.; Wu, R.; Xia, M.; Fei, X.; He, Q.E.; Zha, C.; Wu, F. Identification of eight key miRNAs associated with renal cell carcinoma: A meta-analysis. *Oncol. Lett.* **2018**, *16*, 5847–5855. [CrossRef]
21. Chen, X.; Lou, N.; Ruan, A.; Qiu, B.; Yan, Y.; Wang, X.; Du, Q.; Ruan, H.; Han, W.; Wei, H.; et al. Mir-224/mir-141 ratio as a novel diagnostic biomarker in renal cell carcinoma. *Oncol. Lett.* **2018**, *16*, 1666–1674. [CrossRef] [PubMed]
22. Petillo, D.; Kort, E.; Anema, J.; Furge, K.A.; Yang, X.J.; Teh, B.T. MicroRNA profiling of human kidney cancer subtypes. *Int. J. Oncol.* **2009**, *35*, 109–114. [CrossRef] [PubMed]
23. Edgar, R.; Domrachev, M.; Lash, A.E. Gene Expression Omnibus: NCBI gene expression and hybridization array data repository. *Nucleic Acids Res.* **2002**, *30*, 207–210. [CrossRef] [PubMed]
24. Langmead, B.; Salzberg, S. Bowtie2. *Nat. Methods* **2013**, *9*, 357–359. [CrossRef] [PubMed]
25. Liao, Y.; Smyth, G.K.; Shi, W. FeatureCounts: An efficient general purpose program for assigning sequence reads to genomic features. *Bioinformatics* **2014**, *30*, 923–930. [CrossRef]

26. Pantano, L.; Estivill, X.; Martí, E. SeqBuster, a bioinformatic tool for the processing and analysis of small RNAs datasets, reveals ubiquitous miRNA modifications in human embryonic cells. *Nucleic Acids Res.* **2010**, *38*, e34. [CrossRef]
27. Dobin, A.; Davis, C.A.; Schlesinger, F.; Drenkow, J.; Zaleski, C.; Jha, S.; Batut, P.; Chaisson, M.; Gingeras, T.R. STAR: Ultrafast universal RNA-seq aligner. *Bioinformatics* **2013**, *29*, 15–21. [CrossRef]
28. Zhou, L.; Chen, J.; Li, Z.; Li, X.; Hu, X.; Huang, Y.; Zhao, X.; Liang, C.; Wang, Y.; Sun, L.; et al. Integrated profiling of MicroRNAs and mRNAs: MicroRNAs Located on Xq27.3 associate with clear cell renal cell carcinoma. *PLoS ONE* **2010**, *5*, e15224. [CrossRef]
29. Osanto, S.; Qin, Y.; Buermans, H.P.; Berkers, J.; Lerut, E.; Goeman, J.J.; van Poppel, H. Genome-wide microRNA expression analysis of clear cell renal cell carcinoma by next generation deep sequencing. *PLoS ONE* **2012**, *7*, e38298. [CrossRef]
30. Nientiedt, M.; Deng, M.; Schmidt, D.; Perner, S.; Müller, S.C.; Ellinger, J. Identification of aberrant tRNA-halves expression patterns in clear cell renal cell carcinoma. *Sci. Rep.* **2016**, *6*, 1–9. [CrossRef]
31. Leinonen, R.; Sugawara, H.; Shumway, M. The sequence read archive. *Nucleic Acids Res.* **2011**, *39*, 2010–2012. [CrossRef] [PubMed]
32. Yang, Z.; Wu, L.; Wang, A.; Tang, W.; Zhao, Y.; Zhao, H.; Teschendorff, A.E. DbDEMC 2.0: Updated database of differentially expressed miRNAs in human cancers. *Nucleic Acids Res.* **2017**, *45*, D812–D818. [CrossRef] [PubMed]
33. Zhang, J.; Du, Y.-Y.; Lin, Y.-F.; Chen, Y.-T.; Yang, L.; Wang, H.-J.; Ma, D. The cell growth suppressor, mir-126, targets IRS-1. *Biochem. Biophys. Res. Commun.* **2008**, *377*, 136–140. [CrossRef] [PubMed]
34. Livak, K.J.; Schmittgen, T.D. Analysis of relative gene expression data using real-time quantitative PCR and the 2-ΔΔCT method. *Methods* **2001**, *25*, 402–408. [CrossRef]
35. Warde-Farley, D.; Donaldson, S.L.; Comes, O.; Zuberi, K.; Badrawi, R.; Chao, P.; Franz, M.; Grouios, C.; Kazi, F.; Lopes, C.T.; et al. The GeneMANIA prediction server: Biological network integration for gene prioritization and predicting gene function. *Nucleic Acids Res.* **2010**, *38*, 214–220. [CrossRef]
36. Shannon, P.; Markiel, A.; Ozier, O.; Baliga, N.S.; Wang, J.T.; Ramage, D.; Amin, N.; Schwikowski, B.; Ideker, T. Cytoscape: A Software Environment for Integrated Models of biomolecular interaction networks. *Genome Res.* **1971**, *13*, 426. [CrossRef]
37. Bader, G.D.; Hogue, C.W. An automated method for finding molecular complexes in large protein interaction networks. *BMC Bioinform.* **2003**, *29*, 2. [CrossRef]
38. Chong, J.; Wishart, D.S.; Xia, J. Using MetaboAnalyst 4.0 for Comprehensive and Integrative Metabolomics Data Analysis. *Curr. Protoc. Bioinforma.* **2019**, *68*, 1–128. [CrossRef]
39. Nagy, Á.; Lánczky, A.; Menyhárt, O.; Gyorffy, B. Validation of miRNA prognostic power in hepatocellular carcinoma using expression data of independent datasets. *Sci. Rep.* **2018**, *8*, 1–9. [CrossRef]
40. Kort, E.J.; Farber, L.; Tretiakova, M.; Petillo, D.; Furge, K.A.; Yang, X.J.; Cornelius, A.; Bin, T.T. The E2F3-oncomir-1 axis is activated in Wilms' tumor. *Cancer Res.* **2008**, *68*, 4034–4038. [CrossRef]
41. Jung, M.; Mollenkopf, H.J.; Grimm, C.; Wagner, I.; Albrecht, M.; Waller, T.; Pilarsky, C.; Johannsen, M.; Stephan, C.; Lehrach, H.; et al. MicroRNA profiling of clear cell renal cell cancer identifies a robust signature to define renal malignancy. *J. Cell. Mol. Med.* **2009**, *13*, 3918–3928. [CrossRef] [PubMed]
42. Mathew, L.K.; Lee, S.S.; Skuli, N.; Rao, S.; Keith, B.; Nathanson, K.L.; Lal, P.; Simon, M.C. Restricted expression of miR-30c-2-3p and miR-30a-3p in Clear Cell Renal Carcinomas enhances HIF2α activity. *Cancer Discov.* **2014**, *4*, 53–60. [CrossRef] [PubMed]
43. Wach, S.; Nolte, E.; Theil, A.; Stöhr, C.; Rau, T.T.; Hartmann, A.; Ekici, A.; Keck, B.; Taubert, H.; Wullich, B. MicroRNA profiles classify papillary renal cell carcinoma subtypes. *Br. J. Cancer* **2013**, *109*, 714–722. [CrossRef] [PubMed]
44. Wang, X.; Chen, X.; Han, W.; Ruan, A.; Chen, L.; Wang, R.; Xu, Z.; Xiao, P.; Lu, X.; Zhao, Y.; et al. MiR-200c targets CDK2 and suppresses tumorigenesis in renal cell carcinoma. *Mol. Cancer Res.* **2015**, *13*, 1567–1577. [CrossRef] [PubMed]
45. Telonis, A.G.; Magee, R.; Loher, P.; Chervoneva, I.; Londin, E.; Rigoutsos, I. Knowledge about the presence or absence of miRNA isoforms (isomiRs) can successfully discriminate amongst 32 TCGA cancer types. *Nucleic Acids Res.* **2017**, *45*, 2973–2985. [CrossRef]
46. Hsu, S.D.; Lin, F.M.; Wu, W.Y.; Liang, C.; Huang, W.C.; Chan, W.L.; Tsai, W.T.; Chen, G.Z.; Lee, C.J.; Chiu, C.M.; et al. MiRTarBase: A database curates experimentally validated microRNA-target interactions. *Nucleic Acids Res.* **2011**, *39*, 163–169. [CrossRef]
47. Creighton, C.; Morgan, M.; Gunaratne, P.; Wheeler, D.; Gibbs, R.; Gordon Robertson, A.; Chu, A.; Beroukhim, R.; Cibulskis, K.; Signoretti, S.; et al. Comprehensive molecular characterization of clear cell renal cell carcinoma. *Nature* **2013**, *499*, 43–49. [CrossRef]
48. Hakimi, A.A.; Reznik, E.; Creighton, C.-H.L.C.J.; Brannon, A.R.; Luna, A.; Aksoy, B.A.; Liu, E.M.; Shen, R.; Lee, W.; Chen, Y.; et al. An Integrated Metabolic Atlas of Clear Cell Renal Cell Carcinoma A. *Cancer Cell* **2016**, *29*, 104–116. [CrossRef]
49. Chen, E.; Xu, X.; Liu, R.; Liu, T. Small but Heavy Role: MicroRNAs in Hepatocellular Carcinoma Progression. *Biomed Res. Int.* **2018**, *2018*, 6784607. [CrossRef]
50. Kang, H.; Kim, C.; Lee, H.; Rho, J.G.; Seo, J.W.; Nam, J.W.; Song, W.K.; Nam, S.W.; Kim, W.; Lee, E.K. Downregulation of microRNA-362-3p and microRNA-329 promotes tumor progression in human breast cancer. *Cell Death Differ.* **2016**, *23*, 484–495. [CrossRef]
51. Loh, H.Y.; Norman, B.P.; Lai, K.S.; Rahman, N.M.A.N.A.; Alitheen, N.B.M.; Osman, M.A. The regulatory role of microRNAs in breast cancer. *Int. J. Mol. Sci.* **2019**, *20*, 4940. [CrossRef] [PubMed]
52. Feng, Y.H.; Tsao, C.J. Emerging role of microRNA-21 in cancer (Review). *Biomed. Reports* **2016**, *5*, 395–402. [CrossRef] [PubMed]
53. Hong, L.; Han, Y.; Zhang, Y.; Zhang, H.; Zhao, Q.; Wu, K.; Fan, D. MicroRNA-21: A therapeutic target for reversing drug resistance in cancer. *Expert Opin. Ther. Targets* **2013**, *17*, 1073–1080. [CrossRef] [PubMed]

54. Gan, B.L.; Zhang, L.J.; Gao, L.; Ma, F.C.; He, R.Q.; Chen, G.; Ma, J.; Zhong, J.C.; Hu, X.H. Downregulation of miR-224-5p in prostate cancer and its relevant molecular mechanism via TCGA, GEO database and in silico analyses. *Oncol. Rep.* **2018**, *40*, 3171–3188. [CrossRef]
55. Zhang, G.J.; Zhou, H.; Xiao, H.X.; Li, Y.; Zhou, T. Up-regulation of miR-224 promotes cancer cell proliferation and invasion and predicts relapse of colorectal cancer. *Cancer Cell Int.* **2013**, *13*, 1–10. [CrossRef]
56. Faraoni, I.; Antonetti, F.R.; Cardone, J.; Bonmassar, E. miR-155 gene: A typical multifunctional microRNA. *Biochim. Biophys. Acta Mol. Basis Dis.* **2009**, *1792*, 497–505. [CrossRef]
57. Bayraktar, R.; Van Roosbroeck, K. miR-155 in cancer drug resistance and as target for miRNA-based therapeutics. *Cancer Metastasis Rev.* **2018**, *37*, 33–44. [CrossRef]
58. Lou, W.; Liu, J.; Gao, Y.; Zhong, G.; Chen, D.; Shen, J.; Bao, C.; Xu, L.; Pan, J.; Cheng, J.; et al. MicroRNAs in cancer metastasis and angiogenesis. *Oncotarget* **2017**, *8*, 115787–115802. [CrossRef]
59. Kong, W.; He, L.; Challa, S.; Xu, C.-X.; Permuth-Wey, J.; Lancaster, J.; Coppola, D.; Sellers, T.; Djeu, J.; Cheng, J. Upregulation of miRNA-155 promotes tumour angiogenesis by targeting VHL and is associated with poor prognosis and triple- negative breast cancer. *Oncogene* **2014**, *33*, 679–689. [CrossRef]
60. Humphries, B.; Yang, C. The microRNA-200 family: Small molecules with novel roles in cancer development, progression and therapy. *Oncotarget* **2015**, *6*, 6472–6498. [CrossRef]
61. Li, J.; Tan, Q.; Yan, M.; Liu, L.; Lin, H.; Zhao, F.; Bao, G.; Kong, H. miRNA-200c inhibits invasion and metastasis of human non-small cell lung cancer by directly targeting ubiquitin specific peptidase 25. *Mol. Cancer* **2014**, *13*, 1–14. [CrossRef] [PubMed]
62. O'Brien, S.J.; Carter, J.V.; Burton, J.F.; Oxford, B.G.; Schmidt, M.N.; Hallion, J.C.; Galandiuk, S. The role of the miR-200 family in epithelial—mesenchymal transition in colorectal cancer: A systematic review. *Int. J. Cancer* **2018**, *142*, 2501–2511. [CrossRef] [PubMed]
63. Ni, F.; Zhao, H.; Cui, H.; Wu, Z.; Chen, L.; Hu, Z.; Guo, C.; Liu, Y.; Chen, Z.; Wang, X.; et al. MicroRNA-362-5p promotes tumor growth and metastasis by targeting CYLD in hepatocellular carcinoma. *Cancer Lett.* **2015**, *356*, 809–818. [CrossRef] [PubMed]
64. Shi, C.; Zhang, Z. MicroRNA-362 is downregulated in cervical cancer and inhibits cell proliferation, migration and invasion by directly targeting SIX1. *Oncol. Rep.* **2017**, *37*, 501–509. [CrossRef] [PubMed]
65. Ye, J.; Zhang, W.; Liu, S.; Liu, Y.; Liu, K. MiR-363 inhibits the growth, migration and invasion of hepatocellular carcinoma cells by regulating E2F3. *Oncol. Rep.* **2017**, *38*, 3677–3684. [CrossRef]
66. Wang, Y.; Chen, T.; Huang, H.; Jiang, Y.; Yang, L.; Lin, Z.; He, H.; Liu, T.; Wu, B.; Chen, J.; et al. miR-363-3p inhibits tumor growth by targeting PCNA in lung adenocarcinoma. *Oncotarget* **2017**, *8*, 20133–20144. [CrossRef]
67. Hu, F.; Min, J.; Cao, X.; Liu, L.; Ge, Z.; Hu, J.; Li, X. MiR-363-3p inhibits the epithelial-to-mesenchymal transition and suppresses metastasis in colorectal cancer by targeting Sox4. *Biochem. Biophys. Res. Commun.* **2016**, *474*, 35–42. [CrossRef]
68. Li, T.; Pan, H.; Li, R. The dual regulatory role of miR-204 in cancer. *Tumor Biol.* **2016**, *37*, 11667–11677. [CrossRef]
69. Shi, L.; Zhang, B.; Sun, X.; Lu, S.; Liu, Z.; Liu, Y.; Li, H.; Wang, L.; Wang, X.; Zhao, C. MiR-204 inhibits human NSCLC metastasis through suppression of NUAK1. *Br. J. Cancer* **2014**, *111*, 2316–2327. [CrossRef]
70. Yin, Y.; Zhang, B.; Wang, W.; Fei, B.; Quan, C.; Zhang, J.; Song, M.; Bian, Z.; Wang, Q.; Ni, S.; et al. MiR-204-5p inhibits proliferation and invasion and enhances chemotherapeutic sensitivity of colorectal cancer cells by downregulating RAB22A. *Clin. Cancer Res.* **2014**, *20*, 6187–6699. [CrossRef]
71. Zanette, D.L.; Rivadavia, F.; Molfetta, G.A.; Barbuzano, F.G.; Proto-Siqueira, R.; Falcão, R.P.; Zago, M.A.; Silva, W.A. miRNA expression profiles in chronic lymphocytic and acute lymphocytic leukemia. *Braz. J. Med. Biol. Res.* **2007**, *40*, 1435–1440. [CrossRef] [PubMed]
72. De Luise, M.; Girolimetti, G.; Okere, B.; Porcelli, A.M.; Kurelac, I.; Gasparre, G. Molecular and metabolic features of oncocytomas: Seeking the blueprints of indolent cancers. *Biochim. Biophys. Acta-Bioenerg.* **2017**, *1858*, 591–601. [CrossRef] [PubMed]
73. Porcelli, A.M.; Ghelli, A.; Ceccarelli, C.; Lang, M.; Cenacchi, G.; Capristo, M.; Pennisi, L.F.; Morra, I.; Ciccarelli, E.; Melcarne, A.; et al. The genetic and metabolic signature of oncocytic transformation implicates HIF1α destabilization. *Hum. Mol. Genet.* **2010**, *19*, 1019–1032. [CrossRef] [PubMed]
74. Van der Beek, J.; Jonker, C.; van der Welle, R.; Liv, N.; Klumperman, J. CORVET, CHEVI and HOPS—Multisubunit tethers of the endo-lysosomal system in health and disease. *J. Cell Sci.* **2019**, *132*, 189134. [CrossRef] [PubMed]
75. McCormick, R.I.; Blick, C.; Ragoussis, J.; Schoedel, J.; Mole, D.R.; Young, A.C.; Selby, P.J.; Banks, R.E.; Harris, A.L. MiR-210 is a target of hypoxia-inducible factors 1 and 2 in renal cancer, regulates ISCU and correlates with good prognosis. *Br. J. Cancer* **2013**, *108*, 1133–1142. [CrossRef] [PubMed]
76. Boele, J.; Persson, H.; Shin, J.W.; Ishizu, Y.; Newie, I.S.; Søkilde, R.; Hawkins, S.M.; Coarfa, C.; Ikeda, K.; Takayama, K.I.; et al. PAPD5-mediated 3′ adenylation and subsequent degradation of miR-21 is disrupted in proliferative disease. *Proc. Natl. Acad. Sci. USA* **2014**, *111*, 11467–11472. [CrossRef]
77. Balzeau, J.; Menezes, M.R.; Cao, S.; Hagan, J.P. The LIN28/let-7 pathway in cancer. *Front. Genet.* **2017**, *8*, 1–16. [CrossRef]
78. Silva-Santos, R.M.; Costa-Pinheiro, P.; Luis, A.; Antunes, L.; Lobo, F.; Oliveira, J.; Henrique, R.; Jerónimo, C. MicroRNA profile: A promising ancillary tool for accurate renal cell tumour diagnosis. *Br. J. Cancer* **2013**, *109*, 2646–2653. [CrossRef]
79. Ding, Y.; Miyamoto, H.; Rothberg, P.G. Is Tubulocystic Renal Cell Carcinoma Real?: Genomic Analysis Confirms the World Health Organization Classification. *J. Mol. Diagn.* **2018**, *20*, 28–30. [CrossRef]

80. Lawrie, C.H.; Larrea, E.; Larrinaga, G.; Goicoechea, I.; Arestin, M.; Fernandez-Mercado, M.; Hes, O.; Cáceres, F.; Manterola, L.; Lõpez, J.I. Targeted next-generation sequencing and non-coding RNA expression analysis of clear cell papillary renal cell carcinoma suggests distinct pathological mechanisms from other renal tumour subtypes. *J. Pathol.* **2014**, *232*, 32–42. [CrossRef]
81. Yang, C.S.; Choi, E.; Idrees, M.T.; Chen, S.; Wu, H.H. Percutaneous biopsy of the renal mass: FNA or core needle biopsy? *Cancer Cytopathol.* **2017**, *125*, 407–415. [CrossRef] [PubMed]
82. Liu, J.; Fanning, C.V. Can renal oncocytomas be distinguished from renal cell carcinoma on fine-needle aspiration specimens? A study of conventional smears in conjunction with ancillary studies. *Cancer* **2001**, *93*, 390–397. [CrossRef] [PubMed]
83. Jung, K.O.; Youn, H.; Lee, C.H.; Kang, K.W.; Chung, J.K. Visualization of exosome-mediated miR-210 transfer from hypoxic tumor cells. *Oncotarget* **2017**, *8*, 9899–9910. [CrossRef] [PubMed]

 cancers

Review

Role of MicroRNAs in the Development and Progression of the Four Medulloblastoma Subgroups

Emilia Bevacqua [1], Jasmin Farshchi [2], Maria Victoria Niklison-Chirou [2,*] and Paola Tucci [1,*]

1 Department of Pharmacy, Health and Nutritional Sciences, University of Calabria, 87036 Rende, Italy; emilia.bevacqua@unical.it
2 Centre for Therapeutic Innovation (CTI-Bath), Department of Pharmacy & Pharmacology, University of Bath, Bath BA2 7AY, UK; jkf33@bath.ac.uk
* Correspondence: mvnc20@bath.ac.uk (M.V.N.-C.); paola.tucci@unical.it (P.T.); Tel.: +44-1225-384957 (M.V.N.-C.); +39-0984-493185 (P.T.)

Simple Summary: Medulloblastoma is the most common malignant paediatric brain tumour. Medulloblastoma originates in the cerebellum, a structure located at the base of the brain, affecting movement and balance in patients. Due to DNA alterations, known as mutation, some immature cells acquire new properties, transform from healthy cells into cancer cells and begin multiplying uncontrollably. During carcinogenesis, microRNAs (miRNAs or miRs) play important roles in medulloblastoma, helping cells to proliferate (oncomiRs) or inhibiting cell proliferation and promoting cell differentiation (tumour suppressor miRs). Therefore, in this review, we summarize the role of miRNAs in the four medulloblastoma subgroups and the importance of these non-coding RNAs to provide potential therapeutic applications.

Abstract: Medulloblastoma is the most frequent malignant brain tumour in children. Medulloblastoma originate during the embryonic stage. They are located in the cerebellum, which is the area of the central nervous system (CNS) responsible for controlling equilibrium and coordination of movements. In 2012, medulloblastoma were divided into four subgroups based on a genome-wide analysis of RNA expression. These subgroups are named Wingless, Sonic Hedgehog, Group 3 and Group 4. Each subgroup has a different cell of origin, prognosis, and response to therapies. Wingless and Sonic Hedgehog medulloblastoma are so named based on the main mutation originating these tumours. Group 3 and Group 4 have generic names because we do not know the key mutation driving these tumours. Gene expression at the post-transcriptional level is regulated by a group of small single-stranded non-coding RNAs. These microRNA (miRNAs or miRs) play a central role in several cellular functions such as cell differentiation and, therefore, any malfunction in this regulatory system leads to a variety of disorders such as cancer. The role of miRNAs in medulloblastoma is still a topic of intense clinical research; previous studies have mostly concentrated on the clinical entity of the single disease rather than in the four molecular subgroups. In this review, we summarize the latest discoveries on miRNAs in the four medulloblastoma subgroups.

Keywords: miRNA; medulloblastoma; brain tumour; subgroups; stem cells

Citation: Bevacqua, E.; Farshchi, J.; Niklison-Chirou, M.V.; Tucci, P. Role of MicroRNAs in the Development and Progression of the Four Medulloblastoma Subgroups. *Cancers* **2021**, *13*, 6323. https://doi.org/10.3390/cancers13246323

Academic Editor: Adam E. Frampton

Received: 4 December 2021
Accepted: 14 December 2021
Published: 16 December 2021

1. Introduction

Medulloblastoma (MB) is the most common primary malignant solid tumour of the central nervous system (CNS) in children and originates in a region of the brain known as cerebellum [1]. Embryonic tumours of the CNS account for approximately 4% of childhood cancers [2]. In Italy, according to AIRTUM (Italian Association of Cancer Registries) data, about 7 children per year out of a million are affected by this type of disease [3]. The incidence is slightly higher among males than females and is higher in younger children. Moreover, children with certain genetic diseases, such as Turcot syndrome, Gorlin syndrome, Li-Fraumeni syndrome, are at greater risk of developing medulloblastoma [4,5].

The symptoms related to a medulloblastoma depend on the tumour's size and location [6]. The most common symptoms of medulloblastoma are headache, nausea and vomiting, progressive instability in walking, problems with coordination of the hands, arms, legs or feet, difficulty synchronizing eye movements, and changes in modulation of the voice [7].

Medulloblastoma is caused by different gene mutations, which can transform a healthy cell into a tumour cell [8]. It has been shown, following the discovery of miRNAs, that gene regulation can be altered at different levels, thus leading to tumour formation [9].

In 2005, it was reported for the first time that miRNAs play a central role in brain tumour development [10]. Since then, several studies have been performed in order to shed light on the role of miRNAs in brain tumours such as medulloblastoma in both paediatric and adult populations. Importantly, by using next-generation sequencing in a large cohort of medulloblastoma patients, common driver mutations have been revealed in each medulloblastoma subgroup [11]. However, the role of miRNAs in the framework of the different subgroups is still limited, since most of the studies have concentrated on the clinical entity of the single disease.

In this review, we highlight the major findings on the role of the miRNAs in the development and progression of medulloblastoma, their potential as biomarkers for cancer diagnosis, prognosis and therapeutic applications, with a particular focus on the regulation of the miRNAs in the four different medulloblastoma subgroups.

2. Medulloblastoma's Classification

Medulloblastoma can be classified into different subgroups, which are distinguished based on how they present under the microscope (histological classification) or genetic alterations.

According to the World Health Organization (WHO), histological classification distinguishes four forms [12]:

- Classic, that is the most common subtype;
- Desmoplastic/nodular;
- Extensive nodularity, that is predominantly in infants;
- Anaplastic/large cell.

The classic forms, desmoplastic and with extensive nodularity, generally have a more favourable prognosis, while the anaplastic large cell form is the more aggressive [1] and displays high levels of atypia.

A more recent classification, based on genomics data, also divides medulloblastomas into four subgroups known as Wingless (WNT) and Sonic Hedgehog (SHH), which are better described and Group 3 (Grp3) and Group 4 (Grp4) less characterized. These new medulloblastoma entities are based on the presence of a specific gene mutation or amplification that causes cell proliferation [13,14].

Diagnosis, Prognosis and Therapy

The diagnosis of medulloblastoma is made with imaging techniques such as computed tomography (CT) scan and, subsequently, magnetic resonance imaging (MRI) within one to three months from the appearance of the first symptoms, since medulloblastoma is a rapidly growing tumour [15]. Given the possibility of metastasis to other regions of the CNS, it is always essential to obtain images of both the brain and the spinal cord [13]. A cerebrospinal fluid sampling by lumbar puncture allows to exclude the presence of neoplastic cells at this level. Confirmation by histological examination is obtained after surgery to remove the tumour [16]. Dissemination outside the CNS is very rare.

The evolution of the disease (prognosis) and response to therapies are mainly linked to the medulloblastoma subgroup and to the presence of metastasis at diagnosis, although generally these tumours respond to therapies much better than other neoplasms of embryonic origin. The presence of metastasis in medulloblastoma is a poor prognostic factor. The treatment options for patients with metastases are limited. Unfortunately, it is not uncommon, even if the therapies have worked, for the tumour to reoccur after some time (relapse). In this case, the treatments are generally ineffective [17]. The 5-year survival from diagnosis is around 60–70% [6].

The therapy of choice for medulloblastoma is surgical removal of the tumour followed by chemotherapy and radiotherapy [18]. Ideally, the operation should completely remove all cancer cells, but it may be impractical if the tumour is in an inaccessible area or if there is a risk of damaging an area of vital importance or compromising the physical and cognitive functions of the patient [19]. Complementary therapy to surgery is direct radiotherapy to the head and spine (craniospinal radiotherapy) [20]. Over the decades, radiotherapy techniques and doses, both on the entire CNS and on the site of origin of the disease, have evolved and been modulated in order to make the treatment more effective and less harmful. The introduction of chemotherapy also contributed to this, which, depending on the initial situation, can be used after radiotherapy or before it. In special cases, in relation to the patient's age, histological type and genetic subgroup, it is possible to reduce the total doses of radiotherapy or even omit it. It is important that the treatment plan also includes a rehabilitation path, which improves both the response to treatment and the quality of life of the young patient [21]. Finally, as in all paediatric diseases, an adequate and prolonged follow-up is essential in order to offer the best possible quality of life to the patient treated for cancer.

3. miRNAs

MiRNAs are small, non-coding regions in RNAs of around 22 nucleotides (nt) [22], that induce translational repression or degradation of a target mRNA upon imperfect base pairing to its 3′ untranslated region (3′UTR).

Initially, the biogenesis of miRNAs occurs in the nucleus with the transcription of the miR by an enzyme called RNA-polymerase II. The miRNAs derive from a primary precursor (pri-miRNA) of 100–1000 nt. The formation of mature miRNA occurs in three phases, the first still in the nucleus, the other two in the cytosol: (i) Cropping: cutting performed by RNAse III enzyme Drosha capable of cutting the region flanking the pri-miRNA. Other proteins that confer specificity are associated with the Drosha enzyme (ex. DGCR8). Following the cropping and the action of Drosha, the pre-miRNA composed of 80 nt is released, with a stem-loop structure, it has a 5′P and a 3′OH and 2–3 nt at the 3′OH end single helix protruding; (ii) Export: the pre-miRNA is transported into the cytoplasm by Exportin5/RanGTP, a heterodimer is formed which passes through the nuclear pores; (iii) Dicing: the pre-miRNA undergoes a further cleavage by another RNAse III enzyme Dicer which, together with its partner TRBP (HIV-1 TAR RNA RBP), process the pre-miRNA in a miR duplex of 18–22 nt. [23–25].

Then, while the mature miRNA duplex binds to AGO proteins forming RNA-induced silencing complex (RISC), in some cases, one of the two strands of the duplex is degraded, while the other accumulates as mature miRNA. From the duplex produced by Dicer, the miRNA enters in the protein effector complex RISC, with the presence of proteins belonging to the Argonauts family (AGO), which mediates the degradation or inhibition of mRNA translation of the target gene. In particular, the AGO2 protein, together with other proteins, forms the RISC multiprotein complex with endonuclease activity capable of specifically degrading a target RNA containing sequences complementary to the guide sequence of the miRNA. Eight members of the AGO family have been identified in the man. However, only the enzymatic function of the AGO2 protein is well known [26].

Some miRNAs appear imperfectly with the 3′UTR of the target mRNA and inhibit translation; other miRNAs show a precise complementarity to their target and lead to mRNA degradation. The biogenesis of miRNA and the mechanism by which they silence gene expression are represented in Figure 1.

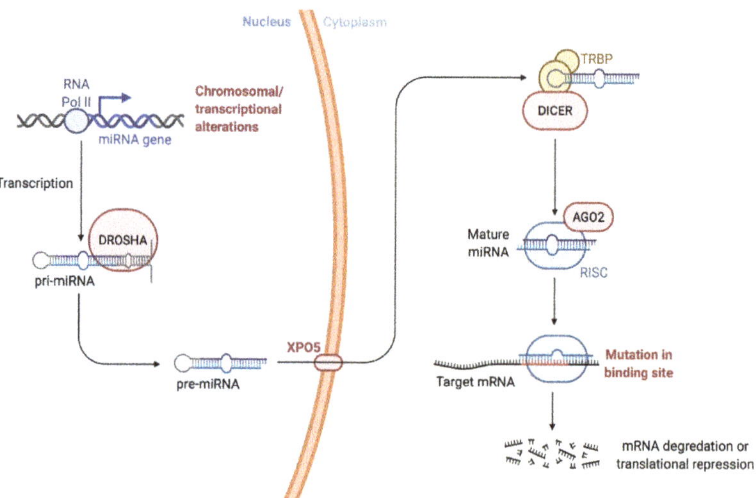

Figure 1. Biogenesis of the microRNA. "Created with BioRender.com".

MiRNAs are essential for the normal development of all tissues, as they control the most important biological processes such as cell growth, differentiation, metabolism and apoptosis [24]. For example, in Drosophila miR-14 prevents cell death and is required for normal lipid metabolism; miR-125b and miR let-7 control cell proliferation; miR-181 is involved in the development of the hematopoietic lineage of B lymphocytes [25]. MiR-15a and miR-16-1 promote the survival of immune B cells; miR-375 is involved in insulin secretion and miR-143 promotes the development of adipocytes [25,27].

3.1. Role of miRNAs in Neuronal Development

Nowadays, it is well accepted that miRNAs play a central role in several physiological processes. In particular, miRNA roles have been described during CNS development-related processes, response to ambient demands and injuries, stress or mental disorders. miRNAs are versatile regulators of gene expression, and they emerge as key players in numerous pathophysiological conditions, including CNS development, adaption and disease. Indeed, the significance of miRNA in development was confirmed by the fact that the loss of Dicer function causes lethal aberrations. It is estimated that over 60% of documented miRNAs are detected in the adult brain, and many of these change their expression as the embryonic brain develops and matures [28]. Recent data have also shown that miRNAs are expressed in the vertebrate nervous system and that their expression is modulated by synaptic activity, essential for learning and memory formation [27]. Altered morphology and neuronal development can result from errors in post-transcriptional processes that are closely regulated by miRNAs. Specific miRNAs are expressed in different compartments of the neural axis, and it has been hypothesized that miRNA pathways play a dominant role in inducing neuronal fate and synaptic plasticity [29]. Since early brain development and later synaptic plasticity are also regulated by miRNAs, it has been hypothesized that neurological disorders are influenced by their expression or alteration [30,31]. Neuronal differentiation, excitability and function are controlled by neuronal-specific miRNAs. For example, the transition from neuronal precursor to mature neurons is caused by the increase in miR-9 and miR-124 and therefore in the differentiation of embryonic stem cells. Scientists have displayed that miR-9 determines an inhibition of neurogenesis along the anterior-posterior axis [32], while miR-124 represses neuron-specific splicing patterns [33]. Neuronal differentiation and neurite growth, on the other hand, is modulated by miR-7 and miR-214 (as compared to miR-1,-16 and -133a) [27]. Neurodegenerative diseases such as Parkinson's, Alzheimer's or cancer also involve a reduction in the function of specific miRNAs [34,35].

3.2. Role of miRNAs in Cancer

The miRNAs are recognized to play a central role in development as well as in cell growth and proliferation, in differentiation, apoptosis, cell cycle, and metabolism controlling the expression levels of many genes [36]. Consequently, the alterations in the expression of these small RNAs play a key role in a wide variety of human diseases, including cancer. The first evidence of the involvement of a miRNA in cancer was demonstrated by Calin et al., in 2002 [37]. Since then, many studies have reported miRNA dysregulation in various human diseases [38]. About 50% of human miRNAs annotated are located in fragile sites of the genome associated with cancer and, moreover, they have been found differentially expressed between tumour cells and normal cells. Some miRNAs are down-regulated while others are overexpressed in cancer, suggesting that miRNAs can act as tumour suppressor genes or oncogenes, respectively [39]. The epigenetic regulation of miRNAs, the hypomethylation of DNA, the increase in DNA methylation and the disruption of histone modification patterns in the miRNA locus, are greater than the genes that encode proteins. The miRNA genes can be silenced in some types of human tumours by aberrant hypermethylation of CpG Island that surrounds it, or is close to the miRNA of histone modification genes [40]. DNA hypermethylation in breast, lung and colon carcinomas was favoured by a decrease in the expression of miR-9-1 [41], miR-124a and miR-145-5p [42].

The aberrant expression of miRNA may be due to mutations in its sequence that cause a reduction in the expression of mature miRNA or an altered regulation of the target gene [43]. The activity of these small regulatory elements can also be altered by genomic rearrangements such as deletions or duplications of the genomic region in which the miRNA is located, or translocations that relocate the miRNA under the control of a new promoter.

The conclusive effects may be an increase in miRNA expression with a consequent decrease in expression of the target gene or a decrease in miR expression with a consequent overexpression of the target (Figure 2).

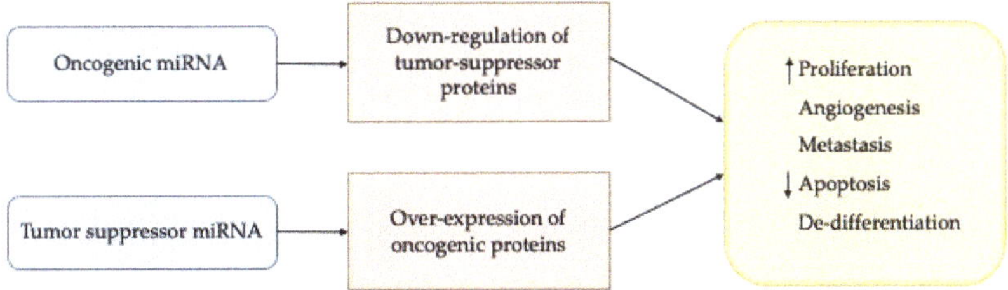

Figure 2. Main effects of deregulation of the miRNA expression.

Despite the huge amount of miRNAs identified to date, their role in tumour processes is not entirely clear. However, the presence of miRNA circulating in the blood of cancer patients has increased the possibility that they could serve as new diagnostic and prognostic biomarkers, either alone or in combination with other well-stablished biomarkers [44].

In fact, some miRNAs are specifically more expressed only in one type of tumour, managing to also characterize malignancy [45,46]. Another reason for the choice of miRNAs as tumour biomarkers is to be found in the non-invasiveness of the analysis. In fact, miRNAs have been isolated from serum, plasma, saliva, urine and other cell fluids [46]. Several studies have shown that in these compartments the expression of miRNAs is correlated with specific tumours.

An early study [47] was concerned with identifying a tumour suppressor on chromosome 13q14, which was involved in chronic lymphocytic leukaemia (CLL), the most common form of leukaemia. They showed that the 13q14 locus does not contain genes encoding tumour suppressor proteins, but two microRNA genes, miR-15a and miR16-a, are expressed in the same polycistronic RNA. This result shows that the deletion of chromosome 13q14 would cause the loss of these two miRNAs, and therefore it is evident that these miRNAs are involved in the pathogenesis of human cancer.

Moreover, the discovery that miRNAs play a vital role in different types of tumour and since they have the advantage of being able to act both as oncogenes and as tumour suppressors they are still considered potential tumour therapeutic targets [48]. Carcinogenesis is favoured by oncogenic miRNAs, which are then over-expressed; on the contrary, the tumour suppressor action is due to a decrease in particular miRNAs (Figure 2). In light of the above, the antagomirs lead to a downregulation of the oncomirs. The concept of "miR replacement therapy" was thus introduced thanks to the observation of the reduction of the pathology following the action of suppressive miRNAs, with the aim of increasing the amount of reduced miRNAs and bringing them to normal values. This approach has great potential to be a more practical strategy than silencing individual genes by siRNAs and represents one of the major commercial areas of interest in today's biotechnology market.

4. miRNAs Involvement in the Different Subgroups of Medulloblastoma

In 2012, it was agreed during an international meeting that medulloblastoma has four distinctive molecular subgroups named: Wingless (WNT-good prognosis), Sonic Hedgehog (SHH-intermedia prognosis), Group 3 (Grp3-bad prognosis) and Group 4 (Grp4-intermedia prognosis) [49]. WNT and SHH are named because these tumours have mutations in the WNT and SHH signalling pathways, respectively. To date, no clear underlying signalling pathways associated with Grp3 and Grp4 have been identified. Emerging evidence suggests that each group may require specific therapeutic strategies [50].

4.1. Wingless (WNT) Subgroup

WNT represent 10% of all medulloblastomas cases [51]. It occurs typically in adolescents and children over the age of 4 and is associated with excellent prognosis (>95% survival at 5 years in paediatric patients) [52]. WNT primary tumours are driven by a mutation in the CTNNB1 gene, which encodes b-catenin [53]. Mutation in this gene causes a constitutively upregulation of gene expression that promotes tumour growth and proliferation. Patients with WNT subgroups harbour TP53 mutations. In fact, WNT with p53 mutation have an excellent prognosis, suggesting that TP53-inactivating mutations on their own do not confer a poor survival [54]. Interestingly, the robust therapeutic response is attributed to an aberrant fenestrated vascular endothelium in the tumour. The fenestrated endothelial surface allows the accumulation of high levels of chemotherapeutic drugs in the tumour, thereby enhancing treatment [55]. However, children with a WNT diagnosis, are predisposed to primary tumour haemorrhage which can lead to severe complications [56]. Due to excellent prognosis of the WNT subgroup, a new clinical trial has been recently created to evaluate the reduction in chemotherapy and radiotherapy doses [57]. It was reported that miR-383, miR-206, miR-183, miR-128a/b and miR-133b are downregulated in this medulloblastoma subgroup [58] and the level expression of miR-449 is also completely different from other MB subtypes. miR-449 is down-regulated by aberrant DNA methylation in the WNT Group [59]. It was found that miR-148a expression is regulated by the NRP1 target. NRP1 target is involved in several pathways promoting tumour growth, invasion and metastasis. The downregulation of this target is due to the tumour suppressive effect of miR-148a expression and the subsequent reduction in tumorigenicity [60].

4.2. Sonic Hedgehog (SHH) Subgroup

SHH represents approximately 30% of MB cases and appears typically in infants and adults, accounting for two thirds of cases in these age groups [61]. The prognosis of this subgroup varies based on age and metastatic status and molecular mutations. It has recently been shown that p53, a tumour suppressor protein, is a prognostic marker for SHH-MB patients. In fact, patients with mutations of *TP53* gene have a worse outcome of the disease than those with wild-type *TP53* [62]. The altered SHH signalling pathway is mainly caused by germline or somatic mutation or copy number alterations in the SHH signalling pathway, which leads to tumour development and proliferation. The most common mutations are protein patched homolog (PTCH) inactivating mutation and smoothened homolog (SMO) activating mutation [63]. In fact, infant (35%) and children (45%) have mutations in the downstream SMO pathway, which makes tumours intrinsically resistant to SMO inhibitors [63]. Therefore, the recent approaches to modulate SHH signalling is focused on SMO inhibition and the mechanisms of acquired resistance in downstream SMO pathway. The metastasis of SHH subgroup happens at the same site of primary tumour. In a recent study, cancer stem cells (CSCs) have been isolated from SHH and expression level of epithelial to mesenchymal transition (EMT) transcript and microRNAs was compared with cerebellar NSCs [5]. Vegfa and its receptor Nrp2 are two molecules up regulated in SHH CSCs and involved in EMT [64]. If these two molecules are inhibited there will be a reduction of the cell viability and the ability of CSCs to self-renew. This mechanism leads to the modulation of two markers involved in EMT, therefore we will see the increase in the epithelial marker (E-Cadherin) and, on the other hand, a reduction of the mesenchymal one (Vimentin). The miRNA identified as an inhibitor of Vegfa and Nrp2 is miR-466-3p [64].

Furthermore, CSCs identified in SHH-MB are controlled by the Sonic Hedgehog/Gli (Hh/Gli) is an aberrant signalling pathway that control CSCs identified in SHH medulloblastoma, regulated by miR-326. More precisely, the downregulation of miR-326 is characteristic of these tumours, therefore an overexpression of miR-326 leads to the inhibition of that signalling pathway [65].

More recently, in vivo and in vitro studies have displayed that SHH MB cells showed a reduction in tumour growth by silencing miR-17/20 and miR-19a/b [66]. Furthermore, miR-17-92 cluster is involved in SHH tumours. Within this cluster belong miR-18a, -19a, -20a, -21, -25 and -106b [67]. Several studies have evaluated the effect of miR-10b on the growth and proliferation of medulloblastoma through the transcriptional induction of BCL-2, a tumour promoter [68]. Potent inhibitors of BCL-2, such as ABT-737 and ABT-199, were evaluated on the expression of miR-10b [69]. Powerful BCL-2 inhibitors significantly inhibit the expression of miR-10b in a dose-dependent manner. This miRNA is strongly associated with tumours, as it plays a crucial role in cell proliferation and survival, moreover miR-10b is not expressed in a normal brain. Several studies suggest that miR-10b is an oncomiR that regulates cell growth and survival of this medulloblastoma subgroup by controlling BCL2 levels [68].

4.3. Group 3 (Grp3) Subgroup

Grp3-MB is the most aggressive paediatric brain tumour and occurs mostly in infants and young children. This subgroup is (40–45%) metastasis at diagnosis and is resistant to combinations of surgery, radiotherapy and chemotherapy [53]. Therefore, it is associated with poor prognosis and the worst survival outcome of any subgroup (under 60% at 5 years). Unlike WNT and SHH subgroups, there is no distinctive altered signalling pathway identified for Grp3. However, amplification of MYC (17%) and hyper activation of the GFI1B oncogene (15–20%) are mostly observed. TP53 mutations were almost never observed in patients with Grp3, in which isochromosome 17q is a common aberration [54]. It was showed that miR-183-96-182 cluster are up-regulated in Group 3 of medulloblastoma. In particular, the expression of the miR-183 cluster in cells was associated with the dysfunction of the DNA damage repair and with the pathways associated with migration, EMT and metastasis [70]. Metastasis of Grp3 happens at a different site of the primary tumour. It was reported that, in medulloblastoma cell lines DAOY, D425 and D283 belonging, respectively, to the SHH subgroup and Grp3, there is an under-expression of miR-30a family [71]. Group 3 and 4 are associated with the highest mortality compared to other MB subgroups. Group 3 of MB displayed a deregulation of miR-1253 expression [72]. They showed that the restoration of miR-1253 expression is linked with a reduction in tumour cell malignancy, which also leads to the activation of apoptotic pathways. A recent study highlighted the expression of seven miRNAs belonging to the miR-30 family in the 4 subtypes of medulloblastoma. These miRNAs are significantly downregulated ($p < 0.00001$) compared to normal cerebellar tissues [71]. The inhibition of the clonogenic potential, proliferation and tumorigenicity of different cell lines of medulloblastoma is notable after the recovery of miR-30a through the lentiviral vector. MiR-30a is known to mediate autophagy through the Beclin1 target. Therefore, it has been shown that the expression of miR-30a leads to a down-regulation of genes implicated in autophagy, such as Beclin1, with consequent inhibition in medulloblastoma cells. Autophagy is a process that allows our cells to recycle and renew themselves. The cells then destroy their components that have become useless and carry them out of the membrane, playing a fundamental role in our defences. This process leads, on the one hand, to cleaning the cell, on the other hand it allows the cell to sustain itself in difficult situations [73]. Additionally, in the medulloblastoma, low levels of miR-4521 leads to an up-regulation of the transcription factor forkhead box M1 (FOXM1). FOXM1 regulates the expression of various genes involved in tumour progression [74]. The Grp3 is the only one with a statistically significant difference in miR-4521 expression reduction compared with the healthy control tissue, while in SHH subgroup there are no particular differences.

4.4. Group 4 (Grp4) Subgroup

Grp4 accounts for 35–40% of all medulloblastoma diagnosis and occurs typically in children and adolescence [75]. This subgroup is (35–40%) metastasis at diagnosis, although the survival outcomes are intermediate, and the recurrences mostly occur late. Grp4 share similar gen amplifications as Grp3 as mentioned above and have not an identified signalling pathway [76]. At the same time, Orthodenticle homeobox 2 (OXT2) amplification and the gain of isochromosome 17q is also seen in Grp4 and Grp3 [77]. Similar to Grp3, TP53 mutations has never been observed in Grp4 and the metastasis is at a different site of the primary tumour. Compared to the other subgroups, Group 4 has a lower expression of miR-181a-2-3p, which is reported to be involved in the formation of glioma acting as tumour suppressors [78,79]. While a high expression was observed for miR-187-3p and could be linked to a poor prognosis of patients with Group 4 MB [80]. Additionally, miR-206 was down-regulated in all four medulloblastoma subgroups. Indeed, miR-206 acts on OTX2, an oncogene which is involved in Grp4 pathogenesis. Overexpression of OTX2 leads to growth and proliferation of medulloblastoma. Therefore, under-expression of miR-206 contributed to the upregulation of OTX2 expression and enhanced growth of G4 cell lines [58]. A recent study showed that the tumour-suppressive let-7 miRNA family is downregulated by gene LIN28B and the expression of these miRNAs is significantly lower in Group 3 and 4 compared with WNT and SHH MB [81]. miR-4521 is located on chromosome 17p13.1. They show that a loss of chromosome 17p is closely associated with Grp3 and Grp4-subgroups [74].

5. Role of miRNAs in Medulloblastoma

The most studied tumours at the level of miRNA deregulation are breast, prostate, colon and leukaemia's; little has been studied regarding the alterations of miRNAs in medulloblastoma. Ferretti et al. conducted one of the first studies on the expression profile of miRNAs in medulloblastoma [82]. A total of 250 miRNAs were screened in 31 primary medulloblastoma specimens and 34 miRNAs differentially expressed between SHH-MB versus WNT-MB, Grp3-MB and Grp4-MB were identified. Additionally, three down-regulated miRNAs were identified in SHH-MB, miR-125b, miR-326 and miR-324-5p. Interestingly, these three miRNAs are known to target Smoothened (SMO), an activator of the Hedgehog signalling pathway [82]. In addition, miR-324-5p also targets Gli1, a committed transcription factor for the Hedgehog signalling pathway. Additionally, it was suggested that a possible genetic anomaly is the cause of the loss of function of miRNA-324-5p in SHH-MB [83]. In fact, this miRNA is contained in a gene region of chromosome 17p, which is one of the most frequent deletions in medulloblastoma. In addition to the miRNA, this chromosomal region also contains the tumour suppressors p53 and HIC1 but also the antagonist of the REN signal pathway [84].

In a different study conducted by Ferretti et al., 248 miRNAs were analysed in medulloblastoma samples, and 248 miRNAs differentially expressed between tumours and normal adult and foetal cerebellar tissue were detected [85]. In this analysis, two miRNAs, miR-9 and miR-125a, were identified as downregulated in medulloblastoma and target the truncated isoform of the neurotropin receptor (TrkC). By comparing medulloblastoma and normal foetal cerebellum it was possible to identify a cluster of upregulated miRNAs in SHH-MB versus non-SHH medulloblastomas known as cluster 17-92 [67]. This miRNA cluster is induced by N-myc in the neuronal cerebellar precursors treated with Sonic Hedgehog; this evidence indicates that the 17-92 miRNAs cluster is a positive effector of the proliferative effects of the Hedgehog signalling pathway.

Additionally, Uziel et al. [86], using medulloblastoma cells from Ink4c $^{-/-}$; Ptch1 $^{+/-}$ and Ink4c $^{-/-}$; p53 $^{-/-}$ genetically modified mice versus mature mice, identified many deregulated miRNAs: 26 upregulated and 24 downregulated. In particular, 9 of these 26 upregulated miRNAs were demonstrated to encode cluster 17-92. To this cluster belong miR-92, miR-19a and miR-20 that are upregulated in the Hedgehog subgroup of medulloblastoma. Thus, demonstrating the close correlation between cluster 17-92 and the Hedgehog signalling pathway [86].

Two miRNAs (miR-30b and miR-30d) were identified located in a commonly amplified region in medulloblastoma, adjacent to the MYC locus on chromosome 8q24. Such miRNAs were found upregulated in a subgroup of primary medulloblastoma [87].

Cluster 183-96-182 was found upregulated in controlled non-Hedgehog medulloblastoma and, in particular, miR-182 was significantly upregulated in metastatic medulloblastoma [88]. It was later shown that this cluster is involved in the suppression of genes associated with apoptosis and the regulation of the PI3K/AKT/mTOR axis [89].

Venkataraman et al. showed that several miRNAs that are downregulated in medulloblastoma have an active role in normal brain development. In particular, miR-128a has been shown to be an antagonist of the Bmi1 oncogene [90].

Many miRNAs can influence tumorigenesis through their tumour suppressor action, such as miR-34a which if overexpressed in medulloblastoma cells and induces apoptosis and restores sensitivity to chemotherapy [89], or miR-199-5p that by its target HES1 regulate the cancer stem cells [91].

Therefore, miRNAs can potentially regulate several pathways involved in the insurgence and progression of the medulloblastoma, acting as both oncogene and tumour suppressor (summarized in Table 1).

Table 1. Summary of deregulated miRNAs involved in the pathogenesis and progression of the four MB subgroups.

SUBGROUP 1: WINGLESS					
miRNAs as Oncogenes	**Cellular Function**	**Ref.**	**miRNAs as Suppressors**	**Cellular Function**	**Ref.**
miR 30b, miR-30d	N/A	[87]	miR-9	Antiproliferation Differentiation Pro-apoptosis	[85,92]
miR-193a	Metastasis Proliferation	[93]	miR-148a	Antiproliferation Invasion Reduces tumorigenicity	[60,85,93]
miR-224	Proliferation Radiation-sensitivity. Anchorage-independent growth	[93,94]			

Table 1. *Cont.*

SUBGROUP 2: SONIC HEDGEHOG					
miRNAs as Oncogenes	**Cellular Function**	**Ref.**	**miRNAs as Suppressors**	**Cellular Function**	**Ref.**
miR-17/92	N-Myc target	[66,67,89]	miR-let-7	Chemoresistance	[82,85,95]
miR-183/96/182	Migration	[70]	miR-34a	Antiproliferation Pro-apoptosis Senescence	[89,96–99]
miR-196b-5p, miR-200b-3p	C-Myc target Proliferation Migration Invasion	[100]	miR-125b	Suppressing progenitor and tumor cell growth	[82]
			miR-128a	Antiproliferation Senescence	[85,90]
			miR-135a	Reduces tumorigenicity	[82,85,101]
			miR-218	Antiproliferation. Reduces clonogenicity Promotes differentiation	[102–104]
			miR-219	Antiproliferation Invasion Migration [85,104,105] (Ferretti et al., 2009, Genovesi et al., 2011, Shi et al., 2014)	[64,101,102]
			miR-324-5p	Proliferation	[82]
			miR-326	Reduces clonogenicity	[82]
SUBGROUP 3					
miRNAs as Oncogenes	**Cellular Function**	**Ref.**	**miRNAs as Suppressors**	**Cellular Function**	**Ref.**
			miR-204	IGF2R and LC3B target Anchorage-indipendent growth Invasion.Autophagy	[106]
			miR-218	Antiproliferation. Reduces clonogenicity Promotes differentiation	[102–104]
			miR-495	Gfi1 target Cell growth inhibition	[107]
			miR-1253	Pro-apoptosis Antiproliferation	[72]
			miR-9	Antiproliferation Differentiation Pro-apoptosis	[85,92]

Table 1. *Cont.*

miRNAs as Oncogenes	Cellular Function	Ref.	miRNAs as Suppressors	Cellular Function	Ref.
			SUBGROUP 4		
miRNAs as Oncogenes	**Cellular Function**	**Ref.**	**miRNAs as Suppressors**	**Cellular Function**	**Ref.**
			miR-9	Antiproliferation Differentiation Pro-apoptosis	[85,92]
			miR-204	IGF2R and LC3B target Anchorage-independent growth Invasion. Autophagy	[106]
			miR-495	Gfi1 target Cell growth inhibition	[107]
			miR-1253	Pro-apoptosis Antiproliferation	[72]
			SUBGROUP NOT SPECIFIED		
miRNAs as Oncogenes	**Cellular Function**	**Ref.**	**miRNAs as Suppressors**	**Cellular Function**	**Ref.**
miR-21	Metastasis	[108]	miR-31	Antiproliferation	[85,109,110]
miR-106a/363	Proliferation Apoptosis Angiogenesis	[111]	miR-124	Differentiation Antiproliferation Pro-apoptosis	[82,112,113]
miR-106b	PTEN target Migration Invasion Tumor-sphere formation	[114]	miR-125a	Antiproliferation	[85]
miR-367	Invasion Proliferation	[115]	miR-199b-5p	Antiproliferation Reduces cancer stem cells	[91,116]
			miR-206	Antiproliferation	[58,93,117,118]
			miR-378	Differentiation Cell growth inhibition	[119]
			miR-383	Pro-apoptosis	[120,121]

Clinical Application of miRNAs in Medulloblastoma

The epigenetic landscape, as well as DNA mutation or miRNAs expression of medulloblastoma, has been investigated for the last 20 years to discover novel biomarkers for diagnosis, treatment, and/or disease progression [122]. miRNAs analysis in medulloblastoma tissue samples, as well as in cerebrospinal fluid (CSF) and in blood has been performed [14,123]. Additionally, miRNA expression in extracellular vesicles isolated from CSF or blood has been investigated. Several miRNAs were found differentially expressed between the different MBs subgroups. Gershanov et al. [76], found three miRNAs differentially express in G4-MB. These miRNAs are miR-20a-5p, 181a-2-3p, and 224-5p. Additionally, Li et al. [59] reported that miR-449a is a very good candidate for WNT-MB. However, Yogi et al. [60] reported that miR-148a is a good candidate for WNT-MBs classification. However, due to the significant variation between samples (primary cells, cell lines, patients) and miRNA expression in these studies is making very difficult to select a miRNA or set of miRNAs to improve medulloblastoma diagnosis and treatment. Fortunately, with the onset of new techniques based on the study of miRNAs and the analysis of patients' samples with medulloblastoma, miRNAs could be drastically improved to select aggressive versus non-aggressive medulloblastoma subgroups for treatment selection. Thus, considering their important roles in medulloblastoma development, miRNAs have been investigated as prognostic and diagnostic biomarkers for cancer detection, and also as useful targets for therapeutic intervention [124]. miRNA-based therapeutic treatments for medulloblastoma may follow the same strategies described above: miRNA over-expression or miRNA repression. However, the use of miRNAs as potential therapeutic targets for medulloblastoma remains controversial with regard to the ability of the miRNA delivery to pass through the blood–brain barrier. In order to overcome this limit, different systems to transport siRNA into the brain have been developed, such as engineered nanoparticles, vector-based, chemically modified, and "packaged" RNA oligonucleotides [95].

6. Conclusions

Medulloblastoma is a tumour of the paediatric population, the second most widespread brain tumour, after astrocytomas, and represents 1% of all cancers of the CNS [6]. A total of 70% of medulloblastomas are diagnosed between the second and tenth year of life. Survival five years from diagnosis (children and adults) is just over 60–70% [125].

Aberrant mechanisms of neuronal and cerebellar development can lead to the formation of medulloblastoma. These genetic and epigenetic changes can cause the abnormal activation of the Hedgehog signal pathway. In recent years, a hierarchical model for the evolution of cancer has been proposed, in which cancer stem cells (CSCs) acquire or maintain the properties of self-renewal, multipotency and tumour generation. This model has also found application in medulloblastoma, as CSCs have been observed in both mice and humans [126]. Furthermore, it was possible to demonstrate the correlation between the Hedgehog signalling pathway and these tumour cells [126], whose presence can lead to greater resistance to classical therapies and probability of relapse.

Most children die within three years due to aggressive treatment or recurrency [27]. Survivors must cope with severe long-term side effects; radiation of the entire developing brain and spinal cord to prevent metastatic recurrence have a devastating effect on intelligence, neurological and endocrine function [27]. Therefore, it is crucial to identify novel and effective therapeutic targets to treat these tumours and improve the quality of life of patients [127].

MiRNAs are known to play vital roles in nervous system development, as well as in various aspects of cancer development, progression, and metastasis. Thus, their involvement in medulloblastoma tumours is not surprising. In this review, we summarized the most important findings present in the literature on the role of miRNAs in influencing the tumorigenesis of medulloblastoma, inducing apoptosis and restoring sensitivity to chemotherapy [89,91]. Such as miR-326 that is absent in brain tumour pathologies and is involved in the modulation of signalling pathways, such as Hedgehog and Notch [82,85]. In particular, it interacts with the Hedgehog signalling pathway by negatively modulating the expression of the SMO activating receptor in cerebellar granules [82].

Furthermore, as we have seen, miRNAs are able not only to distinguish normal tissue from tumour, but also to characterize the different subgroups of medulloblastoma. Therefore, they can be used as biomarkers of tumour early diagnosis, prognosis, and provide new opportunities to treat the different clinical and biological features between subgroups.

In conclusion, it is crucial to know the functional and disease-associated mechanisms causing the deregulation of these small RNAs in medulloblastoma. Even though substantial questions must be answered, such as the role of the miRNAs in the development and progression of the different tumoral subgroups, they still represent a suitable target for the future medical treatment of medulloblastoma therapy, able to change the medical practice in the foreseeable future.

Author Contributions: E.B., J.F., M.V.N.-C. and P.T. wrote the paper. E.B. prepared the figures. M.V.N.-C. and P.T. correct the manuscript. All authors have read and agreed to the published version of the manuscript.

Funding: This work was funded by a Children with Cancer UK fellowship (No. 2014-178), Barts Charity (No. MGU0473), and Life Sciences Institute Fellowship Seed Award-QMUL awarded to M.V.N.-C. and by ERASMUS fellowship to J.F.

Conflicts of Interest: The authors declare no conflict of interest.

References

1. Packer, R.J.; Vezina, G. Management of and prognosis with medulloblastoma: Therapy at a crossroads. *Arch. Neurol.* **2008**, *65*, 1419–1424. [CrossRef] [PubMed]
2. Bondy, M.L.; Scheurer, M.E.; Malmer, B.; Barnholtz-Sloan, J.S.; Davis, F.G.; Il'yasova, D.; Kruchko, C.; McCarthy, B.J.; Rajaraman, P.; Schwartzbaum, J.A.; et al. Brain tumor epidemiology: Consensus from the Brain Tumor Epidemiology Consortium. *Cancer* **2008**, *113*, 1953–1968. [CrossRef] [PubMed]
3. Group, A.W.; Busco, S.; Buzzoni, C.; Mallone, S.; Trama, A.; Castaing, M.; Bella, F.; Amodio, R.; Bizzoco, S.; Cassetti, T.; et al. Italian cancer figures—Report 2015: The burden of rare cancers in Italy. *Epidemiol. Prev.* **2016**, *40*, 1–120. [CrossRef]
4. Northcott, P.A.; Robinson, G.W.; Kratz, C.P.; Mabbott, D.J.; Pomeroy, S.L.; Clifford, S.C.; Rutkowski, S.; Ellison, D.W.; Malkin, D.; Taylor, M.D.; et al. Medulloblastoma. *Nat. Rev. Dis. Primers* **2019**, *5*, 11. [CrossRef] [PubMed]
5. Huang, M.; Tailor, J.; Zhen, Q.; Gillmor, A.H.; Miller, M.L.; Weishaupt, H.; Chen, J.; Zheng, T.; Nash, E.K.; McHenry, L.K.; et al. Engineering Genetic Predisposition in Human Neuroepithelial Stem Cells Recapitulates Medulloblastoma Tumorigenesis. *Cell Stem Cell* **2019**, *25*, 433–446.e7. [CrossRef] [PubMed]
6. Quinlan, A.; Rizzolo, D. Understanding medulloblastoma. *J. Am. Acad. Phys. Assist.* **2017**, *30*, 30–36. [CrossRef] [PubMed]
7. Wells, E.M.; Packer, R.J. Pediatric brain tumors. *Continuum* **2015**, *21*, 373–396. [CrossRef] [PubMed]
8. Waszak, S.M.; Northcott, P.A.; Buchhalter, I.; Robinson, G.W.; Sutter, C.; Groebner, S.; Grund, K.B.; Brugieres, L.; Jones, D.T.W.; Pajtler, K.W.; et al. Spectrum and prevalence of genetic predisposition in medulloblastoma: A retrospective genetic study and prospective validation in a clinical trial cohort. *Lancet Oncol.* **2018**, *19*, 785–798. [CrossRef]
9. Kumar, V.; Kumar, V.; McGuire, T.; Coulter, D.W.; Sharp, J.G.; Mahato, R.I. Challenges and Recent Advances in Medulloblastoma Therapy. *Trends Pharmacol. Sci.* **2017**, *38*, 1061–1084. [CrossRef] [PubMed]
10. Buruiana, A.; Florian, S.I.; Florian, A.I.; Timis, T.L.; Mihu, C.M.; Miclaus, M.; Osan, S.; Hrapsa, I.; Cataniciu, R.C.; Farcas, M.; et al. The Roles of miRNA in Glioblastoma Tumor Cell Communication: Diplomatic and Aggressive Negotiations. *Int. J. Mol. Sci.* **2020**, *21*, 1950. [CrossRef] [PubMed]
11. Archer, T.C.; Mahoney, E.L.; Pomeroy, S.L. Medulloblastoma: Molecular Classification-Based Personal Therapeutics. *Neurotherapeutics* **2017**, *14*, 265–273. [CrossRef] [PubMed]
12. Louis, D.N.; Ohgaki, H.; Wiestler, O.D.; Cavenee, W.K.; Burger, P.C.; Jouvet, A.; Scheithauer, B.W.; Kleihues, P. The 2007 WHO classification of tumours of the central nervous system. *Acta Neuropathol.* **2007**, *114*, 97–109. [CrossRef] [PubMed]
13. Cavalli, F.M.G.; Remke, M.; Rampasek, L.; Peacock, J.; Shih, D.J.H.; Luu, B.; Garzia, L.; Torchia, J.; Nor, C.; Morrissy, A.S.; et al. Intertumoral Heterogeneity within Medulloblastoma Subgroups. *Cancer Cell* **2017**, *31*, 737–754.e6. [CrossRef] [PubMed]
14. Laneve, P.; Caffarelli, E. The Non-coding Side of Medulloblastoma. *Front. Cell Dev. Biol.* **2020**, *8*, 275. [CrossRef] [PubMed]
15. Orr, B.A. Pathology, diagnostics, and classification of medulloblastoma. *Brain Pathol.* **2020**, *30*, 664–678. [CrossRef] [PubMed]
16. Garzia, L.; Kijima, N.; Morrissy, A.S.; De Antonellis, P.; Guerreiro-Stucklin, A.; Holgado, B.L.; Wu, X.; Wang, X.; Parsons, M.; Zayne, K.; et al. A Hematogenous Route for Medulloblastoma Leptomeningeal Metastases. *Cell* **2018**, *172*, 1050–1062.e14. [CrossRef] [PubMed]
17. Khatua, S.; Song, A.; Citla Sridhar, D.; Mack, S.C. Childhood Medulloblastoma: Current Therapies, Emerging Molecular Landscape and Newer Therapeutic Insights. *Curr. Neuropharmacol.* **2018**, *16*, 1045–1058. [CrossRef] [PubMed]
18. Gottardo, N.G.; Gajjar, A. Current therapy for medulloblastoma. *Curr. Treat. Options Neurol.* **2006**, *8*, 319–334. [CrossRef] [PubMed]
19. Frisira, E.; Rashid, F.; Varma, S.N.; Badodi, S.; Benjamin-Ombo, V.A.; Michod, D.; Niklison-Chirou, M.V. NPI-0052 and gamma-radiation induce a synergistic apoptotic effect in medulloblastoma. *Cell Death Dis.* **2019**, *10*, 785. [CrossRef]
20. Ashley, D.M.; Merchant, T.E.; Strother, D.; Zhou, T.; Duffner, P.; Burger, P.C.; Miller, D.C.; Lyon, N.; Bonner, M.J.; Msall, M.; et al. Induction chemotherapy and conformal radiation therapy for very young children with nonmetastatic medulloblastoma: Children's Oncology Group study P9934. *J. Clin. Oncol.* **2012**, *30*, 3181–3186. [CrossRef]
21. Gupta, T.; Sarkar, C.; Rajshekhar, V.; Chatterjee, S.; Shirsat, N.; Muzumdar, D.; Pungavkar, S.; Chinnaswamy, G.; Jalali, R. Indian Society of Neuro-Oncology consensus guidelines for the contemporary management of medulloblastoma. *Neurol. India* **2017**, *65*, 315–332. [CrossRef] [PubMed]
22. Garzon, R.; Calin, G.A.; Croce, C.M. MicroRNAs in Cancer. *Annu. Rev. Med.* **2009**, *60*, 167–179. [CrossRef] [PubMed]
23. Vaucheret, H.; Vazquez, F.; Crete, P.; Bartel, D.P. The action of ARGONAUTE1 in the miRNA pathway and its regulation by the miRNA pathway are crucial for plant development. *Genes Dev.* **2004**, *18*, 1187–1197. [CrossRef] [PubMed]
24. Esquela-Kerscher, A.; Slack, F.J. Oncomirs—MicroRNAs with a role in cancer. *Nat. Rev. Cancer* **2006**, *6*, 259–269. [CrossRef] [PubMed]

25. Harfe, B.D. MicroRNAs in vertebrate development. *Curr. Opin. Genet. Dev.* **2005**, *15*, 410–415. [CrossRef]
26. Fabian, M.R.; Sonenberg, N. The mechanics of miRNA-mediated gene silencing: A look under the hood of miRISC. *Nat. Struct. Mol. Biol.* **2012**, *19*, 586–593. [CrossRef] [PubMed]
27. Chen, H.; Shalom-Feuerstein, R.; Riley, J.; Zhang, S.D.; Tucci, P.; Agostini, M.; Aberdam, D.; Knight, R.A.; Genchi, G.; Nicotera, P.; et al. miR-7 and miR-214 are specifically expressed during neuroblastoma differentiation, cortical development and embryonic stem cells differentiation, and control neurite outgrowth in vitro. *Biochem. Biophys. Res. Commun.* **2010**, *394*, 921–927. [CrossRef] [PubMed]
28. Feng, W.; Feng, Y. MicroRNAs in neural cell development and brain diseases. *Sci. China Life Sci.* **2011**, *54*, 1103–1112. [CrossRef] [PubMed]
29. Bak, M.; Silahtaroglu, A.; Moller, M.; Christensen, M.; Rath, M.F.; Skryabin, B.; Tommerup, N.; Kauppinen, S. MicroRNA expression in the adult mouse central nervous system. *RNA* **2008**, *14*, 432–444. [CrossRef]
30. Agostini, M.; Tucci, P.; Killick, R.; Candi, E.; Sayan, B.S.; Rivetti di Val Cervo, P.; Nicotera, P.; McKeon, F.; Knight, R.A.; Mak, T.W.; et al. Neuronal differentiation by TAp73 is mediated by microRNA-34a regulation of synaptic protein targets. *Proc. Natl. Acad. Sci. USA* **2011**, *108*, 21093–21098. [CrossRef] [PubMed]
31. Agostini, M.; Tucci, P.; Steinert, J.R.; Shalom-Feuerstein, R.; Rouleau, M.; Aberdam, D.; Forsythe, I.D.; Young, K.W.; Ventura, A.; Concepcion, C.P.; et al. microRNA-34a regulates neurite outgrowth, spinal morphology, and function. *Proc. Natl. Acad. Sci. USA* **2011**, *108*, 21099–21104. [CrossRef]
32. Bonev, B.; Pisco, A.; Papalopulu, N. MicroRNA-9 reveals regional diversity of neural progenitors along the anterior-posterior axis. *Dev. Cell* **2011**, *20*, 19–32. [CrossRef] [PubMed]
33. Makeyev, E.V.; Zhang, J.; Carrasco, M.A.; Maniatis, T. The MicroRNA miR-124 promotes neuronal differentiation by triggering brain-specific alternative pre-mRNA splicing. *Mol. Cell* **2007**, *27*, 435–448. [CrossRef] [PubMed]
34. Titze-de-Almeida, S.S.; Soto-Sanchez, C.; Fernandez, E.; Koprich, J.B.; Brotchie, J.M.; Titze-de-Almeida, R. The Promise and Challenges of Developing miRNA-Based Therapeutics for Parkinson's Disease. *Cells* **2020**, *9*, 841. [CrossRef] [PubMed]
35. Gui, Y.; Liu, H.; Zhang, L.; Lv, W.; Hu, X. Altered microRNA profiles in cerebrospinal fluid exosome in Parkinson disease and Alzheimer disease. *Oncotarget* **2015**, *6*, 37043–37053. [CrossRef] [PubMed]
36. Bruce, J.P.; Hui, A.B.; Shi, W.; Perez-Ordonez, B.; Weinreb, I.; Xu, W.; Haibe-Kains, B.; Waggott, D.M.; Boutros, P.C.; O'Sullivan, B.; et al. Identification of a microRNA signature associated with risk of distant metastasis in nasopharyngeal carcinoma. *Oncotarget* **2015**, *6*, 4537–4550. [CrossRef] [PubMed]
37. Calin, G.A.; Dumitru, C.D.; Shimizu, M.; Bichi, R.; Zupo, S.; Noch, E.; Aldler, H.; Rattan, S.; Keating, M.; Rai, K.; et al. Frequent deletions and down-regulation of micro- RNA genes miR15 and miR16 at 13q14 in chronic lymphocytic leukemia. *Proc. Natl. Acad. Sci. USA* **2002**, *99*, 15524–15529. [CrossRef]
38. Mollashahi, B.; Aghamaleki, F.S.; Movafagh, A. The Roles of miRNAs in Medulloblastoma: A Systematic Review. *J. Cancer Prev.* **2019**, *24*, 79–90. [CrossRef] [PubMed]
39. Fasoulakis, Z.; Daskalakis, G.; Diakosavvas, M.; Papapanagiotou, I.; Theodora, M.; Bourazan, A.; Alatzidou, D.; Pagkalos, A.; Kontomanolis, E.N. MicroRNAs Determining Carcinogenesis by Regulating Oncogenes and Tumor Suppressor Genes During Cell Cycle. *Microrna* **2020**, *9*, 82–92. [CrossRef] [PubMed]
40. Zhang, L.; Huang, J.; Yang, N.; Greshock, J.; Megraw, M.S.; Giannakakis, A.; Liang, S.; Naylor, T.L.; Barchetti, A.; Ward, M.R.; et al. microRNAs exhibit high frequency genomic alterations in human cancer. *Proc. Natl. Acad. Sci. USA* **2006**, *103*, 9136–9141. [CrossRef]
41. Lehmann, U.; Hasemeier, B.; Christgen, M.; Muller, M.; Romermann, D.; Langer, F.; Kreipe, H. Epigenetic inactivation of microRNA gene hsa-mir-9-1 in human breast cancer. *J. Pathol.* **2008**, *214*, 17–24. [CrossRef]
42. Kano, M.; Seki, N.; Kikkawa, N.; Fujimura, L.; Hoshino, I.; Akutsu, Y.; Chiyomaru, T.; Enokida, H.; Nakagawa, M.; Matsubara, H. miR-145, miR-133a and miR-133b: Tumor-suppressive miRNAs target FSCN1 in esophageal squamous cell carcinoma. *Int. J. Cancer* **2010**, *127*, 2804–2814. [CrossRef] [PubMed]
43. Boscaino, V.; Fiannaca, A.; La Paglia, L.; La Rosa, M.; Rizzo, R.; Urso, A. MiRNA therapeutics based on logic circuits of biological pathways. *BMC Bioinform.* **2019**, *20*, 344. [CrossRef]
44. Lu, J.; Getz, G.; Miska, E.A.; Alvarez-Saavedra, E.; Lamb, J.; Peck, D.; Sweet-Cordero, A.; Ebert, B.L.; Mak, R.H.; Ferrando, A.A.; et al. MicroRNA expression profiles classify human cancers. *Nature* **2005**, *435*, 834–838. [CrossRef]
45. Di Leva, G.; Garofalo, M.; Croce, C.M. MicroRNAs in cancer. *Annu. Rev. Pathol.* **2014**, *9*, 287–314. [CrossRef] [PubMed]
46. Cortez, M.A.; Bueso-Ramos, C.; Ferdin, J.; Lopez-Berestein, G.; Sood, A.K.; Calin, G.A. MicroRNAs in body fluids—The mix of hormones and biomarkers. *Nat. Rev. Clin. Oncol.* **2011**, *8*, 467–477. [CrossRef]
47. Iorio, M.V.; Croce, C.M. MicroRNAs in cancer: Small molecules with a huge impact. *J. Clin. Oncol.* **2009**, *27*, 5848–5856. [CrossRef] [PubMed]
48. Gambari, R.; Brognara, E.; Spandidos, D.A.; Fabbri, E. Targeting oncomiRNAs and mimicking tumor suppressor miRNAs: Nuew trends in the development of miRNA therapeutic strategies in oncology (Review). *Int. J. Oncol.* **2016**, *49*, 5–32. [CrossRef] [PubMed]
49. Northcott, P.A.; Shih, D.J.; Peacock, J.; Garzia, L.; Morrissy, A.S.; Zichner, T.; Stutz, A.M.; Korshunov, A.; Reimand, J.; Schumacher, S.E.; et al. Subgroup-specific structural variation across 1,000 medulloblastoma genomes. *Nature* **2012**, *488*, 49–56. [CrossRef] [PubMed]

50. Taylor, M.D.; Northcott, P.A.; Korshunov, A.; Remke, M.; Cho, Y.J.; Clifford, S.C.; Eberhart, C.G.; Parsons, D.W.; Rutkowski, S.; Gajjar, A.; et al. Molecular subgroups of medulloblastoma: The current consensus. *Acta Neuropathol.* **2012**, *123*, 465–472. [CrossRef]
51. Ellison, D.W.; Dalton, J.; Kocak, M.; Nicholson, S.L.; Fraga, C.; Neale, G.; Kenney, A.M.; Brat, D.J.; Perry, A.; Yong, W.H.; et al. Medulloblastoma: Clinicopathological correlates of SHH, WNT, and non-SHH/WNT molecular subgroups. *Acta Neuropathol.* **2011**, *121*, 381–396. [CrossRef] [PubMed]
52. Goschzik, T.; Zur Muhlen, A.; Kristiansen, G.; Haberler, C.; Stefanits, H.; Friedrich, C.; von Hoff, K.; Rutkowski, S.; Pfister, S.M.; Pietsch, T. Molecular stratification of medulloblastoma: Comparison of histological and genetic methods to detect Wnt activated tumours. *Neuropathol. Appl. Neurobiol.* **2015**, *41*, 135–144. [CrossRef] [PubMed]
53. Juraschka, K.; Taylor, M.D. Medulloblastoma in the age of molecular subgroups: A review. *J. Neurosurg. Pediatrics* **2019**, *24*, 353–363. [CrossRef]
54. Ramaswamy, V.; Nor, C.; Taylor, M.D. p53 and Meduloblastoma. *Cold Spring Harb. Perspect. Med.* **2015**, *6*, a026278. [CrossRef] [PubMed]
55. Phoenix, T.N.; Patmore, D.M.; Boop, S.; Boulos, N.; Jacus, M.O.; Patel, Y.T.; Roussel, M.F.; Finkelstein, D.; Goumnerova, L.; Perreault, S.; et al. Medulloblastoma Genotype Dictates Blood Brain Barrier Phenotype. *Cancer Cell* **2016**, *29*, 508–522. [CrossRef] [PubMed]
56. Manoranjan, B.; Venugopal, C.; Bakhshinyan, D.; Adile, A.A.; Richards, L.; Kameda-Smith, M.M.; Whitley, O.; Dvorkin-Gheva, A.; Subapanditha, M.; Savage, N.; et al. Wnt activation as a therapeutic strategy in medulloblastoma. *Nat. Commun.* **2020**, *11*, 4323. [CrossRef] [PubMed]
57. Dai, J.; Li, Q.; Bing, Z.; Zhang, Y.; Niu, L.; Yin, H.; Yuan, G.; Pan, Y. Comprehensive analysis of a microRNA expression profile in pediatric medulloblastoma. *Mol. Med. Rep.* **2017**, *15*, 4109–4115. [CrossRef] [PubMed]
58. Panwalkar, P.; Moiyadi, A.; Goel, A.; Shetty, P.; Goel, N.; Sridhar, E.; Shirsat, N. MiR-206, a Cerebellum Enriched miRNA Is Down-regulated in All Medulloblastoma Subgroups and Its Overexpression Is Necessary for Growth Inhibition of Medulloblastoma Cells. *J. Mol. Neurosci.* **2015**, *56*, 673–680. [CrossRef] [PubMed]
59. Li, Y.; Jiang, T.; Shao, L.; Liu, Y.; Zheng, C.; Zhong, Y.; Zhang, J.; Chang, Q. Mir-449a, a potential diagnostic biomarker for WNT group of medulloblastoma. *J. Neurooncol.* **2016**, *129*, 423–431. [CrossRef] [PubMed]
60. Yogi, K.; Sridhar, E.; Goel, N.; Jalali, R.; Goel, A.; Moiyadi, A.; Thorat, R.; Panwalkar, P.; Khire, A.; Dasgupta, A.; et al. MiR-148a, a microRNA upregulated in the WNT subgroup tumors, inhibits invasion and tumorigenic potential of medulloblastoma cells by targeting Neuropilin 1. *Oncoscience* **2015**, *2*, 334–348. [CrossRef]
61. Kool, M.; Jones, D.T.; Jager, N.; Northcott, P.A.; Pugh, T.J.; Hovestadt, V.; Piro, R.M.; Esparza, L.A.; Markant, S.L.; Remke, M.; et al. Genome sequencing of SHH medulloblastoma predicts genotype-related response to smoothened inhibition. *Cancer Cell* **2014**, *25*, 393–405. [CrossRef] [PubMed]
62. Zhukova, N.; Ramaswamy, V.; Remke, M.; Pfaff, E.; Shih, D.J.; Martin, D.C.; Castelo-Branco, P.; Baskin, B.; Ray, P.N.; Bouffet, E.; et al. Subgroup-specific prognostic implications of TP53 mutation in medulloblastoma. *J. Clin. Oncol.* **2013**, *31*, 2927–2935. [CrossRef] [PubMed]
63. Rausch, T.; Jones, D.T.; Zapatka, M.; Stutz, A.M.; Zichner, T.; Weischenfeldt, J.; Jager, N.; Remke, M.; Shih, D.; Northcott, P.A.; et al. Genome sequencing of pediatric medulloblastoma links catastrophic DNA rearrangements with TP53 mutations. *Cell* **2012**, *148*, 59–71. [CrossRef] [PubMed]
64. Besharat, Z.M.; Sabato, C.; Po, A.; Gianno, F.; Abballe, L.; Napolitano, M.; Miele, E.; Giangaspero, F.; Vacca, A.; Catanzaro, G.; et al. Low Expression of miR-466f-3p Sustains Epithelial to Mesenchymal Transition in Sonic Hedgehog Medulloblastoma Stem Cells Through Vegfa-Nrp2 Signaling Pathway. *Front. Pharmacol.* **2018**, *9*, 1281. [CrossRef]
65. Miele, E.; Po, A.; Begalli, F.; Antonucci, L.; Mastronuzzi, A.; Marras, C.E.; Carai, A.; Cucchi, D.; Abballe, L.; Besharat, Z.M.; et al. beta-arrestin1-mediated acetylation of Gli1 regulates Hedgehog/Gli signaling and modulates self-renewal of SHH medulloblastoma cancer stem cells. *BMC Cancer* **2017**, *17*, 488. [CrossRef] [PubMed]
66. Murphy, B.L.; Obad, S.; Bihannic, L.; Ayrault, O.; Zindy, F.; Kauppinen, S.; Roussel, M.F. Silencing of the miR-17~92 cluster family inhibits medulloblastoma progression. *Cancer Res.* **2013**, *73*, 7068–7078. [CrossRef] [PubMed]
67. Northcott, P.A.; Fernandez, L.A.; Hagan, J.P.; Ellison, D.W.; Grajkowska, W.; Gillespie, Y.; Grundy, R.; Van Meter, T.; Rutka, J.T.; Croce, C.M.; et al. The miR-17/92 polycistron is up-regulated in sonic hedgehog-driven medulloblastomas and induced by N-myc in sonic hedgehog-treated cerebellar neural precursors. *Cancer Res.* **2009**, *69*, 3249–3255. [CrossRef] [PubMed]
68. Tantawy, M.; Elzayat, M.G.; Yehia, D.; Taha, H. Identification of microRNA signature in different pediatric brain tumors. *Genet. Mol. Biol.* **2018**, *41*, 27–34. [CrossRef] [PubMed]
69. Pal, R.; Greene, S. microRNA-10b Is Overexpressed and Critical for Cell Survival and Proliferation in Medulloblastoma. *PLoS ONE* **2015**, *10*, e0137845. [CrossRef]
70. Weeraratne, S.D.; Amani, V.; Teider, N.; Pierre-Francois, J.; Winter, D.; Kye, M.J.; Sengupta, S.; Archer, T.; Remke, M.; Bai, A.H.; et al. Pleiotropic effects of miR-183~96~182 converge to regulate cell survival, proliferation and migration in medulloblastoma. *Acta Neuropathol.* **2012**, *123*, 539–552. [CrossRef]
71. Singh, S.V.; Dakhole, A.N.; Deogharkar, A.; Kazi, S.; Kshirsagar, R.; Goel, A.; Moiyadi, A.; Jalali, R.; Sridhar, E.; Gupta, T.; et al. Restoration of miR-30a expression inhibits growth, tumorigenicity of medulloblastoma cells accompanied by autophagy inhibition. *Biochem. Biophys. Res. Commun.* **2017**, *491*, 946–952. [CrossRef] [PubMed]

72. Kanchan, R.K.; Perumal, N.; Atri, P.; Chirravuri Venkata, R.; Thapa, I.; Klinkebiel, D.L.; Donson, A.M.; Perry, D.; Punsoni, M.; Talmon, G.A.; et al. MiR-1253 exerts tumor-suppressive effects in medulloblastoma via inhibition of CDK6 and CD276 (B7-H3). *Brain Pathol.* **2020**, *30*, 732–745. [CrossRef]
73. Glick, D.; Barth, S.; Macleod, K.F. Autophagy: Cellular and molecular mechanisms. *J. Pathol.* **2010**, *221*, 3–12. [CrossRef] [PubMed]
74. Senfter, D.; Samadaei, M.; Mader, R.M.; Gojo, J.; Peyrl, A.; Krupitza, G.; Kool, M.; Sill, M.; Haberler, C.; Ricken, G.; et al. High impact of miRNA-4521 on FOXM1 expression in medulloblastoma. *Cell Death Dis.* **2019**, *10*, 696. [CrossRef]
75. Kool, M.; Korshunov, A.; Remke, M.; Jones, D.T.; Schlanstein, M.; Northcott, P.A.; Cho, Y.J.; Koster, J.; Schouten-van Meeteren, A.; van Vuurden, D.; et al. Molecular subgroups of medulloblastoma: An international meta-analysis of transcriptome, genetic aberrations, and clinical data of WNT, SHH, Group 3, and Group 4 medulloblastomas. *Acta Neuropathol.* **2012**, *123*, 473–484. [CrossRef] [PubMed]
76. Gershanov, S.; Toledano, H.; Michowiz, S.; Barinfeld, O.; Pinhasov, A.; Goldenberg-Cohen, N.; Salmon-Divon, M. MicroRNA-mRNA expression profiles associated with medulloblastoma subgroup 4. *Cancer Manag. Res.* **2018**, *10*, 339–352. [CrossRef] [PubMed]
77. Stromecki, M.; Tatari, N.; Morrison, L.C.; Kaur, R.; Zagozewski, J.; Palidwor, G.; Ramaswamy, V.; Skowron, P.; Wolfl, M.; Milde, T.; et al. Characterization of a novel OTX2-driven stem cell program in Group 3 and Group 4 medulloblastoma. *Mol. Oncol.* **2018**, *12*, 495–513. [CrossRef] [PubMed]
78. Shi, L.; Cheng, Z.; Zhang, J.; Li, R.; Zhao, P.; Fu, Z.; You, Y. hsa-mir-181a and hsa-mir-181b function as tumor suppressors in human glioma cells. *Brain Res.* **2008**, *1236*, 185–193. [CrossRef]
79. Zhao, J.; Lei, T.; Xu, C.; Li, H.; Ma, W.; Yang, Y.; Fan, S.; Liu, Y. MicroRNA-187, down-regulated in clear cell renal cell carcinoma and associated with lower survival, inhibits cell growth and migration though targeting B7-H3. *Biochem. Biophys. Res. Commun.* **2013**, *438*, 439–444. [CrossRef]
80. Mulrane, L.; Madden, S.F.; Brennan, D.J.; Gremel, G.; McGee, S.F.; McNally, S.; Martin, F.; Crown, J.P.; Jirstrom, K.; Higgins, D.G.; et al. miR-187 is an independent prognostic factor in breast cancer and confers increased invasive potential in vitro. *Clin. Cancer Res.* **2012**, *18*, 6702–6713. [CrossRef]
81. Hovestadt, V.; Jones, D.T.; Picelli, S.; Wang, W.; Kool, M.; Northcott, P.A.; Sultan, M.; Stachurski, K.; Ryzhova, M.; Warnatz, H.J.; et al. Decoding the regulatory landscape of medulloblastoma using DNA methylation sequencing. *Nature* **2014**, *510*, 537–541. [CrossRef]
82. Ferretti, E.; De Smaele, E.; Miele, E.; Laneve, P.; Po, A.; Pelloni, M.; Paganelli, A.; Di Marcotullio, L.; Caffarelli, E.; Screpanti, I.; et al. Concerted microRNA control of Hedgehog signalling in cerebellar neuronal progenitor and tumour cells. *EMBO J.* **2008**, *27*, 2616–2627. [CrossRef]
83. Tang, B.; Xu, A.; Xu, J.; Huang, H.; Chen, L.; Su, Y.; Zhang, L.; Li, J.; Fan, F.; Deng, J.; et al. MicroRNA-324-5p regulates stemness, pathogenesis and sensitivity to bortezomib in multiple myeloma cells by targeting hedgehog signaling. *Int. J. Cancer* **2018**, *142*, 109–120. [CrossRef] [PubMed]
84. De Smaele, E.; Di Marcotullio, L.; Ferretti, E.; Screpanti, I.; Alesse, E.; Gulino, A. Chromosome 17p deletion in human medulloblastoma: A missing checkpoint in the Hedgehog pathway. *Cell Cycle* **2004**, *3*, 1263–1266. [CrossRef] [PubMed]
85. Ferretti, E.; De Smaele, E.; Po, A.; Di Marcotullio, L.; Tosi, E.; Espinola, M.S.; Di Rocco, C.; Riccardi, R.; Giangaspero, F.; Farcomeni, A.; et al. MicroRNA profiling in human medulloblastoma. *Int. J. Cancer* **2009**, *124*, 568–577. [CrossRef]
86. Uziel, T.; Karginov, F.V.; Xie, S.; Parker, J.S.; Wang, Y.D.; Gajjar, A.; He, L.; Ellison, D.; Gilbertson, R.J.; Hannon, G.; et al. The miR-17~92 cluster collaborates with the Sonic Hedgehog pathway in medulloblastoma. *Proc. Natl. Acad. Sci. USA* **2009**, *106*, 2812–2817. [CrossRef]
87. Lu, Y.; Ryan, S.L.; Elliott, D.J.; Bignell, G.R.; Futreal, P.A.; Ellison, D.W.; Bailey, S.; Clifford, S.C. Amplification and overexpression of Hsa-miR-30b, Hsa-miR-30d and KHDRBS3 at 8q24.22-q24.23 in medulloblastoma. *PLoS ONE* **2009**, *4*, e6159. [CrossRef] [PubMed]
88. Bai, A.H.; Milde, T.; Remke, M.; Rolli, C.G.; Hielscher, T.; Cho, Y.J.; Kool, M.; Northcott, P.A.; Jugold, M.; Bazhin, A.V.; et al. MicroRNA-182 promotes leptomeningeal spread of non-sonic hedgehog-medulloblastoma. *Acta Neuropathol.* **2012**, *123*, 529–538. [CrossRef] [PubMed]
89. Weeraratne, S.D.; Amani, V.; Neiss, A.; Teider, N.; Scott, D.K.; Pomeroy, S.L.; Cho, Y.J. miR-34a confers chemosensitivity through modulation of MAGE-A and p53 in medulloblastoma. *Neuro Oncol.* **2011**, *13*, 165–175. [CrossRef] [PubMed]
90. Venkataraman, S.; Alimova, I.; Fan, R.; Harris, P.; Foreman, N.; Vibhakar, R. MicroRNA 128a increases intracellular ROS level by targeting Bmi-1 and inhibits medulloblastoma cancer cell growth by promoting senescence. *PLoS ONE* **2010**, *5*, e10748. [CrossRef] [PubMed]
91. Garzia, L.; Andolfo, I.; Cusanelli, E.; Marino, N.; Petrosino, G.; De Martino, D.; Esposito, V.; Galeone, A.; Navas, L.; Esposito, S.; et al. MicroRNA-199b-5p impairs cancer stem cells through negative regulation of HES1 in medulloblastoma. *PLoS ONE* **2009**, *4*, e4998. [CrossRef] [PubMed]
92. Fiaschetti, G.; Abela, L.; Nonoguchi, N.; Dubuc, A.M.; Remke, M.; Boro, A.; Grunder, E.; Siler, U.; Ohgaki, H.; Taylor, M.D.; et al. Epigenetic silencing of miRNA-9 is associated with HES1 oncogenic activity and poor prognosis of medulloblastoma. *Br. J. Cancer* **2014**, *110*, 636–647. [CrossRef] [PubMed]
93. Gokhale, A.; Kunder, R.; Goel, A.; Sarin, R.; Moiyadi, A.; Shenoy, A.; Mamidipally, C.; Noronha, S.; Kannan, S.; Shirsat, N.V. Distinctive microRNA signature of medulloblastomas associated with the WNT signaling pathway. *J. Cancer Res. Ther.* **2010**, *6*, 521–529. [CrossRef] [PubMed]

94. Lu, S.; Wang, S.; Geng, S.; Ma, S.; Liang, Z.; Jiao, B. Upregulation of microRNA-224 confers a poor prognosis in glioma patients. *Clin. Transl. Oncol.* **2013**, *15*, 569–574. [CrossRef] [PubMed]
95. Kim, D.H.; Rossi, J.J. Strategies for silencing human disease using RNA interference. *Nat. Rev. Genet.* **2007**, *8*, 173–184. [CrossRef] [PubMed]
96. De Antonellis, P.; Medaglia, C.; Cusanelli, E.; Andolfo, I.; Liguori, L.; De Vita, G.; Carotenuto, M.; Bello, A.; Formiggini, F.; Galeone, A.; et al. MiR-34a targeting of Notch ligand delta-like 1 impairs CD15+/CD133+ tumor-propagating cells and supports neural differentiation in medulloblastoma. *PLoS ONE* **2011**, *6*, e24584. [CrossRef] [PubMed]
97. Morrissy, A.S.; Garzia, L.; Shih, D.J.; Zuyderduyn, S.; Huang, X.; Skowron, P.; Remke, M.; Cavalli, F.M.; Ramaswamy, V.; Lindsay, P.E.; et al. Divergent clonal selection dominates medulloblastoma at recurrence. *Nature* **2016**, *529*, 351–357. [CrossRef] [PubMed]
98. Fan, Y.N.; Meley, D.; Pizer, B.; See, V. Mir-34a mimics are potential therapeutic agents for p53-mutated and chemo-resistant brain tumour cells. *PLoS ONE* **2014**, *9*, e108514. [CrossRef]
99. Thor, T.; Kunkele, A.; Pajtler, K.W.; Wefers, A.K.; Stephan, H.; Mestdagh, P.; Heukamp, L.; Hartmann, W.; Vandesompele, J.; Sadowski, N.; et al. MiR-34a deficiency accelerates medulloblastoma formation in vivo. *Int. J. Cancer* **2015**, *136*, 2293–2303. [CrossRef] [PubMed]
100. Visani, M.; Marucci, G.; Biase, D.; Giangaspero, F.; Buttarelli, F.R.; Brandes, A.A.; Franceschi, E.; Acquaviva, G.; Ciarrocchi, A.; Rhoden, K.J.; et al. miR-196B-5P and miR-200B-3P Are Differentially Expressed in Medulloblastomas of Adults and Children. *Diagnostics* **2020**, *10*, 265. [CrossRef] [PubMed]
101. Hemmesi, K.; Squadrito, M.L.; Mestdagh, P.; Conti, V.; Cominelli, M.; Piras, I.S.; Sergi, L.S.; Piccinin, S.; Maestro, R.; Poliani, P.L.; et al. miR-135a Inhibits Cancer Stem Cell-Driven Medulloblastoma Development by Directly Repressing Arhgef6 Expression. *Stem Cells* **2015**, *33*, 1377–1389. [CrossRef] [PubMed]
102. Venkataraman, S.; Birks, D.K.; Balakrishnan, I.; Alimova, I.; Harris, P.S.; Patel, P.R.; Handler, M.H.; Dubuc, A.; Taylor, M.D.; Foreman, N.K.; et al. MicroRNA 218 acts as a tumor suppressor by targeting multiple cancer phenotype-associated genes in medulloblastoma. *J. Biol. Chem.* **2013**, *288*, 1918–1928. [CrossRef] [PubMed]
103. Shi, J.; Yang, L.; Wang, T.; Zhang, J.; Guo, X.; Huo, X.; Niu, H. miR-218 is downregulated and directly targets SH3GL1 in childhood medulloblastoma. *Mol. Med. Rep.* **2013**, *8*, 1111–1117. [CrossRef] [PubMed]
104. Shi, J.A.; Lu, D.L.; Huang, X.; Tan, W. miR-219 inhibits the proliferation, migration and invasion of medulloblastoma cells by targeting CD164. *Int. J. Mol. Med.* **2014**, *34*, 237–243. [CrossRef] [PubMed]
105. Genovesi, L.A.; Carter, K.W.; Gottardo, N.G.; Giles, K.M.; Dallas, P.B. Integrated analysis of miRNA and mRNA expression in childhood medulloblastoma compared with neural stem cells. *PLoS ONE* **2011**, *6*, e23935. [CrossRef]
106. Bharambe, H.S.; Paul, R.; Panwalkar, P.; Jalali, R.; Sridhar, E.; Gupta, T.; Moiyadi, A.; Shetty, P.; Kazi, S.; Deogharkar, A.; et al. Downregulation of miR-204 expression defines a highly aggressive subset of Group 3/Group 4 medulloblastomas. *Acta Neuropathol. Commun.* **2019**, *7*, 52. [CrossRef] [PubMed]
107. Wang, C.; Yun, Z.; Zhao, T.; Liu, X.; Ma, X. MiR-495 is a Predictive Biomarker that Downregulates GFI1 Expression in Medulloblastoma. *Cell Physiol. Biochem.* **2015**, *36*, 1430–1439. [CrossRef] [PubMed]
108. Grunder, E.; D'Ambrosio, R.; Fiaschetti, G.; Abela, L.; Arcaro, A.; Zuzak, T.; Ohgaki, H.; Lv, S.Q.; Shalaby, T.; Grotzer, M. MicroRNA-21 suppression impedes medulloblastoma cell migration. *Eur. J. Cancer* **2011**, *47*, 2479–2490. [CrossRef]
109. Jin, Y.; Xiong, A.; Zhang, Z.; Li, S.; Huang, H.; Yu, T.T.; Cao, X.; Cheng, S.Y. MicroRNA-31 suppresses medulloblastoma cell growth by inhibiting DNA replication through minichromosome maintenance 2. *Oncotarget* **2014**, *5*, 4821–4833. [CrossRef] [PubMed]
110. Ma, H.; Cao, W.; Ding, M. MicroRNA-31 weakens cisplatin resistance of medulloblastoma cells via NF-kappaB and PI3K/AKT pathways. *Biofactors* **2020**, *46*, 831–838. [CrossRef] [PubMed]
111. Gruszka, R.; Zakrzewski, K.; Liberski, P.P.; Zakrzewska, M. mRNA and miRNA Expression Analyses of the MYC/E2F/miR-17-92 Network in the Most Common Pediatric Brain Tumors. *Int. J. Mol. Sci.* **2021**, *22*, 543. [CrossRef] [PubMed]
112. Pierson, J.; Hostager, B.; Fan, R.; Vibhakar, R. Regulation of cyclin dependent kinase 6 by microRNA 124 in medulloblastoma. *J. Neurooncol.* **2008**, *90*, 1–7. [CrossRef] [PubMed]
113. Silber, J.; Hashizume, R.; Felix, T.; Hariono, S.; Yu, M.; Berger, M.S.; Huse, J.T.; VandenBerg, S.R.; James, C.D.; Hodgson, J.G.; et al. Expression of miR-124 inhibits growth of medulloblastoma cells. *Neuro Oncol.* **2013**, *15*, 83–90. [CrossRef]
114. Li, K.K.; Xia, T.; Ma, F.M.; Zhang, R.; Mao, Y.; Wang, Y.; Zhou, L.; Lau, K.M.; Ng, H.K. miR-106b is overexpressed in medulloblastomas and interacts directly with PTEN. *Neuropathol. Appl. Neurobiol.* **2015**, *41*, 145–164. [CrossRef] [PubMed]
115. Kaid, C.; Silva, P.B.; Cortez, B.A.; Rodini, C.O.; Semedo-Kuriki, P.; Okamoto, O.K. miR-367 promotes proliferation and stem-like traits in medulloblastoma cells. *Cancer Sci.* **2015**, *106*, 1188–1195. [CrossRef] [PubMed]
116. Andolfo, I.; Liguori, L.; De Antonellis, P.; Cusanelli, E.; Marinaro, F.; Pistollato, F.; Garzia, L.; De Vita, G.; Petrosino, G.; Accordi, B.; et al. The micro-RNA 199b-5p regulatory circuit involves Hes1, CD15, and epigenetic modifications in medulloblastoma. *Neuro Oncol.* **2012**, *14*, 596–612. [CrossRef] [PubMed]
117. Kunder, R.; Jalali, R.; Sridhar, E.; Moiyadi, A.; Goel, N.; Goel, A.; Gupta, T.; Krishnatry, R.; Kannan, S.; Kurkure, P.; et al. Real-time PCR assay based on the differential expression of microRNAs and protein-coding genes for molecular classification of formalin-fixed paraffin embedded medulloblastomas. *Neuro Oncol.* **2013**, *15*, 1644–1651. [CrossRef]
118. Pan, X.; Wang, Z.; Wan, B.; Zheng, Z. MicroRNA-206 inhibits the viability and migration of medulloblastoma cells by targeting LIM and SH3 protein 1. *Exp. Ther. Med.* **2017**, *14*, 3894–3900. [CrossRef] [PubMed]

119. Zhang, Z.Y.; Zhu, B.; Zhao, X.W.; Zhan, Y.B.; Bao, J.J.; Zhou, J.Q.; Zhang, F.J.; Yu, B.; Liu, J.; Wang, Y.M.; et al. Regulation of UHRF1 by microRNA-378 modulates medulloblastoma cell proliferation and apoptosis. *Oncol. Rep.* **2017**, *38*, 3078–3084. [CrossRef] [PubMed]
120. Li, K.K.; Pang, J.C.; Lau, K.M.; Zhou, L.; Mao, Y.; Wang, Y.; Poon, W.S.; Ng, H.K. MiR-383 is downregulated in medulloblastoma and targets peroxiredoxin 3 (PRDX3). *Brain Pathol.* **2013**, *23*, 413–425. [CrossRef] [PubMed]
121. Cui, Y.; Chen, L.G.; Yao, H.B.; Zhang, J.; Ding, K.F. Upregulation of microRNA-383 inhibits the proliferation, migration and invasion of colon cancer cells. *Oncol Lett.* **2018**, *15*, 1184–1190. [CrossRef] [PubMed]
122. Pickles, J.C.; Fairchild, A.R.; Stone, T.J.; Brownlee, L.; Merve, A.; Yasin, S.A.; Avery, A.; Ahmed, S.W.; Ogunbiyi, O.; Gonzalez Zapata, J.; et al. DNA methylation-based profiling for paediatric CNS tumour diagnosis and treatment: A population-based study. *Lancet Child. Adolesc. Health* **2020**, *4*, 121–130. [CrossRef]
123. Shalaby, T.; Grotzer, M.A. Tumor-Associated CSF MicroRNAs for the Prediction and Evaluation of CNS Malignancies. *Int. J. Mol. Sci.* **2015**, *16*, 29103–29119. [CrossRef] [PubMed]
124. Wang, X.; Holgado, B.L.; Ramaswamy, V.; Mack, S.; Zayne, K.; Remke, M.; Wu, X.; Garzia, L.; Daniels, C.; Kenney, A.M.; et al. miR miR on the wall, who's the most malignant medulloblastoma miR of them all? *Neuro Oncol.* **2018**, *20*, 313–323. [CrossRef] [PubMed]
125. Shih, D.J.; Northcott, P.A.; Remke, M.; Korshunov, A.; Ramaswamy, V.; Kool, M.; Luu, B.; Yao, Y.; Wang, X.; Dubuc, A.M.; et al. Cytogenetic prognostication within medulloblastoma subgroups. *J. Clin. Oncol.* **2014**, *32*, 886–896. [CrossRef] [PubMed]
126. Po, A.; Ferretti, E.; Miele, E.; De Smaele, E.; Paganelli, A.; Canettieri, G.; Coni, S.; Di Marcotullio, L.; Biffoni, M.; Massimi, L.; et al. Hedgehog controls neural stem cells through p53-independent regulation of Nanog. *EMBO J.* **2010**, *29*, 2646–2658. [CrossRef] [PubMed]
127. Niklison-Chirou, M.V.; Erngren, I.; Engskog, M.; Haglof, J.; Picard, D.; Remke, M.; McPolin, P.H.R.; Selby, M.; Williamson, D.; Clifford, S.C.; et al. TAp73 is a marker of glutamine addiction in medulloblastoma. *Genes Dev.* **2017**, *31*, 1738–1753. [CrossRef] [PubMed]

Review

Insulin Resistance and Endometrial Cancer: Emerging Role for microRNA

Iwona Sidorkiewicz [1,*], Maciej Jóźwik [2], Magdalena Niemira [1] and Adam Krętowski [1,3]

[1] Clinical Research Centre, Medical University of Białystok, M. Skłodowskiej-Curie 24a, 15-276 Białystok, Poland; magdalena.niemira@umb.edu.pl (M.N.); adamkretowski@wp.pl (A.K.)
[2] Department of Gynecology and Gynecologic Oncology, Medical University of Białystok, M. Skłodowskiej-Curie 24a, 15-276 Białystok, Poland; jozwikmc@interia.pl
[3] Department of Endocrinology, Diabetology and Internal Medicine, Medical University of Białystok, M. Skłodowskiej-Curie 24a, 15-276 Białystok, Poland
* Correspondence: iwona.sidorkiewicz@umb.edu.pl; Tel.: +48-85-831-8893

Received: 30 July 2020; Accepted: 7 September 2020; Published: 8 September 2020

Simple Summary: Insulin resistance is one of the risk factors of endometrial cancer. Hyperinsulinemia can trigger many physiological effects that drive carcinogenesis, which is also modulated by epigenetic dysregulation including miRNAs expression. Our working hypothesis was that there must be a more pronounced relationship between insulin resistance and alterations in miRNA profiles of endometrial cancer patients. Consequently, this work was undertaken to better clarify this assumption. Our careful literature search indicated that miRNA could represent a potential molecular link between the metabolic alterations related to insulin resistance and endometrial cancer. Additionally, by reporting the known relationships between miRNA and both insulin resistance and endometrial cancer, we highlighted their potential role as predictive factors of future endometrial cancer in insulin resistant patients.

Abstract: Endometrial cancer (EC) remains one of the most common cancers of the female reproductive system. Epidemiological and clinical data implicate insulin resistance (IR) and its accompanying hyperinsulinemia as key factors in the development of EC. MicroRNAs (miRNAs) are short molecules of non-coding endogenous RNA that function as post-transcriptional regulators. Accumulating evidence has shown that the miRNA expression pattern is also likely to be associated with EC risk factors. The aim of this work was the verification of the relationships between IR, EC, and miRNA, and, as based on the literature data, elucidation of miRNA's potential utility for EC prevention in IR patients. The pathways affected in IR relate to the insulin receptors, insulin-like growth factors and their receptors, insulin-like growth factor binding proteins, sex hormone-binding globulin, and estrogens. Herein, we present and discuss arguments for miRNAs as a plausible molecular link between IR and EC development. Specifically, our careful literature search indicated that dysregulation of at least 13 miRNAs has been ascribed to both conditions. We conclude that there is a reasonable possibility for miRNAs to become a predictive factor of future EC in IR patients.

Keywords: adipokines; endometrial cancer; estrogens; hyperinsulinemia; insulin; insulin resistance; insulin signaling; insulin-like growth factors; microRNA

---

1. Introduction

Endometrial cancer (EC) is the most common gynecological cancer in developed countries, with annual rates continuing to increase. It is estimated that more than 60,000 new EC diagnoses and 11,000 deaths from the disease occur in the United States alone every year [1]. However, the etiology of this disease is still not fully understood [1,2]. EC has been generally divided into two clinical categories. The first is classified as type I, which represents the vast majority (80–90%) of cases and is associated with a hyperplastic, low-grade, estrogen-related endometrium. It occurs primarily in obese pre-, peri-, and early postmenopausal women, and is associated with a good prognosis. Type II is characterized by a non-estrogenic, high-grade atrophic endometrium, which is also less well-differentiated. It occurs mostly in postmenopausal women and has a high risk of relapse and metastatic disease [3]. There are several histologic types of EC, and the most common endometrioid carcinoma tends to have a favorable prognosis. Other histotypes (such as serous or clear cell carcinoma) of EC are associated with a poor prognosis [4,5]. It was initially noted that type I EC generally presents an endometrioid morphology, whereas type II cancers are characterized by non-endometrioid histology, predominantly serous (Table 1). However, this classical distinction into two EC types has been challenged by long-term follow-up of patients with cancer of endometrioid histology and grades 2 and 3 of differentiation, whose survival turned out to be worse than expected [6,7]. In line with this, Setiawan et al. observed that the risk factor patterns of high-grade endometrioid tumors and type II tumors were similar [3]. Currently, only endometrioid grade 1 (well differentiated) EC is considered to be type I, with the remainder of EC cases being included into type II.

Table 1. Conventional comparison between type I and type II endometrial cancer [8–12].

Characteristic Feature	Type I	Type II
Frequency	~80% of cases	Up to 20% of cases
Estrogenic status	Estrogen-dependent	Estrogen-independent
Histology	Mostly endometrioid adenocarcinomas	Non-endometrioid carcinoma
Precursor lesion	Atypical hyperplasia	Endometrial intraepithelial carcinoma
Growth	Slow growth	Rapid growth
Risk factors	Imbalance between estrogen and progesterone exposures (such as the use of unopposed estrogen therapy)	Early age at menarche, low parity, tobacco smoking
Obesity	Often present	Often absent
Type 2 diabetes mellitus	Often present	Often absent
Estrogen and progesterone receptors	Usually ER (+), PR (+)	Usually ER (−), PR (−)
Prognosis	Usually good prognosis	Poor prognosis
PTEN Mutations	Yes	No
P53 Overexpression	No	Yes
Other Frequent Mutations	*ARID1A* *PIK3CA* *CTNNB1* *FGFR2*	*PPP2R1A* *FBXW7* *HER2*

ARID1A: AT-Rich Interaction Domain 1A; *CTNNB1*: Catenin Beta 1; ER: Estrogen Receptor; *FBXW7*: F-Box and WD Repeat Domain-containing 7; *FGFR2*: Fibroblast Growth Factor Receptor 2; *HER2*: Human Epidermal Growth Factor Receptor 2; *PIK3CA*: Phosphatidylinositol-4,5-Bisphosphate 3-Kinase Catalytic Subunit Alpha; *PPP2R1A*: Protein Phosphatase 2 Scaffold Subunit A alpha; PR: Progesterone Receptor; *PTEN*: Phosphatase and tensin homolog deleted on chromosome 10; P53: Protein 53.

In 2013, The Cancer Genome Atlas (or TCGA) Research Network classification for EC applied four molecular subgroups: DNA polymerase epsilon (POLE)-mutated (ultra-mutated), microsatellite-instable (MSI-high, hypermutated), copy-number-low/protein 53 (P53)-wild-type (CNL), and copy-number-high/P53-mutant (CNH) [11]. The POLE and MSI groups suggest better prognosis of

EC patients, and CNL and CNH groups are coupled to worse prognosis [13,14]. The implementation of this novel EC classification laid the grounds for the refined differential diagnosis of particular cancer subtypes based on molecular signatures and provided a precision approach for both research and clinical management [5]. A long-term follow-up of patients with these specific cancer subtypes is now mandatory and of utmost importance.

Although EC is generally considered to be hormone-sensitive, its development is widely considered to also be regulated by environmental and lifestyle factors. One of this cancer's risk factors is insulin resistance (IR), a prominent component in many metabolic disorders, including prediabetes, type 2 diabetes mellitus (T2DM), metabolic syndrome, and polycystic ovary syndrome (PCOS) [15–18]. IR is a condition of reduced sensitivity of insulin-responsive tissues to insulin, which leads to an increase in blood insulin and glucose concentrations. According to the International Diabetes Federation Diabetes Atlas, the global prevalence of T2DM developed from IR continues to be on the increase [19]. Hyperinsulinemia can trigger many physiological effects that drive carcinogenesis, as insulin is a major anabolic hormone that can stimulate cell proliferation [15]. Reduced receptor binding and decreased insulin receptor-mediated transduction lead to hyperinsulinemia which, in turn, triggers the deregulation of many metabolic pathways [20]. The exact molecular mechanisms linking IR and EC are still uncertain. However, the direct effect(s) on endometrial cells of insulin and insulin-like growth factors (IGFs), as well as of alterations in the mitogen-activated protein kinase (MAPK)/extracellular-signal-regulated kinase (ERK) and in the complex of phosphatidylinositol 3-kinase (P13K)–phosphatase and tensin homolog deleted on chromosome 10 (PTEN)–protein kinase B (Akt) signaling pathways, may play crucial roles [16,21,22].

Cancer development is also associated with epigenetic dysregulation, occurring at the earliest stage of cancer [23]. The most common epigenetic modifications are DNA methylation, histone methylation and acetylation, and the actions of non-coding RNAs, including microRNAs (miRNAs). All of them can regulate multiple genes and are involved in various important signaling pathways [24]. miRNAs belong to a class of highly conserved, sequence-specific, single-stranded, endogenous small non-coding RNAs, which bind to the 3′ end of the target mRNAs to induce their destabilization, degradation, and/or translation inhibition [25]. Deregulation of miRNA profiles has been implicated in a variety of cellular processes, including cancer development. Therefore, miRNAs have been drawing attention for their potential usefulness as diagnostic and/or prognostic markers [26,27].

To date, numerous studies have focused on the miRNAs' role in endometrial carcinogenesis or IR, albeit no possible reciprocal interactions of miRNA and IR on EC have been taken into account [28]. Our working hypothesis was that there must be a more pronounced relationship between IR and alterations in miRNA profiles of EC patients. Consequently, this work was undertaken to better clarify this assumption. Furthermore, we discuss the known relationships between miRNA and both clinical conditions.

2. Clinical Importance of the Association between Insulin Resistance and Dndometrial Cancer

Generally, IR is a principal pathophysiological process that relates not only to diabetes but also to prediabetes, as well as preclinical hyperinsulinemia and dysglycemia of varied degrees. IR has been defined as the resistance of target organs to the actions of insulin so that increased concentrations of this hormone are necessary to obtain a normal biological effect [29]. Accordingly, IR is the primary cause of T2DM and occurs years before its clinical manifestation [30]. This prediabetic state plays an important role in the development and progression of some types of cancers, including breast, prostate, colorectal, and endometrial neoplasia [31]. There is accumulating evidence that the risk factors for IR are also risk factors for EC, which strongly suggests that the development of IR and EC may be parallelly promoted at the same time. A meta-analysis conducted by Saed et al. demonstrated that diabetes increases the risk of EC by 72% [32]. Another work, a meta-analysis of 16 studies (3 cohort and 13 case-control studies), found that diabetes is associated with a 2.1-fold increase in the relative risk for EC [33]. Notably, a higher prevalence of EC was demonstrated in non-diabetic women with IR [34].

Decreased serum adiponectin (a polypeptide hormone increasing the cell's insulin sensitivity and a surrogate marker for IR) concentration was found to be independently and inversely correlated with EC occurrence [35,36]. It has also been established that the EC risk increases quite shortly following the diagnosis of IR and diabetes; that is, approximately past 6 months after their detection [37]. Elevated levels of insulin in prediabetic and diabetic patients seem to affect their cancer risk rather quickly [38]. Similarly, epidemiological evidence shows that the presence of accompanying diseases substantially influences EC risk estimations [17]. For instance, the relationship between diabetes and EC incidence can be largely promoted by increased body weight [31]. In their pooled analysis of cohort studies, Stocks et al. found direct linear relationships of body mass index (BMI), blood pressure, blood glucose, triglycerides, and total cholesterol concentrations with EC risk [39]. This finding is particularly worrying in the present era of widespread overweight and obesity.

3. Insulin Resistance as a Driving Force for Endometrial Cancer

Over past decades, hyperinsulinemia and IR have been implicated as playing a major role in diabetes-promoted cancers. Multiple studies were able to demonstrate a direct association between IR and the incidence of EC with several biological mechanisms as a result of their common regulation by molecular factors (such as mediators of inflammation and adipokines) [40–42]. Figure 1 presents a model of links between metabolic alterations in the development of this malignancy, highlighting the roles of changes in the insulin and IGF system and mediators of inflammation.

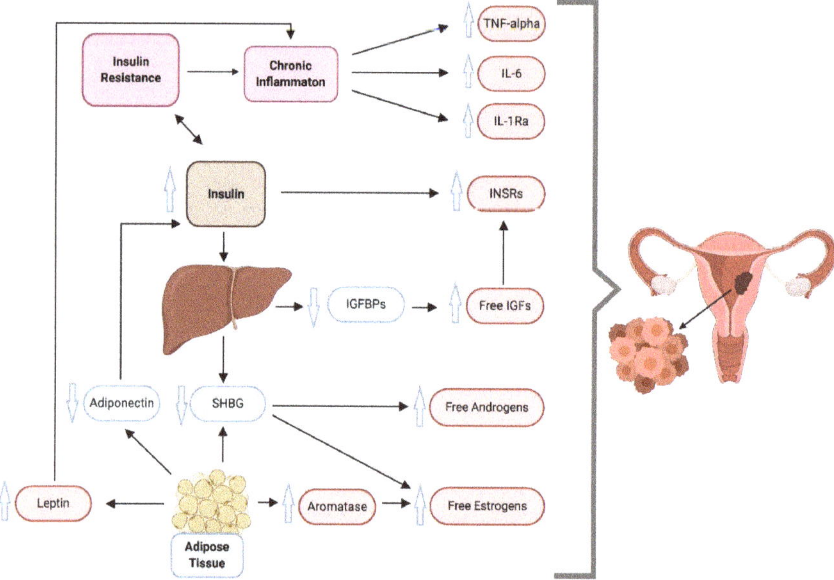

Figure 1. A proposed multidimensional model of endometrial cancer development, which suggests insulin resistance, inflammation, and overweight/obesity as driving forces behind cancer. IGFs, insulin-like growth factors; IGFBPs, insulin-like growth factor binding proteins; IL-1Ra, interleukin 1 receptor antagonist; IL-6, interleukin 6; INSR, insulin receptor; SHBG, sex hormone-binding globulin; TNF-alpha, tumor necrosis factor-alpha. Compiled from References [35,43–47].

3.1. Insulin Receptor

Molecular signaling downstream of insulin receptor (INSR) is tightly regulated by a large number of factors. This control system supervises energy homeostasis in peripheral target tissues for insulin. Both insulin and IGF1 activate a specific tyrosine kinase and the two main pathways of insulin signaling are the PI3K/Akt and the MAPK/ERK pathways. These two metabolic cascades contain several points of regulation, crosstalk with other signaling pathways, and control proliferation, differentiation, and survival at the cellular level [48,49].

Insulin signal transduction occurs through two INSR isoforms resulting from transcriptional alternative splicing: INSR-A and INSR-B, differing by the absence (INSR-A) or presence (INSR-B) of exon 11. INSR-A is the principal receptor during fetal development, recognizes both insulin and IGFs, demonstrates a greater affinity for IGF2 than IGF1, and is responsible for intracellular signaling that results in mitogenic responses. INSR-B, expressed in mature insulin-sensitive tissues, is quite insulin-specific and primarily involved in glucose homeostasis [50]. The differences in the effects exerted by INSR-A and INSR-B could be due to their varying abilities to bind IGF2 [51]. INSR-A overexpression was found in many cancer cells and tissues, suggesting that INSR-A-mediated signaling pathways may contribute to cancer pathogenesis [52]. Wang et al. demonstrated that the total *INSR* and *INSR-A* mRNA levels and the ratio of *INSR-A*/total *INSR* were significantly higher in EC than in control endometrium [53]. However, no comparisons between clinical types I and II of EC, or between particular EC histotypes, were made, making the interpretation of these results somewhat difficult. On the contrary, Flannery et al. found that *INSR-B* expression was increased in non-diabetic patients both in complex endometrial hyperplasia and EC, relative to normal tissue [54].

3.2. Insulin-Like Growth Factors and Their Receptors

Combined, insulin, IGF1, and IGF2 critically control many aspects of metabolism, growth, and survival. IGFs are predominantly produced in the liver by hepatocytes in response to growth hormone (GH) stimulation [55]. IGF1 displays significant amino acid sequence homology with insulin and enhances insulin sensitivity [56]. To date, IGF2 actions have been insufficiently characterized; however, some relevant roles have been determined for fetal development and cerebral protection [51]. Both ligands IGF1 and IGF2 activate the IGF1 receptor (IGF1R), making it their common receptor. The IGF system plays a central role in human carcinogenesis. Interestingly, it has been hypothesized that IGF2 is more closely linked to the etiology of EC than IGF1 [57].

At the pathophysiological level, insulin can also bind to IGF1R, which is a cell-surface tyrosine kinase receptor coupled to several intracellular secondary messenger pathways, including the PI3K/Akt signaling cascade. IGF1R plays a pivotal role in cell survival by regulating somatic growth, development, and metabolism, as demonstrated by using *IGF1R* knock-out mice that displayed severe growth deficiency, lethal neonatal lung hypoplasia, and muscle hypoplasia [58]. Although INSR and IGF1R are highly homologous and are coupled to similar intracellular pathways, insulin and IGFs stimulate distinct and specific functions, such as glucose metabolism for insulin and cell growth and proliferation for IGFs [59]. Additionally, the functional specificity of insulin/INSR signaling can be affected by: (1) crosstalk between INSR-A and IGF1 because of the abundant synthesis of INSR-A in tissues and its increased binding affinity for IGF1, (2) enhanced formation of an INSR-A/IGF1R hybrid receptor, and (3) autocrine and/or paracrine IGF production [60]. There is convincing evidence for a direct effect of insulin and IGF1 on EC cells, with the activation of the INSR resulting in both increased cell proliferation and inhibition of apoptosis [55]. From Dai et al., although serum concentrations of IGF1 and IGF2, as well as the degree of activation of IGF1R in endometrial cells did not differ between diabetic patients with or without EC, both the degree of activation of IGF2R and of PI3K were significantly higher in endometrial cells in T2DM patients with EC [61]. These same authors suggested that increased IGF2R protein expression in endometrial cells in T2DM patients could increase PI3K/cyclin D1 (CCND1)-dependent cell growth through the loss of competitive binding of IGF2 to IGF1R, a possible explanation for the higher risk of developing EC in T2DM. Moreover, this study indicated that IGF1 and IGF2 compete for binding to IGF1R, whereas binding of IGF2 to IGF1R may cause alternative phosphorylation of IGF1R with the resultant suppression of downstream PI3K and CCND1 signaling cascades [61]. Somewhat similar conclusions were obtained by Petridou et al., who suggested a more pronounced role of IGF2 than IGF1 in the etiology of EC [57]. Of clinical relevance, Price et al. noted that increased IGF1R expression is linked to higher BMI and better overall survival and disease-free survival in EC [62]. Gunter et al. observed that free IGF1 levels were inversely associated with the incidence of EC [63]. Of interest, Roy et al. suggested the existence of two different mechanisms that activate IGF1R in EC: ligand-dependent in type I and ligand-independent in type II [64]. These authors found that *IGF1* mRNA expression was increased in type I compared with type II.

Further interesting findings were reported by Ding et al., who found that higher protein expression of IGF1, IGF1R, and INSR in colorectal cancer was associated with a history of diabetes, suggesting that IGF1/INSR signaling may play an important role in the development of this cancer in diabetic patients [65]. Unfortunately, many clinical trials with anti-IGF1R showed only limited responses in small proportions of cancer patients. Despite promising preclinical data, anti-IGF1R/INSR-targeted therapies lacked overall efficacy and the multitude of side effects led to their discontinuation [50,66,67]. In contrast, metformin (an oral antidiabetic drug of the biguanide family used for the treatment of T2DM) is known to interact with the IGF pathway, induces apoptosis, and inhibits proliferation and migration of EC cells [68]. Three separate systematic reviews and meta-analyses confirmed a beneficial role of metformin in improving overall survival and progression-free survival in EC [69–71]. In an in vitro study, metformin even inhibited proliferation and migration of endometrial serous carcinoma cell lines. The authors suggested that this drug could be a novel and attractive therapeutic approach for the treatment of this highly aggressive variant of EC [68].

3.3. *Insulin-Like Growth Factor Binding Proteins*

Although IGF1 is structurally related to insulin, unlike insulin, it circulates in the blood bound to specific carrier proteins, called IGF binding proteins (IGFBPs), with variable affinity. IGFBPs tightly regulate IGF1 availability by increasing its half-life, usually by forming a tertiary complex that blocks IGF1 from binding to IGF1R. Six IGFBPs (labeled IGFBP-1 to -6) have been identified so far [72]. Hyperinsulinemia has been shown to increase hepatic production and bioavailability of IGF1, in part by inhibiting hepatic production of IGFBP-1 and -2. This surplus IGF1 may excessively activate IGF1R, INSR/IGF1R, and proliferative and anti-apoptotic signaling in both premalignant and malignant tissues [45]. Insulin-sensitizing, blood pressure-lowering, and antiatherosclerotic properties of IGFBP 1 have been demonstrated, raising the possibility that increasing IGFBP-1 levels may be a therapeutic option to protect individuals from IR, arterial hypertension, and atherosclerosis [73]. However, high IGFBP-1 concentrations seem to be associated with EC risk in older women and women with an elevated BMI [74]. One study, by Weiderpass et al., found an increased risk for EC based on serum levels of IGFBP-1 solely in women who had ever used hormonal replacement therapy [75].

Recent evidence suggests beneficial effects of IGFBP-2 on systemic metabolism by indirect interacting with IGF1 signaling, including inhibition of adipogenesis and enhanced long-term insulin sensitivity [76,77]. Besides binding to IGFs, IGFBP-2 interacts with cellular components and exerts other key functions within the nucleus, directly or indirectly promoting transcriptional activation of specific genes. IGFBP-2 activities, both IGF-dependent and IGF-independent, contribute to the protein's functional roles in growth, development, metabolism, and malignancy [78]. The overexpression of IGFBP-2 has been shown to correlate with tumor progression in a number of cancers, including ovarian, lung, and pancreatic cancer [79–82]. However, the mechanisms by which IGFBP-2 contributes to the progression of cancer are still unclear [83].

As for IGFBP-3, this protein inhibits adipocyte differentiation and impacts the peroxisome proliferator-activated receptor-gamma (PPARγ) system, suggesting a role for IGFBP-3 in the pathogenesis of obesity and IR. Apart from physiological IGF-dependent effects, this carrier protein has been demonstrated to regulate cell proliferation independently of binding to IGFs [84]. Mochizuki et al. found that the anti-proliferative and proapoptotic activities of IGFBP-3 are IGF-independent and attenuate epidermal growth factor (EGF)-induced EC cell proliferation. However, the exact details of action by which IGFBP-3 inhibits the EGF-mediated survival pathway require elucidation [85]. Recently, the binding of IGFBP-3 to a variety of growth factors was shown to improve the efficacy of anti-cancer precision therapy, counteract numerous mechanisms of tumor resistance, and combat tumor heterogeneity [86].

3.4. Estrogens

Estrogens are important participants in the metabolic regulation, playing a mitogenic role in the normal endometrium [87]. The effects of estrogen are mostly mediated by three receptors: two cytosolic estrogen receptors (ER), α (ERα) and β (ERβ), and transmembrane G protein-coupled estrogen receptor 1 (or GPER). ERs can carry out both genomic (transcription and gene expression regulation) and nongenomic (regulatory protein modifications) signal transduction. Estradiol (E2), the principal biologically active form of estrogen, controls insulin activity directly via actions on insulin-sensitive tissues or indirectly by regulating factors responsible for oxidative stress, with both outcomes contributing to IR [88]. Of note, some data suggest opposed effects of ERα and ERβ on glucose tolerance, and that ERβ ligands exert diabetogenic actions [89].

Estrogen signaling causes proliferation of EC [90]. Extensive crosstalk between estrogen signaling and the insulin/IGF axis was recently thoroughly discussed [91]. Research on ER-positive breast cancer (cell line MCF-7 xenografts) demonstrated that tamoxifen, a selective ER modulator, effectively inhibits classical ER-dependent transcription, including the transcription of *IGF1R* gene product [92]. In vivo studies have shown that E2 improves insulin sensitivity and glucose tolerance via activation of ERα/PI3K/Akt signaling [93,94]. However, E2 has been demonstrated to inhibit the in vitro binding of insulin to INSR by binding to both insulin and its receptor instead, an observation strongly suggesting the ability of E2 to induce IR either directly or indirectly [95]. Attention should be paid to tissue-specific roles of E2 [96]. Estrogens mediate the expression of IGFs in the uterus, but IGFBPs also interfere with this process [55,97]. Merritt et al. observed lower expression of *IGF1*, but higher expression of *IGFBP-1* and *IGFBP-3*, coupled with higher protein expression for ER, INSR, and IGF1R in postmenopausal endometrium as compared to premenopausal proliferative phase endometrium [98].

Several EC risk factors provide strong support for the hypothesis of the causative role of unopposed estrogen, stating that EC risks are increased in women with high plasma estrogens and/or low progesterone, so that estrogenic effects are not sufficiently counterbalanced by the latter [99–101]. Insulin and IGF1 have been shown to stimulate ovarian steroid synthesis, resulting in cellular proliferation and inhibition of apoptosis in breast epithelium and endometrium [102]. Estrogens, both internal and from external provision, play a significant role in EC development. The majority of type I EC cases express ER, and higher ER expression has been associated with better clinical outcomes [8]. A recent analysis by Tian et al. underlined that insulin and estrogens could exert combined or even synergistic effects on the progression of type I EC [103]. Estrogens are thought to trigger proliferation and growth in cancer cells through the activation of ERα and the subsequent activation of PI3K and MAPK pathways [90]. Furthermore, an in vitro study showed that insulin directly stimulates aromatase activity in both endometrial glands and stroma, which strongly suggests that hyperinsulinemia caused by IR predisposes to EC by enhancing endogenous endometrial estrogen production [104]. Work by Galvão Wolff et al. indicated that IR stimulates endometrial expression of ERs and progesterone receptors (PRs), thereby contributing to the increased occurrence of endometrial proliferative lesions [105].

3.5. Sex Hormone-Binding Globulin

The activity of estrogens depends on their bioavailability, which is primarily determined by sex hormone-binding globulin (SHBG). The synthesis of SHBG occurs in the liver and is increased by sex steroids (mostly estrogens) and thyroxine, whereas insulin is a known important inhibitor of its production. Winters et al. found lower serum SHBG and lower hepatic *SHBG* expression with increasing IR, together with a weak association between fasting insulin concentration and *SHBG* mRNA expression [106]. Not only insulin but also IGF1 inhibits the hepatic production of SHBG [102].

In premenopausal women, hyperinsulinemia promotes the stimulation of ovarian androgen synthesis and decreases hepatic production of SHBG. This leads to increased circulating levels of free androgens. In contrast, in postmenopausal women, adipose tissue is the main source of estrogens through the aromatization of androgens. Obesity and hyperinsulinemia, if present, further lead to increased aromatization of androgens and decreased production of SHBG, the results of which are increased levels of bioavailable estrogen [107]. Excessive estrogens promote the development of EC, as described by the unopposed estrogen hypothesis.

3.6. Adipokines

The adipose tissue secretes a wide variety of bioactive molecules, including adipokines and hormones, such as adiponectin, leptin, resistin, visfatin, and chemerin, as well as proinflammatory cytokines, such as tumor necrosis factor-α (TNF-α) [108]. Among adipokines, adiponectin proved to be particularly important. Decreased adiponectin production exerts a key role in the pathogenesis of obesity-associated disorders: arterial hypertension, metabolic syndrome, atherosclerosis, and cancer [109–111]. Two separate studies demonstrated a significant relationship between high circulating adiponectin levels and reduced EC risk, an observation largely independent of other obesity-related risk factors [112,113].

Leptin has contrasting biological functions to adiponectin: it decreases tissue sensitivity to insulin and increases plasma insulin concentration. Hyperinsulinemia and obesity are therefore linked with high leptin and low adiponectin levels [114]. The role of leptin in IR is still not fully clarified, but solid evidence indicates that leptin is a major metabolic regulator of circulating IGFBP-2 [115,116]. A relationship between serum leptin and insulin concentrations has been confirmed, regardless of body fatness [117–119]. Since obesity and adipokines are independent risk factors for EC, this notion supports the roles of two distinct mechanisms involved in endometrial carcinogenesis: excess estrogen and IR [120,121]. Resistance to leptin is considered a hallmark of obesity and has been shown to lead to hepatic IR [122]. Importantly, leptin plays a proinflammatory role, contributing to the generation and maintenance of low-grade inflammation, recently also linked to EC [42]. However, literature data regarding circulating levels of adiponectin and leptin in EC report conflicting results [111,123,124]. Increased circulating adiponectin and adiponectin/leptin ratio and decreased leptin concentration were shown to be associated with reduced risk for EC [125]. On the other hand, Ma et al. observed increased leptin and decreased adiponectin levels in EC [126]. Unfortunately, adiponectin and leptin concentrations and insulin pathway receptor expressions were not found useful for defining molecular subtypes of EC [127]. Moreover, molecular links between adipokines and cancer cells are complex and as yet, are not fully understood [128]. It has been hypothesized that circulating levels of adiponectin and leptin, together with insulin pathway molecules, exert oncogenic effects on endometrial tissue not only through their impact on the expression of tumor cell receptors but also by the activation of multiple epigenetic pathways within neoplastic cells and their microenvironment [129,130].

Other adipokines/cytokines of adipose origin such as visfatin, vaspin, and omentin display proinflammatory properties and affect insulin sensitivity and secretion. Recent research indicates that serum visfatin concentration is elevated in patients with EC and that combined serum visfatin and resistin levels could be used to predict the risk of advanced stages of EC [131]. Hlavna et al. showed increased circulating levels of resistin in EC patients compared with control subjects [132]. Unraveling the pathophysiological roles of adipokines in IR and EC should be prioritized in future research [133].

4. miRNAs in Both Insulin Resistance and Endometrial Cancer

Post-transcriptional regulation by miRNAs is of interest as a mechanism to silence gene expression [134]. Aberrant expression of the miRNA profile plays a key role in a wide variety of physiological processes, including cell proliferation, apoptosis, and tissue differentiation [135]. Yet, deregulation in miRNA biogenesis and function have been shown to modulate many fundamental signaling pathways, including insulin synthesis, secretion, and signal transduction, and therefore, specific miRNA patterns are likely to play a role in the development of IR and related metabolic complications [136]. Importantly, miRNA-mediated insulin signaling modulation is tissue- and cell-specific, with distinct miRNAs modulating components of the insulin transduction pathway only in some tissues or cells. The basis for IR is multifactorial and includes obesity, inflammation, mitochondrial dysfunction, endoplasmic reticulum stress, oxidative stress, lipotoxicity/hyperlipidemia, genetic background, and hypoxia. These factors contribute quite differently to the disruption of insulin signaling [137].

Various conditions are caused by dysregulation of gene networks due to changes in miRNA expression, and the association between miRNAs and cancer is currently under vivid investigation [138]. miRNAs regulate cell metabolic processes either directly by targeting key molecules of metabolic pathways (transporters and enzymes, including kinases), or indirectly by modulating the expression of important transcription factors [139].

On the one hand, EC molecular subtypes have been shown to demonstrate distinct miRNA signatures. These miRNA signatures are reduced, and particular levels of depletion are characteristic for particular EC subtypes [140]. In summary, many miRNAs, either circulating or of tissue origin, have been found dysregulated in EC. Table 2 presents their comprehensive list.

Table 2. MicroRNAs (miRNAs) found as dysregulated in endometrial cancer (EC).

miRNA	Form of Dysregulation	Studied Specimen	Reference Specimen	Reference
hsa-miR-1307-3p; hsa-miR-183-3p; hsa-miR-183-5p; hsa-miR-200b-3p; hsa-miR-429	up	EC	Normal endometrium	[141]
hsa-miR-152-3p; hsa-miR-24-1-5p; hsa-miR-374b-5p; hsa-miR-542-3p	down	EC	Normal endometrium	[141]
miR-650; miR-168; miR-572; miR-200a; miR-182; miR-622; miR-34a; miR-205	up	Endometrioid EC	Benign endometrium	[142]
miR-411; miR-487b	down	Endometrioid EC	Benign endometrium	[142]
let-7c-5p; miR-125b-5p; miR-23b-3p; miR-99a-5p	down	Endometrioid EC	Non-neoplastic endometrium	[143]
let-7g-5p; miR-195-5p; miR-34a-5p; miR-497-5p	down	Endometrioid EC (grade 1+2) and serous EC	Non-neoplastic endometrium	[143]
miR-205; miR-182; miR-325; miR-183; miR-203; miR-210; miR-223; miR-194; miR-95; miR-151; miR-200a; miR-301; miR-141; miR-215; miR-103; miR-106a; miR-191; miR-184; miR-326; miR-34a; miR-200c; miR-23a	up	Endometrioid EC	Normal endometrium	[144]
miR-1; miR-101; miR-10b*; miR-127–3p; miR-132*; miR-133a; miR-133b; miR-136; miR-136*; miR-139-5p; miR-140-3p; miR-140-5p; miR-142-3p; miR-142-5p; miR-143; miR-143*; miR-145; miR-145*; miR-152; miR-195; miR-196b; miR-199a-5p; miR-199b-3p; miR-199b-5p; miR-214; miR-214*; miR-23b; miR-24-1*; miR-27b; miR-299-3p; miR-299-5p; miR-29b; miR-33a; miR-337-5p; miR-34b; miR-34b*; miR-34c-5p; miR-376a; miR-376c; miR-377; miR-379; miR-381; miR-410; miR-411; miR-424; miR-450a; miR-455-3p; miR-455-5p; miR-497; miR-503; miR-542-3p; miR-542-5p; miR-654-3p; miR-873	down	Serous EC	Normal endometrium	[145]

Table 2. *Cont.*

miRNA	Form of Dysregulation	Studied Specimen	Reference Specimen	Reference
miR-222; miR-223; miR-186; miR-204	up	Serum of endometrioid EC patients	Serum of healthy controls	[146]
miR-186; miR-222; miR-223	up	Serum of EC patients	Serum of healthy controls	[147]
miR-204	down	Serum of EC patients	Serum of healthy controls	[147]
let-7g*; miR-181c*; miR-516a-3p; miR-9; miR-203; miR-375; miR-652; miR-146a; miR-9*; miR-210; miR-32; miR-148a; miR-425; miR-592; miR-21; miR-7-1*; miR-107	up	Endometrioid EC	Normal endometrium	[148]
miR-502-3p; miR-130a; miR-214; miR-218; miR-99a; miR-410; miR-100; miR-199a-3p; miR-424; miR-199a-5p; miR-214*; miR-99a*; let-7c; miR-212, miR-130a*; miR-495; miR-100*; miR-125b*; miR-218-2*; miR-502-5p; miR-532-5p	down	Endometrioid EC	Normal endometrium	[148]
miR-31; miR-995-5p; miR-490-3p; miR-644; miR-522; miR-519d; miR-98; miR-425; miR-518e; miR-155	up	Serous EC	Normal endometrium	[148]
miR-370; miR-423-5p	down	Serous EC	Normal endometrium	[148]
miR-516; let-7a; miR-424; miR-496; miR-409; miR-451; miR-431; miR-516; miR-503; miR-369; miR-032; miR-032b; miR-425; miR-181c; miR-19b; miR-009; miR-205; miR-423; miR-223; miR-183; miR-146; miR-200c	up	Endometrioid EC	Normal endometrium	[149]

Nomenclature was given as provided by cited references. EC: endometrial cancer. *: the less expressed strand.

On the other hand, at least several miRNAs are known to be involved in the pathogenesis of cancer. As for endometrial neoplasia, a 4-miRNA signature (miR-4758, miR-876, miR-142, miR-190b) has been established as an independent prognostic factor for overall survival in EC patients (area under the curve (AUC) of receiver operating characteristic (ROC) curve was 0.7 at 5-year overall survival) [150]. By contrast, based on their systematic review, Donkers et al. proposed miR-205, the whole miR-200 family, miR-135b, miR-182, miR-183, and miR-223 as promising diagnostic biomarkers in EC [151]. Such studies were performed in the hope that the expression pattern of miRNA would become an early diagnostic and prognostic biomarker, whilst particular miRNAs could be identified as novel therapeutic targets.

Although the pathophysiology that underlies the association of IR with EC requires further investigation, miRNAs may be a missing link. Of interest, our careful literature search indicated that dysregulation of at least 13 miRNAs is actually shared by or has been ascribed to both IR and EC. Table 3 substantiates these findings.

Table 3. miRNAs found as dysregulated in both insulin resistance (IR) and EC.

miRNA	Form of Dysregulation	Target Genes Involved in IR	Target Genes Involved in EC	Reference
let-7	down	*IGF1R, IGF2BP-2, INSR, IRS-1, IRS-2*	*HMGA2, c-myc, JAK*, Aurora B kinase, *STAT3*	[152–154]
miR-9	up	*OC-2, SIRT1*	*FOXO1*	[155,156]
miR-29a	up	*PPARδ*	*TPX2*	[157,158]
miR-29b	up	*CAV2, INSIG1, PIK3R1*	*PTEN*	[159–161]

Table 3. Cont.

miRNA	Form of Dysregulation	Target Genes Involved in IR	Target Genes Involved in EC	Reference
miR-29c	down	*HK2, GLUT1, IRS-1*	*COL4A1*	[162,163]
miR-103	up	*CAV1*	*TIMP-3*	[164,165]
miR-107	up	*CAV1*	*ERα*	[165,166]
miR-126	down	*IRS-1*	*IRS-1*	[167,168]
miR-141	up	*FOXA2*	*PLA2R*	[169,170]
miR-200	up	*ZEB1*	*PTEN*	[171,172]
miR-221	up	*SIRT1*	*SLUG*	[173,174]
miR-222	up	*IRS-1*	*ERα*	[175,176]
miR-320a	up	*PI3Kp85*	*IGR1R*	[177,178]

CAV1, caveolin-1; *CAV2*, caveolin-2; *ERα*, estrogen receptor alpha; *FOXA2*, forkhead box A2; *FOXO1*, forkhead transcription factor 1; *GLUT1*, glucose transporter 1; *HK2*, hexokinase 2; *HMGA2*, high mobility group AT-hook 2; *IGFBP-2*, insulin-like growth factor binding protein 2; *IGF1R*, insulin-like growth factor receptor; *INSIG1*, Insulin-Induced Gene 1; *INSR*, insulin receptor; *IRS-1*, Insulin receptor substrate 1; *IRS-2*, Insulin receptor substrate 2; *JAK*, Janus protein tyrosine kinase; *OC-2*, Transcription Factor Onecut-2; *PI3Kp85*, phosphoinositide 3-kinase regulatory subunit p85; *PIK3R1*, phosphatidylinositol 3-kinase regulatory subunit; *PLA2R*, phospholipase A2 receptor; *PTEN*, phosphatase and tensin homolog deleted on chromosome 10; *SIRT1*, Sirtuin 1; *SLUG*, Zinc Finger Protein SNAI2 (Snail Family Transcriptional Repressor 2); *STAT3*, signal transducer and activator of transcription 3; *TIMP-3*, tissue inhibitor of metalloproteinase 3; *ZEB1*, zinc finger E-box-binding homeobox 1.

The let-7 miRNA family, whose decreased expression in EC tissues has been demonstrated, is also involved in the development of IR [179]. A let-7 loss contributes to carcinogenesis via an increase in its target oncogenes (such as high-mobility group AT-hook 2 (*HMGA2*), *c-Myc*, Janus protein tyrosine kinase (*JAK*), Aurora B kinase, and signal transducer and activator of transcription 3 (*STAT3*)) and stemness factors [153,154,180]. However, the overexpression of let-7 in mouse skeletal muscles is related to the impairment of glucose tolerance and enhancement of IR [181]. The lin-28/let-7 axis regulates the insulin/PI3K/mammalian target of rapamycin (*mTOR*) pathway via multiple targets, such as *IGF1R*, *INSR*, and insulin receptor substrates 1 and 2 (*IRS-1*, *IRS-2*), thereby directly regulating glucose metabolism [152,182]. Understanding the tissue-specific regulation of let-7 may fill the current data gap and result in its potential use as a therapeutic for an array of metabolic diseases [183].

Increased expression of miR-9 in EC versus normal endometrial tissue has been shown [184–186]. Myatt et al. demonstrated that miR-9 was increased in EC tissue but lower in HEC-1B (type II EC cell line) compared with Ishikawa cells (type I EC line). Moreover, miR-9 expression was inversely correlated with forkhead transcription factor 1 (*FOXO1*) expression both in EC in vivo and in Ishikawa cells [155]. Two studies reported that the gene for sirtuin 1 (*SIRT1*), together with *FOXO1* and the gene for sterol regulatory element-binding protein 1 (*SREBP-1*) act as a pathway involved in tumorigenesis suppression and play a role in the development of progestin resistance in EC cells [187,188]. Since *FOXO1* and *SREBP-1* are targets of insulin, their role in IR can be hypothesized. miR-9 can regulate insulin secretion by inhibiting transcription factor onecut-2 (*OC-2*) and *SIRT1* in vivo in pancreatic β-islets [156]. In turn, this decrease in *OC-2* in insulin-secreting cells results in an increase in the expression of its target gene, granuphilin, a key player in insulin secretion and known to negatively regulate insulin exocytosis [189]. mir-9 expression was first thought to be restricted to the brain and pancreatic islets, yet recent studies emphasize the need to focus on its precise functional role in cancer [190].

Members of the miR-29 family (i.e., miR-29a, b, and c) have been shown to be involved in the EC development [191]. Specifically, miR-29b was found to play important roles in proliferation and progression in EC cells by direct regulation of *PTEN*, whose involvement in inhibiting cell migration, invasion, and cytoskeleton rearrangement has been proven [161]. Chen et al. demonstrated that miR-29b contributes to EC angiogenesis by targeting both MAPK/ERK and PI3K/Akt signaling pathways [160]. Significant downregulation of miR-29c was observed to occur in EC, possibly resulting in increased cell proliferation and collagen type IV alpha 1 synthesis [163]. The miR-29 family is involved in IR, as its in vivo suppression in adult mice led to a significant reduction of fasting blood glucose concentration and improvement in insulin sensitivity [192]. A study by Massart et al. reported that miR-29a and miR-29c expression are increased in skeletal muscle from patients with T2DM, playing a pivotal role in glucose and fatty acid metabolism [162]. In line with this, silencing miR-29a resulted in decreased glucose transport and altered lipid metabolism in myotube cells, indicating the involvement of this miRNA in IR by targeting peroxisome proliferator-activated receptor δ (*PPARδ*) in skeletal muscle [157]. The various mechanisms of action by the miR-29 family suggest its dichotomous role as a tumor suppressor and oncogene based on tissue specificity [193,194].

Other miRNAs involved in both IR and EC are miR-103 and miR-107. An in vitro study by Du et al. showed that miR-103 overexpression significantly promoted EC cell proliferation, whereas downregulation significantly suppressed EC cell proliferation [195]. miR-103 has been demonstrated to directly target tissue inhibitor of metalloproteinase 3, leading to an imbalance between matrix metalloproteinases and their tissue inhibitors, well known to play a critical role in tumor development [164]. Hepatic miR-103 overexpression in obese mice promotes glucose intolerance and IR [165]. In turn, miR-107-5p promotes EC progression and invasion by targeting *ERα* [166]. miR-103/107 inactivation leads to increased expression of caveolin-1 (*CAV1*) in adipocytes, thereby reducing downstream insulin signaling and decreasing adipocyte size [165]. Interestingly, miR-103 and miR-107 target RNase III-like enzyme named DICER, which is a key component of the miRNA processing machinery, resulting in global miRNA inhibition. However, these inhibiting effects may also be mediated by other miRNAs [140].

miR-126 has been reported to directly target *IRS-1* in SK-Hep1 hepatocytes [167]. This miRNA was found to be frequently downregulated in EC. Moreover, *IRS-1* is involved in miR-126-mediated EC cell migration and invasion, thus raising a possibility of miR-126-based molecular targeted therapy for EC [168].

High expression of the miRNA-200 family (including miR-141, miR-200a, miR-200b, miR-200c, and miR-429) has been demonstrated in endometrioid EC compared with normal endometrium, suggestive of their substantial role in cancer growth [196]. Importantly, miR-200 has been implicated in IR by inducing pancreatic β-cell proinflammatory state and damage, and by downmodulating *IRS-2* [197,198]. An in vitro study by Lu et al. showed that the expression of the miR-200 family in Ishikawa cells (type I EC cell line) was increased when compared with HEC-1B cells (type II EC line). There is convincing evidence that dynamic expression changes during transition from the normal to cancerous state reflect a link between ovarian steroids and the miRNA expression pattern [199]. Zinc finger E-box binding homeobox 1 (*ZEB1*) is a target gene of miR-200. The product of this gene is involved in epithelial–mesenchymal transition (EMT), which contributes to cancer invasion, metastasis, recurrence, and therapeutic resistance [200–202]. Therefore, there should be a role for miR-200 in EMT. Upregulation of miR-141 has been demonstrated in IR, as well as in EC [169,170,196]. The increased expression of miR-141 resulted in impaired glucose-stimulated insulin secretion and pancreatic β-cell proliferation. In addition, a positive correlation was observed in diabetic patients between miR-141 expression and blood glucose concentration. Forkhead box A2 (*FOXA2*) was identified as a direct miR-141 target gene [170]. Separate work demonstrated that *FOXA2* must be important in tumorigenesis based on its role in the inhibition of EMT in cancer [203,204].

Interestingly, miR-221 and miR-222 are also related to both IR and EC. Ramon et al. showed a significant downregulation of miR-221/222 in endometrioid EC in comparison with control endometrium. miR-221 and -222 were negatively correlated with the vascular endothelial growth factor A (VEGF-A) protein level, an observation suggesting their involvement in the mechanism of increased VEGF-A ratios observed in EC. miR-222-3p expression was found lower in ERα-positive EC tissues as compared with ERα-negative ones [205]. Consequently, the level of miR-222-3p expression was lower in tumors of lower grades and earlier stages. Further, regulation of *ERα* expression by miR-222-3p was confirmed in RL95-2 EC cells [176]. Overexpression of miR-221 caused inflammation and IR in differentiated 3T3-L1 adipocytes through the suppression of *SIRT1* [173]. An in vitro study on preadipocytes demonstrated that leptin and TNF-α downregulate miR-221, which inversely affects the adiponectin receptor 1 (*ADIPOR1*) and transcription factor v-ets erythroblastosis virus E26 oncogene homolog 1 (*ETS1*) expression. Adiponectin signaling promotes insulin sensitivity, and *ETS1* is known to regulate the expression of cytokines, chemokines, and other genes involved in angiogenesis [206]. The association of miR-221 in the complex interplay between *ERα*, *PR*, hypoxia-inducible factor 1-alpha (*HIF1-α*), and zinc finger protein SNAI2 (snail family transcriptional repressor 2) (*SLUG*) has been demonstrated and ascribed to EMT in endometrioid EC [174].

A recent communication reported on downregulated expression of miR-320a in EC [177]. From Abbas et al., miR-320a induces proliferation inhibition in EC cells, *IGF1R* is a direct functional mediator for miR-320a, and IGF1R is a critical negative regulator of insulin sensitivity in the endothelium [207]. Additionally, a study by Ling et al. showed that miR-320 increases insulin sensitivity of insulin-resistant 3T3-L1 adipocytes [178]. miR-320 may inhibit insulin/PI3K signaling in adipocytes, leading to IR; thus, anti-miR-320 oligo has been proposed as a potentially new therapeutic strategy to control IR [178].

Another aspect of the EC relationship with IR is the pleiotropic function of adipokines. Adipokine-regulated miRNAs can act as either oncogenic or anti-tumoral factors [208]. Moreover, adipose tissue is a major source of circulating miRNAs and they constitute a novel class of adipokines that can act as regulators of metabolism in tissues other than fat [209,210]. The discovery of cell signaling mechanisms followed by the appreciation of a wide network of miRNA-target genes' expression patterns has been crucial to identify the adipokine-regulated miRNAs in the development of EC. However, the data from human patients are limited and large in vivo studies are needed [211].

It has been hypothesized that the role of miRNAs in the metabolic crosstalk is not only between cellular and non-cellular components within the tumor microenvironment but also between cancerous and other cells, such as adipocytes [174]. Another suggestion is that transfer of specific adipose cell- or other cell-derived miRNAs may be involved in the regulation of endometrial tumor progression, providing a new form of intercellular communication. Overexpressed miRNAs are included in exosomes released from cells and play a functional role in cell-to-cell communication [212].

5. Perspective and Future Directions

Although diabetes has long been known to be an independent risk factor for EC, little is known about the relationships between IR and EC [213]. Literature data were unable to indicate clear associations between insulin, IGFs, and sex steroid hormones with EC incidence because of the multitude of dysregulated pathways that lead to EC progression. Yet, meta-analyses support the theory about the association between IR and EC. Currently, there is still a need for new precise molecular tools for the early diagnosis, risk assessment, and prediction of EC development, and miRNA may be a promising marker [214].

Epidemiological and cohort studies should determine the risk of EC in patients with IR based on miRNA expression pattern. That would allow timely intervention(s) to prevent cancer development. Patients with T2DM, prediabetes, metabolic syndrome, and PCOS should be included.

Taking into account that EC is a hormone-dependent cancer, studies on epigenetic mechanisms, including miRNA and sex steroid pathway profiling, both in cancer and IR, are worth undertaking. EC molecular subtypes have been shown to demonstrate distinct miRNA signatures [141]. These miRNA signatures are reduced, and particular levels of depletion are characteristic for particular EC subtypes [140]. A long-term follow-up of patients with these specific cancer subtypes is now mandatory to unveil the clinical significance of miRNA signatures. Similarly, long-term studies should reveal the significance of miRNAs in reference to type I and II EC. The thirteen miRNAs found by us to be dysregulated in both IR and EC are worth special attention.

6. Conclusions

This review highlights changes in miRNA involved in both IR and EC. In support of the possible role of miRNA in both conditions, our careful literature search found that dysregulation of at least 13 miRNAs has been ascribed to both IR and EC. Therefore, miRNA could represent a potential molecular link between the metabolic alterations related to IR and EC. There is a reasonable possibility for miRNAs to become a predictive factor of future EC in IR patients.

Author Contributions: Conceptualization, I.S.; writing—original draft preparation, I.S.; writing—review and editing, M.J., M.N. and A.K.; visualization, I.S.; supervision, M.J. and A.K. All authors have read and agreed to the published version of the manuscript.

Funding: This research was funded by a Grant No. SUB/1/DN/20/005/1196 from the Medical University of Białystok, Poland.

Conflicts of Interest: The authors declare no conflict of interest.

References

1. Morice, P.; Leary, A.; Creutzberg, C.; Abu-Rustum, N.; Darai, E. Endometrial cancer. *Lancet* **2016**, *387*, 1094–1108. [CrossRef]
2. Memon, A.; Paudyal, P. Epidemiology of endometrial cancer. In *Endometrial Cancer: Current Epidemiology, Detection and Management*; Nova Science Pub Inc.: Hauppauge, NY, USA, 2014. [CrossRef]
3. Setiawan, V.W.; Yang, H.P.; Pike, M.C.; McCann, S.E.; Yu, H.; Xiang, Y.B.; Wolk, A.; Wentzensen, N.; Weiss, N.S.; Webb, P.M.; et al. Type i and II endometrial cancers: Have they different risk factors? *J. Clin. Oncol.* **2013**, *31*, 2607–2618. [CrossRef] [PubMed]
4. Burke, W.M.; Orr, J.; Leitao, M.; Salom, E.; Gehrig, P.; Olawaiye, A.B.; Brewer, M.; Boruta, D.; Villella, J.; Herzog, T.; et al. Endometrial cancer: A review and current management strategies: Part I. *Gynecol. Oncol.* **2014**, *134*, 385–392. [CrossRef] [PubMed]
5. McAlpine, J.; Leon-Castillo, A.; Bosse, T. The rise of a novel classification system for endometrial carcinoma; integration of molecular subclasses. *J. Pathol.* **2018**, *244*, 538–549. [CrossRef]
6. Voss, M.A.; Ganesan, R.; Ludeman, L.; McCarthy, K.; Gornall, R.; Schaller, G.; Wei, W.; Sundar, S. Should grade 3 endometrioid endometrial carcinoma be considered a type 2 cancer-a clinical and pathological evaluation. *Gynecol. Oncol.* **2012**, *124*, 15–20. [CrossRef]
7. Murali, R.; Soslow, R.A.; Weigelt, B. Classification of endometrial carcinoma: More than two types. *Lancet Oncol.* **2014**, *15*, e268–e278. [CrossRef]
8. Malik, T.Y.; Chishti, U.; Aziz, A.B.; Sheikh, I. Comparison of risk factors and survival of type-I and type-II endometrial cancers. *Pak. J. Med. Sci.* **2016**, *32*, 886–890. [CrossRef]
9. Amant, F.; Moerman, P.; Neven, P.; Timmerman, D.; Van Limbergen, E.; Vergote, I. Endometrial cancer. *Lancet* **2005**, *366*, 491–505. [CrossRef]
10. Remmerie, M.; Janssens, V. Targeted therapies in type II endometrial cancers: Too little, but not too late. *Int. J. Mol. Sci.* **2018**, *19*, 2380. [CrossRef] [PubMed]
11. Getz, G.; Gabriel, S.B.; Cibulskis, K.; Lander, E.; Sivachenko, A.; Sougnez, C.; Lawrence, M.; Kandoth, C.; Dooling, D.; Fulton, R.; et al. Integrated genomic characterization of endometrial carcinoma. *Nature* **2013**, *497*, 67–73. [CrossRef]
12. Raglan, O.; Kalliala, I.; Markozannes, G.; Cividini, S.; Gunter, M.J.; Nautiyal, J.; Gabra, H.; Paraskevaidis, E.; Martin-Hirsch, P.; Tsilidis, K.K.; et al. Risk factors for endometrial cancer: An umbrella review of the literature. *Int. J. Cancer* **2019**, *145*, 1719–1730. [CrossRef] [PubMed]
13. Coll-de la Rubia, E.; Martinez-Garcia, E.; Dittmar, G.; Gil-Moreno, A.; Cabrera, S.; Colas, E. Prognostic biomarkers in endometrial cancer: A systematic review and meta-analysis. *J. Clin. Med.* **2020**, *9*, 1900. [CrossRef] [PubMed]
14. León-Castillo, A.; Gilvazquez, E.; Nout, R.; Smit, V.T.H.B.M.; McAlpine, J.N.; McConechy, M.; Kommoss, S.; Brucker, S.Y.; Carlson, J.W.; Epstein, E.; et al. Clinicopathological and molecular characterisation of 'multiple-classifier' endometrial carcinomas. *J. Pathol.* **2020**, *250*, 312–322. [CrossRef] [PubMed]
15. Orgel, E.; Mittelman, S.D. The links between insulin resistance, diabetes, and cancer. *Curr. Diabetes Rep.* **2013**, *13*, 213–222. [CrossRef] [PubMed]

16. Shafiee, M.N.; Khan, G.; Ariffin, R.; Abu, J.; Chapman, C.; Deen, S.; Nunns, D.; Barrett, D.A.; Seedhouse, C.; Atiomo, W. Preventing endometrial cancer risk in polycystic ovarian syndrome (PCOS) women: Could metformin help? *Gynecol. Oncol.* **2014**, *132*, 248–253. [CrossRef] [PubMed]
17. Kitson, S.J.; Gareth Evans, D.; Crosbie, E.J. Identifying high-risk women for endometrial cancer prevention strategies: Proposal of an endometrial cancer risk prediction model. *Cancer Prev. Res.* **2017**, *10*, 1–13. [CrossRef]
18. Mu, N.; Zhu, Y.; Wang, Y.; Zhang, H.; Xue, F. Insulin resistance: A significant risk factor of endometrial cancer. *Gynecol. Oncol.* **2012**, *125*, 751–757. [CrossRef]
19. Saeedi, P.; Petersohn, I.; Salpea, P.; Malanda, B.; Karuranga, S.; Unwin, N.; Colagiuri, S.; Guariguata, L.; Motala, A.A.; Ogurtsova, K.; et al. Global and regional diabetes prevalence estimates for 2019 and projections for 2030 and 2045: Results from the International Diabetes Federation Diabetes Atlas, 9th edition. *Diabetes Res. Clin. Pract.* **2019**, *157*, 107843. [CrossRef]
20. Thomas, D.D.; Corkey, B.E.; Istfan, N.W.; Apovian, C.M. Hyperinsulinemia: An early indicator of metabolic dysfunction. *J. Endocr. Soc.* **2019**, *3*, 1727–1747. [CrossRef]
21. Shafiee, M.N.; Chapman, C.; Barrett, D.; Abu, J.; Atiomo, W. Reviewing the molecular mechanisms which increase endometrial cancer (EC) risk in women with polycystic ovarian syndrome (PCOS): Time for paradigm shift? *Gynecol. Oncol.* **2013**, *131*, 489–492. [CrossRef]
22. Shafiee, M.N.; Seedhouse, C.; Mongan, N.; Chapman, C.; Deen, S.; Abu, J.; Atiomo, W. Up-regulation of genes involved in the insulin signalling pathway (IGF1, PTEN and IGFBP1) in the endometrium may link polycystic ovarian syndrome and endometrial cancer. *Mol. Cell. Endocrinol.* **2016**, *424*, 94–101. [CrossRef]
23. Banno, K.; Kisu, I.; Yanokura, M.; Masuda, K.; Kobayashi, Y.; Ueki, A.; Tsuji, K.; Yamagami, W.; Nomura, H.; Susumu, N.; et al. Endometrial cancer and hypermethylation: Regulation of DNA and MicroRNA by epigenetics. *Biochem. Res. Int.* **2012**, *2012*, 738274. [CrossRef]
24. Wang, H.; Peng, R.; Wang, J.; Qin, Z.; Xue, L. Circulating microRNAs as potential cancer biomarkers: The advantage and disadvantage. *Clin. Epigenet.* **2018**, *10*, 59. [CrossRef]
25. Valihrach, L.; Androvic, P.; Kubista, M. Circulating miRNA analysis for cancer diagnostics and therapy. *Mol. Asp. Med.* **2019**, *72*, 100825. [CrossRef]
26. Hayes, J.; Peruzzi, P.P.; Lawler, S. MicroRNAs in cancer: Biomarkers, functions and therapy. *Trends Mol. Med.* **2014**, *20*, 460–469. [CrossRef]
27. MacFarlane, L.-A.; Murphy, P.R. MicroRNA: Biogenesis, Function and Role in Cancer. *Curr. Genom.* **2010**, *11*, 537–561. [CrossRef]
28. Vasilatou, D.; Sioulas, V.D.; Pappa, V.; Papageorgiou, S.G.; Vlahos, N.F. The role of miRNAs in endometrial cancer. *Epigenomics* **2015**, *7*, 951–959. [CrossRef]
29. Petersen, M.C.; Shulman, G.I. Mechanisms of insulin action and insulin resistance. *Physiol. Rev.* **2018**, *98*, 2133–2223. [CrossRef]
30. Tabak, A.G.; Herder, C.; Rathman, W.; Brunner, E.J.; Kivimaki, M. Prediabetes: A high-risk state for developing diabetes. *Lancet* **2014**, *379*, 2279–2290. [CrossRef]
31. Lai, Y.; Sun, C. Association of abnormal glucose metabolism and insulin resistance in patients with atypical and typical endometrial cancer. *Oncol. Lett.* **2018**, *15*, 2173–2178. [CrossRef]
32. Saed, L.; Varse, F.; Baradaran, H.R.; Moradi, Y.; Khateri, S.; Friberg, E.; Khazaei, Z.; Gharahjeh, S.; Tehrani, S.; Sioofy-Khojine, A.B.; et al. The effect of diabetes on the risk of endometrial Cancer: An updated a systematic review and meta-analysis. *BMC Cancer* **2019**, *19*, 527. [CrossRef] [PubMed]
33. Friberg, E.; Orsini, N.; Mantzoros, C.S.; Wolk, A. Diabetes mellitus and risk of endometrial cancer: A meta-analysis. *Diabetologia* **2007**, *50*, 1365–1374. [CrossRef] [PubMed]
34. Sun, W.; Lu, J.; Wu, S.; Bi, Y.; Mu, Y.; Zhao, J.; Liu, C.; Chen, L.; Shi, L.; Li, Q.; et al. Association of insulin resistance with breast, ovarian, endometrial and cervical cancers in non-diabetic women. *Am. J. Cancer Res.* **2016**, *6*, 2334–2344. [PubMed]
35. Soliman, P.T.; Wu, D.; Tortolero-Luna, G.; Schmeler, K.M.; Slomovitz, B.M.; Bray, M.S.; Gershenson, D.M.; Lu, K.H. Association between adiponectin, insulin resistance, and endometrial cancer. *Cancer* **2006**, *106*, 1376–2381. [CrossRef] [PubMed]
36. Zeng, F.; Shi, J.; Long, Y.; Tian, H.; Li, X.; Zhao, A.Z.; Li, R.F.; Chen, T. Adiponectin and Endometrial Cancer: A Systematic Review and Meta-Analysis. *Cell. Physiol. Biochem.* **2015**, *36*, 1670–1678. [CrossRef] [PubMed]

37. Lukanova, A.; Zeleniuch-Jacquotte, A.; Lundin, E.; Micheli, A.; Arslan, A.A.; Rinaldi, S.; Muti, P.; Lenner, P.; Koenig, K.L.; Biessy, C.; et al. Prediagnostic levels of C-peptide, IGF-I, IGFBP-1, -2 and -3 and risk of endometrial cancer. *Int. J. Cancer* **2004**, *108*, 262–268. [CrossRef]
38. Burzawa, J.K.; Schmeler, K.M.; Soliman, P.T.; Meyer, L.A.; Bevers, M.W.; Pustilnik, T.L.; Anderson, M.L.; Ramondetta, L.M.; Tortolero-Luna, G.; Urbauer, D.L.; et al. Prospective evaluation of insulin resistance among endometrial cancer patients. *Am. J. Obstet. Gynecol.* **2011**, *204*, e1–e7. [CrossRef]
39. Stocks, T.; Bjørge, T.; Ulmer, H.; Manjer, J.; Häggström, C.; Nagel, G.; Engeland, A.; Johansen, D.; Hallmans, G.; Selmer, R.; et al. Metabolic risk score and cancer risk: Pooled analysis of seven cohorts. *Int. J. Epidemiol.* **2015**, *44*, 1353–1363. [CrossRef]
40. Nead, K.T.; Sharp, S.J.; Thompson, D.J.; Painter, J.N.; Savage, D.B.; Semple, R.K.; Barker, A.; Perry, J.R.B.; Attia, J.; Dunning, A.M.; et al. Evidence of a causal sssociation between insulinemia and endometrial cancer: A mendelian randomization analysis. *J. Natl. Cancer Inst.* **2015**, *107*, djv178. [CrossRef]
41. Byrne, F.L.; Martin, A.R.; Kosasih, M.; Caruana, B.T.; Farrell, R. The role of hyperglycemia in endometrial cancer pathogenesis. *Cancers Basel* **2020**, *12*, 1191. [CrossRef]
42. Garikapati, K.K.; Ammu, V.V.V.R.K.; Krishnamurthy, P.T.; Chintamaneni, P.K.; Pindiprolu, S.K.S.S. Type-II endometrial cancer: Role of adipokines. *Arch. Gynecol. Obstet.* **2019**, *300*, 239–249. [CrossRef] [PubMed]
43. Kaya, S.; Kaya, B.; Keskin, H.L.; Kayhan Tetik, B.; Yavuz, F.A. Is there any relationship between benign endometrial pathologies and metabolic status? *J. Obstet. Gynaecol. Lahore* **2019**, *39*, 176–183. [CrossRef]
44. Dossus, L.; Rinaldi, S.; Becker, S.; Lukanova, A.; Tjonneland, A.; Olsen, A.; Stegger, J.; Overvad, K.; Chabbert-Buffet, N.; Jimenez-Corona, A.; et al. Obesity, inflammatory markers, and endometrial cancer risk: A prospective case—Control study. *Endocr. Relat. Cancer* **2010**, *17*, 1007–1019. [CrossRef] [PubMed]
45. Dossus, L.; Lukanova, A.; Rinaldi, S.; Allen, N.; Cust, A.E.; Becker, S.; Tjonneland, A.; Hansen, L.; Overvad, K.; Chabbert-Buffet, N.; et al. Hormonal, metabolic, and inflammatory profiles and endometrial cancer risk within the EPIC cohort—A factor analysis. *Am. J. Epidemiol.* **2013**, *177*, 787–799. [CrossRef]
46. Di Zazzo, E.; Polito, R.; Bartollino, S.; Nigro, E.; Porcile, C.; Bianco, A.; Daniele, A.; Moncharmont, B. Adiponectin as link factor between adipose tissue and cancer. *Int. J. Mol. Sci.* **2019**, *20*, 839. [CrossRef] [PubMed]
47. Calle, E.E.; Kaaks, R. Overweight, obesity and cancer: Epidemiological evidence and proposed mechanisms. *Nat. Rev. Cancer* **2004**, *4*, 579–591. [CrossRef]
48. Boucher, J.; Kleinridders, A.; Kahn, C.R. Insulin receptor signaling in normal and insulin-resistant states. *Cold Spring Harb. Perspect. Biol.* **2014**, *6*, a009191. [CrossRef]
49. Haeusler, R.A.; McGraw, T.E.; Accili, D. Metabolic signalling: Biochemical and cellular properties of insulin receptor signalling. *Nat. Rev. Mol. Cell Biol.* **2018**, *19*, 31–44. [CrossRef]
50. Belfiore, A.; Malaguarnera, R. Insulin receptor and cancer. *Endocr. Relat. Cancer* **2011**, *18*, R125–R147. [CrossRef]
51. Kadakia, R.; Josefson, J. The relationship of insulin-like growth factor 2 to fetal growth and adiposity. *Horm. Res. Paediatr.* **2016**, *85*, 75–82. [CrossRef]
52. Denley, A.; Wallace, J.C.; Cosgrove, L.J.; Forbes, B.E. The insulin receptor isoform exon 11-(IR-A) in cancer and other diseases: A review. *Horm. Metab. Res.* **2003**, *35*, 778–785. [CrossRef] [PubMed]
53. Wang, C.F.; Zhang, G.; Zhao, L.J.; Qi, W.J.; Li, X.P.; Wang, J.L.; Wei, L.H. Overexpression of the Insulin Receptor Isoform A Promotes Endometrial Carcinoma Cell Growth. *PLoS ONE* **2013**, *8*, e69001. [CrossRef] [PubMed]
54. Flannery, C.A.; Saleh, F.L.; Choe, G.H.; Selen, D.J.; Kodaman, P.H.; Kliman, H.J.; Wood, T.L.; Taylor, H.S. Differential expression of IR-A, IR-B and IGF-1R in endometrial physiology and distinct signature in adenocarcinoma. *J. Clin. Endocrinol. Metab.* **2016**, *101*, 2883–2891. [CrossRef] [PubMed]
55. Bruchim, I.; Sarfstein, R.; Werner, H. The IGF hormonal network in endometrial cancer: Functions, regulation, and targeting approaches. *Front. Endocrinol. Lausanne* **2014**, *5*, 76. [CrossRef]
56. Djiogue, S.; Kamdje, A.H.N.; Vecchio, L.; Kipanyula, M.J.; Farahna, M.; Aldebasi, Y.; Etet, P.F.S. Insulin resistance and cancer: The role of insulin and IGFs. *Endocr. Relat. Cancer* **2013**, *20*, R1–R17. [CrossRef]
57. Petridou, E.; Koukoulomatis, P.; Alexe, D.M.; Voulgaris, Z.; Spanos, E.; Trichopoulos, D. Endometrial cancer and the IGF system: A case-control study in Greece. *Oncology* **2003**, *64*, 341–345. [CrossRef]

58. Epaud, R.; Aubey, F.; Xu, J.; Chaker, Z.; Clemessy, M.; Dautin, A.; Ahamed, K.; Bonora, M.; Hoyeau, N.; Fléjou, J.F.; et al. Knockout of insulin-like growth factor-1 receptor impairs distal lung morphogenesis. *PLoS ONE* **2012**, *7*, e48071. [CrossRef]
59. Alvino, C.L.; Ong, S.C.; McNeil, K.A.; Delaine, C.; Booker, G.W.; Wallace, J.C.; Forbes, B.E. Understanding the mechanism of insulin and insulin-like growth factor (IGF) receptor activation by IGF-II. *PLoS ONE* **2011**, *6*, e27488. [CrossRef]
60. Wang, C.; Su, K.; Zhang, Y.; Zhang, W.; Zhao, Q.; Chu, D.; Guo, R. IR-A/IGF-1R-mediated signals promote epithelial-mesenchymal transition of endometrial carcinoma cells by activating PI3K/AKT and ERK pathways. *Cancer Biol. Ther.* **2019**, *20*, 295–306. [CrossRef]
61. Dai, C.; Li, N.; Song, G.; Yang, Y.; Ning, X. Insulin-like growth factor 1 regulates growth of endometrial carcinoma through PI3k signaling pathway in insulin-resistant type 2 diabetes. *Am. J. Transl. Res.* **2016**, *8*, 3329–3336.
62. Joehlin-Price, A.S.; Stephens, J.A.; Zhang, J.; Backes, F.J.; Cohn, D.E.; Suarez, A.A. Endometrial cancer insulin-like growth factor 1 receptor (IGF1R) expression increases with body mass index and is associated with pathologic extent and prognosis. *Cancer Epidemiol. Biomark. Prev.* **2016**, *25*, 438–445. [CrossRef] [PubMed]
63. Gunter, M.J.; Hoover, D.R.; Yu, H.; Wassertheil-Smoller, S.; Manson, J.E.; Li, J.; Harris, T.G.; Rohan, T.E.; Xue, X.N.; Ho, G.Y.F.; et al. A prospective evaluation of insulin and insulin-like growth factor-I as risk factors for endometrial cancer. *Cancer Epidemiol. Biomark. Prev.* **2008**, *89*, 921–929. [CrossRef] [PubMed]
64. Roy, R.N.; Gerulath, A.H.; Cecutti, A.; Bhavnani, B.R. Discordant expression of insulin-like growth factors and their receptor messenger ribonucleic acids in endometrial carcinomas relative to normal endometrium. *Mol. Cell. Endocrinol.* **1999**, *153*, 19–27. [CrossRef]
65. Ding, J.; Li, C.; Tang, J.; Yi, C.; Liu, J.Y.; Qiu, M. Higher Expression of Proteins in IGF/IR Axes in Colorectal Cancer is Associated with Type 2 Diabetes Mellitus. *Pathol. Oncol. Res.* **2016**, *22*, 773–779. [CrossRef]
66. Werner, H.; Sarfstein, R.; Bruchim, I. Investigational IGF1R inhibitors in early stage clinical trials for cancer therapy. *Expert Opin. Investig. Drugs* **2019**, *28*, 1101–1112. [CrossRef]
67. Cohen, D.H.; LeRoith, D. Obesity, type 2 diabetes, and cancer: The insulin and IGF connection. *Endocr. Relat. Cancer* **2012**, *19*, F25–F45. [CrossRef]
68. Sarfstein, R.; Friedman, Y.; Attias-Geva, Z.; Fishman, A.; Bruchim, I.; Werner, H. Metformin downregulates the insulin/IGF-I signaling pathway and inhibits different uterine serous carcinoma (USC) cells proliferation and migration in p53-dependent or -independent manners. *PLoS ONE* **2013**, *8*, e61537. [CrossRef]
69. Xie, W.; Li, T.; Yang, J.; Shang, M.; Xiao, Y.; Li, Q.; Yang, J. Metformin use and survival outcomes in endometrial cancer: A systematic review and meta-analysis. *Oncotarget* **2017**, *8*, 73079–73086. [CrossRef]
70. Chu, D.; Wu, J.; Wang, K.; Zhao, M.; Wang, C.; Li, L.; Guo, R. Effect of metformin use on the risk and prognosis of endometrial cancer: A systematic review and meta-analysis. *BMC Cancer* **2018**, *18*, 438. [CrossRef]
71. Guo, J.; Xu, K.; An, M.; Zhao, Y. Metformin and endometrial cancer survival: A quantitative synthesis of observational studies. *Oncotarget* **2017**, *8*, 66169–66177. [CrossRef]
72. Allard, J.B.; Duan, C. IGF-binding proteins: Why do they exist and why are there so many? *Front. Endocrinol. Lausanne* **2018**, *9*, 117. [CrossRef]
73. Rajwani, A.; Ezzat, V.; Smith, J.; Yuldasheva, N.Y.; Duncan, E.R.; Gage, M.; Cubbon, R.M.; Kahn, M.B.; Imrie, H.; Abbas, A.; et al. Increasing circulating IGFBP1 levels improves insulin sensitivity, promotes nitric oxide production, lowers blood pressure, and protects against atherosclerosis. *Diabetes* **2012**, *61*, 915–924. [CrossRef] [PubMed]
74. Poloz, Y.; Stambolic, V. Obesity and cancer, a case for insulin signaling. *Cell Death Dis.* **2015**, *6*, e2037. [CrossRef] [PubMed]
75. Weiderpass, E.; Brismar, K.; Bellocco, R.; Vainio, H.; Kaaks, R. Serum levels of insulin-like growth factor-I, IGF-binding protein 1 and 3, and insulin and endometrial cancer risk. *Br. J. Cancer* **2003**, *89*, 1697–1704. [CrossRef] [PubMed]
76. Russo, V.C.; Azar, W.J.; Yau, S.W.; Sabin, M.A.; Werther, G.A. IGFBP-2: The dark horse in metabolism and cancer. *Cytokine Growth Factor Rev.* **2015**, *26*, 329–346. [CrossRef] [PubMed]
77. Wheatcroft, S.B.; Kearney, M.T.; Shah, A.M.; Ezzat, V.A.; Miell, J.R.; Modo, M.; Williams, S.C.R.; Cawthorn, W.P.; Medina-Gomez, G.; Vidal-Puig, A.; et al. IGF-binding protein-2 protects against the development of obesity and insulin resistance. *Diabetes* **2007**, *56*, 285–294. [CrossRef] [PubMed]

78. Yau, S.W.; Azar, W.J.; Sabin, M.A.; Werther, G.A.; Russo, V.C. IGFBP-2—Taking the lead in growth, metabolism and cancer. *J. Cell Commun. Signal.* **2015**, *9*, 125–142. [CrossRef] [PubMed]
79. Lee, E.J.; Mircean, C.; Shmulevich, I.; Wang, H.; Liu, J.; Niemistö, A.; Kavanagh, J.J.; Lee, J.H.; Zhang, W. Insulin-like growth factor binding protein 2 promotes ovarian cancer cell invasion. *Mol. Cancer* **2005**, *4*, 7. [CrossRef]
80. Guo, C.; Lu, H.; Gao, W.; Wang, L.; Lu, K.; Wu, S.; Pataer, A.; Huang, M.; El-Zein, R.; Lin, T.; et al. Insulin-like growth factor binding protein-2 level is increased in blood of lung cancer patients and associated with poor survival. *PLoS ONE* **2013**, *8*, e74973. [CrossRef]
81. Liu, H.; Li, L.; Chen, H.; Kong, R.; Pan, S.; Hu, J.; Wang, Y.; Li, Y.; Sun, B. Silencing IGFBP-2 decreases pancreatic cancer metastasis and enhances chemotherapeutic sensitivity. *Oncotarget* **2017**, *8*, 61674–61686. [CrossRef]
82. Liu, Y.; Song, C.; Shen, F.; Zhang, J.; Song, S.W. IGFBP2 promotes immunosuppression associated with its mesenchymal induction and FcγRIIB phosphorylation in glioblastoma. *PLoS ONE* **2019**, *14*, e0222999. [CrossRef] [PubMed]
83. Pickard, A.; McCance, D.J. IGF-binding protein 2—Oncogene or tumor suppressor? *Front. Endocrinol. Lausanne* **2015**, *6*, 25. [CrossRef] [PubMed]
84. Jogie-Brahim, S.; Feldman, D.; Oh, Y. Unraveling insulin-like growth factor binding protein-3 actions in human disease. *Endocr. Rev.* **2009**, *30*, 417–437. [CrossRef]
85. Mochizuki, T.; Sakai, K.; Iwashita, M. Effects of insulin-like growth factor (IGF) binding protein-3 (IGFBP-3) on endometrial cancer (HHUA) cell apoptosis and EGF stimulated cell proliferation in vitro. *Growth Horm. IGF Res.* **2006**, *16*, 202–210. [CrossRef] [PubMed]
86. Wang, E.A.; Chen, W.Y.; Wong, C.H. Multiple growth factor targeting by engineered insulin-like rowth factor binding protein-3 augments EGF receptor tyrosine kinase inhibitor efficacy. *Sci. Rep.* **2020**, *10*, 2735. [CrossRef] [PubMed]
87. Groothuis, P.G.; Dassen, H.H.N.M.; Romano, A.; Punyadeera, C. Estrogen and the endometrium: Lessons learned from gene expression profiling in rodents and human. *Hum. Reprod. Update* **2007**, *13*, 405–417. [CrossRef]
88. Gupte, A.A.; Pownall, H.J.; Hamilton, D.J. Estrogen: An emerging regulator of insulin action and mitochondrial function. *J. Diabetes Res.* **2015**, *2015*, 916585. [CrossRef]
89. Barros, R.P.A.; Machado, U.F.; Warner, M.; Gustafsson, J.Å. Muscle GLUT4 regulation by estrogen receptors ERβ and ERα. *Proc. Natl. Acad. Sci. USA* **2006**, *103*, 1605–1608. [CrossRef]
90. Rodriguez, A.C.; Blanchard, Z.; Maurer, K.A.; Gertz, J. Estrogen signaling in endometrial cancer: A key oncogenic pathway with several open questions. *Horm. Cancer* **2019**, *10*, 51–63. [CrossRef]
91. De Marco, P.; Cirillo, F.; Vivacqua, A.; Malaguarnera, R.; Belfiore, A.; Maggiolini, M. Novel aspects concerning the functional cross-talk between the insulin/IGF-I system and estrogen signaling in cancer cells. *Front. Endocrinol. Lausanne* **2015**, *6*, 30. [CrossRef]
92. Massarweh, S.; Osborne, C.K.; Creighton, C.J.; Qin, L.; Tsimelzon, A.; Huang, S.; Weiss, H.; Rimawi, M.; Schiff, R. Tamoxifen resistance in breast tumors is driven by growth factor receptor signaling with repression of classic estrogen receptor genomic function. *Cancer Res.* **2008**, *68*, 826–833. [CrossRef] [PubMed]
93. Lundholm, L.; Bryzgalova, G.; Gao, H.; Portwood, N.; Fält, S.; Berndt, K.D.; Dicker, A.; Galuska, D.; Zierath, J.R.; Gustafsson, J.Å.; et al. The estrogen receptor α-selective agonist propyl pyrazole triol improves glucose tolerance in ob/ob mice; potential molecular mechanisms. *J. Endocrinol.* **2008**, *199*, 275–286. [CrossRef] [PubMed]
94. Yan, H.; Yang, W.; Zhou, F.; Li, X.; Pan, Q.; Shen, Z.; Han, G.; Newell-Fugate, A.; Tian, Y.; Majeti, R.; et al. Estrogen improves insulin sensitivity and suppresses gluconeogenesis via the transcription factor Foxo1. *Diabetes* **2019**, *68*, 291–304. [CrossRef]
95. Root-Bernstein, R.; Podufaly, A.; Dillon, P.F. Estradiol binds to insulin and insulin receptor decreasing insulin binding in vitro. *Front. Endocrinol. Lausanne* **2014**, *5*, 118. [CrossRef] [PubMed]
96. Tee, M.K.; Rogatsky, I.; Tzagarakis-Foster, C.; Cvoro, A.; An, J.; Christy, R.J.; Yamamoto, K.R.; Leitman, D.C. Estradiol and selective estrogen receptor modulators differentially regulate target genes with estrogen receptors α and β. *Mol. Biol. Cell* **2004**, *15*, 1262–1272. [CrossRef]
97. Yu, H. Role of the Insulin-Like Growth Factor Family in Cancer Development and Progression. *J. Natl. Cancer Inst.* **2000**, *92*, 1472–1489. [CrossRef]

98. Merritt, M.A.; Strickler, H.D.; Einstein, M.H.; Yang, H.P.; Sherman, M.E.; Wentzensen, N.; Brouwer-Visser, J.; Cossio, M.J.; Whitney, K.D.; Yu, H.; et al. Insulin/IGF and sex hormone axes in human endometrium and associations with endometrial cancer risk factors. *Cancer Causes Control.* **2016**, *27*, 737–748. [CrossRef]
99. Ito, K.; Utsunomiya, H.; Yaegashi, N.; Sasano, H. Biological roles of estrogen and progesterone in human endometrial carcinoma—New developments in potential endocrine therapy for endometrial cancer. *Endocr. J.* **2007**, *54*, 667–679. [CrossRef]
100. Trabert, B.; Wentzensen, N.; Yang, H.P.; Sherman, M.E.; Hollenbeck, A.R.; Park, Y.; Brinton, L.A. Is estrogen plus progestin menopausal hormone therapy safe with respect to endometrial cancer risk? *Int. J. Cancer* **2013**, *132*, 417–426. [CrossRef]
101. Brinton, L.A.; Trabert, B.; Anderson, G.L.; Falk, R.T.; Felix, A.S.; Fuhrman, B.J.; Gass, M.L.; Kuller, L.H.; Pfeiffer, R.M.; Rohan, T.E.; et al. Serum estrogens and estrogen metabolites and endometrial cancer risk among postmenopausal women. *Cancer Epidemiol. Biomark. Prev.* **2016**, *25*, 1081–1089. [CrossRef]
102. Arcidiacono, B.; Iiritano, S.; Nocera, A.; Possidente, K.; Nevolo, M.T.; Ventura, V.; Foti, D.; Chiefari, E.; Brunetti, A. Insulin resistance and cancer risk: An overview of the pathogenetic mechanisms. *Exp. Diabetes Res.* **2012**, *2012*, 789174. [CrossRef]
103. Tian, W.; Teng, F.; Zhao, J.; Gao, J.; Gao, C.; Sun, D.; Liu, G.; Zhang, Y.; Yu, S.; Zhang, W.; et al. Estrogen and insulin synergistically promote type 1 endometrial cancer progression. *Cancer Biol. Ther.* **2017**, *18*, 1000–1010. [CrossRef]
104. Randolph, J.F.; Kipersztok, S.; Ayers, J.W.T.; Ansbacher, R.; Peegel, H.; Menon, K.M.J. The effect of insulin on aromatase activity in isolated human endometrial glands and stroma. *Am. J. Obstet. Gynecol.* **1987**, *157*, 1534–1539. [CrossRef]
105. Wolff, L.G.; Vassallo, J.; Pinto, C.B.; Yela, D.A.; Monteiro, I.U. Correlation between insulin resistance and steroid endometrial receptors, KI-67 and BCL-2 after menopause. *Women's Health Sci. J.* **2017**, *1*, 000104. [CrossRef]
106. Winters, S.J.; Gogineni, J.; Karegar, M.; Scoggins, C.; Wunderlich, C.A.; Baumgartner, R.; Ghooray, D.T. Sex hormone-binding globulin gene expression and insulin resistance. *J. Clin. Endocrinol. Metab.* **2014**, *99*, E2780–E2788. [CrossRef]
107. Gallagher, E.J.; LeRoith, D. The proliferating role of insulin and insulin-like growth factors in cancer. *Trends Endocrinol. Metab.* **2010**, *21*, 610–618. [CrossRef]
108. Makki, K.; Froguel, P.; Wolowczuk, I. Adipose tissue in obesity-related inflammation and insulin resistance: Cells, cytokines, and chemokines. *ISRN Inflamm.* **2013**, *2013*, 139239. [CrossRef]
109. Parida, S.; Siddharth, S.; Sharma, D. Adiponectin, obesity, and cancer: Clash of the bigwigs in health and disease. *Int. J. Mol. Sci.* **2019**, *20*, 2519. [CrossRef]
110. Prakash, J.; Mittal, B.; Awasthi, S.; Agarwal, C.G.; Srivastava, N. Hypoadiponectinemia in obesity: Association with insulin resistance. *Indian J. Clin. Biochem.* **2013**, *28*, 158–163. [CrossRef]
111. Gelsomino, L.; Naimo, G.D.; Catalano, S.; Mauro, L.; Andò, S. The emerging role of adiponectin in female malignancies. *Int. J. Mol. Sci.* **2019**, *20*, 2127. [CrossRef]
112. Cust, A.E.; Kaaks, R.; Friedenreich, C.; Bonnet, F.; Laville, M.; Lukanova, A.; Rinaldi, S.; Dossus, L.; Slimani, N.; Lundin, E.; et al. Plasma adiponectin levels and endometrial cancer risk in pre- and postmenopausal women. *J. Clin. Endocrinol. Metab.* **2007**, *92*, 255–263. [CrossRef] [PubMed]
113. Dal Maso, L.; Augustin, L.S.A.; Karalis, A.; Talamini, R.; Franceschi, S.; Trichopoulos, D.; Mantzoros, C.S.; La Vecchia, C. Circulating adiponectin and endometrial cancer risk. *J. Clin. Endocrinol. Metab.* **2004**, *89*, 1160–1163. [CrossRef] [PubMed]
114. Piya, M.K.; McTernan, P.G.; Kumar, S. Adipokine inflammation and insulin resistance: The role of glucose, lipids and endotoxin. *J. Endocrinol.* **2013**, *216*, T1–T15. [CrossRef] [PubMed]
115. Hedbacker, K.; Birsoy, K.; Wysocki, R.W.; Asilmaz, E.; Ahima, R.S.; Farooqi, I.S.; Friedman, J.M. Antidiabetic effects of IGFBP2, a leptin-regulated gene. *Cell Metab.* **2010**, *11*, 11–22. [CrossRef] [PubMed]
116. Neumann, U.H.; Chen, S.; Tam, Y.Y.C.; Baker, R.K.; Covey, S.D.; Cullis, P.R.; Kieffer, T.J. IGFBP2 is neither sufficient nor necessary for the physiological actions of leptin on glucose homeostasis in male ob/ob mice. *Endocrinology* **2014**, *155*, 716–725. [CrossRef] [PubMed]
117. Osegbe, I.; Okpara, H.; Azinge, E. Relationship between serum leptin and insulin resistance among obese Nigerian women. *Ann. Afr. Med.* **2016**, *15*, 14–19. [CrossRef]

118. Wang, T.N.; Chang, W.T.; Chiu, Y.W.; Lee, C.Y.; Lin, K.D.; Cheng, Y.Y.; Su, Y.J.; Chung, H.F.; Huang, M.C. Relationships between changes in leptin and insulin resistance levels in obese individuals following weight loss. *Kaohsiung J. Med. Sci.* **2013**, *29*, 436–443. [CrossRef]
119. Zuo, H.; Shi, Z.; Yuan, B.; Dai, Y.; Wu, G.; Hussain, A. Association between serum leptin concentrations and insulin resistance: A population-based study from China. *PLoS ONE* **2013**, *8*, e54615. [CrossRef]
120. Ding, S.; Madu, C.O.; Lu, Y. The impact of hormonal imbalances associated with obesity on the incidence of endometrial cancer in postmenopausal women. *J. Cancer* **2020**, *11*, 5456–5465. [CrossRef]
121. Schmandt, R.E.; Iglesias, D.A.; Co, N.N.; Lu, K.H. Understanding obesity and endometrial cancer risk: Opportunities for prevention. *Am. J. Obstet. Gynecol.* **2011**, *205*, 518–525. [CrossRef]
122. Wang, J.; Obici, S.; Morgan, K.; Barzilai, N.; Feng, Z.; Rossetti, L. Overfeeding rapidly induces leptin and insulin resistance. *Diabetes* **2001**, *50*, 2786–2791. [CrossRef]
123. Uzan, J.; Laas, E.; Alsamad, I.A.; Skalli, D.; Mansouri, D.; Haddad, B.; Touboul, C. Supervised clustering of adipokines and hormonal receptors predict prognosis in a population of obese women with type 1 endometrial cancer. *Int. J. Mol. Sci.* **2017**, *185*, 1055. [CrossRef]
124. Lin, T.; Zhao, X.; Kong, W.M. Association between adiponectin levels and endometrial carcinoma risk: Evidence from a dose-response meta-analysis. *BMJ Open* **2015**, *5*, e008541. [CrossRef]
125. Gong, T.T.; Wu, Q.J.; Wang, Y.L.; Ma, X.X. Circulating adiponectin, leptin and adiponectin-leptin ratio and endometrial cancer risk: Evidence from a meta-analysis of epidemiologic studies. *Int. J. Cancer* **2015**, *137*, 1967–1978. [CrossRef]
126. Ma, Y.; Liu, Z.; Zhang, Y.; Lu, B. Serum leptin, adiponectin and endometrial cancer risk in Chinese women. *J. Gynecol. Oncol.* **2013**, *24*, 336–341. [CrossRef]
127. Busch, E.L.; Crous-Bou, M.; Prescott, J.; Downing, M.J.; Rosner, B.A.; Mutter, G.L.; De Vivo, I. Adiponectin, leptin, and insulin-pathway receptors as endometrial cancer subtyping markers. *Horm. Cancer* **2018**, *9*, 33–39. [CrossRef]
128. Hebbard, L.; Ranscht, B. Multifaceted roles of Adiponectin in cancer. *Best Pract. Res. Clin. Endocrinol. Metab.* **2014**, *28*, 56–59. [CrossRef]
129. VanSaun, M.N. Molecular pathways: Adiponectin and leptin signaling in cancer. *Clin. Cancer Res.* **2013**, *19*, 1926–1932. [CrossRef]
130. Tumminia, A.; Vinciguerra, F.; Parisi, M.; Graziano, M.; Sciacca, L.; Baratta, R.; Frittitta, L. Adipose tissue, obesity and adiponectin: Role in endocrine cancer risk. *Int. J. Mol. Sci.* **2019**, *20*, 2863. [CrossRef]
131. Ilhan, T.T.; Kebapcilar, A.; Yilmaz, S.A.; Ilhan, T.; Kerimoglu, O.S.; Pekin, A.T.; Akyurek, F.; Unlu, A.; Celik, C. Relations of serum visfatin and resistin levels with endometrial cancer and factors associated with its prognosis. *Asian Pac. J. Cancer Prev.* **2015**, *16*, 4503–4508. [CrossRef]
132. Hlavna, M.; Kohut, L.; Lipkova, J.; Bienertova-Vasku, J.; Dostalova, Z.; Chovanec, J.; Vasku, A. Relationship of resistin levels with endometrial cancer risk. *Neoplasma* **2011**, *58*, 124–128. [CrossRef] [PubMed]
133. Rabe, K.; Lehrke, M.; Parhofer, K.G.; Broedl, U.C. Adipokines and insulin resistance. *Mol. Med.* **2008**, *14*, 741–751. [CrossRef] [PubMed]
134. Catalanotto, C.; Cogoni, C.; Zardo, G. MicroRNA in control of gene expression: An overview of nuclear functions. *Int. J. Mol. Sci.* **2016**, *17*, 1712. [CrossRef] [PubMed]
135. O'Brien, J.; Hayder, H.; Zayed, Y.; Peng, C. Overview of microRNA biogenesis, mechanisms of actions, and circulation. *Front. Endocrinol. Lausanne* **2018**, *9*, 402. [CrossRef] [PubMed]
136. Nigi, L.; Grieco, G.E.; Ventriglia, G.; Brusco, N.; Mancarella, F.; Formichi, C.; Dotta, F.; Sebastiani, G. MicroRNAs as regulators of insulin signaling: Research updates and potential therapeutic perspectives in type 2 diabetes. *Int. J. Mol. Sci.* **2018**, *19*, 3705. [CrossRef]
137. Khan, S.; Wang, C.H. ER stress in adipocytes and insulin resistance: Mechanisms and significance (Review). *Mol. Med. Rep.* **2014**, *10*, 2234–2240. [CrossRef]
138. Wang, J.; Chen, J.; Sen, S. MicroRNA as Biomarkers and Diagnostics. *J. Cell. Physiol.* **2016**, *231*, 25–30. [CrossRef]
139. La Ferlita, A.; Battaglia, R.; Andronico, F.; Caruso, S.; Cianci, A.; Purrello, M.; Di Pietro, C. Non-coding RNAs in endometrial physiopathology. *Int. J. Mol. Sci.* **2018**, *19*, 2120. [CrossRef]
140. Martello, G.; Rosato, A.; Ferrari, F.; Manfrin, A.; Cordenonsi, M.; Dupont, S.; Enzo, E.; Guzzardo, V.; Rondina, M.; Spruce, T.; et al. A microRNA targeting dicer for metastasis control. *Cell* **2010**, *141*, 1195–1207. [CrossRef]

141. Wang, Q.; Xu, K.; Tong, Y.; Dai, Z.; Xu, T.; He, D.; Ying, J. Novel miRNA markers for the diagnosis and prognosis of endometrial cancer. *J. Cell. Mol. Med.* **2020**, *24*, 4533–4546. [CrossRef]
142. Ratner, E.S.; Tuck, D.; Richter, C.; Nallur, S.; Patel, R.M.; Schultz, V.; Hui, P.; Schwartz, P.E.; Rutherford, T.J.; Weidhaas, J.B. MicroRNA signatures differentiate uterine cancer tumor subtypes. *Gynecol. Oncol.* **2010**, *118*, 251–257. [CrossRef] [PubMed]
143. Kalinkova, L.; Kajo, K.; Karhanek, M.; Wachsmannova, L.; Suran, P.; Zmetakova, I.; Fridrichova, I. Discriminating miRNA profiles between endometrioid well- and poorly-differentiated tumours and endometrioid and serous subtypes of endometrial cancers. *Int. J. Mol. Sci.* **2020**, *21*, 6071. [CrossRef] [PubMed]
144. Chung, T.K.H.; Cheung, T.H.; Huen, N.Y.; Wong, K.W.Y.; Lo, K.W.K.; Yim, S.F.; Siu, N.S.S.; Wong, Y.M.; Tsang, P.T.; Pang, M.W.; et al. Dysregulated microRNAs and their predicted targets associated with endometrioid endometrial adenocarcinoma in Hong Kong women. *Int. J. Cancer* **2009**, *124*, 1358–1365. [CrossRef] [PubMed]
145. Hiroki, E.; Akahira, J.I.; Suzuki, F.; Nagase, S.; Ito, K.; Suzuki, T.; Sasano, H.; Yaegashi, N. Changes in microRNA expression levels correlate with clinicopathological features and prognoses in endometrial serous adenocarcinomas. *Cancer Sci.* **2010**, *101*, 241–249. [CrossRef]
146. Jia, W.; Wu, Y.; Zhang, Q.; Gao, G.; Zhang, C.; Xiang, Y. Identification of four serum microRNAs from a genome-wide serum microRNA expression profile as potential non-invasive biomarkers for endometrioid endometrial cancer. *Oncol. Lett.* **2013**, *6*, 261–267. [CrossRef]
147. Montagnana, M.; Benati, M.; Danese, E.; Giudici, S.; Perfranceschi, M.; Ruzzenenete, O.; Salvagno, G.L.; Bassi, A.; Gelati, M.; Paviati, E.; et al. Aberrant microRNA epression in patients with endometrial cancer. *Int. J. Gynecol. Cancer* **2017**, *27*, 459–466. [CrossRef]
148. Devor, E.J.; Hovey, A.M.; Goodheart, M.J.; Ramachandran, S.; Leslie, K.K. MicroRNA expression profiling of endometrial endometrioid adenocarcinomas and serous adenocarcinomas reveals profiles containing shared, unique and differentiating groups of microRNAs. *Oncol. Rep.* **2011**, *26*, 995–1002. [CrossRef]
149. Cohn, D.E.; Fabbri, M.; Valeri, N.; Alder, H.; Ivanov, I.; Liu, C.G.; Croce, C.M.; Resnick, K.E. Comprehensive miRNA profiling of surgically staged endometrial cancer. *Am. J. Obstet. Gynecol.* **2010**, *202*, 656.e1–656.e8. [CrossRef]
150. Wu, Y.S.; Lin, H.; Chen, D.; Yi, Z.; Zeng, B.; Jiang, Y.; Ren, G. A four-miRNA signature as a novel biomarker for predicting survival in endometrial cancer. *Gene* **2019**, *697*, 86–93. [CrossRef]
151. Donkers, H.; Bekkers, R.; Galaal, K. Diagnostic value of microRNA panel in endometrial cancer: A systematic review. *Oncotarget* **2020**, *11*, 2010–2023. [CrossRef]
152. Zhu, H.; Ng, S.C.; Segr, A.V.; Shinoda, G.; Shah, S.P.; Einhorn, W.S.; Takeuchi, A.; Engreitz, J.M.; Hagan, J.P.; Kharas, M.G.; et al. The Lin28/let-7 axis regulates glucose metabolism. *Cell* **2011**, *147*, 81–94. [CrossRef] [PubMed]
153. Liu, P.; Qi, M.; Ma, C.; Lao, G.; Liu, Y.; Liu, Y.; Liu, Y. Let7a inhibits the growth of endometrial carcinoma cells by targeting Aurora-B. *FEBS Lett.* **2013**, *587*, 2523–2529. [CrossRef] [PubMed]
154. Chirshev, E.; Oberg, K.C.; Ioffe, Y.J.; Unternaehrer, J.J. Let-7 as biomarker, prognostic indicator, and therapy for precision medicine in cancer. *Clin. Transl. Med.* **2019**, *8*, 24. [CrossRef] [PubMed]
155. Myatt, S.S.; Wang, J.; Monteiro, L.J.; Christian, M.; Ho, K.-K.; Fusi, L.; Dina, R.E.; Brosens, J.J.; Ghaem-Maghami, S.; Lam, E.W.-F. Repression of FOXO1 expression by microRNAs in endometrial cancer. *Cancer Res.* **2010**, *70*, 367–377. [CrossRef] [PubMed]
156. Ramachandran, D.; Roy, U.; Garg, S.; Ghosh, S.; Pathak, S.; Kolthur-Seetharam, U. Sirt1 and mir-9 expression is regulated during glucose-stimulated insulin secretion in pancreatic β-islets. *FEBS J.* **2011**, *287*, 1167–1174. [CrossRef]
157. Wu, P.; Wang, Q.; Jiang, C.; Chen, C.; Liu, Y.; Chen, Y.; Zeng, Y. MicroRNA-29a is involved lipid metabolism dysfunction and insulin resistance in C2C12 myotubes by targeting PPARδ. *Mol. Med. Rep.* **2018**, *17*, 8493–8501. [CrossRef]
158. Jiang, T.; Sui, D.; You, D.; Yao, S.; Zhang, L.; Wang, Y.; Zhao, J.; Zhang, Y. MiR-29a-5p inhibits proliferation and invasion and induces apoptosis in endometrial carcinoma via targeting TPX2. *Cell Cycle* **2018**, *17*, 1268–1278. [CrossRef] [PubMed]

159. He, A.; Zhu, L.; Gupta, N.; Chang, Y.; Fang, F. Overexpression of micro ribonucleic acid 29, highly up-regulated in diabetic rats, leads to insulin resistance in 3T3-L1 adipocytes. *Mol. Endocrinol.* **2007**, *21*, 2785–2794. [CrossRef]
160. Chen, H.X.; Xu, X.X.; Tan, B.Z.; Zhang, Z.; Zhou, X.D. MicroRNA-29b inhibits angiogenesis by targeting VEGFA through the MAPK/ERK and PI3K/Akt signaling pathways in endometrial carcinoma. *Cell. Physiol. Biochem.* **2017**, *41*, 933–946. [CrossRef]
161. Kong, J.; He, X.; Wang, Y.; Li, J. Effect of microRNA-29b on proliferation, migration, and invasion of endometrial cancer cells. *J. Int. Med. Res.* **2019**, *47*, 3803–3817. [CrossRef]
162. Massart, J.; Sjögren, R.J.O.; Lundell, L.S.; Mudry, J.M.; Franck, N.; O'Gorman, D.J.; Egan, B.; Zierath, J.R.; Krook, A. Altered miR-29 expression in type 2 diabetes influences glucose and lipid metabolism in skeletal muscle. *Diabetes* **2017**, *66*, 1807–1818. [CrossRef]
163. van Sinderen, M.; Griffiths, M.; Menkhorst, E.; Niven, K.; Dimitriadis, E. Restoration of microRNA-29c in type I endometrioid cancer reduced endometrial cancer cell growth. *Oncol. Lett.* **2019**, *18*, 2684–2693. [CrossRef] [PubMed]
164. Yu, D.; Zhou, H.; Xun, Q.; Xu, X.; Ling, J.; Hu, Y. microRNA-103 regulates the growth and invasion of endometrial cancer cells through the downregulation of tissue inhibitor of metalloproteinase 3. *Oncol. Lett.* **2012**, *3*, 1221–1226. [CrossRef]
165. Trajkovski, M.; Hausser, J.; Soutschek, J.; Bhat, B.; Akin, A.; Zavolan, M.; Heim, M.H.; Stoffel, M. MicroRNAs 103 and 107 regulate insulin sensitivity. *Nature* **2011**, *474*, 649–654. [CrossRef] [PubMed]
166. Bao, W.; Zhang, Y.; Li, S.; Fan, Q.; Qiu, M.; Wang, Y.; Li, Y.; Ji, X.; Yang, Y.; Sang, Z.; et al. MiR-107-5p promotes tumor proliferation and invasion by targeting estrogen receptor-α in endometrial carcinoma. *Oncol. Rep.* **2019**, *41*, 1575–1585. [CrossRef]
167. Ryu, H.S.; Park, S.Y.; Ma, D.; Zhang, J.; Lee, W. The induction of microrna targeting IRS-1 is involved in the development of insulin resistance under conditions of mitochondrial dysfunction in hepatocytes. *PLoS ONE* **2011**, *6*, e17343. [CrossRef]
168. Zhao, X.; Zhu, D.; Lu, C.; Yan, D.; Li, L.; Chen, Z. MicroRNA-126 inhibits the migration and invasion of endometrial cancer cells by targeting insulin receptor substrate 1. *Oncol. Lett.* **2016**, *11*, 1207–1212. [CrossRef]
169. Liu, Y.; Hua, T.; Chi, S.; Wang, H. Identification of key pathways and genes in endometrial cancer using bioinformatics analyses. *Oncol. Lett.* **2019**, *17*, 897–906. [CrossRef]
170. Yu, X.; Zhong, L. Pioglitazone/microRNA-141/FOXA2: A novel axis in pancreatic β-cells proliferation and insulin secretion. *Mol. Med. Rep.* **2018**, *17*, 7931–7938. [CrossRef]
171. Yoneyama, K.; Ishibashi, O.; Kawase, R.; Kurose, K.; Takeshita, T. MiR-200a, miR-200b and miR-429 are onco-miRs that target the PTEN gene in endometrioid endometrial carcinoma. *Anticancer Res.* **2015**, *35*, 1401–1410.
172. Filios, S.R.; Xu, G.; Chen, J.; Hong, K.; Jing, G.; Shalev, A. MicroRNA-200 is induced by thioredoxin-interacting protein and regulates Zeb1 protein signaling and beta cell. *J. Biol. Chem.* **2014**, *289*, 36275–36283. [CrossRef] [PubMed]
173. Peng, J.; Zhou, Y.; Deng, Z.; Zhang, H.; Wu, Y.; Song, T.; Yang, Y.; Wei, H.; Peng, J. miR-221 negatively regulates inflammation and insulin sensitivity in white adipose tissue by repression of sirtuin-1 (SIRT1). *J. Cell. Biochem.* **2018**, *119*, 6418–6428. [CrossRef] [PubMed]
174. Penolazzi, L.; Bonaccorsi, G.; Gafà, R.; Ravaioli, N.; Gabriele, D.; Bosi, C.; Lanza, G.; Greco, P.; Piva, R. SLUG/HIF1-α/miR-221 regulatory circuit in endometrial cancer. *Gene* **2019**, *711*, 143938. [CrossRef] [PubMed]
175. Ono, K.; Igata, M.; Kondo, T.; Kitano, S.; Takaki, Y.; Hanatani, S.; Sakaguchi, M.; Goto, R.; Senokuchi, T.; Kawashima, J.; et al. Identification of microRNA that represses IRS-1 expression in liver. *PLoS ONE* **2018**, *13*, e0191553. [CrossRef]
176. Liu, B.; Che, Q.; Qiu, H.; Bao, W.; Chen, X.; Lu, W.; Li, B.; Wan, X. Elevated miR-222-3p promotes proliferation and invasion of endometrial carcinoma via targeting ERα. *PLoS ONE* **2014**, *9*, e87563. [CrossRef]
177. Shu, S.; Liu, X.; Xu, M.; Gao, X.; Chen, S.; Zhang, L.; Li, R. MicroRNA-320a acts as a tumor suppressor in endometrial carcinoma by targeting IGF-1R. *Int. J. Mol. Med.* **2019**, *43*, 1505–1512. [CrossRef]
178. Ling, H.Y.; Ou, H.S.; Feng, S.D.; Zhang, X.Y.; Tuo, Q.H.; Chen, L.X.; Zhu, B.Y.; Gao, Z.P.; Tang, C.K.; Yin, W.D.; et al. Changes in microrna (mir) profile and effects of mir-320 in insulin-resistant 3t3-l1 adipocytes. *Clin. Exp. Pharmacol. Physiol.* **2009**, *36*, e32–e39. [CrossRef]

179. Jayaraman, M.; Radhakrishnan, R.; Mathews, C.A.; Yan, M.; Husain, S.; Moxley, K.M.; Song, Y.S.; Dhanasekaran, D.N. Identification of novel diagnostic and prognostic miRNA signatures in endometrial cancer. *Genes Cancer* **2017**, *8*, 566–576. [CrossRef]
180. Cinkornpumin, J.; Roos, M.; Nguyen, L.; Liu, X.; Gaeta, X.; Lin, S.; Chan, D.N.; Liu, A.; Gregory, R.I.; Jung, M.; et al. A small molecule screen to identify regulators of let-7 targets. *Sci. Rep.* **2017**, *7*, 15973. [CrossRef]
181. Frost, R.J.A.; Olson, E.N. Control of glucose homeostasis and insulin sensitivity by the Let-7 family of microRNAs. *Proc. Natl. Acad. Sci. USA* **2011**, *108*, 21075–21080. [CrossRef]
182. Zhang, B.H.; Shen, C.A.; Zhu, B.W.; An, H.Y.; Zheng, B.; Xu, S.B.; Sun, J.C.; Sun, P.C.; Zhang, W.; Wang, J.; et al. Insight into miRNAs related with glucometabolic disorder. *Biomed. Pharmacother.* **2019**, *111*, 657–665. [CrossRef] [PubMed]
183. Jiang, S. A regulator of metabolic reprogramming: MicroRNA Let-7. *Transl. Oncol.* **2019**, *12*, 1005–1013. [CrossRef] [PubMed]
184. Delangle, R.; De Foucher, T.; Larsen, A.K.; Sabbah, M.; Azaïs, H.; Bendifallah, S.; Daraï, E.; Ballester, M.; Mehats, C.; Uzan, C.; et al. The use of microRNAs in the management of endometrial cancer: A meta-analysis. *Cancers* **2019**, *11*, 832. [CrossRef]
185. Srivastava, S.K.; Ahmad, A.; Zubair, H.; Miree, O.; Singh, S.; Rocconi, R.P.; Scalici, J.; Singh, A.P. MicroRNAs in gynecological cancers: Small molecules with big implications. *Cancer Lett.* **2017**, *407*, 123–138. [CrossRef] [PubMed]
186. Jurcevic, S.; Olsson, B.; Klinga-Levan, K. MicroRNA expression in human endometrial adenocarcinoma. *Cancer Cell Int.* **2014**, *14*, 88. [CrossRef]
187. Zhang, Y.; Zhang, L.; Sun, H.; Lv, Q.; Qiu, C.; Che, X.; Liu, Z.; Jiang, J. Forkhead transcription factor 1 inhibits endometrial cancer cell proliferation via sterol regulatory element-binding protein 1. *Oncol. Lett.* **2017**, *13*, 731–737. [CrossRef]
188. Wang, Y.; Zhang, L.; Che, X.; Li, W.; Liu, Z.; Jiang, J. Roles of SIRT1/FoxO1/SREBP-1 in the development of progestin resistance in endometrial cancer. *Arch. Gynecol. Obstet.* **2018**, *298*, 961–969. [CrossRef]
189. Plaisance, V.; Abderrahmani, A.; Perret-Menoud, V.; Jacquemin, P.; Lemaigre, F.; Regazzi, R. MicroRNA-9 controls the expression of Granuphilin/Slp4 and the secretory response of insulin-producing cells. *J. Biol. Chem.* **2006**, *281*, 26932–26942. [CrossRef]
190. Nowek, K.; Wiemer, E.A.C.; Jongen-Lavrencic, M. The versatile nature of miR-9/9* in human cancer. *Oncotarget* **2018**, *9*, 20838–20854. [CrossRef]
191. Slusarz, A.; Pulakat, L. The two faces of miR-29. *J. Cardiovasc. Med.* **2015**, *16*, 480–490. [CrossRef]
192. Yu-Han, H.; Kanke, M.; Kurtz, C.L.; Cubitt, R.; Bunaciu, R.P.; Miao, J.; Zhou, L.; Graham, J.L.; Hussain, M.M.; Havel, P.; et al. Acute suppression of insulin resistance-associated hepatic miR-29 in vivo improves glycemic control in adult mice. *Physiol. Genom.* **2019**, *51*, 379–389. [CrossRef]
193. Kwon, J.J.; Factora, T.D.; Dey, S.; Kota, J. A systematic review of miR-29 in cancer. *Mol. Ther. Oncolytics* **2019**, *12*, 173–194. [CrossRef] [PubMed]
194. Jiang, H.; Zhang, G.; Wu, J.H.; Jiang, C.P. Diverse roles of miR-29 in cancer (Review). *Oncol. Rep.* **2014**, *31*, 1509–1516. [CrossRef] [PubMed]
195. Du, J.; Zhang, F.; Zhang, L.; Jia, Y.; Chen, H. MicroRNA-103 regulates the progression in endometrial carcinoma through ZO-1. *Int. J. Immunopathol. Pharmacol.* **2019**, *33*, 1–8. [CrossRef] [PubMed]
196. Snowdon, J.; Zhang, X.; Childs, T.; Tron, V.A.; Feilotter, H. The microRNA-200 family is upregulated in endometrial carcinoma. *PLoS ONE* **2011**, *6*, e22828. [CrossRef] [PubMed]
197. Belgardt, B.F.; Ahmed, K.; Spranger, M.; Latreille, M.; Denzler, R.; Kondratiuk, N.; Von Meyenn, F.; Villena, F.N.; Herrmanns, K.; Bosco, D.; et al. The microRNA-200 family regulates pancreatic beta cell survival in type 2 diabetes. *Nat. Med.* **2015**, *21*, 619–627. [CrossRef]
198. Magenta, A.; Ciarapica, R.; Capogrossi, M.C. The emerging role of miR-200 family in cardiovascular diseases. *Circ. Res.* **2017**, *120*, 1399–1402. [CrossRef]
199. Lu, J.; Zhang, X.; Zhang, R.; Ge, Q. MicroRNA heterogeneity in endometrial cancer cell lines revealed by deep sequencing. *Oncol. Lett.* **2015**, *6*, 3457–3465. [CrossRef]
200. Panda, H.; Pelakh, L.; Chuang, T.D.; Luo, X.; Bukulmez, O.; Chegini, N. Endometrial miR-200c is altered during transformation into cancerous states and targets the expression of ZEBs, VEGFA, FLT1, IKKβ, KLF9, and FBLN5. *Reprod. Sci.* **2012**, *19*, 786–796. [CrossRef]

201. Liu, Y.; El-Naggar, S.; Darling, D.S.; Higashi, Y.; Dean, D.C. Zeb1 links epithelial-mesenchymal transition and cellular senescence. *Development* **2008**, *135*, 579–588. [CrossRef]
202. Wellner, U.; Schubert, J.; Burk, U.C.; Schmalhofer, O.; Zhu, F.; Sonntag, A.; Waldvogel, B.; Vannier, C.; Darling, D.; Zur Hausen, A.; et al. The EMT-activator ZEB1 promotes tumorigenicity by repressing stemness-inhibiting microRNAs. *Nat. Cell Biol.* **2009**, *11*, 1487–1495. [CrossRef] [PubMed]
203. Tang, Y.; Shu, G.; Yuan, X.; Jing, N.; Song, J. FOXA2 functions as a suppressor of tumor metastasis by inhibition of epithelial-to-mesenchymal transition in human lung cancers. *Cell Res.* **2011**, *21*, 316–326. [CrossRef] [PubMed]
204. Bach, D.H.; Long, N.P.; Luu, T.T.T.; Anh, N.H.; Kwon, S.W.; Lee, S.K. The dominant role of forkhead box proteins in cancer. *Int. J. Mol. Sci.* **2018**, *19*, 3279. [CrossRef] [PubMed]
205. Ramón, L.A.; Braza-Boïls, A.; Gilabert, J.; Chirivella, M.; España, F.; Estellés, A.; Gilabert-Estellés, J. MicroRNAs related to angiogenesis are dysregulated in endometrioid endometrial cancer. *Hum. Reprod.* **2012**, *27*, 3036–3045. [CrossRef] [PubMed]
206. Meerson, A.; Traurig, M.; Ossowski, V.; Fleming, J.M.; Mullins, M.; Baier, L.J. Human adipose microRNA-221 is upregulated in obesity and affects fat metabolism downstream of leptin and TNF-α. *Diabetologia* **2013**, *56*, 1971–1979. [CrossRef]
207. Abbas, A.; Imrie, H.; Viswambharan, H.; Sukumar, P.; Rajwani, A.; Cubbon, R.M.; Gage, M.; Smith, J.; Galloway, S.; Yuldeshava, N.; et al. The insulin-like growth factor-1 receptor is a negative regulator of nitric oxide bioavailability and insulin sensitivity in the endothelium. *Diabetes* **2011**, *60*, 2169–2178. [CrossRef]
208. Jasinski-Bergner, S.; Kielstein, H. Adipokines regulate the expression of tumor-relevant microRNAs. *Obes. Facts* **2019**, *12*, 211–225. [CrossRef]
209. Thomou, T.; Mori, M.A.; Dreyfuss, J.M.; Konishi, M.; Sakaguchi, M.; Wolfrum, C.; Rao, T.N.; Winnay, J.N.; Garcia-Martin, R.; Grinspoon, S.K.; et al. Adipose-derived circulating miRNAs regulate gene expression in other tissues. *Nature* **2017**, *542*, 450–455. [CrossRef]
210. Withers, S.B.; Dewhurst, T.; Hammond, C.; Topham, C.H. MiRNAs as novel adipokines: Obesity-related circulating MiRNAs influence chemosensitivity in cancer patients. *Non-Coding RNA* **2020**, *6*, 5. [CrossRef]
211. Ge, Q.; Brichard, S.; Yi, X.; Li, Q. MicroRNAs as a new mechanism regulating adipose tissue inflammation in obesity and as a novel therapeutic strategy in the metabolic syndrome. *J. Immunol. Res.* **2014**, *2014*, 987285. [CrossRef]
212. Valadi, H.; Ekström, K.; Bossios, A.; Sjöstrand, M.; Lee, J.J.; Lötvall, J.O. Exosome-mediated transfer of mRNAs and microRNAs is a novel mechanism of genetic exchange between cells. *Nat. Cell Biol.* **2007**, *9*, 654–659. [CrossRef] [PubMed]
213. Constantine, G.D.; Kessler, G.; Graham, S.; Goldstein, S.R. Increased incidence of endometrial cancer following the women's health initiative: An assessment of risk factors. *J. Women's Health* **2019**, *28*, 237–243. [CrossRef] [PubMed]
214. Hernandez, A.V.; Pasupuleti, V.; Benites-Zapata, V.A.; Thota, P.; Deshpande, A.; Perez-Lopez, F.R. Insulin resistance and endometrial cancer risk: A systematic review and meta-analysis. *Eur. J. Cancer* **2015**, *51*, 2747–2758. [CrossRef] [PubMed]

 *cancers*

Article

Tumor Suppressor miR-584-5p Inhibits Migration and Invasion in Smoking Related Non-Small Cell Lung Cancer Cells by Targeting YKT6

Saet Byeol Lee [1,2], Young Soo Park [1], Jae Sook Sung [3], Jong Won Lee [1,4], Boyeon Kim [1,2] and Yeul Hong Kim [1,2,3,5,*]

1. Cancer Research Institute, Korea University College of Medicine, Goryeodae-ro 73, Seongbuk-gu, Seoul 02841, Korea; akdltmxkf@korea.ac.kr (S.B.L.); difco@korea.ac.kr (Y.S.P.); potato23@korea.ac.kr (J.W.L.); alice_1989@korea.ac.kr (B.K.)
2. BK21 Plus Program, Korea University College of Medicine, Goryeodae-ro 73, Seongbuk-gu, Seoul 02841, Korea
3. K-MASTER Cancer Precision Medicine Diagnosis and Treatment Enterprise, Korea University Anam Hospital, Goryeodae-ro 73, Seongbuk-gu, Seoul 02841, Korea; jsssung@korea.ac.kr
4. Department of Biomedical Sciences, Korea University College of Medicine, Goryeodae-ro 73, Seongbuk-gu, Seoul 02841, Korea
5. Department of Oncology/Hematology, Korea University Anam Hospital, Korea University College of Medicine, Goryeodae-ro 73, Seongbuk-gu, Seoul 02841, Korea

* Correspondence: yhk0215@korea.ac.kr; Tel.: +82-2-920-5569; Fax: +82-2-926-4534

Simple Summary: Cigarette smoke is a major carcinogen that causes lung cancer and induces DNA methylation. DNA methylation regulates the expression of microRNA (miRNAs), which are important regulators of cancer biology. However, the association between smoking and miRNAs has not been fully elucidated in smoking-related lung carcinogenesis. In this study, we found that miR-584-5p expression was downregulated with cancer progression using a lung carcinogenesis model cell line. Moreover, we demonstrated that miR-584-5p is downregulated by the methylation of its promoter region and that it suppresses migration and invasion by targeting YKT6 in smoking-related non-small cell lung cancer (NSCLC) cells. Our results provide a better understanding of the underlying changes in miRNA expression in smoking-related lung carcinogenesis and suggest that miR-584-5p is a potential molecular biomarker for smoking-related NSCLC.

Abstract: Cigarette smoke (CS) affects the expression of microRNAs (miRNAs), which are important regulators of gene expression by inducing DNA methylation. However, the effects of smoking on miRNA expression have not been fully elucidated in smoking-related lung carcinogenesis. Therefore, in this study, to investigate the change of miRNA expression pattern and to identify tumor suppressor miRNAs by smoking in lung carcinogenesis, we used lung carcinogenesis model cell lines that, derived from a murine xenograft model with human bronchial epithelial cells (BEAS-2B), exposed CS or not. The microarray analysis revealed that miR-584-5p expression was downregulated with cancer progression in lung carcinogenesis model cell lines. We confirmed by pyrosequencing that the methylation level of the miR-584-5p promoter increased with cancer progression. In vitro and in vivo experiments showed that miR-584-5p suppressed migration and invasion in non-small cell lung cancer (NSCLC) cells by targeting YKT6. Furthermore, we showed that high level of YKT6 was associated with a poor survival rate in NSCLC patients with a history of smoking. These results suggest that miR-584-5p acts as a tumor suppressor and is a potential molecular biomarker for smoking-related NSCLC.

Keywords: smoking; non-small cell lung cancer; methylation; miR-584-5p; YKT6

Citation: Lee, S.B.; Park, Y.S.; Sung, J.S.; Lee, J.W.; Kim, B.; Kim, Y.H. Tumor Suppressor miR-584-5p Inhibits Migration and Invasion in Smoking Related Non-Small Cell Lung Cancer Cells by Targeting YKT6. *Cancers* **2021**, *13*, 1159. https://doi.org/10.3390/cancers13051159

Academic Editor: Paola Tucci

Received: 28 January 2021
Accepted: 4 March 2021
Published: 8 March 2021

1. Introduction

Lung cancer (LC) is one of the most common cancers worldwide [1,2] and is classified into two main categories: small cell lung cancer (SCLC) and non-small cell lung cancer (NSCLC). NSCLC accounts for about 85% of all lung cancers and is classified into three main histological subtypes: adenocarcinoma, squamous cell carcinoma, and large cell carcinoma [3,4]. LC is caused by factors such as smoking, exposure to radon, exposure to asbestos, and air pollution. Among these, smoking is the major risk factor for development of LC. In addition, the lungs are the major organ affected by cigarette smoke, and smoking accounts for 87% of deaths from LC [5]. Cigarette smoke (CS) contains a complex mixture of about 5000 chemicals, including nicotine, tar, benzo(a)pyrene (BaP), acetaldehyde, and nitric oxide (NO). Of these, more than 60 CS compounds are well known carcinogens [6]. Despite the proven relationship between smoking and increased risk of LC, the underlying mechanisms of how smoking contributes to lung carcinogenesis are not completely understood [7].

Smoking is known to contribute to carcinogenesis by causing epigenetic changes, such as DNA methylation and histone modification [8]. The DNA methylation of promotor regions regulates gene expression by suppressing the transcription of protein-coding genes and microRNA-coding genes [9]. Although DNA methylation is essential for the normal functioning of cells, abnormal hypermethylation and hypomethylation can contribute to cancer [10]. For example, DNA methylation of the promoter regions of tumor suppressor genes can contribute to tumor formation [11,12]. Therefore, an assessment of the methylation status of the promoter regions of specific genes has been proposed as a method for the early detection of cancer [13].

MicroRNAs (miRNAs) are small non-coding RNAs that regulate target gene expression by binding to complementary bases in the 3′ untranslated region (UTR) of their target mRNAs [14,15]. miRNAs regulate various biological processes, including those involved in critical pathways related to cell proliferation, apoptosis, metastasis, and invasion [16]. Several studies have shown that the expression level of specific miRNAs varies according to disease stage [17]. However, no prior studies have examined miRNA expression pattern in lung carcinogenesis due to smoking. Therefore, we analyzed changes in the miRNA expression pattern and degree of methylation using a lung carcinogenesis model cell line (Table 1). As a result, we identified a tumor suppressor miRNA that plays an important role in smoking-related lung cancer and investigated its biological role in smoking-related NSCLC cells.

Table 1. Characteristics of the lung carcinogenesis model cell lines.

Cell Lines	Histological Stage [a]	CSC Exposure	Tumorigenicity [b]	Histology
1799	Immortalized	-	-	Adenocarcinoma
1198	Transformed	+	-	Adenocarcinoma
1170I	Tumorigenic	+	+	Adenocarcinoma

Abbreviations: CSC (Cigarette Smoke Condensate). [a] Lacroix et al. [18] and [b] Klein-Szanto et al. [19].

2. Materials and Methods

2.1. Cell Culture and Transfection

Human lung cancer cell lines (H1703, A549, H522, H1299, and H358) were acquired from American Type Culture Collection (ATCC) (Manassas, VA, USA) and cultured in Roswell Park Memorial Institute (RPMI)—1640 with 1% of antibiotics and 10% of fetal bovine serum (FBS) (HyClone, Logan, UT, USA). To establish stable cell lines overexpressing miR-584-5p, A549 cells were infected with Lv12-u6/miR-584-5p or the negative control (Genepharma, Shanghai, China). Forty-eight hours after infection, cells were treated with 1-μg/mL puromycin to select transformed cells. BEAS-2B (human bronchial epithelial cells) and lung carcinogenesis model cell lines (1799 cells, normal immortalized cells; 1198 cells, transformed cells; and 1170I, tumorigenic cells) were gifts from Dr. Curtis Harris (National Institutes of Health, Bethesda, MD, USA) [20] and Dr. Andres Klein-Szanto (Fox

Chase Cancer Center, Philadelphia, PA, USA) [19]. Culture conditions for BEAS-2B and lung carcinogenesis model cell lines were described in a previous study [21]. All cells were incubated at 5% CO_2 and 37 °C. miR-584-5p mimics and negative controls were generated by Ambion (Thermo Fisher Scientific, Austin, TX, USA). Negative control small interfering RNA (siRNA) and YKT6 siRNA were purchased from Santa Cruz Biotechnology (Santa Cruz, CA, USA). miRNA and siRNA were used at 20 nM and transfected into cells with Lipofectamine RNAiMAX (Invitrogen, Carlsbad, CA, USA), according to the manufacturer's instructions. pCMV6 control plasmid and pCMV6-YKT6 were purchased from Origene (Rockville, MD, USA). In rescue experiments, cells were co-transfected with miR-NC or miR-584-5p mimics and pCMV6 control vector or YKT6 overexpression vector using Lipofectamine 3000 (Invitrogen, Carlsbad, CA, USA).

2.2. Treatment with Demethylation Agent

Cells were treated with 5-aza-2′-deoxycytidine (5-Aza-dC) or DMSO (Sigma-Aldrich, St. Louis, MO, USA) as the control for 72 h. Then, cells were seeded for migration and invasion assays. Cell pellets were stored at −80 °C for DNA and RNA experiments.

2.3. Microarray Analysis

MicroRNA expression in lung carcinogenesis model cell lines (1799, 1198, and 1170I) was analyzed using Affymetrix's Gene Chip miRNA Array 4.0 (Affymetrix, Santa Clara, CA, USA). Total RNA (500 ng), including miRNA, was biotin HSR-labeled using FlashTag. Samples were hybridized to the Affymetrix miRNA microarray (DNA link, Seoul, Korea) in a hybridization oven according to the protocols provided by the manufacturer.

2.4. Pyrosequencing Analysis

Genomic DNA was extracted from cell pellets using the QIAamp DNA Blood Mini kit (QIAGEN, Hilden, Germany) and quantified by Nanodrop (NanoDrop Technologies, Wilmington, DE, USA). Genomic DNA (300 ng) was used in bisulfite conversion reactions with the Lightning kit (Zymo Research, Irvine, CA, USA), according to the manufacturer's protocols. Pyrosequencing was performed according to the manufacturer's protocols (PyroGold Reagent kit, QIAGEN) by a service provider (Genomictree, Daejeon, Korea). PCR conditions consisted of incubation at 95 °C for 10 min, followed by 45 cycles of 95 °C for 30 s, 55 °C for 30 s, and 72 °C for 30 s and then a final annealing and extension step at 72 °C for 5 min. Primer sequences were as follows: primer 1, miR-584-5p (-730)-F: ATTAAAGGTTGTATTGTGTATTGA, miR-584-5p (-730)-R: biotin- CACCCATATATATACCATCCTAC, and miR-584-5p (-730)-S: TTGTGTATTGAGTAGGTT and primer2, miR-584-5p (-730)-F: ATTAAAGGTTGTATTGTG TATTGA, miR-584-5p (-730)-R: biotin-CACCCATATATATACCATCCTAC, miR-584-5p (-730)-S: TTGTGTATTGAGTAGGTT, primer 3, miR-584-5p (-583)-F: GGTTAGGGTA GGATGGTATATATATGG, miR-584-5p (-583)-R: biotin-CCCAACAAATCCCTAAAC CTCTA, and miR-584-5p (-583)-S: GGTGGTTGTTTTTGTAT.

2.5. Gene Expression Omnibus (GEO) Database Analysis

Microarray data (https://www.ncbi.nlm.nih.gov/geo/; accession numbers GSE74190 and GSE19945) were used in this study to evaluate the expression levels of miR-584-5p in various tumor types. The GSE74190 dataset includes data from 18 small cell lung carcinoma (SCLC) tissue samples, 29 squamous cell carcinoma (SQ) tissue samples, and 44 adjacent normal tissue samples, while the GSE19945 dataset contains microarray data from 35 SCLC tissue samples, five SQ tissue samples, and eight adjacent normal tissue samples. We also used microarray data (accession number GSE31210) to evaluate the association between the expression level of YKT6 mRNA and survival rate. The GSE31210 dataset contains data from lung tumor and normal tumor tissue samples. An independent Student's t-test was performed to determine the significant difference in the miR-584-5p expression between lung cancer tissues and adjacent normal control tissues. $p < 0.05$ was considered statistically significant.

2.6. RNA Isolation and Real-Time RT-PCR

Total RNA was isolated using Qiazol reagent (QIAGEN). miRNA was purified and extracted using the miRNeasy Mini kit (QIAGEN) according to the manufacturer's recommendations. Complementary DNA (cDNA) synthesis was performed using the TaqMan™ MicroRNA reverse transcription kit (Applied Biosystems, Foster City, CA, USA), and TaqMan real time-PCR was carried out according to the manufacturer's instructions (Applied Biosystems). The expression of YKT6 mRNA was measured by SYBR Green quantitative PCR (Applied Biosystems). The expression of miR-584-5p was normalized to that of RNU6B, and the mRNA expression of YKT6 was normalized to that of β-actin.

2.7. Dual-Luciferase Reporter Assay

For the dual-luciferase reporter assay, the cells were seeded in 96-well plates at a number that reached confluency after a 72-h incubation. Then, the cells were co-transfected with 20-nM miRNA mimics or negative control miRNAs and 500 ng of pGL3-wt-YKT6 3′UTR or pGL3-mut-YKT6 3′UTR using Lipofectamine 3000. Forty-eight hours after transfection, luciferase activities were measured using the Dual-Glo Luciferase Assay System (Promega, Madison, WI, USA).

2.8. Wound-Healing Assay

For the dual-luciferase reporter assay, cells were seeded in 96-well plates at a number that ensured confluency after a 72-h incubation. Then, cells were co-transfected with 20-nM miRNA mimics or negative control miRNAs and 500 ng of pGL3-wt-YKT6 3′UTR or pGL3-mut-YKT6 3′UTR using Lipofectamine 3000. Forty-eight hours after transfection, luciferase activities were measured using the Dual-Glo Luciferase Assay System (Promega).

2.9. Trans-Well Assays

Invasion assays were carried out in 24-well Transwell chambers (Corning Costar Corp, Corning, NY, USA). Forty-eight hours after transfection, 1×10^5 cells were seeded in the upper chamber in 200 μL serum-free medium, whereas the bottom chamber was filled with 750 μL 10% FBS medium. Twenty-four and 48 h later, respectively, the two chambers were washed and wiped off, and cells were fixed with 4% paraformaldehyde and 100% methanol. Next, chambers were stained with 0.1% crystal violet. Cells were counted and photographed in five randomly selected fields.

2.10. Western Blot Analysis

Cells were harvested and lysed in RIPA buffer with a protease inhibitor cocktail (1183170001, Roche, Hvidovre, Denmark) and phosphatase inhibitor cocktail (04906837001, Roche, New York, NY, USA). Total protein lysates were separated by 10% SDS-PAGE and transferred to Nitrocellulose membranes (66485, Pall Corporation, Port Washington, NY, USA). Membranes were then incubated with primary antibodies at 4 °C overnight. Subsequently, membranes were incubated with secondary antibodies at room temperature. Western blot analyses were carried out using the following antibodies: YKT6 (cat# sc-365732, Santa Cruz Biotechnology), matrix metalloproteinase 9 (MMP-9) (cat# 13667, Cell Signaling Technology, Danvers, MA, USA), and β-actin (cat# A5316, Sigma-Aldrich, St. Louis, MO, USA). Signals were visualized by enhanced chemiluminescence assays (Bio-Rad, Hercules, CA, USA).

2.11. Animal Studies

BALB/c athymic nude mice (3 to 4 weeks old) were purchased from Orient Bio Animal Center (Seongnam, Korea). All animal experiments were performed in accordance with the International Animal Care Use Committee (IACUC) of Korea University College of Medicine (IACUC approval No. KOREA-2019-0122-C1, date 8 January 2020).

2.11.1. In Vivo Tumorigenicity Assays

Ten mice were randomly divided into two groups ($n = 5$): an Lv-miR-NC group and an Lv-miR-584-5p-overexpression group. A total of 1×10^6 A549 cells transfected with Lv-miR-584-5p (or Lv-miR-NC) in 200 µL of PBS and Matrigel (356231, Corning Costar Corp) (1:1) mixture was injected subcutaneously into the flanks of mice to generate xenograft tumors. Tumor growth and weights were monitored every day. After thirty-two days, mice were sacrificed, and all tumor tissues were fixed with 4% paraformaldehyde. The expression of YKT6 in tumors was detected by qRT-PCR and immunohistochemistry assays.

2.11.2. In Vivo Metastasis Assays

A549 cells were transfected with Lv-miR-NC or Lv-miR584-5p. Sixteen mice were randomly divided into two groups ($n = 8$). A total of 1×10^6 transfected cells in PBS was injected through the tail vein. Seventy-nine days after injection, mice were sacrificed. Lung tissues were collected and fixed in 4% formalin.

2.12. Immunohistochemistry

Paraffin sections from tumor tissue samples in mice were prepared to detect the Ki-67(Dako, Glostrup, Denmark), YKT6 (cat# PA5-56565, Invitrogen). Immunostaining was performed using the Polink-2 Plus HRP Broad Kit with DAB detection system (GBI Labs, Mukilteo, WA, USA) according to the manufacturer's protocols. The stained sections slides were visualized using the Slide Scanner (Axio Scan Z1, ZEISS, Oberkochen, Germany).

2.13. TUNEL Assay

We detected apoptosis in tumor tissue samples using an ApopTag Peroxidase in situ apoptosis detection kit (Merck Millipore, Darmstadt, Germany) according to the manufacturer's instructions. The number of apoptosis-positive cells was counted under microscopy.

2.14. Kaplan–Meier Plot Analysis

Kaplan–Meier plot (https://kmplot.com/analysis/) analysis was utilized to determine the correlation between the mRNA expression of *YKT6* and survival outcomes in lung cancer patients with smoking experience. The Affymetrix ID corresponding to YKT6 is 217784_at.

2.15. Statistical Analysis

Each of our experiments was performed at least three times to ensure reproducibility. Data are expressed as mean ± standard deviation. A p-value < 0.05 was considered statistically significant. The significance of differences between groups was determined by Student's t-test in GraphPad Prism 5.0 (GraphPad Inc., San Diego, CA, USA).

3. Results

3.1. miR-584-5p Expression Is Downregulated in Lung Carcinogenesis Model Cell Lines

We hypothesized that the expression of tumor suppressor miRNAs would decrease with progression of lung carcinogenesis caused by smoking. Thus, we analyzed the following cell lines, all of which were derived from a single cell line but represent different lung cancer development stages: (Cigarette Smoke Condensate or CSC-nonexposed, immortalized) 1799 cells vs. (CSC-exposed, transformed) 1198 cells and (CSC-nonexposed, immortalized) 1799 cells vs. (CSC-exposed, tumorigenic) 1170I cells. (Figure 1A,B). The microarray data analysis showed that the expression of eight miRNAs (miR-183-5p, miR-424-5p, miR-29c-5p, miR-4448, miR-584-3p, miR-3180-3p, miR-584-5p, and miR-1183) was significantly downregulated in these lung carcinogenesis model cell lines as the cancer progressed (Figure 1C). We then validated the expression of these eight miRNAs and confirmed four of the eight (miR-183-5p, miR-424-5p, miR -29c-5p, and miR-584-5p) were consistent with the analysis results of the microarray (Figure 1D). Furthermore, we evaluated the methylation levels of the promoter regions of these miRNAs in the lung carcinogenesis model cell lines. As shown in Figure 1E, among the four miRNAs, the methylation level of only the miR-584-5p promoter region increased significantly according to stage of cancer progression. These results indicate that the CpG regions methylation level of the promoter of miR-584-5p was negatively correlated with the expression of this miRNA in 1799, 1198, and 1170I cells. Additionally, we evaluated expression level of miR-584-5p in the Gene Expression Omnibus (GEO) database. We found that miR-584-5p expression was significantly downregulated in the lung tissue of patients with smoking-related lung cancer compared to adjacent normal tissues (GSE74190 and GSE19945) (Figure 1F).

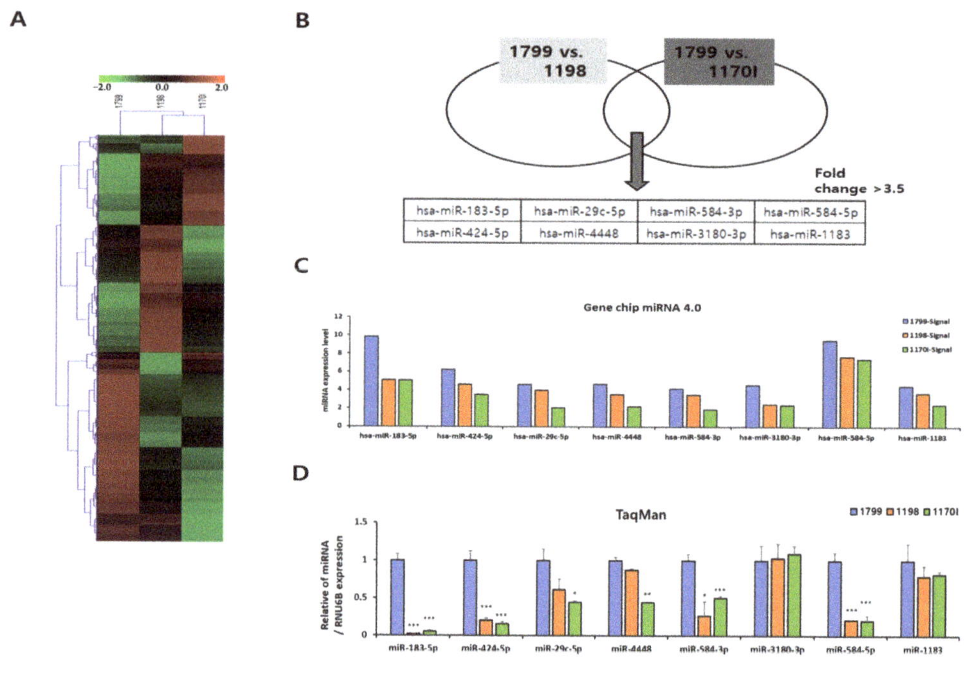

Figure 1. *Cont.*

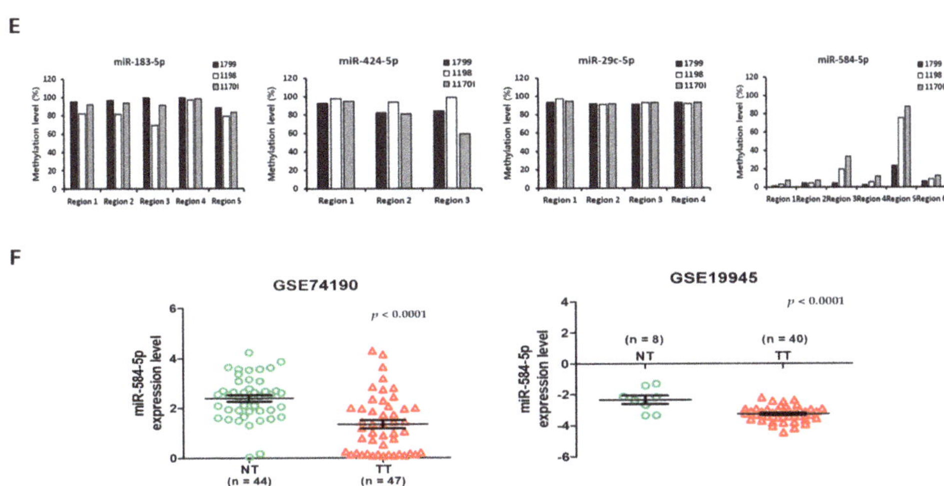

Figure 1. miR-584-5p was downregulated in lung carcinogenesis model cell lines. (**A**) Heat map clustering of differentially expressed microRNAs (miRNAs) in lung carcinogenesis model cell lines. Columns: lung carcinogenesis model cell lines (1799, 1198, and 1170I) and rows: miRNA (**B**) The Venn diagram shows the categories of the analysis group. Cigarette Smoke Condensate (CSC)-nonexposed, immortalized 1799 cells were compared with CSC-exposed, transformed 1198 cells, and CSC-nonexposed, immortalized 1799 cells were compared with CSC-exposed, tumorigenic 1170I cells. The eight miRNAs shown below the diagram are common miRNAs that showed decreased expression in 1198 cells compared to 1799 cells and in 1170I cells compared to 1799 cells. Threshold of >3.5-fold change and $p < 0.05$ were used to determine the significantly regulated miRNAs. (**C**) Microarray analysis of the expression of the eight miRNAs in lung carcinogenesis model cell lines. (**D**) qRT-PCR analysis of the eight miRNAs in lung carcinogenesis model cell lines; * $p < 0.05$, ** $p < 0.01$, and *** $p < 0.001$. (**E**) Methylation levels of the eight miRNAs in lung carcinogenesis model cell lines as determined by pyrosequencing. (**F**) Microarray analysis of miR-584-5p expression in tissues of lung cancer patients (https://www.ncbi.nlm.nih.gov/geo/; Gene Expression Omnibus (GEO) database accession numbers GSE74190 and GSE19945); *** $p < 0.001$.

3.2. miR-584-5p Regulates Migration and Invasion in Lung Carcinogenesis Model Cell Lines

We analyzed the methylation levels of six regions of the miR-584-5p promoter in lung carcinogenesis model cell lines. The 1170I cell line, which was the most highly methylated, showed the highest level of demethylation after treatment with the demethylation agent, 5-aza-2′-deoxycytidine (Figure 2A,B). In particular, among the CpGs, region 5 was the most demethylated after 5-aza-dC treatment (83% to 25%) in the 1170I cell line, followed by the 1198 cell line (80% to 42%), whereas the methylation levels of the 1799 cell line were not affected by the 5-aza-dC treatment (Figure 2C). These results indicate that downregulation of miR-584-5p was due to methylation of the promoter region of this miRNA and that region 5 is the major methylated CpG region in the promoter regions. In addition, the 5-aza-dC treatment restored miR-584-5p expression in 1198 and 1170I cell lines but not in the 1799 cell line (Figure 2D). Next, we investigated the migration and invasion ability of 1198 and 1170I cell lines in response to miR-584-5p overexpression. The overexpression of this miRNA in 1170I cells significantly reduced their migration and invasion ability, whereas miR-584-5p overexpression in 1198 cells decreased their migration ability only (Figure 2E,F).

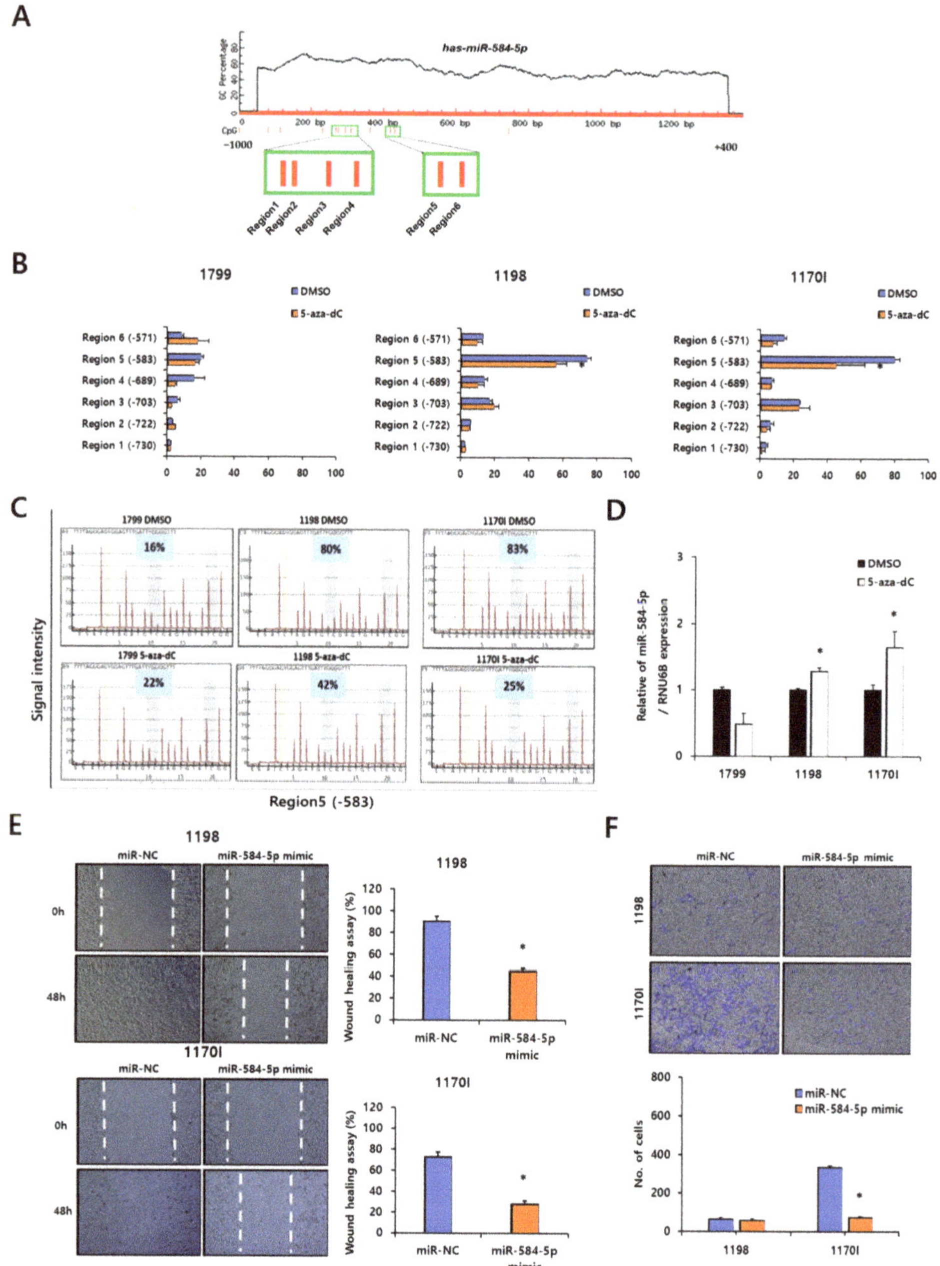

Figure 2. miR-584-5p downregulation is associated with hypermethylation in lung carcinogenesis model cell lines. (**A**) The CpG regions of miR-584-5p were predicted by MethPrimer (http://www.urogene.org), and the analyzed area is indicated with a green box. (**B**,**C**) The methylation levels and methylation signal intensity in lung carcinogenesis model cell lines with or without 5-aza-2′-deoxycytidine (5-aza-dC) treatment for 72 h by pyrosequencing. * $p < 0.05$. (**D**) The level of miR-584-5p expression in lung carcinogenesis model cell lines with or without 5-aza-2′-deoxycytidine (5-aza-dC) treatment for 72 h by qRT-PCR. * $p < 0.05$. Representative images and quantification of wound-healing (**E**) and Transwell (**F**) assays in 1198 and 1170I cells transfected with miR-584-5p mimic and the negative control. * $p < 0.05$.

3.3. Downregulation of miR-584-5p Expression Is Associated with Hypermethylation of Promoter CpG Island in Smoking-Related NSCLC Cells

We examined endogenous miR-584-5p levels in smoking-related NSCLC cell lines (H1703, H522, A549, H1299, and H358) and a normal human bronchial epithelial cell line (BEAS-2B). As shown in Figure 3A, miR-584-5p was significantly downregulated in the H1703, H522, and A549 cell lines but upregulated in the H1299 and H358 cell lines compared with the BEAS-2B cell line. Next, we investigated the methylation levels of those NSCLC cell lines. Most miR-584-5p promoter CpGs were more highly methylated in H1703 and A549 cells than in the other cell lines evaluated (Figure 3B). Consistent with our previous findings, region 5 of CpGs was the most strongly affected by 5-aza-dC treatment (Figure 3C). In addition, a significant dose-dependent increase of miR-584-5p expression was detected in H1703 and A549 cells (Figure 3D).

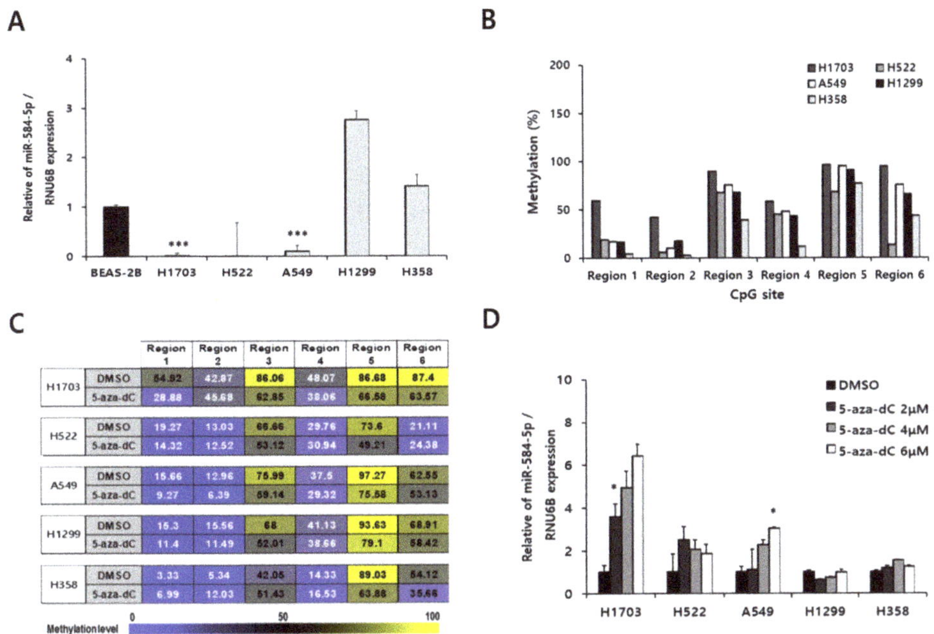

Figure 3. miR-584-5p expression was regulated by hypermethylation in smoking-related non-small cell lung cancer (NSCLC) cells. (**A**) qRT-PCR analysis of miR-584-5p expression level in smoking-related NSCLC cells; *** $p < 0.001$. (**B**) Basal methylation level of six CpGs of the miR-584-5p promoter in smoking-related NSCLC cells, as determined by pyrosequencing. (**C**) Methylation level of six CpGs of the miR-584-5p promoter in smoking-related NSCLC cells after 5-aza-2′-deoxycytidine (5-aza-dC) treatment, as determined by pyrosequencing. (**D**) qRT-PCR analysis of the miR-584-5p expression level in smoking-related NSCLC cells after 5-aza-2′-deoxycytidine (5-aza-dC) treatment; * $p < 0.05$.

3.4. miR-584-5p Suppresses Migration and Invasion in Smoking-Related NSCLC Cell Lines

We next investigated the role of miR-584-5p in smoking-related-NSCLC cell migration and invasion in H1703 and A549 cells (two cell lines with a low expression of miR-584-5p and elevated methylation level). Consistent with our previous findings (Figure 2E,F), the migration and invasion abilities of H1703 and A549 cells were significantly decreased by miR-584-5p overexpression (Figure 4A,B). Additionally, we investigated changes in metastatic ability after 5-aza-dC treatment. Interestingly, 5-aza-dC treatment suppressed the migration and invasion of H1703 and A549 cells (Figure 4C,D). Several studies have reported that the inhibition of matrix metalloproteinases (MMPs) is involved in decreased cancer cell migration and invasion in NSCLC [22,23]. Thus, we evaluated the protein levels of the lung cancer metastasis-related factors, including MMP-9. As shown in Figure 4E, the expression of MMP-9 in smoking-related-NSCLC cells was decreased by an overexpression of miR-584-5p.

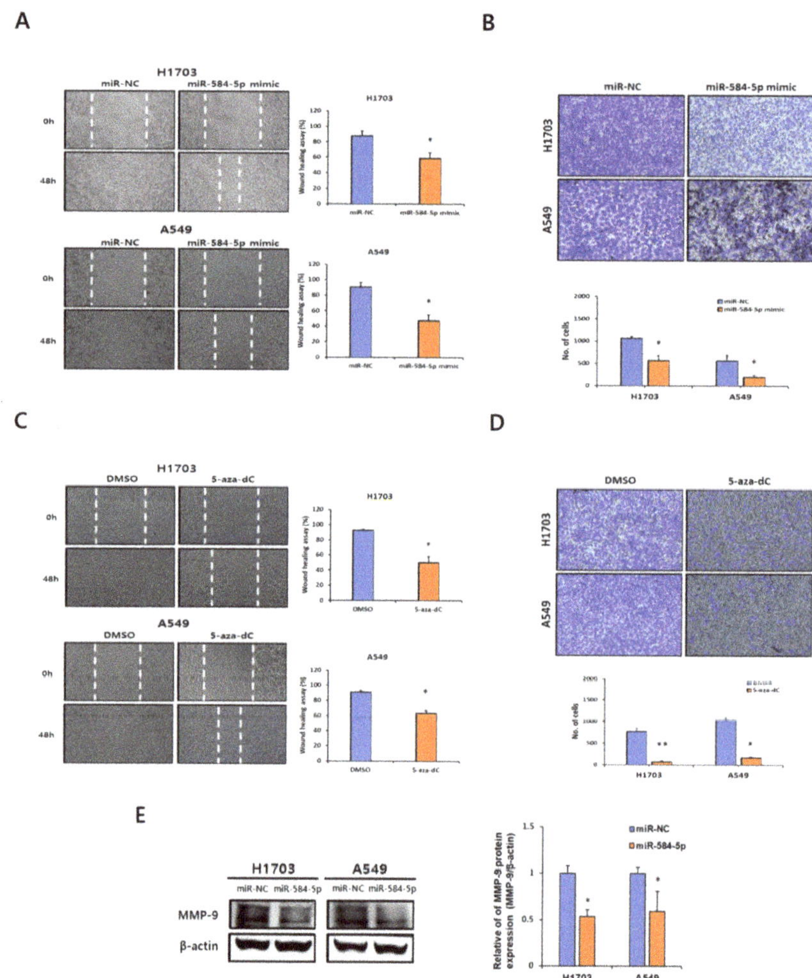

Figure 4. miR-584-5p inhibits the migration and invasion of smoking-related NSCLC cells in vitro. Representative images and quantification of the wound-healing assay (**A**) and Transwell assay (**B**) in H1703 and A549 cells transfected with miR-584-5p mimic or the negative control; * $p < 0.05$. Representative images and quantification of the wound-healing assay (**C**) and Transwell assay (**D**) in H1703 and A549 cells after the treatment with 5-aza-2′-deoxycytidine (5-aza-dC); * $p < 0.05$ and ** $p < 0.01$. (**E**) Effects of miR-584-5p mimic on the matrix metalloproteinase 9 (MMP-9) protein expression level as determined by the Western blot assay. Full Western Blot images can be found in Figure S1.

3.5. miR-584-5p Inhibits Tumor Growth and Lung Metastasis Abilities of NSCLC Cells In Vivo

To examine the effects of miR-584-5p on tumor growth and lung metastasis abilities in vivo, we generated a stable cell line, which were infected with Lv-miR-584-5p in A549 cells and verified the expression of miR-584-5p (Figure 5A). Then, Lv-miR-NC-A549 and Lv-miR-584-5p-A549 cells were subcutaneously injected into nude mice. After 32 days, we identified that the average tumor volumes and weights were significantly reduced in the miR-584-5p-overexpressing group (Figure 5B–D). We also confirmed that the expression of miR-584-5p was more upregulated in the tumors of the miR-584-5p-overexpressing group than in the tumors of the miR-NC group (Figure 5E).

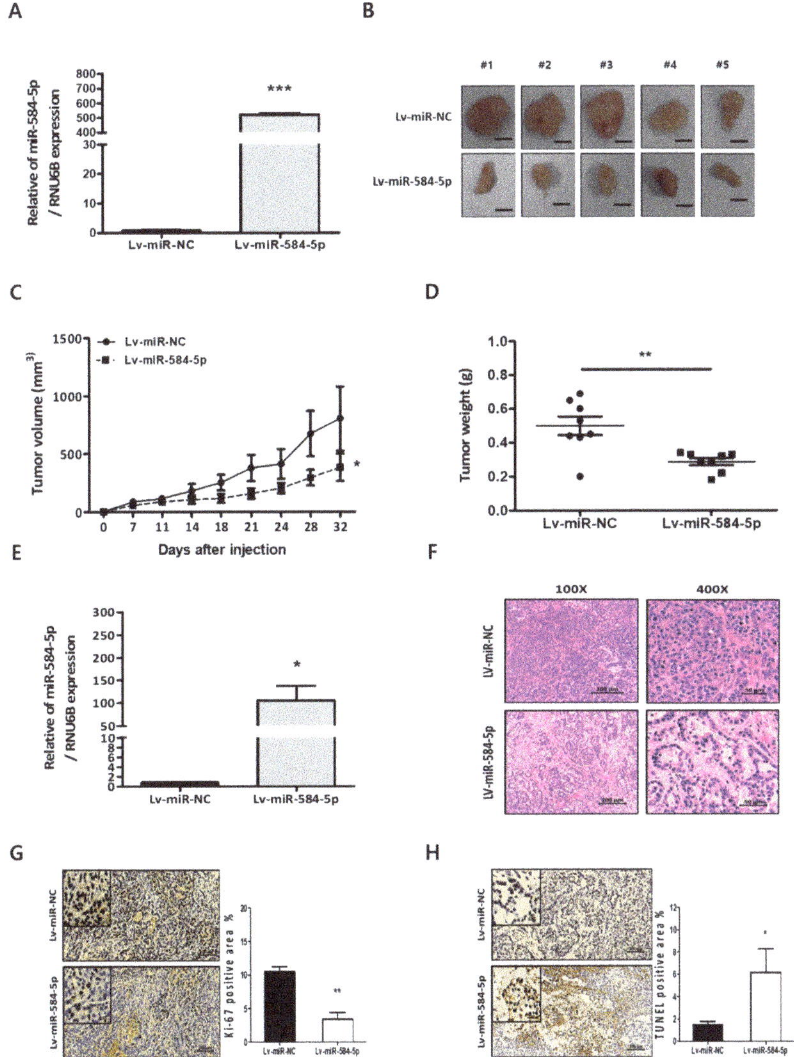

Figure 5. miR-584-5p suppresses tumor growth of smoking-related NSCLC cells in vivo. (**A**) Verification of the expression level of miR-584-5p by qRT-PCR in A549 cells infected with miR-584-5p-overexpressing lentivirus (Lv-miR-584-5p) or miR-negative control lentivirus (Lv-miR-NC); *** $p < 0.001$. (**B**) BALB/c nude mice ($n = 5$ mice per group) were subcutaneously injected with A549 cells infected with Lv-miR-584-5p or Lv-miR-NC. Representative images of tumors from BALB/c nude mice from the two groups. Scale bar represents 100 mm. (**C**,**D**) Tumor volumes and weights were measured in the two groups; * $p < 0.05$ and ** $p < 0.01$. (**E**) Verification of the expression level of miR-584-5p by qRT-PCR from xenograft tumors in the two groups; * $p < 0.05$. (**F**) Representative images of hematoxylin and eosin (H&E)-stained xenograft tumor sections from the two groups; magnification 100× and 400×, scale bar = 200 µm. Representative images are immunohistochemical (IHC) staining results for Ki-67 (**G**) and apoptotic cells (**H**) in xenograft tumor sections from the two groups; * $p < 0.05$ and ** $p < 0.01$. scale bar = 100 µm.

Hematoxylin and eosin (H&E) staining of tumor samples showed that miR-584-5p expression impaired tumor formation in the miR-584-5p-overexpressing group compared with that of the miR-NC group (Figure 5F). Ki-67 expression as a proliferation marker was decreased in miR-584-5p-overexpressing group tumors. Conversely, TUNEL staining as a marker of apoptosis was increased in miR-584-5p-overexpressing tumors compared to miR-NC tumors (Figure 5G,H). For the in vivo metastasis experiments, Lv-miR-NC-A549 and Lv-miR-584-5p-A549 cells were injected into mice through their tail veins (Figure 6A). Seventy-nine days after injection, these mice were sacrificed, and the metastatic foci in their lungs were assessed. We found that the overexpression of miR-584-5p significantly attenuated the incidence of lung metastasis and decreased the number of metastatic nodules (Figure 6B–D). H&E staining showed that lungs from miR-584-5p-overexpressing mice had fewer tumor nodules than lungs of miR-NC mice (Figure 6E).

A

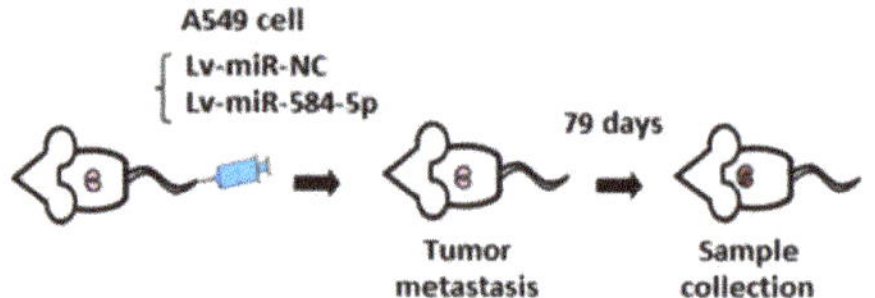

B

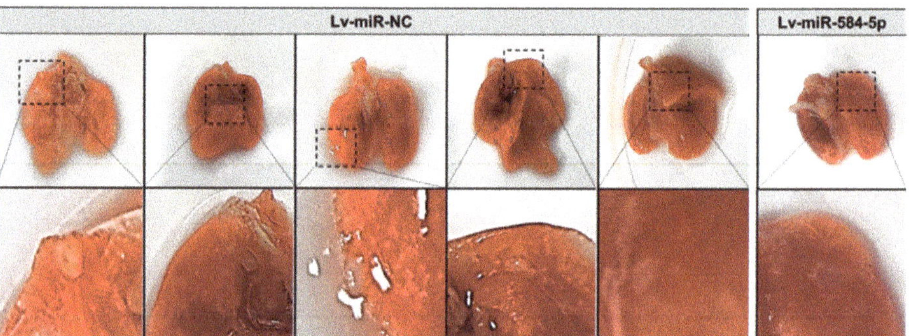

Figure 6. *Cont.*

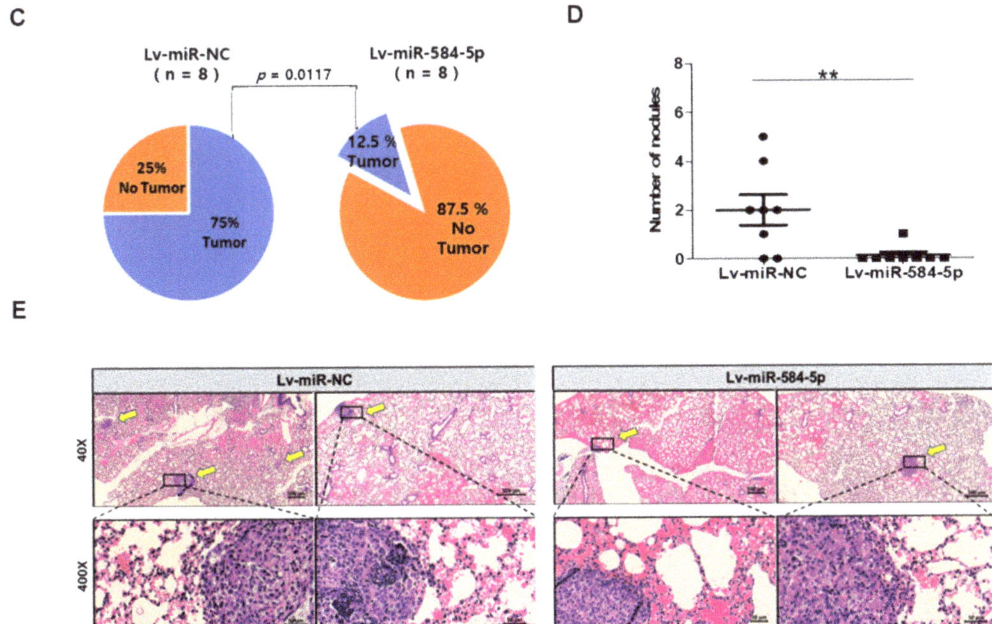

Figure 6. miR-584-5p inhibits the lung metastasis abilities of smoking-related NSCLC cells in vivo. (**A**) A549 cells transfected with miR-584-5p-overexpressing lentivirus or miR-control lentivirus were injected through the tail veil into BALB/c nude mice ($n = 8$ mice per group). Mice were sacrificed at 79 days after cell injection, and their lungs were collected. (**B**) Representative images of the lungs of the two groups of mice. (**C**) Pie graphs showing the incidence of metastatic nodules in lungs from the two groups of mice. (**D**) Numbers of metastatic lung nodules in the two groups; ** $p < 0.01$. (**E**) Representative images of H&E-stained tumor sections from lungs of mice in the two groups. Magnification, 40× and 400×. Arrows indicate metastatic lung nodules.

3.6. YKT6 Is a Direct Target of miR-584-5p

We searched two public bioinformatic databases (Target scan and miR DB) to predict the target genes of miR-584-5p and found 11 candidate genes with a target score of 80 or higher (Figure 7A). Among the candidate target genes, the endogenous mRNA expression of HDAC1, YKT6, RAP2A, and ENAH was reduced by more than 40% by overexpressed miR-584-5p in H1703 and A549 cells (Figure 7B). We also confirmed that the protein expression of YKT6 was significantly downregulated in H1703 and A549 cells by miR-584-5p (Figure 7C). Next, we searched for interaction sites between miR-584-5p and the 3′UTR of the YKT6 mRNA using miRbase (Figure 7D) and constructed a YKT6 3′UTR reporter luciferase assay system. Co-transfection experiments showed that miR-584-5p overexpression significantly decreased the luciferase activity of the YKT6 wild-type vector but not that of the mutant vector in H1703 and A549 cells (Figure 7E). Furthermore, we validated the results in the tumor samples from nude mice. As expected, the protein expression of YKT6 was slightly decreased in tumor tissues of the Lv-miR-584-5p group compared to the Lv-miR-NC group but did not reach statistical significance (Figure 7F). Consistent with these results, the IHC staining results showed that YKT6 expression was downregulated in tumor tissues of the Lv-miR-584-5p mice group (Figure 7G).

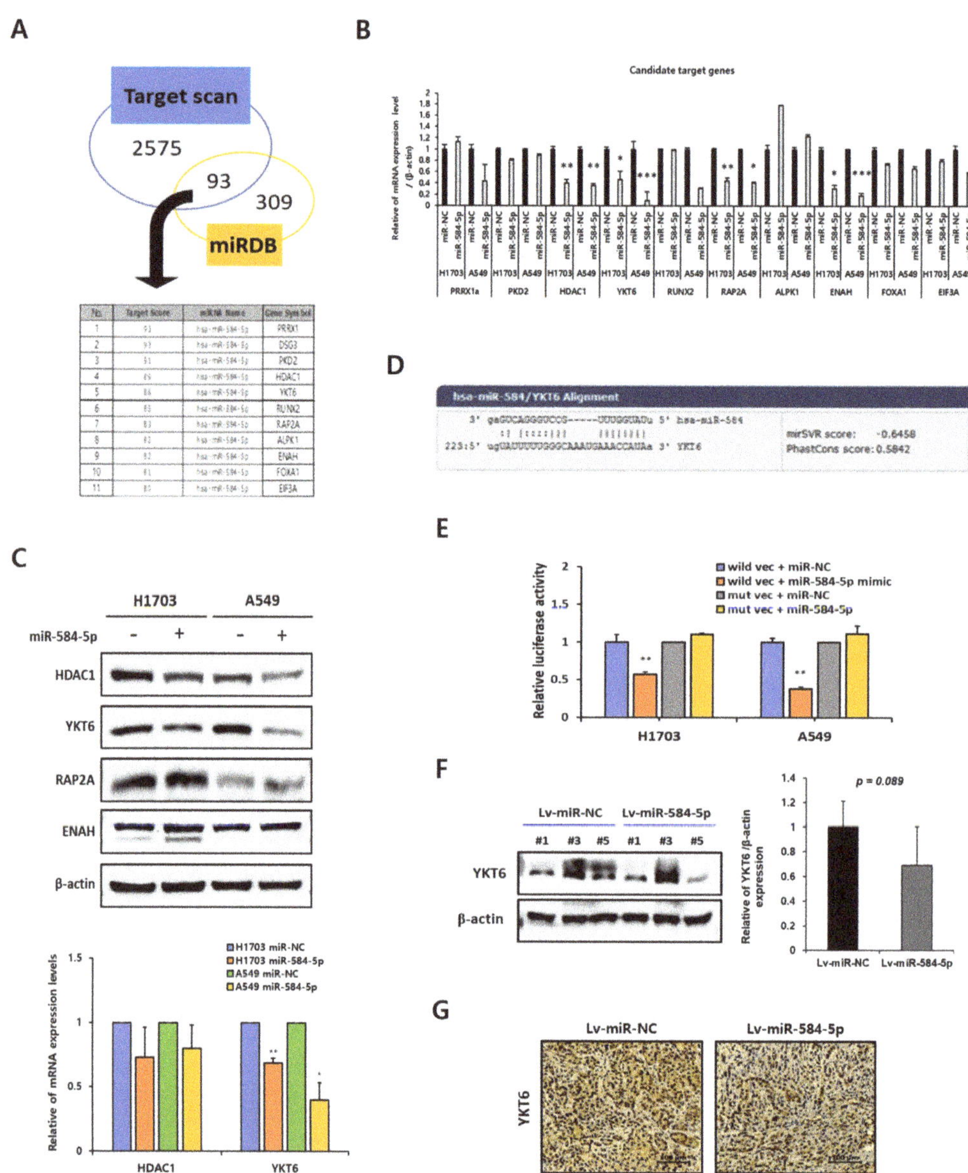

Figure 7. YKT6 is a direct target gene of miR-584-5p in smoking-related NSCLC cells. (**A**) miR DB (target score ≥ 80) and Targetscan database were used to analyze the target gene of miR-584-5p. Venn diagrams show groups of predicted target genes. (**B**) Verification of the mRNA expression of candidate target genes by qRT-PCR in H1703 and A549 cells transfected with miR-584-5p mimic or miR control; * $p < 0.05$, ** $p < 0.01$, and *** $p < 0.001$. (**C**) Verification of the protein expression of candidate target genes (HDAC1, YKT6, RAP2A, and ENAH) by Western blot assays in H1703 and A549 cells transfected with the miR-584-5p mimic or miR control; * $p < 0.05$ and ** $p < 0.01$. Full Western Blot images can be found in Figure S2. (**D**) Representative images of the has-miR-584/YKT6 binding sequence. (**E**) Relative luciferase activity was measured in H1703 and A549 cells transfected with the reporter vector containing wild-type or mutant-type YKT6 3′UTR (untranslated region), along with miR-584-5p mimic or miR control; ** $p < 0.01$. (**F**,**G**) Expression of YKT6 in xenograft tumor tissue was analyzed by Western blot and immunohistochemical (IHC) staining. Full Western Blot images can be found in Figure S3.

3.7. YKT6 Regulates Migration and Invasion in Smoking-Related NSCLC Cell Lines

We found that the depletion of YKT6 inhibited the migration and invasion abilities of smoking-related NSCLC cells (Figure 8A,B). Next, we assessed the protein level of MMP-9. Consistent with our earlier observations (Figure 4E), the protein level of MMP-9 was markedly decreased by the depletion of YKT6 in H1703 and A549 cells (Figure 8C). Rescue experiments revealed that YKT6 overexpression attenuated the inhibitory effect of miR-584-5p mimics on the YKT6 expression in H1703 and A549 cells (Figure 8D). We also investigated the effects of YKT6 overexpression on smoking-related NSCLC cell migration and invasion. The inhibitory effects of miR-584-5p overexpression on H1703 and A549 cell migration and invasion were attenuated by the expression of exogenous YKT6 (Figure 8E,F).

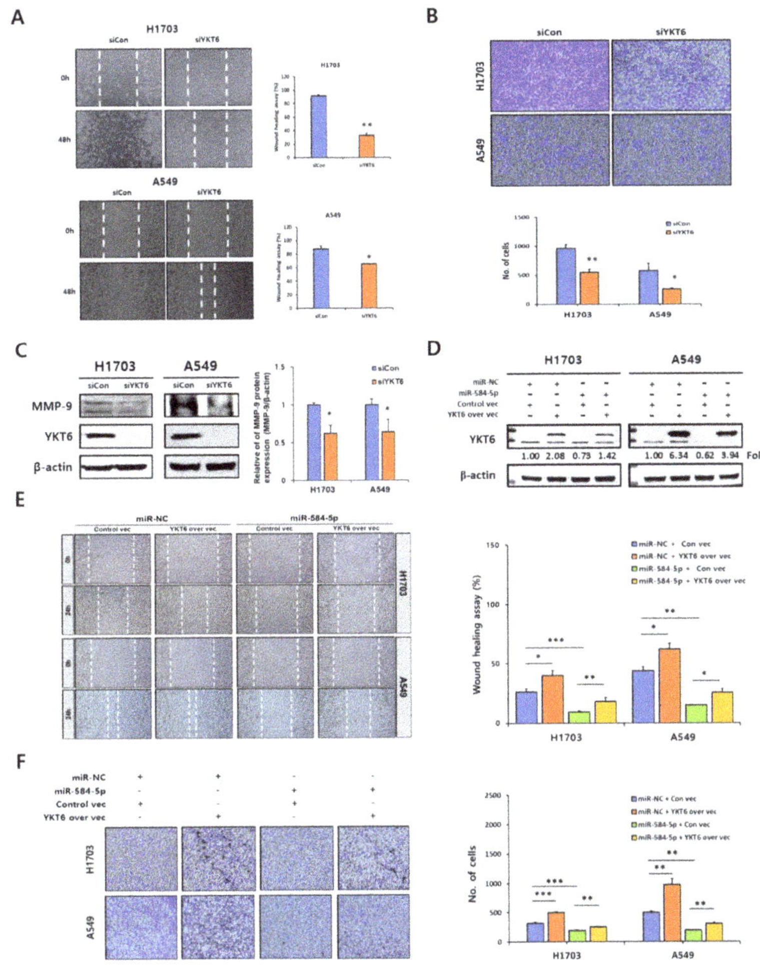

Figure 8. Depletion of YKT6 inhibited migration and invasion of smoking-related NSCLC cells. Representative images and quantification of the wound-healing assay (**A**) and Transwell assay (**B**) in H1703 and A549 cells transfected with siYKT6 or Control siRNA; * $p < 0.05$ and ** $p < 0.01$. (**C**) Verification of MMP-9 expression by Western blot assay in H1703 and A549 cells transfected with siYKT6 or Control siRNA. Full Western Blot images can be found in Figure S4 (**D**) Confirmation of transfection efficiency of H1703 and A549 cells by Western blot analysis. Full Western Blot images can be found in Figure S5. (**E**,**F**) Rescue of the migration and invasion abilities of H1703 and A549 cells by the exogenous expression of YKT6; * $p < 0.05$, ** $p < 0.01$, and *** $p < 0.001$.

3.8. YKT6 Expression Level Is Associated with Survival Rate in Smoking-Related NSCLC Patients

We hypothesized that the expression of YKT6, the target gene of miR-584-5p, would be associated with smoking-related lung cancer. In the GSE31210 dataset, YKT6 mRNA expression was significantly upregulated in tumor tissues of ever-smokers compared to the tumor tissues of never-smokers in lung cancer patients (Figure 9A). Additionally, we analyzed the overall survival rate of lung cancer patients according to the YKT6 expression level using Kaplan–Meier plots. The group with elevated YKT6 expression had a low survival rate, and a significant *p*-value was obtained for smoking-related patients (Figure 9B).

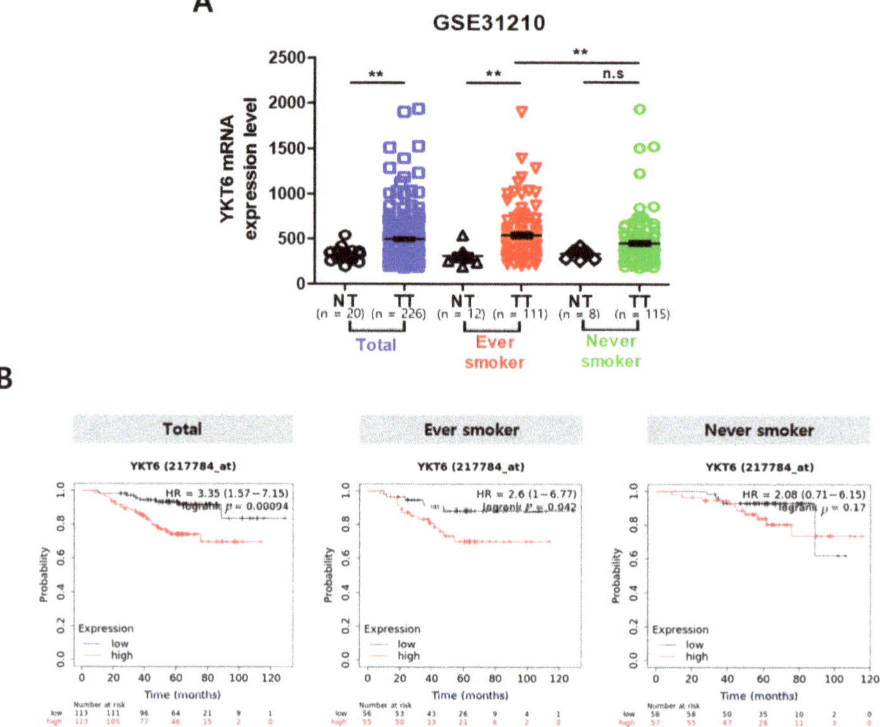

Figure 9. YKT6 expression level is associated with survival rate in NSCLC patients who smoked. (**A**) YKT6 expression level in NSCLC patient tumor tissue and adjacent normal tissue from the GSE31210 (GEO Database); ** $p < 0.01$. (**B**) Kaplan–Meier survival analysis shows that patients with a higher expression of YKT6 (217784_at) had a poorer overall survival than those with low YKT6 expression; ** $p < 0.01$, and n.s. = not significant.

4. Discussion

Many studies have demonstrated a correlation between smoking and miRNA expression in various cancers, including lung cancer [24–26]. In addition, previous reports have shown that cigarette smoking induces genetic changes through its effects on miRNAs [27]. Thus, miRNAs appear to play an important role in the development of smoking-related cancer. For example, cigarette smoking can induce miR-994 expression in oral cancer [28]. The expression of miR-21 was increased by cigarette smoke extract exposure in colorectal cancer [29], and the expression of this miRNA was elevated in esophageal cancers patients with consistent cigarette smoking [30]. miR-205 and miR-99a were shown to be downregulated in bladder cancer in smokers [31]. The expression of miR-486-5p is associated with the smoking-induced development of lung adenocarcinoma [25]. However, it is still insufficient to elucidate the specific miRNAs involved with the process of lung carcinogenesis caused by smoking. This is because most studies selected miRNAs that change with CSC exposure in normal or cancer cells or chose miRNAs based on smoking history [32,33]. In smoking-related carcinogenesis studies, cigarette smoke extract (CSE), cigarette smoke condensate (CSC), and whole cigarette smoke (WCS) are used to mimic the impacts of smoking [34]. CSE and CSC are mainly used in cell culture-based studies. CSE refers to the aqueous solution obtained by dissolving WCS in a cell culture medium or phosphate-buffered saline (PBS), while CSC refers to the solution obtained by collecting WCS in a filter pad and then dissolving it in organic solvents such as methanol or dimethyl sulfide side (DMSO) [35]. However, CSE or CSC does not reflect the composition of the gas released by smoking cigarettes, and these experimental methods do not reflect that CS affects the human lung by inhalation through the bronchus. In this study, we used three previously established lung carcinogenesis model cell lines—namely, 1799, 1198, and 1170I cells [19]. Although all these cell lines are derived from BEAS-2B cells, which are immortalized human normal bronchial epithelial cell lines, they represent different histological stages. For example, 1799 cells (immortalized) were derived from BEAS-2B cells exposed in vivo to beeswax only as a control, while 1198 (transformed) and 1170I (tumorigenic) cells were derived from BEAS-2B cells exposed in vivo to a beeswax pellet containing CSC. Importantly, the lung carcinogenesis model cell lines that we evaluated mimic the gradual changes that occur during human lung carcinogenesis induced by smoking [19]. Therefore, these cell lines are the best available in vitro model of lung carcinogenesis caused by smoking.

In the present study, we found that the expression of miR-584-5p is decreased during the process of lung carcinogenesis induced by cigarette smoking. Previous studies have reported that miR-584-5p expression is dysregulated in a variety of cancers, including hepatocellular carcinoma [36], medulloblastoma [37], gastric cancer [38], lung adenocarcinoma [39], and non-small cell lung cancer [40]. For example, Wei et al. [40] reported that miR-584-5p expression was downregulated in tissues of NSCLC patients and that the overexpression of miR-584-5p inhibited migration and invasion by targeting MMP-14. However, the mechanism by which miR-584-5p is downregulated in several cancers, including NSCLC cells, has not been identified. In this report, we showed that miR-584-5p expression is regulated by methylation in lung carcinogenesis model cell lines and smoking-related NSCLC cells. For the first time, our results demonstrated a link between the expression of miR-584-5p and smoking-induced methylation. Furthermore, we verified that miR-584-5p is controlled by methylation, as its expression was increased by demethylation. As shown in Figure 2D, the expression level of miR-584-5p was not increased by the 5-aza-dC treatment in 1799 cells, unlike what we observed in the 1198 and 1170I cells. This might be because the CpG sites of the miR-584-5p promoter were hardly methylated in 1799 cells exposed only to beeswax without CSC exposure [41]. Further, when we confirmed the effect of miR-584-5p on the invasion of the lung carcinogenesis model cell lines, an overexpression of miR-584-5p inhibited the invasion of 1170I cells but not 1198 cells (Figure 2F). We suspect that this is because 1198 and 1170I cells have different morphological characteristics. In addition, when we examined CpGs regulating the expression of miR-584-5p, we found that region 5 is a major site for regulation through methylation.

We also demonstrated that the overexpression of miR-584-5p could inhibit the migration and invasion of smoking-related NSCLC cells through its inhibitory effects on MMP-9. Additionally, this study is the first to show that miR-584-5p inhibits smoking-related NSCLC cell migration and invasion both in vitro and in vivo. To understand the mechanisms by which miR-584-5p inhibits migration and invasion in smoking-related NSCLC cells, we investigated a candidate target gene of miR-584-5p. According to previous reports, a single miRNA can regulate multiple target genes [42]. Likewise, we identified several candidate target genes for miR-584-5p, including Ras-related protein Rap-2a (RAP2A), ENAH, histone deacetylase 1 (HDAC1), and YKT6. However, among these candidate target genes, only YKT6 expression was significantly regulated by miR-584-5p at both the mRNA and protein levels in smoking-related NSCLC cells. Here, we showed that miR-584-5p could only inhibit migration and invasion but confirmed that miR-584-5p could also inhibit the proliferation of NSCLC cells (data not shown). However, YKT6 depletion could not significantly inhibit cell proliferation. Therefore, it is assumed that miR-584-5p may have other targets that inhibit the proliferation of NSCLC cells. Although this was not a prospective study, we found that a high expression level of YKT6 was associated with low survival rates in lung cancer patients with smoking experience by an analysis of the GEO database. Several previous studies have reported that YKT6 controls cell migration and invasion [43], and the level of this protein is significantly upregulated in p53-mutated tumors and in breast cancer cells resistant to docetaxel [44]. YKT6 is the target of miR-134 and miR-135b in NSCLC cells, and a low YKT6 expression has been reported to be associated with an improved survival of NSCLC patients [45]. YKT6, which is a SNARE protein recognition molecule, is involved in vesicular transport between secretory compartments [46] and is located in the membrane, cytosol, and perinuclear regions of cells. Due to its likely participation in various stages of intracellular vesicle trafficking, this SNARE protein might play essential roles in controlling the membrane dynamics during cell adhesion and migration [43]. Taken together, our results indicate that miR-584-5p, which functions as a tumor suppressor, was downregulated by methylation in smoking-related NSCLC cells. In addition, we demonstrated that the overexpression of miR-584-5p inhibited smoking-related NSCLC cell migration and invasion by targeting YKT6 both in vitro and in vivo. Therefore, these novel findings suggest that miR-584-5p is a potential molecular biomarker for smoking-related NSCLC.

5. Conclusions

In our study, we demonstrated that tumor suppressor miR-584-5p is an important factor in the developmental stage of smoking-related lung carcinogenesis. We found that miR-584-5p expression was downregulated by methylation. Moreover, overexpressed miR-584-5p suppresses the migration and invasion of smoking-related NSCLC cells by targeting YKT6 both in vitro and in vivo. In conclusion, tumor suppressor miR-584-5p can be used as molecular biomarker for smoking-related NSCLC.

Supplementary Materials: The following are available online at https://www.mdpi.com/2072-6694/13/5/1159/s1, Figure S1: Original images in Figure 4E for Western blots, Figure S2: Original images in Figure 7C for Western blots, Figure S3: Original images in Figure 7F for Western blots, Figure S4: Original images in Figure 8C for Western blots, Figure S5: Original images in Figure 8D for Western blots.

Author Contributions: Conceptualization, S.B.L. and J.S.S.; methodology, S.B.L. and Y.S.P.; data analysis, S.B.L., B.K., J.W.L., Y.S.P. and Y.H.K.; writing—original draft preparation and writing—review and editing, Y.S.P. and Y.H.K.; supervision, Y.H.K.; and funding acquisition, J.S.S. and Y.H.K. All authors have read and agreed to the published version of the manuscript.

Funding: This research was funded by the Ministry of Education, Science and Technology, Republic of Korea (NRF-2013R1A1A2062146) and Korea University Grant (K1904521 and K1922321).

Institutional Review Board Statement: This study was approved by Institutional Animal Care and Use Committee of Korea University College of Medicine, approval number KOREA-2019-0122-C1, date 8 January 2020.

Data Availability Statement: Data is contained within the article or Supplementary Materials.

Conflicts of Interest: The authors declare no conflict of interest. The funders had no role in the design of the study; in the collection, analyses, or interpretation of data; in the writing of the manuscript; or in the decision to publish the results.

References

1. Khuder, S.A. Effect of cigarette smoking on major histological types of lung cancer: A meta-analysis. *Lung Cancer* **2001**, *31*, 139–148. [CrossRef]
2. Nasim, F.; Sabath, B.F.; Eapen, G.A. Lung Cancer. *Med. Clin. N. Am.* **2019**, *103*, 463–473. [CrossRef]

3. Pikor, L.A.; Ramnarine, V.R.; Lam, S.; Lam, W.L. Genetic alterations defining NSCLC subtypes and their therapeutic implications. *Lung Cancer* **2013**, *82*, 179–189. [CrossRef] [PubMed]
4. Inamura, K. Lung Cancer: Understanding Its Molecular Pathology and the 2015 WHO Classification. *Front. Oncol.* **2017**, *7*, 193. [CrossRef]
5. Bakulski, K.M.; Dou, J.; Lin, N.; London, S.J.; Colacino, J.A. DNA methylation signature of smoking in lung cancer is enriched for exposure signatures in newborn and adult blood. *Sci. Rep.* **2019**, *9*, 4576. [CrossRef] [PubMed]
6. Jeon, S.Y.; Go, R.E.; Heo, J.R.; Kim, C.W.; Hwang, K.A.; Choi, K.C. Effects of cigarette smoke extracts on the progression and metastasis of human ovarian cancer cells via regulating epithelial-mesenchymal transition. *Reprod. Toxicol.* **2016**, *65*, 1–10. [CrossRef] [PubMed]
7. Lee, M.K.; Hong, Y.; Kim, S.Y.; London, S.J.; Kim, W.J. DNA methylation and smoking in Korean adults: Epigenome-wide association study. *Clin. Epigenetics* **2016**, *8*, 103. [CrossRef]
8. Peluso, M.E.; Munnia, A.; Bollati, V.; Srivatanakul, P.; Jedpiyawongse, A.; Sangrajrang, S.; Ceppi, M.; Giese, R.W.; Boffetta, P.; Baccarelli, A.A. Aberrant methylation of hypermethylated-in-cancer-1 and exocyclic DNA adducts in tobacco smokers. *Toxicol. Sci.* **2014**, *137*, 47–54. [CrossRef]
9. Heller, G.; Altenberger, C.; Steiner, I.; Topakian, T.; Ziegler, B.; Tomasich, E.; Lang, G.; End-Pfützenreuter, A.; Zehetmayer, S.; Döme, B.; et al. DNA methylation of microRNA-coding genes in non-small-cell lung cancer patients. *J. Pathol.* **2018**, *245*, 387–398. [CrossRef] [PubMed]
10. Kulis, M.; Esteller, M. DNA methylation and cancer. *Adv. Genet.* **2010**, *70*, 27–56. [CrossRef] [PubMed]
11. Shui, I.M.; Wong, C.J.; Zhao, S.; Kolb, S.; Ebot, E.M.; Geybels, M.S.; Rubicz, R.; Wright, J.L.; Lin, D.W.; Klotzle, B.; et al. Prostate tumor DNA methylation is associated with cigarette smoking and adverse prostate cancer outcomes. *Cancer* **2016**, *122*, 2168–2177. [CrossRef] [PubMed]
12. Yang, Z.; Qi, W.; Sun, L.; Zhou, H.; Zhou, B.; Hu, Y. DNA methylation analysis of selected genes for the detection of early-stage lung cancer using circulating cell-free DNA. *Adv. Clin. Exp. Med.* **2019**, *28*, 355–360. [CrossRef]
13. Pan, Y.; Liu, G.; Zhou, F.; Su, B.; Li, Y. DNA methylation profiles in cancer diagnosis and therapeutics. *Clin. Exp. Med.* **2018**, *18*, 1–14. [CrossRef]
14. Yan, F.; Shen, N.; Pang, J.; Xie, D.; Deng, B.; Molina, J.R.; Yang, P.; Liu, S. Restoration of miR-101 suppresses lung tumorigenesis through inhibition of DNMT3a-dependent DNA methylation. *Cell Death Dis.* **2014**, *5*, e1413. [CrossRef]
15. Iqbal, M.A.; Arora, S.; Prakasam, G.; Calin, G.A.; Syed, M.A. MicroRNA in lung cancer: Role, mechanisms, pathways and therapeutic relevance. *Mol. Asp. Med.* **2019**, *70*, 3–20. [CrossRef] [PubMed]
16. Wu, K.L.; Tsai, Y.M.; Lien, C.T.; Kuo, P.L.; Hung, A.J. The Roles of MicroRNA in Lung Cancer. *Int. J. Mol. Sci.* **2019**, *20*, 1611. [CrossRef]
17. Petrovic, N.; Ergün, S.; Isenovic, E.R. Levels of MicroRNA Heterogeneity in Cancer Biology. *Mol. Diagn.* **2017**, *21*, 511–523. [CrossRef]
18. Lacroix, L.; Feng, G.; Lotan, R. Identification of genes expressed differentially in an in vitro human lung carcinogenesis model. *Cancer Biol.* **2006**, *5*, 665–673. [CrossRef] [PubMed]
19. Klein-Szanto, A.J.; Iizasa, T.; Momiki, S.; Garcia-Palazzo, I.; Caamano, J.; Metcalf, R.; Welsh, J.; Harris, C.C. A tobacco-specific N-nitrosamine or cigarette smoke condensate causes neoplastic transformation of xenotransplanted human bronchial epithelial cells. *Proc. Natl. Acad. Sci. USA* **1992**, *89*, 6693–6697. [CrossRef]
20. Reddel, R.R.; Ke, Y.; Gerwin, B.I.; McMenamin, M.G.; Lechner, J.F.; Su, R.T.; Brash, D.E.; Park, J.B.; Rhim, J.S.; Harris, C.C. Transformation of human bronchial epithelial cells by infection with SV40 or adenovirus-12 SV40 hybrid virus, or transfection via strontium phosphate coprecipitation with a plasmid containing SV40 early region genes. *Cancer Res.* **1988**, *48*, 1904–1909. [PubMed]
21. Whang, Y.M.; Jo, U.; Sung, J.S.; Ju, H.J.; Kim, H.K.; Park, K.H.; Lee, J.W.; Koh, I.S.; Kim, Y.H. Wnt5a is associated with cigarette smoke-related lung carcinogenesis via protein kinase C. *PLoS ONE* **2013**, *8*, e53012. [CrossRef] [PubMed]
22. Yin, P.; Peng, R.; Peng, H.; Yao, L.; Sun, Y.; Wen, L.; Wu, T.; Zhou, J.; Zhang, Z. MiR-451 suppresses cell proliferation and metastasis in A549 lung cancer cells. *Mol. Biotechnol.* **2015**, *57*, 1–11. [CrossRef] [PubMed]
23. Rajasinghe, L.D.; Pindiprolu, R.H.; Gupta, S.V. Delta-tocotrienol inhibits non-small-cell lung cancer cell invasion via the inhibition of NF-κB, uPA activator, and MMP-9. *Onco Targets Ther.* **2018**, *11*, 4301–4314. [CrossRef] [PubMed]
24. Zhang, L.; Wang, H.; Wang, C. Persistence of smoking induced non-small cell lung carcinogenesis by decreasing ERBB pathway-related microRNA expression. *Thorac. Cancer* **2019**, *10*, 890–897. [CrossRef]
25. Tessema, M.; Yingling, C.M.; Picchi, M.A.; Wu, G.; Ryba, T.; Lin, Y.; Bungum, A.O.; Edell, E.S.; Spira, A.; Belinsky, S.A. ANK1 Methylation regulates expression of MicroRNA-486-5p and discriminates lung tumors by histology and smoking status. *Cancer Lett.* **2017**, *410*, 191–200. [CrossRef]
26. Wang, B.; Liu, Y.; Luo, F.; Xu, Y.; Qin, Y.; Lu, X.; Xu, W.; Shi, L.; Liu, Q.; Xiang, Q. Epigenetic silencing of microRNA-218 via EZH2-mediated H3K27 trimethylation is involved in malignant transformation of HBE cells induced by cigarette smoke extract. *Arch. Toxicol.* **2016**, *90*, 449–461. [CrossRef] [PubMed]
27. Fujii, T.; Shimada, K.; Nakai, T.; Ohbayashi, C. MicroRNAs in Smoking-Related Carcinogenesis: Biomarkers, Functions, and Therapy. *J. Clin. Med.* **2018**, *7*, 98. [CrossRef]

28. Peng, H.Y.; Hsiao, J.R.; Chou, S.T.; Hsu, Y.M.; Wu, G.H.; Shieh, Y.S.; Shiah, S.G. MiR-944/CISH mediated inflammation via STAT3 is involved in oral cancer malignance by cigarette smoking. *Neoplasia* **2020**, *22*, 554–565. [CrossRef]
29. Dino, P.; D'Anna, C.; Sangiorgi, C.; Di Sano, C.; Di Vincenzo, S.; Ferraro, M.; Pace, E. Cigarette smoke extract modulates E-Cadherin, Claudin-1 and miR-21 and promotes cancer invasiveness in human colorectal adenocarcinoma cells. *Toxicol. Lett.* **2019**, *317*, 102–109. [CrossRef]
30. Zhang, Y.; Pan, T.; Zhong, X.; Cheng, C. Nicotine upregulates microRNA-21 and promotes TGF-β-dependent epithelial-mesenchymal transition of esophageal cancer cells. *Tumour Biol.* **2014**, *35*, 7063–7072. [CrossRef]
31. Ganji, S.M.; Saidijam, M.; Amini, R.; Mousavi-Bahar, S.H.; Shabab, N.; Seyedabadi, S.; Mahdavinezhad, A. Evaluation of MicroRNA-99a and MicroRNA-205 Expression Levels in Bladder Cancer. *Int. J. Mol. Cell. Med.* **2017**, *6*, 87–95. [CrossRef] [PubMed]
32. Huang, J.; Wu, J.; Li, Y.; Li, X.; Yang, T.; Yang, Q.; Jiang, Y. Deregulation of serum microRNA expression is associated with cigarette smoking and lung cancer. *Biomed Res. Int.* **2014**, *2014*, 364316. [CrossRef]
33. Xi, S.; Xu, H.; Shan, J.; Tao, Y.; Hong, J.A.; Inchauste, S.; Zhang, M.; Kunst, T.F.; Mercedes, L.; Schrump, D.S. Cigarette smoke mediates epigenetic repression of miR-487b during pulmonary carcinogenesis. *J. Clin. Investig.* **2013**, *123*, 1241–1261. [CrossRef]
34. Dalle-Donne, I.; Colombo, G.; Gornati, R.; Garavaglia, M.L.; Portinaro, N.; Giustarini, D.; Bernardini, G.; Rossi, R.; Milzani, A. Protein Carbonylation in Human Smokers and Mammalian Models of Exposure to Cigarette Smoke: Focus on Redox Proteomic Studies. *Antioxid. Redox Signal.* **2017**, *26*, 406–426. [CrossRef]
35. Kim, Y.H.; An, Y.J.; Jo, S.; Lee, S.H.; Lee, S.J.; Choi, S.J.; Lee, K. Comparison of volatile organic compounds between cigarette smoke condensate (CSC) and extract (CSE) samples. *Environ. Health Toxicol.* **2018**, *33*, e2018010–e2018012. [CrossRef] [PubMed]
36. Wei, H.; Wang, J.; Xu, Z.; Lu, Y.; Wu, X.; Zhuo, C.; Tan, C.; Tang, Q.; Pu, J. miR-584-5p regulates hepatocellular carcinoma cell migration and invasion through targeting KCNE2. *Mol. Genet. Genom. Med.* **2019**, *7*, e702. [CrossRef] [PubMed]
37. Abdelfattah, N.; Rajamanickam, S.; Panneerdoss, S.; Timilsina, S.; Yadav, P.; Onyeagucha, B.C.; Garcia, M.; Vadlamudi, R.; Chen, Y.; Brenner, A.; et al. MiR-584-5p potentiates vincristine and radiation response by inducing spindle defects and DNA damage in medulloblastoma. *Nat. Commun.* **2018**, *9*, 4541. [CrossRef]
38. Li, Q.; Li, Z.; Wei, S.; Wang, W.; Chen, Z.; Zhang, L.; Chen, L.; Li, B.; Sun, G.; Xu, J.; et al. Overexpression of miR-584-5p inhibits proliferation and induces apoptosis by targeting WW domain-containing E3 ubiquitin protein ligase 1 in gastric cancer. *J. Exp. Clin. Cancer Res.* **2017**, *36*, 59. [CrossRef] [PubMed]
39. Zhou, X.; Wen, W.; Shan, X.; Zhu, W.; Xu, J.; Guo, R.; Cheng, W.; Wang, F.; Qi, L.W.; Chen, Y.; et al. A six-microRNA panel in plasma was identified as a potential biomarker for lung adenocarcinoma diagnosis. *Oncotarget* **2017**, *8*, 6513–6525. [CrossRef] [PubMed]
40. Guo, T.; Zheng, C.; Wang, Z.; Zheng, X. miR-584-5p regulates migration and invasion in non-small cell lung cancer cell lines through regulation of MMP-14. *Mol. Med. Rep.* **2019**, *19*, 1747–1752. [CrossRef] [PubMed]
41. Poli, E.; Zhang, J.; Nwachukwu, C.; Zheng, Y.; Adedokun, B.; Olopade, O.I.; Han, Y.J. Molecular Subtype-Specific Expression of MicroRNA-29c in Breast Cancer Is Associated with CpG Dinucleotide Methylation of the Promoter. *PLoS ONE* **2015**, *10*, e0142224. [CrossRef] [PubMed]
42. Xu, P.; Wu, Q.; Yu, J.; Rao, Y.; Kou, Z.; Fang, G.; Shi, X.; Liu, W.; Han, H. A Systematic Way to Infer the Regulation Relations of miRNAs on Target Genes and Critical miRNAs in Cancers. *Front. Genet.* **2020**, *11*, 278. [CrossRef] [PubMed]
43. Naydenov, N.G.; Joshi, S.; Feygin, A.; Saini, S.; Litovchick, L.; Ivanov, A.I. A membrane fusion protein, Ykt6, regulates epithelial cell migration via microRNA-mediated suppression of Junctional Adhesion Molecule A. *Cell Cycle* **2018**, *17*, 1812–1831. [CrossRef] [PubMed]
44. Ooe, A.; Kato, K.; Noguchi, S. Possible involvement of CCT5, RGS3, and YKT6 genes up-regulated in p53-mutated tumors in resistance to docetaxel in human breast cancers. *Breast Cancer Res. Treat.* **2007**, *101*, 305–315. [CrossRef] [PubMed]
45. Ruiz-Martinez, M.; Navarro, A.; Marrades, R.M.; Viñolas, N.; Santasusagna, S.; Muñoz, C.; Ramírez, J.; Molins, L.; Monzo, M. YKT6 expression, exosome release, and survival in non-small cell lung cancer. *Oncotarget* **2016**, *7*, 51515–51524. [CrossRef] [PubMed]
46. Fukasawa, M.; Varlamov, O.; Eng, W.S.; Söllner, T.H.; Rothman, J.E. Localization and activity of the SNARE Ykt6 determined by its regulatory domain and palmitoylation. *Proc. Natl. Acad. Sci. USA* **2004**, *101*, 4815–4820. [CrossRef] [PubMed]

20, 12, 2008

Article

Chemically Induced Hypoxia Enhances miRNA Functions in Breast Cancer

Emma Gervin †, Bonita Shin †, Reid Opperman ‡, Mackenzie Cullen ‡, Riley Feser ‡, Sujit Maiti and Mousumi Majumder *

Department of Biology, Brandon University, 3rd Floor, John R. Brodie Science Centre, 270-18th Street, Brandon, MB R7A6A9, Canada; GERVINEE02@BrandonU.CA (E.G.); SHINBW60@brandonu.ca (B.S.); OPPERMRM72@brandonu.ca (R.O.); CULLENMW01@brandonu.ca (M.C.); FESERRJ36@brandonu.ca (R.F.); MaitiS@BrandonU.CA (S.M.)

* Correspondence: majumderm@brandonu.ca; Tel.: +1(204)-727-7324

† These authors contributed equally to this work.

‡ These authors contributed equally to this work.

Received: 29 June 2020; Accepted: 19 July 2020; Published: 22 July 2020

Abstract: In aggressively growing tumors, hypoxia induces HIF-1α expression promoting angiogenesis. Previously, we have shown that overexpression of oncogenic microRNAs (miRNAs, miRs) miR526b/miR655 in poorly metastatic breast cancer cell lines promotes aggressive cancer phenotypes in vitro and in vivo. Additionally, miR526b/miR655 expression is significantly higher in human breast tumors, and high miR526b/miR655 expression is associated with poor prognosis. However, the roles of miR526b/miR655 in hypoxia are unknown. To test the relationship between miR526b/miR655 and hypoxia, we used various in vitro, in silico, and in situ assays. In normoxia, miRNA-high aggressive breast cancer cell lines show higher HIF-1α expression than miRNA-low poorly metastatic breast cancer cell lines. To test direct involvement of miR526b/miR655 in hypoxia, we analyzed miRNA-high cell lines (MCF7-miR526b, MCF7-miR655, MCF7-COX2, and SKBR3-miR526b) compared to controls (MCF7 and SKBR3). $CoCl_2$-induced hypoxia in breast cancer further promotes *HIF-1α* mRNA and protein expression while reducing *VHL* expression (a negative HIF-1α regulator), especially in miRNA-high cell lines. Hypoxia enhances oxidative stress, epithelial to mesenchymal transition, cell migration, and vascular mimicry more prominently in MCF7-miR526b/MCF7-miR655 cell lines compared to MCF7 cells. Hypoxia promotes inflammatory and angiogenesis marker (*COX-2, EP4, NFκB1, VEGFA*) expression in all miRNA-high cells. Hypoxia upregulates miR526b/miR655 expression in MCF7 cells, thus observed enhancement of hypoxia-induced functions in MCF7 could be attributed to miR526b/miR655 upregulation. In silico bioinformatics analysis shows miR526b/miR655 regulate *PTEN* (a negative regulator of *HIF-1α*) and *NFκB1* (positive regulator of *COX-2* and *EP4*) expression by downregulation of transcription factors *NR2C2, SALL4*, and *ZNF207*. Hypoxia-enhanced functions in miRNA-high cells are inhibited by COX-2 inhibitor (Celecoxib), EP4 antagonist (ONO-AE3-208), and irreversible PI3K/Akt inhibitor (Wortmannin). This establishes that hypoxia enhances miRNA functions following the COX-2/EP4/PI3K/Akt pathways and this pathway can serve as a therapeutic target to abrogate hypoxia and miRNA induced functions in breast cancer. In situ, *HIF-1α* expression is significantly higher in human breast tumors (n = 96) compared to non-cancerous control tissues (n = 20) and is positively correlated with miR526b/miR655 expression. In stratified tumor samples, *HIF-1α* expression was significantly higher in ER-positive, PR-positive, and HER2-negative breast tumors. Data extracted from the TCGA database also show a strong correlation between *HIF-1α* and miRNA-cluster expression in breast tumors. This study, for the first time, establishes the dynamic roles of miR526b/miR655 in hypoxia.

Keywords: Breast cancer; Hypoxia inducible factor 1-alpha (HIF-1α); MicroRNA (miRNA); miR526b; miR655; Oxidative stress; Migration; Cyclooxygenase-2 (COX-2); Prostaglandin E2 receptor 4 (EP4); PI3K/Akt

1. Introduction

Breast cancer is the most common form of cancer, as well as the second leading cancer-related death among Canadian women [1]. According to Canadian Cancer Statistics, 1 in 8 Canadian women will develop breast cancer in their lifetime, while 1 in 33 will die from it [1]. Understanding the complexity of the disease is urgently required to find personalized therapy for various kinds (i.e., estrogen receptor (ER)-positive and ER-negative; progesterone receptor (PR)-positive and PR-negative; human epidermal growth factor receptor 2 (HER2)-positive and HER2-negative; and triple-negative (ER-PR-HER2-negative)) of breast cancer, as there is no single target for treating such a complex malignancy. One of the factors that contribute to the complexity of tumor growth, metastasis, and patient survival in breast cancer is the level of hypoxia (oxygen deficiency) within the tumor microenvironment [2]. Due to their rapid proliferation, cancer cells outgrow the available blood supply. This limits the delivery of oxygen and nutrients to the cells, making the center of the aggressively growing tumor largely hypoxic [2]. To counteract hypoxia, cancerous cells secrete growth factors and stimulants that facilitate tumor-associated angiogenesis in the tumor microenvironment to deliver the required oxygen and nutrients to dividing tumor cells [3].

Hypoxia influences multiple signaling pathways in cells, including the hypoxia-inducible factor (HIF), NFκB, ERK, and PI3K/Akt/mTOR pathways, which regulate apoptosis, migration, proliferation, and inflammation in cancer [4–6]. HIF-1 is a heterodimer composed of the HIF-1α and HIF-1β subunits. Under normoxia (physiologically normal oxygen levels), both the HIF-1α and -1β subunits are constitutively expressed, but the HIF-1 dimer is not formed as the HIF-1α subunit is degraded in the presence of oxygen [4]. Under normoxic condition, the oxygen-dependent degradation domain of HIF-1α is hydroxylated by the PHD (prolyl hydroxylase domain) enzyme, which further allows the tumor suppressor pVHL (Von Hippel-Lindau) to catalyze the ubiquitin-dependent degradation of the HIF-1α protein [7]. In hypoxic conditions, this hydroxylation does not occur, and pVHL does not catalyze the ubiquitination of the HIF-1α protein, allowing it to avoid degradation. The HIF-1α subunit can then dimerize with the HIF-1β subunit to form HIF-1 [5]. HIF-1 is a transcription factor that binds to promoter regions and regulates the expression of multiple genes, including vascular endothelial growth factors (*VEGFs*) (a pro-angiogenic agent), anaerobic respiration enzymes, glucose metabolism, and regulates microRNA (miRNA, miR) biogenesis and functions [8–10]. The net effects of these changes increase the amount of ATP available to the tumor cell, promoting rapid growth.

miRNAs are defined as a group of endogenously-produced, small, non-coding RNAs that can downregulate gene expression of target messenger RNAs (mRNAs) at the post-transcription level by complete or partial complementary base pairing. Dysregulated miRNA expression has been associated with various cancers, including breast cancer [3,11–14]. Using gene expression and miRNA microarray assays, we have identified that the overexpression of COX-2 in a poorly metastatic MCF7 cells (an ER/PR-positive and HER2-negative breast cancer cell line) upregulates two miRNAs, miR526b and miR655, which have been classified as oncogenic miRNAs in human breast cancer [13,14]. We found that miR526b and miR655 collectively target a total of 13 genes in COX-2 overexpressing MCF7 cells (MCF7-COX2), 12 of which are classified as tumor-suppressor-like genes [15]. The single gene targeted by both miRNAs was identified as cytoplasmic polyadenylation element-binding protein 2 (*CPEB2*). Recently, it was identified that *CPEB2* is a tumor suppressor gene, further validating miR526b and miR655 as oncogenic miRNAs promoting breast cancer by collectively targeting this gene [15]. We have previously shown that in SKBR3, MDA-MB-231, and MCF7-COX2 cell lines COX-2, miR526b, and miR655 were upregulated, while *CPEB2* was downregulated [15]. miR526b is located on a large cluster of miRNAs on chromosome 19 with the chromosomal location 19q13.42, in the gene family miR515 [16,17]. miR655 is located on a large cluster of miRNA on chromosome 14 on the host gene miR381HG in the chromosomal location 14q32.31 and belongs to the miR154 gene family [17,18].

We have also shown that miR526b and miR655 overexpression in ER-positive breast cancer cell line MCF7 and an ER-negative HER2-positive breast cancer cell line SKBR3 promotes epithelial-to-mesenchymal transition (EMT), cell migration, invasion, induction of stem-like cells (SLCs) phenotype, tumor growth, and metastasis in vivo [13,14]. In growing tumors, the core of the mass becomes hypoxic and requires a new means for oxygen delivery. This is achieved through tumor-associated angiogenesis, a phenotype that can be induced by the expression of certain miRNAs [3,19,20]. We have identified that overexpression of miR526b/miR655 in MCF7 cells enhances tumor-associated angiogenesis and lymphangiogenesis by the production of VEGFA and that miRNA cell secretion enhances tube formation in vascular endothelial cells [3]. Cancer cells can also mimic the properties of vascular endothelial cells to induce tumor-associated angiogenesis, known as vascular mimicry [21,22]. We have shown that in human breast tumors, miR526b and miR655 expression is highly correlated with angiogenesis and lymphangiogenesis markers (VEGFA, VEGFC, and VEGFD) [3]. In this article, we investigated the roles of miRNA in promoting angiogenic marker expression and vascular mimicry in hypoxia.

EMT is an important biological process characterized by the progressive loss of cell-to-cell adhesion, alterations in cellular polarity, and actin cytoskeletal rearrangements leading to the formation of filopodia and upregulation of mesenchymal phenotypes and markers [23]. Tumor cells lose intercellular junction proteins such as E-Cadherin (CDH1) and are able to travel through the extracellular matrix, in a process known as cell migration [24]. EMT is necessary for the migration of embryonic cells to establish the development of an embryo, and to complete wound healing in adult tissues. However, EMT in cancer leads to the promotion of aggressive phenotypes, such as migration, invasion, angiogenesis, stem-like phenotypes in cancer cells, and resistance to chemo-radiotherapy [25]. Previously, we have shown that miR526b and miR655 induce EMT in breast cancer, promote tumor cell migration and invasion [13,14], and that miRNA cell secretions enhance the migration of vascular endothelial cells to enhance angiogenesis [3]. However, hypoxia's influence on miRNA-induced EMT is not clear.

Another known phenotype in hypoxic tumors is the formation of reactive oxygen species (ROS), such as superoxide (SO), which are byproducts of cellular metabolism. Cellular inability to neutralize and eliminate these ROS leads to oxidative stress. Furthermore, increased levels of SO have shown regulation of signaling cascades for cell proliferation and survival [26]. We have shown that a dynamic relationship exists between oxidative stress and miR526b/miR655 expression, where an increase in miRNA leads to an increase in ROS and SO. Likewise, an increase in ROS was shown to significantly increase miR526b and miR655 expression, suggesting that a positive feedback loop relationship between both miRNAs and oxidative stress is present in human breast cancer [27]. We have previously shown that when we treat poorly metastatic breast cancer tumor cell line MCF7 and primary endothelial cell line human umbilical vein endothelial cells (HUVECs) with cell secretions from miR526b and miR655-overexpressing cells, there is an increase in ROS, SO, and oxidative stress marker thioredoxin reductase 1 (TXNRD1) expression. This suggests that miR526b/miR655-high cells' metabolites induce oxidative stress in the tumor microenvironment. Thus, we wanted to investigate the effect of hypoxia on miRNA-induced oxidative stress in breast cancer cells.

For the first time, with this specific research, we investigate the capability of miR526b, miR655, and hypoxia collaborating to promote aggressive breast cancer phenotypes. First, we show that highly metastatic and miRNA-high cell lines show high expression of *HIF-1α* in normoxia, while poorly metastatic, miRNA-low cell lines show low expression. Next, we used $CoCl_2$ to induce hypoxia in ER-positive MCF7, MCF7-miR526b, and MCF7-miR655 cells, as well as HER2-positive SKBR3 and SKBR3-miR526b cells, since $CoCl_2$ has long been used as a chemical inducer of hypoxia and has been shown to induce *HIF-1α* expression [28,29]. We further verified the effects of hypoxia enhancing miRNA-induced oxidative stress, cell migration, induction of EMT, expression of hypoxia-linked genes such as *VHL*, *HIF-1α*, and *NFκB1*, and expression of inflammation-associated genes such as *VEGFA*, *COX-2*, and *EP4* in breast cancer cell lines. Here we demonstrated that hypoxia enhances oncogenic miRNA functions in breast cancer, which can be inhibited by COX-2, EP4, and PI3K/Akt signaling pathway inhibitors. In silico bioinformatics analysis further confirms that miRNA functions in hypoxia are regulated by COX-2/EP4/PI3K/Akt pathways and that miRNA has a negative correlation with transcription factors that regulate the expression of *NFκB1* and *PTEN*. In human breast tumors, *HIF-1α* expression is significantly high and we estimated the highest expression in the ER-positive, PR-positive, and HER2-negative breast tumors. Both miR526b and miR655 expression in breast tumors is positively and significantly correlated with *HIF-1α* expression in the set of tumor samples we used in this study and also data extracted from The Cancer Genome Atlas (TCGA) cBioPortal database, which includes data from 16 different breast cancer studies, strongly suggesting that hypoxia and miRNAs collaborate to promote breast cancer progression. This is a novel function of miR526b and miR655 in breast cancer.

2. Results

We do not have access to a hypoxic chamber, thus, we used $CoCl_2$ to induce hypoxia. $CoCl_2$ increases the expression of hypoxic marker HIF-1α and induces hypoxia in MCF7 cells [29]. First, we conducted a dose-response assay of *HIF-1α* expression with various concentrations of $CoCl_2$ (Figure S1).

To investigate the interaction between miRNA and hypoxia, we used various breast cancer cell lines with differential levels of miR526b and miR655 expression [13,14]. We used the breast epithelial cell line MCF10A, poorly-metastatic breast cancer cell lines MCF7, T47D (ER-positive, PR-positive, and HER2-negative) and SKBR3 (ER-negative, PR-negative, and HER2-positive); highly metastatic breast cancer cell lines Hs578T, MDA-MB-231 (ER-negative, PR-negative, and HER2-negative), and MCF7-COX2 (ER-positive, PR-positive, and HER2-negative); as well as the highly metastatic stable miRNA-overexpression cell lines MCF7-miR526b, MCF7-miR655, and SKBR3-miR526b. We used empty vector transfected cells MCF7-Mock as a control for miRNA-overexpressing cell lines. We have previously shown that there is no significant difference in miRNA expression between MCF7 and MCF7-Mock cells [13,14]. Thus, for experiments in this article, we used MCF7 as a low miRNA-expressing control cell line and MCF7-COX2 as a high miRNA-expressing cell line. We also used human breast tumor tissues to test the correlation of miRNA with *HIF-1α* expression in tumors.

2.1. *HIF-1α Gene and Protein Expression in Normoxia*

2.1.1. HIF-1α Gene Expression in Various Breast Cancer Cell Lines

We measured gene expression of *HIF-1α* in a variety of breast cancer cell lines in comparison to the mammary epithelial cell line MCF10A. Poorly metastatic and miRNA-low breast cancer cell lines MCF7, SKBR3, and T47D; and highly metastatic MCF7-COX2, MDA-MB-231, and Hs578T cell lines were used. miRNA expressions in these cell lines are presented in Figure S2A,B [13,14]. MDA-MB-231, MCF7-COX2, and Hs578T cell lines show very high and significant upregulation of *HIF-1α* compared to MCF10A and T47D, MCF7, and SKBR3 cell lines showed lower expressions of *HIF-1α* (Figure 1A). We observed that MCF7 and SKBR3 cell lines had the lowest levels of *HIF-1α*, and also miRNA expression. In normoxia, MCF7-miR526b and MCF7-miR655 cell lines show an extremely significant upregulation of *HIF-1α* gene expression compared to MCF7 (Figure 1B) and SKBR3-miR526b cells showed an extremely significant increase in *HIF-1α* gene expression compared to SKBR3 (Figure 1C).

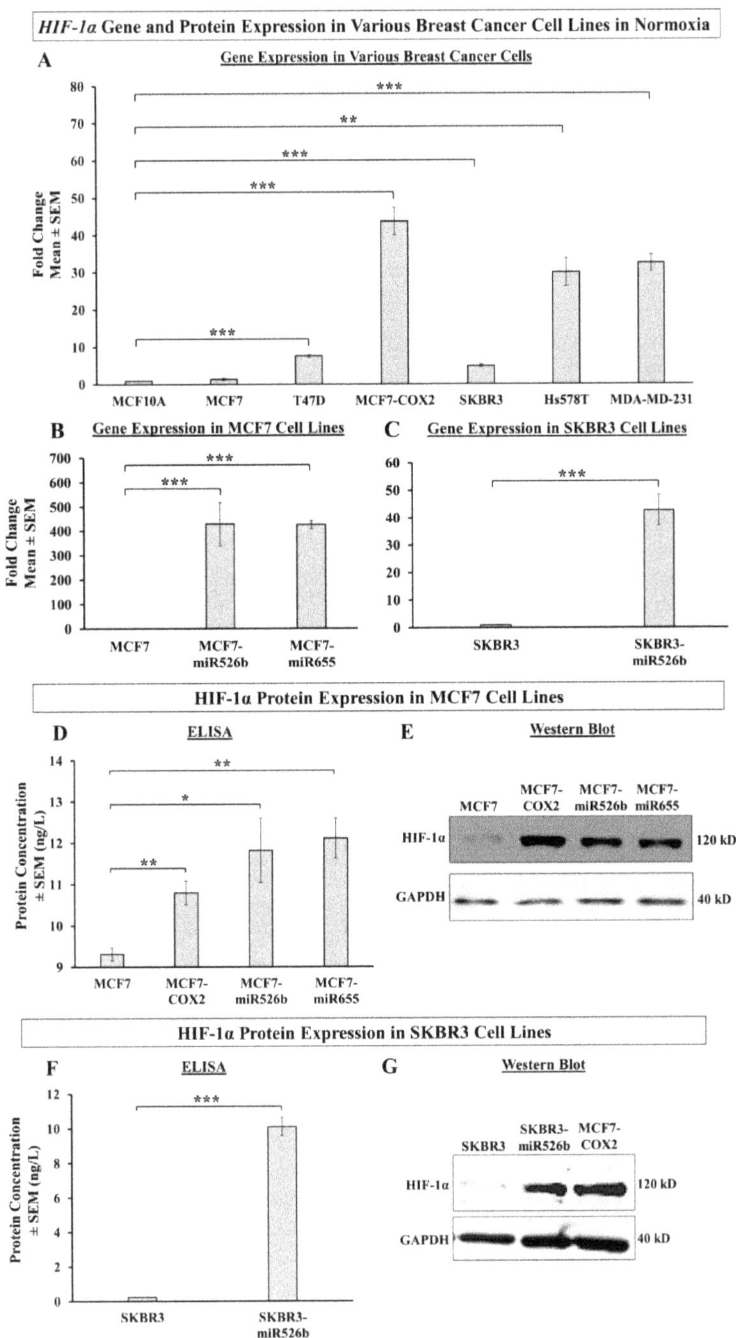

Figure 1. HIF-1α mRNA and protein expression in various breast cancer cell lines in normoxic conditions: (**A**) *HIF-1α* mRNA expression in various breast cancer cell lines (T47D, SKBR3, MCF7-COX2, Hs578T, and MDA-MB-231) in comparison to mammary epithelial cell line MCF10A. (**B**,**C**) *HIF-1α* mRNA

expression in miRNA-high cell lines MCF7-miR526b, MCF7-miR655, and SKBR3-miR526b compared to their respective control cell lines MCF7 and SKBR3. (**D**) HIF-1α protein expression measured with ELISA in miRNA-overexpressed cell lines (MCF7-miR526b, MCF7-miR655) and miRNA-high cell line MCF7-COX2 compared to control cell line MCF7. (**E**) Western blot analysis of total endogenous HIF-1α protein expression in MCF7, MCF7-COX2, MCF7-miR526b, and MCF7-miR655 cell lines. (**F**) HIF-1α protein expression in SKBR3 and SKBR3-miR526b measured with ELISA. (**G**) Western blot analysis showing total expression of endogenous HIF-1α protein in SKBR3 and SKBR3-miR526b cell lines, MCF7-COX2 as a positive control. Full western blots are provided in Figure S3A–C. Data are presented as the mean ± SEM of triplicate replicates; * $p < 0.05$, ** $p < 0.01$, and *** $p < 0.001$.

2.1.2. HIF-1α Protein Expression in Various Breast Cancer Cell Lines

We performed an enzyme-linked immunosorbent assay (ELISA) to test HIF-1α protein expression in the stable miRNA-overexpressed cell lines MCF7-miR526b, MCF7-miR655, and SKBR3-miR526b, as well as the naturally miRNA-high cell line MCF7-COX2, in comparison to their respective controls. HIF-1α protein levels were significantly increased in MCF7-miRNA-high cell lines compared to MCF7 cells, with MCF7-COX2 cells showing a significant but moderate increase and MCF7-miR526b and MCF7-miR655 cell lines showing high upregulation (Figure 1D). SKBR3-miR526b cells also show a significant increase in HIF-1α protein expression compared to SKBR3 cells (Figure 1F). miRNA-overexpression very significantly enhances HIF-1α expression in both ER-positive MCF7 cells and HER2-positive SKBR3 cells.

Total endogenous HIF-1α protein expression was measured with western blot analysis, data showing high expression of HIF-1α total protein in the MCF7-miR526b and MCF7-miR655 (Figure 1E) and SKBR3-miR526b (Figure 1G) cell lines compared to control miRNA low MCF7 and SKBR3 cell lines, respectively. Endogenous HIF-1α protein expression further supports results recorded with HIF-1α ELISA.

2.2. *Induction of Hypoxia Using* $CoCl_2$

To mimic the effect of a hypoxia chamber, we used $CoCl_2$ to induce hypoxia as described in other publications [28–31]. We conducted a $CoCl_2$ treatment dose-response assay using *HIF-1α* gene expression fold changes (Figure S1) and selected 150 μM for further experiments. It should also be noted that during $CoCl_2$ treatment, we observed changes in cell density. We seeded an average of 6000 cells per well in a six-well plate and observed an increase in cell density in $CoCl_2$-treated cells (Figure S4), showing that the $CoCl_2$ treatment we selected was not toxic to the cells.

2.2.1. HIF-1α Gene Expression in Hypoxia

We used qRT-PCR to analyze *HIF-1α* gene expression in MCF7, MCF7-miR526b, MCF7-miR655, SKBR3-miR526b, and MCF7-COX2 cell lines in hypoxia, and considered sterile H_2O treatment as the control or "normoxia." *HIF-1α* gene expression was significantly upregulated in all cell lines except MCF7 in hypoxia compared to normoxia (Figure 2A). It should be noted, however, that miRNA-overexpressing MCF7 cell lines (MCF7-miR526b, MCF7-miR655) showed the greatest upregulation of *HIF-1α*. Thus, we decided to test hypoxia-enhanced functions in miRNA-overexpressing MCF7-miR526b and MCF7-miR655 cell lines compared to MCF7 cell lines. MCF7-COX2 showed the highest expression of *HIF-1α* expression in normoxia, thus, $CoCl_2$ treatment could only moderately, but very significantly increase *HIF-1α* expression. The increase in *HIF-1α* expression in the SKBR3-miR526b cell line from normoxia to hypoxia was also modest; however, this could be the effect of HER2-positivity, which warrants further investigation.

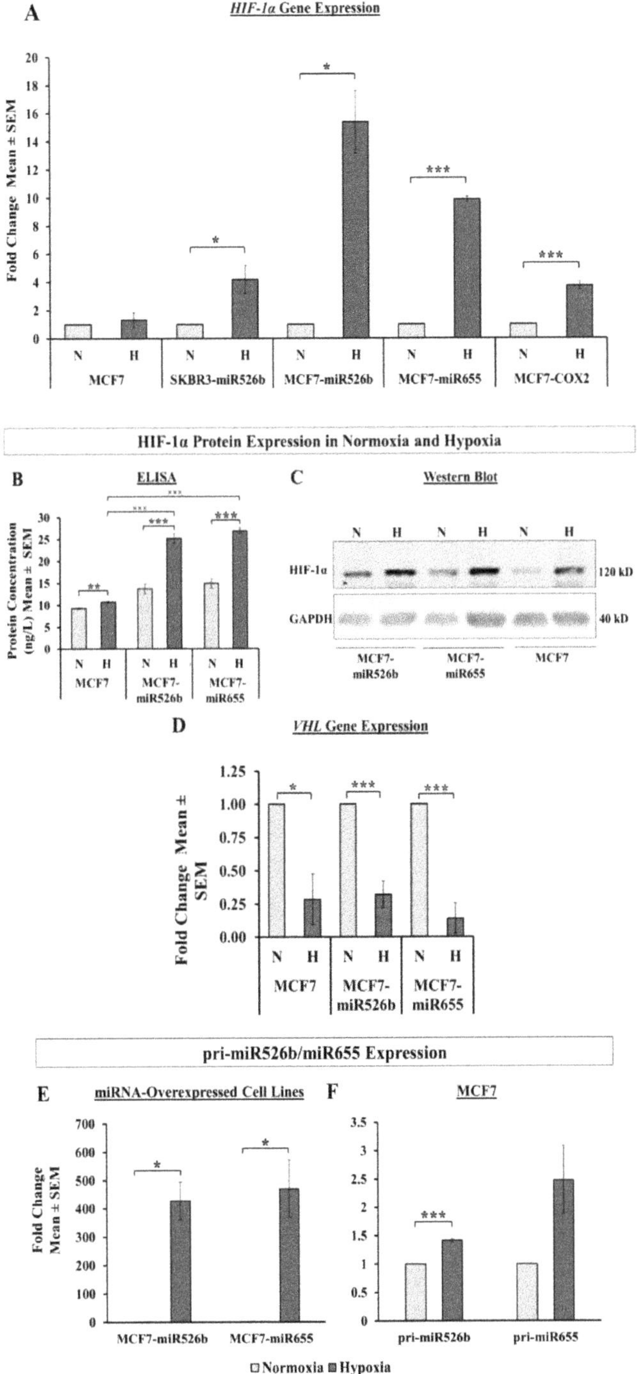

Figure 2. Induction of hypoxia using $CoCl_2$: In all figures, 'N' indicates normoxia and 'H' indicates hypoxia. (**A**) Gene expression of *HIF-1α* in MCF7, SKBR3-miR526b, MCF7-miR526b, MCF7-miR655, and MCF7-COX2 cell lines under normoxic and hypoxic conditions measured using qRT-PCR. (**B**) Protein

levels of HIF-1α in MCF7, MCF7-miR526b, and MCF7-miR655 cell lines measured using ELISA. (**C**) Total HIF-1α protein expression in both hypoxia and normoxia were measured with western blots. Complete western blots are presented in Figure S3D. (**D**) *VHL* gene expression in MCF7, MCF7-miR526b, and MCF7-miR655 cell lines measured via qRT-PCR. (**E**) Pri-miRNA expression in MCF7-miR526b and MCF7-miR655 cells in normoxia and hypoxia. (**F**) Pri-miRNA expression in MCF7 cells in normoxia and hypoxia. Data are presented as the mean ± SEM of triplicate biological replicates; * $p < 0.05$, ** $p < 0.01$, *** $p < 0.001$, and xxx indicates $p < 0.001$. * Also indicates comparison between normoxia and hypoxia of the same cell line and x indicates comparison between cell lines only in a hypoxic condition.

2.2.2. HIF-1α Protein Expression in Hypoxia

We performed an enzyme-linked immunosorbent assay (ELISA) to specifically test HIF-1α protein expression in MCF7, MCF7-miR526b, and MCF7-miR655 cells in hypoxia and normoxia. Microplate ELISA analysis showed that there was a significant increase in HIF-1α protein levels in all ER-positive cell lines in hypoxia compared to normoxia; however, this increase was most significant in MCF7-miR526b and MCF7-miR655 cell lines (Figure 2B). Within the hypoxic condition, we compared MCF7 and miRNA-high cell lines and both miRNA-high cell lines demonstrated a very significant increase in HIF-1α protein expression compared to MCF7 cells, indicating some form of direct involvement of miRNA in hypoxia (Figure 2B).

We also conducted a western blot analysis of HIF-1α protein expression in MCF7, MCF7-miR526b, and MCF7-miR655 cell lines in hypoxia to measure change in total HIF-1α protein expression. For all cell lines, $CoCl_2$ treatment enhanced HIF-1α expression in hypoxia compared to normoxia (Figure 2C). This enhancement of HIF-1α protein expression further confirms that $CoCl_2$ treatment induces hypoxia in breast cancer.

2.2.3. Analysis of VHL Gene Expression in Hypoxia

HIF-1α protein stability is dependent on *VHL*, a tumor suppressor gene that downregulates HIF-1α. We identified that *VHL* gene expression was significantly decreased in hypoxia compared to normoxia in all cell lines, with the most significant change occurring in the miRNA-high cell lines (Figure 2D). Thus, $CoCl_2$ treatment successfully increased *HIF-1α* expression and downregulated *VHL* expression.

2.2.4. Hypoxia Enhances miRNA Expression

Both MCF7-miR526b and MCF7-miR655 cells showed a very significant increase in primary miRNA (pri-miRNA) expression in normoxia compared to MCF7 cells (Figure S2C,D). Since we observed an increase in HIF-1α expression in hypoxia, we wanted to determine if miRNA expression was also increased in hypoxia. Pri-miR526b expression was significantly increased in hypoxia compared to normoxia in MCF7-miR526b cells. Similarly, pri-miR655 expression was significantly increased in hypoxia in MCF7-miR655 cells compared to normoxia (Figure 2E). Most prominent changes were recorded in MCF7 cells, which showed an extremely significant increase in pri-miR526b expression and a marginal increase in pri-miR655 expression in hypoxia compared to normoxia (Figure 2F).

2.3. Hypoxia Induces Oxidative Stress

Previously, we have shown that miRNA overexpression in MCF7 cells and cell-free conditioned media from miRNA-high cells (MCF7-miR526b and MCF7-miR655) induces oxidative stress [27]. In the current study, we tested if hypoxia can further stimulate oxidative stress in miRNA-high cells. Here we show data for only 150 μM $CoCl_2$ treatment, since we found that the 150 μM concentration of $CoCl_2$ induced maximum hypoxia (Figure S1).

2.3.1. Fluorescence Microscopy Assay to Measure Cellular Fluorescence

Fluorescence microscopy images showing ROS (green) and SO (red) production in MCF7, MCF7-miR526b, and MCF7-miR655 cells in hypoxia and normoxia, with quantification presented in Figure 3. Negative controls of MCF7 (Figure 3A,O), MCF7-miR526b (Figure 3B,P), and MCF7-miR655 (Figure 3C,Q) as well as positive controls of MCF7 (Figure 3D,R), MCF7-miR526b (Figure 3E,S), and MCF7-miR655 (Figure 3F,T) were used to normalize fluorescence-positive cell quantifications. Only bright fluorescent cells normalized to the negative control of respective cell lines were considered for quantifications. We observed a significant increase in ROS and SO in MCF7-miR526b cells in hypoxia (Figure 3K,Y) compared to normoxia (Figure 3H,V). There was an increase in ROS and SO producing cells in MCF7 cells in hypoxia (Figure 3J,X) compared to normoxia (Figure 3G,U) as well. Quantitative data show that ROS production in hypoxia was increased three-fold in both MCF7 and MCF7-miR526b cells (Figure 3M); however, SO production in hypoxia was enhanced 1.8- and 2.4-fold in MCF7 and MCF7-miR526b cells, respectively, compared to normoxia (Figure 3AA). Images of MCF7-miR655 cells in hypoxia (Figure 3L,Z) and normoxia (Figure 3I,W) evidently show higher expression of ROS and SO in hypoxia; however, quantitative data for MCF7-miR655 are not presented.

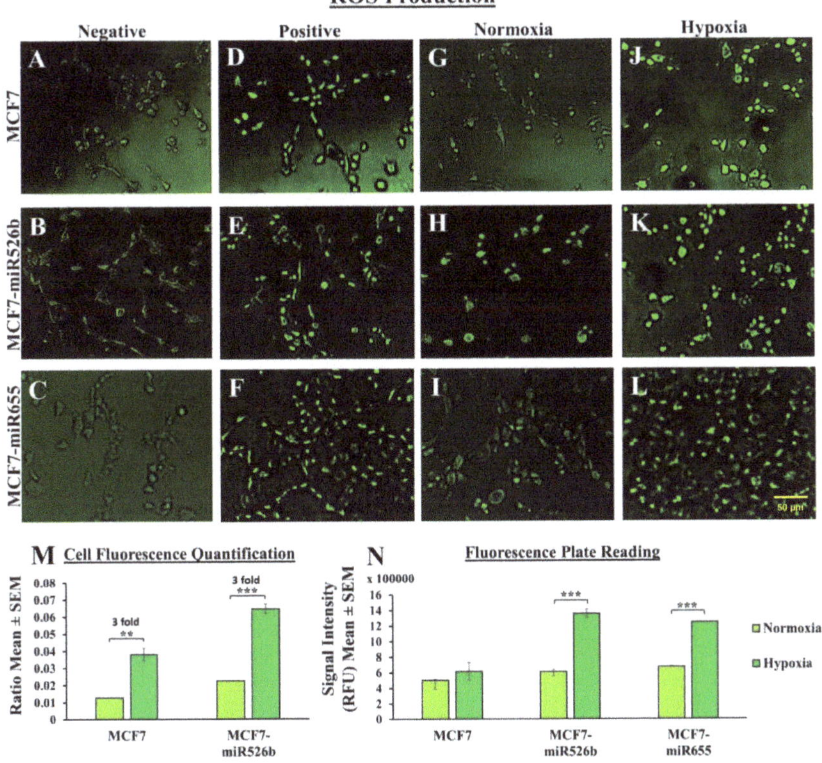

Figure 3. *Cont.*

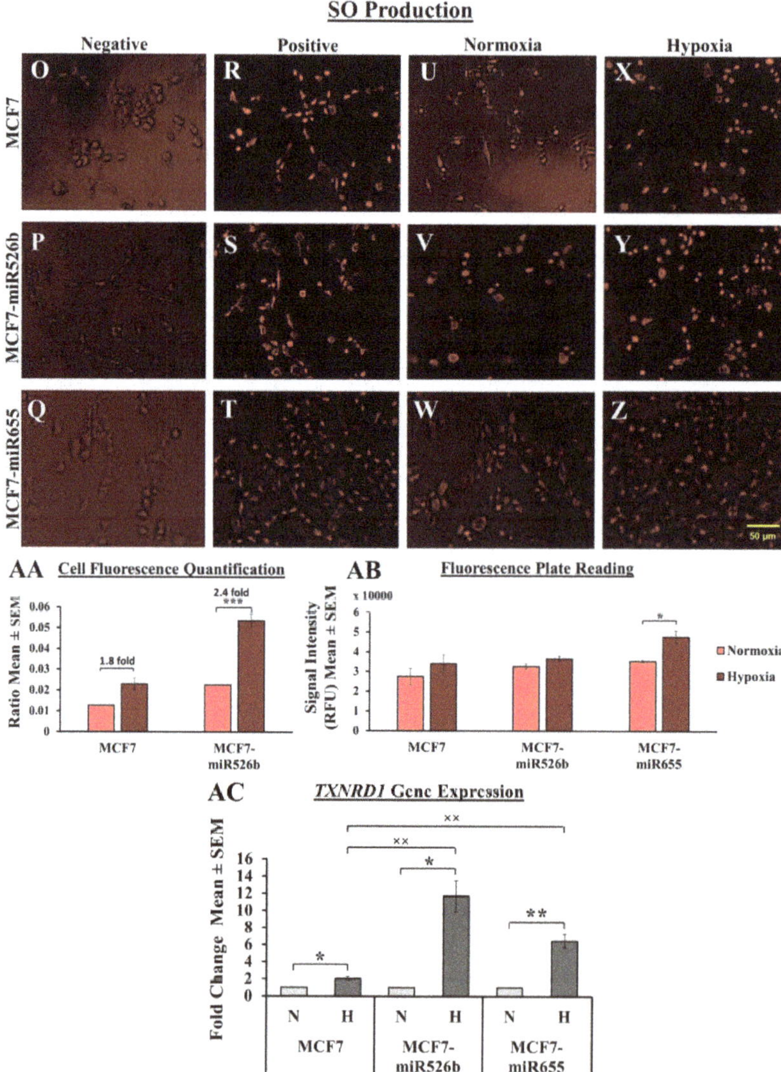

Figure 3. Fluorescence microscopy and fluorescence microplate assays to quantify ROS and SO production: MCF7, MCF7-miR526b, and MCF7-miR655 cells in (**A–C**) negative control, (**D–F**) positive control, (**G–I**) in normoxia, and (**J–L**) in hypoxia under the green filter for total ROS detection. (**M**) Quantification ratios of MCF7 and MCF7-miR526b cells positive for ROS in normoxia and hypoxia. (**N**) Fluorescence microplate assay to quantify total ROS production in MCF7, MCF7-miR526b, and MCF7-miR655 cells. Fluorescent SO-positive MCF7, MCF7-miR526b, and MCF7-miR655 cells (**O–Q**) in negative control, (**R–T**) in positive control, (**U–W**) in normoxia, and (**X–Z**) in hypoxia under the red filter for total SO detection. (**AA**) Quantification ratios of MCF7 and MCF7-miR526b cells positive for SO in normoxic and hypoxic conditions. (**AB**) Fluorescence microplate assay to quantify total SO production in MCF7, MCF7-miR526b, and MCF7-miR655 cells. (**AC**) Gene expression of *TXNRD1* measured with qRT-PCR. 'N' indicates normoxia, and 'H' indicates hypoxia. Scale bar represents 50 μM. (**M**,**N**,**AA**,**AB**,**AC**) Data are presented as the mean ± SEM of triplicate biological replicates; * $p < 0.05$, ** $p < 0.01$, *** $p < 0.001$, and xx indicates $p < 0.01$. * Also indicates comparison between normoxia and hypoxia of the same cell line and x, indicates comparison between cell lines only in a hypoxic condition.

2.3.2. Fluorescence Microplate Assay to Measure Total Fluorescence

After finding cellular fluorescence in the microscopy assay, we measured total ROS and SO production using fluorescence microplate assays as described in previous studies [27]. Total ROS production was marginally but non-significantly increased in MCF7 in hypoxia compared to normoxia, whereas we recorded a significant increase in ROS production by both MCF7-miR526b and MCF7-miR655 in hypoxia (Figure 3N). Additionally, there was a marginal increase in total SO production by MCF7 cells and MCF7-miR526b cells, but only in MCF7-miR655 we observed a significant increase in SO production in hypoxia compared to normoxia (Figure 3A,B). Fluorescence microplate assays evidently show that hypoxia only enhances total ROS (Figure 3N) and SO production (Figure 3A,B) in miRNA-high cells.

2.3.3. Overexpression of TXNRD1

TXNRD1 is a marker associated with oxidative stress. We previously showed that MCF7-miR526b and MCF7-miR655 cell lines overexpress *TXNRD1* compared to MCF7 cells in normoxia [27]. In the current study, we measured changes in *TXNRD1* expression in hypoxia in MCF7, MCF7-miR526b, and MCF7-miR655 cell lines to determine if hypoxia further enhances *TXNRD1* expression. We observed that hypoxia promotes *TXNRD1* expression in all three cell lines; however, MCF7-miR526b and MCF7-miR655 cells showed a very significant increase compared to normoxia (Figure 3AC). Additionally, we compared *TXNRD1* expression in hypoxia between cell lines and found that *TXNRD1* expression is very significantly higher in miRNA-high cells in comparison to MCF7 (Figure 3AC). Collectively, our results strongly suggest that hypoxia further enhances oxidative stress induction in miRNA-high cells.

2.4. Hypoxia Promotes EMT in miRNA-High Cells

Previously, we have indicated that the overexpression of miR526b and miR655 induces EMT phenotypes in MCF7 and SKBR3 cells [13,14]. Furthermore, we have shown cell-free secretions from miRNA-high cells induce migration of vascular endothelial cells [3]. Thus, we wanted to investigate the role of hypoxia in promoting the EMT of miRNA-high cell lines. We used qRT-PCR to measure the gene expressions of mesenchymal markers (*VIM*, *TWIST1*, *SNAIL*) and the epithelial marker *CDH1*, and proceeded to perform a migration assay on miRNA-high cells in normoxic and hypoxic conditions. In hypoxia, we observed miRNA-high cell lines mimicking vascular properties, forming tube-like structures on growth factor-reduced Matrigel.

2.4.1. Hypoxic Condition Regulates EMT Markers Expression in Cancer Cells

We measured mRNA expression of the epithelial marker *CDH1* and the mesenchymal markers *VIM*, *TWIST1*, and *SNAIL* in MCF7 and miRNA-high cells in both normoxic and hypoxic conditions using qRT-PCR. We observed a significant downregulation of the epithelial marker *CDH1* in all cell lines in hypoxia compared to normoxia (Figure 4A). Moreover, we observed an extremely significant upregulation of the mesenchymal markers *VIM* and *TWIST1* in MCF7-miR526b and MCF7-miR655 cells in hypoxia compared to normoxia (Figure 4B,C). For all cell lines, there was a marginal, but non-significant increase in *SNAIL* expression in hypoxia compared to normoxia (Figure 4D). Although there was a marginal increase in mesenchymal marker expression in MCF7 cells in hypoxia compared to normoxia, these changes were not significant (Figure 4B–D). We also compared *CDH1*, *VIM*, and *TWIST1* expressions in MCF7-miR526b and MCF7-miR655 cell lines compared to MCF7 only in hypoxia. We found a significant downregulation of *CDH1* in MCF7-miR655 cells in hypoxia compared to MCF7 cells in hypoxia (Figure 4A) and an extremely significant upregulation of *VIM* and *TWIST1* in miRNA-high cells (Figure 4B,C).

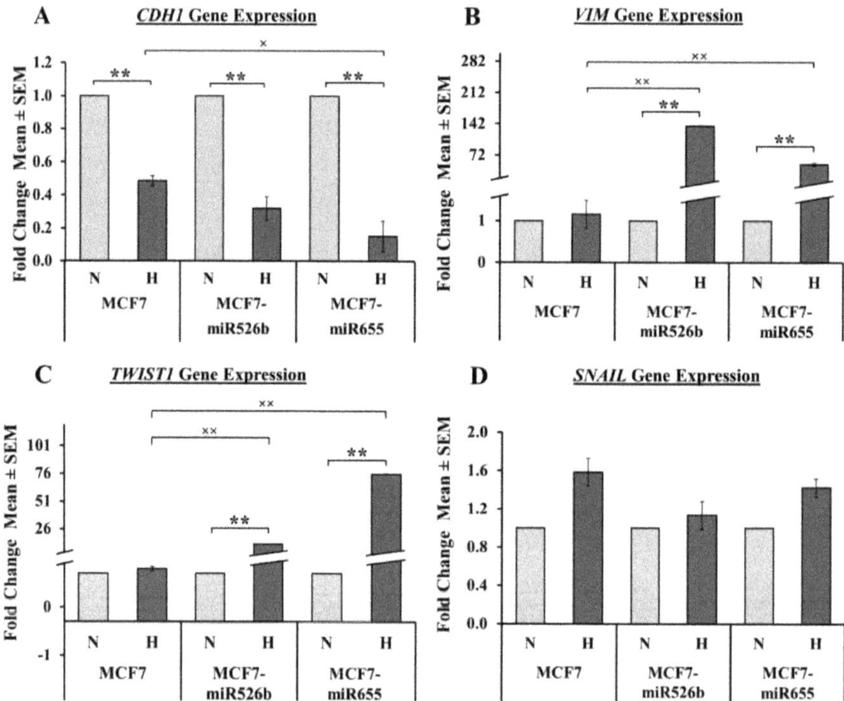

Figure 4. Expression of EMT markers in MCF7, MCF7-miR526b, and MCF7-miR655 cell lines: 'N' indicates normoxia, and 'H' indicates hypoxia. (**A**) Epithelial marker *CDH1* gene expression. (**B**) Gene expression of mesenchymal marker *VIM*. (**C**) Gene expression of mesenchymal marker *TWIST1*. (**D**) Gene expression of mesenchymal marker *SNAIL*. Data are presented as the mean ± SEM of quadruplicate replicates; ** $p < 0.001$, x indicates $p < 0.05$ and xx indicates $p < 0.001$. * Also indicates comparison between normoxia and hypoxia of the same cell line and x, indicates comparison between cell lines only in a hypoxic condition.

2.4.2. Hypoxic Condition Promotes Migration of miRNA-High Cells

miRNA overexpression induces cell migration and invasion of both MCF7 and SKBR3 cell lines [13,14]. We previously showed that cell-free secretions from MCF7-miR526b and MCF7-miR655 cells promote migration of HUVECs [3]. Here, we tested changes in cell migration in hypoxia of MCF7, MCF7-miR526b, and MCF7-miR655 cell lines by conducting a scratch-wound cell migration assay over 48 h. In normoxia, both MCF7 (Figure 5A–D) and MCF7-miR526b (Figure 5I–L) migrated; however, MCF7-miR526b cells migrated faster through the various time points (Figure 5Q). In hypoxia, MCF7-miR526b cells (Figure 5M–P) significantly migrated and closed the wound by 24 h and completely sealed the wound by 48 h, whereas this movement was limited for MCF7 cells (Figure 5E–H). Quantitative data are presented in Figure 5R for hypoxia. MCF7-miR655 showed similar phenotypes, image data are presented in Figure S5 and quantitative data are presented in Figure 5R. Hypoxic conditions very significantly increased miRNA-induced cell migration in miRNA-high cells.

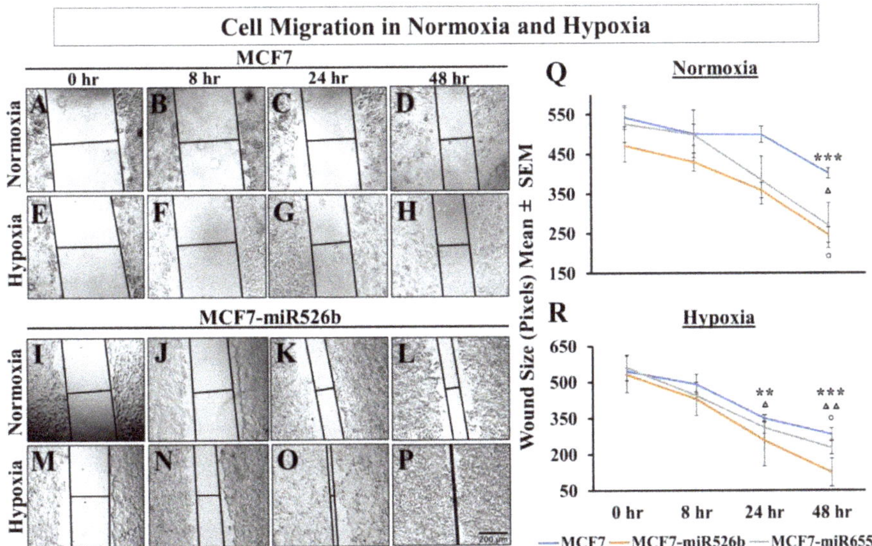

Figure 5. Cell migration in normoxia and hypoxia: baseline scratches represented by black lines at 0, 8, 24, and 48 h time points. Representative images of MCF7 are presented in (**A–D**) normoxia and (**E–H**) hypoxia. Representative images for MCF7-miR526b are presented in (**I–L**) normoxia and (**M–P**) hypoxia. Scale bar represents 200 μM. Wound size measured in pixels. (**Q**) Mean wound size in normoxic conditions over 0–48 h. (**R**) Mean wound size in hypoxic conditions over 0–48 h. Data are presented as the mean ± SEM of quadruplicate biological replicates; ** $p < 0.01$ and *** $p < 0.001$. * = MCF7, ○ = MCF7-miR526b, Δ = MCF7-miR655.

2.5. Hypoxia Promotes Inflammatory Gene Expression and Vascular Mimicry in miRNA-High Cells

We have previously shown that miR655 overexpression in MCF7 cells promotes COX-2 expression (Figure S2E) [14] and proposed that this could be via *NFκB1* upregulation in the ER-positive breast cancer cell line [13,14]. We have also shown that COX-2 stimulates the production of PGE2 (prostaglandin E_2), which activates EP4 and consequently activates the PI3K/Akt pathway and promotes breast cancer angiogenesis and lymphangiogenesis [32–35]. Furthermore, we have indicated the overexpression of miR526b and miR655 downregulates *PTEN* [3], resulting in the upregulation of *VEGF*s. Here we investigated if hypoxia can regulate miRNA functions following the same signaling pathways. To establish a link between miRNA, HIF-1α, and the COX-2/EP4/PI3K/Akt pathway, we measured *NFκB1*, *COX-2*, *EP4*, and *VEGFA* gene expression in breast cancer cell lines, in both normoxia and hypoxia. We used qRT-PCR to measure gene expression of *NFκB1*, *COX-2*, and *EP4*.

2.5.1. Hypoxia Promotes Expression of NFκB1, COX-2, and EP4

While *NFκB1* expression significantly increased in all three MCF7 cell lines in hypoxia compared to normoxia, this increase was more prominent and significant in MCF7-miR526b and MCF7-miR655 cell lines (Figure 6A). We also compared the gene expression of *NFκB1* in MCF7-miR526b and MCF7-miR655 cell lines in hypoxia with MCF7 cells and observed that both miRNA-high cell lines show a significant increase in *NFκB1* expression compared to MCF7 cells (Figure 6A). We tested the change in *NFκB1* expression in two other miRNA high cell lines SKBR3-miR526b and MCF7-COX2. In SKBR3-miR526b cells under hypoxia compared to normoxia there was no change in *NFκB1* expression. However, we noted a significant increase in *NFκB1* expression in the ER-positive, miRNA-high MCF7-COX2 cells in hypoxia compared to normoxia (Figure 6E). *COX-2* gene expression was significantly increased in hypoxia compared to normoxia in all cell lines except MCF7-COX2; data for MCF7, MCF7-miR526b, MCF7-miR655 are presented in Figure 6B and data for SKBR3-miR526b and MCF7-COX2 are presented in Figure 6F. It should be noted, however, that *COX-2* overexpression in hypoxia was larger in miRNA-high cell lines (Figure 6B,F) compared to that of MCF7-COX2, which only showed a marginal increase in *COX-2* expression. This could be due to the fact that MCF7-COX2 cells are already high in COX-2, so hypoxia could only marginally enhance *COX-2* expression. However, miRNA-overexpression enhances *COX-2* expression, and hypoxia further enhances this in both MCF7 and SKBR3 miRNA-overexpressed cell lines. MCF7-miR655 cells in hypoxia exhibited a very significant increase in *COX-2* expression compared to MCF7 cells in hypoxia (Figure 6B). *EP4* gene expression was significantly higher in MCF7, MCF7-miR526b, and MCF7-miR655 cell lines (Figure 6C), as well as in SKBR3-miR526b and MCF7-COX2 cell lines (Figure 6G) in hypoxia compared to normoxia.

2.5.2. Hypoxia Promotes VEGFA Gene Expression

We previously showed that miRNA overexpression in MCF7 cell lines enhanced *VEGFA* mRNA and protein production in both miRNA-overexpressed cell lines [3]. Here we analyzed mRNA expression of *VEGFA* in MCF7, MCF7-miR526b, MCF7-miR655, SKBR3-miR526b, and MCF7-COX2 cell lines in hypoxia and normoxia using qRT-PCR. *VEGFA* expression was increased in hypoxia for all cell lines, but this increase was highest in miRNA-high cell lines. Data for MCF7, MCF7-miR526b, and MCF7-miR655 are presented in Figure 6D and data for MCF7-COX2 and SKBR3-miR526b are presented in Figure 6H. In hypoxic conditions, miRNA-high cells show a significant increase in *VEGFA* expression in MCF7-miR526b and MCF7-miR655 cell lines compared to MCF7 cells (Figure 6D). Thus, hypoxia further enhanced vascular gene expression in miRNA-high cells.

2.5.3. Hypoxia Promotes Vascular Mimicry

Tumor cells mimic the properties of vascular endothelial cells and form tube-like vascular structures in a process called vascular mimicry, which show an overexpression of VEGF. MCF7 cells are poorly metastatic cell lines with no vascular properties and cannot form tubes on growth factor-reduced Matrigel. We previously showed that cell-free secretions from MCF7-miR526b and MCF7-miR655 cell lines promote tube formation in HUVECs and produce VEGFs [3]; however, we have never tested the tube formation abilities of miRNA-overexpressing cells in hypoxia. Here we tested the vascular mimicry properties of MCF7, MCF7-miR526b, and MCF7-miR655 cell lines in hypoxia and normoxia. In normoxia, we found that only MCF7-miR526b and MCF7-miR655 cell lines can form tube-like structures at 24 and 48 h, but MCF7 cannot (images in Figure S6A–F, data in Figure S6G–L). In hypoxic conditions, we observed tube-like structures in MCF7 and miRNA-high cells, but miRNA-high cells produced a significantly higher number of complete tubes compared to MCF7 (images in Figure S6M,Q,U, data in Figure S6G–L). These results further confirmed that hypoxic conditions enhance vascular properties in ER-positive breast cancer cells and that hypoxia enhances vascular mimicry properties of miRNA-high cells.

2.6. *Inhibition of Hypoxia-Enhanced Functions in miRNA-High Cells*

The above results indicate that hypoxia enhances *COX-2*, *EP4*, and *NFκB1* expression. We have previously shown that miRNA expression and miRNA-induced functions can be abrogated with a COX-2 inhibitor (Celecoxib, CEL), an EP4 antagonist (ONO-AE3-208, ONO), and an irreversible PI3K/Akt inhibitor (Wortmannin, WM) [3,13,14]. Here we wanted to investigate the effect of inhibition of COX-2/EP4/PI3K/Akt signaling pathways on hypoxia-enhanced miRNA functions and miRNA expression. To investigate the direct involvement of miR526b and miR655 in hypoxia, we would also need to knockdown miR526b/miR655 in aggressive breast cancer cells in normoxia and hypoxia and test if that would inhibit miRNA induced functions. However, we were unable to conduct miRNA-knockdown experiments.

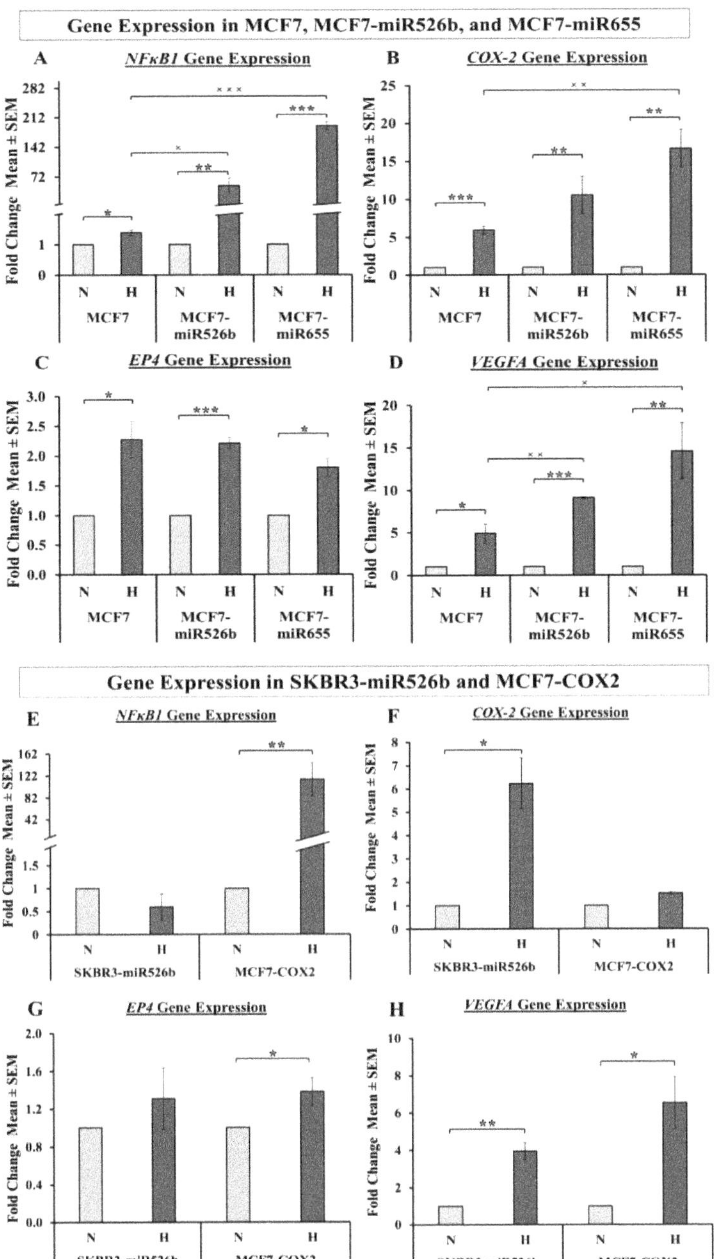

Figure 6. *NFκB1*, *COX-2*, *EP4*, and *VEGFA* gene expression: 'N' indicates normoxia, and 'H' indicates hypoxia. (**A**) *NFκB1*, (**B**) *COX-2*, (**C**) *EP4*, and (**D**) *VEGFA* represents gene expression in MCF7, MCF7-miR526b, and MCF7-miR655 cell lines. (**E**) *NFκB1*, (**F**) *COX-2*, (**G**) *EP4*, and (**H**) *VEGFA* gene expression in SKBR3-miR526b and MCF7-COX2 cell lines. Data are presented as the mean ± SEM of quadruplicate replicates; * $p < 0.05$, ** $p < 0.01$ and *** $p < 0.001$. x indicates $p < 0.05$, xx indicates $p < 0.01$ and xxx indicates $p < 0.001$. * Also indicates comparison between normoxia and hypoxia of the same cell line and x indicates comparison between cell lines only in hypoxia.

2.6.1. Pri-miRNA and HIF-1α Gene Expression Abrogated with COX-2 Inhibitor and EP4 Antagonist

We wanted to determine if pri-miR526b, pri-miR655, and *HIF-1α* gene expression could be reduced in hypoxia with CEL and ONO treatments. Expression of pri-miR526b in hypoxic MCF7-miR526b cells was significantly reduced by non-specific COX-2 inhibitor Celecoxib (CEL), and very significantly reduced by EP4 receptor-specific antagonist ONO-AE3-208 (ONO) treatments. Expression of pri-miR655 in hypoxic MCF7-miR655 cells was very significantly downregulated by both CEL and ONO compared to cells in hypoxic condition only (Figure 7A). We then measured *HIF-1α* expression in cells in hypoxia that had been treated with inhibitors. Both MCF7-miR526b and MCF7-miR655 cells in hypoxia show a very significant decrease in *HIF-1α* gene expression when treated with CEL and ONO, in comparison to hypoxic cells without inhibitor treatment (Figure 7B). These results strongly suggested that enhanced functions and increase in inflammatory gene expression in miRNA-high cell lines during hypoxia is following COX-2/EP4 signaling [3,13,14]. To test the involvement of PI3K/Akt cell signaling in hypoxia-induced functions, we use an irreversible inhibitor WM to block miRNA functions, which we previously showed to strongly regulate miRNA functions in breast cancer [3].

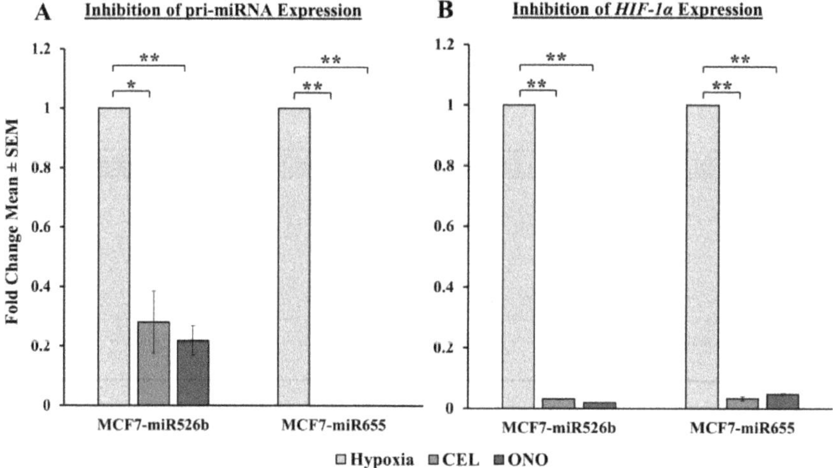

Figure 7. Pri-miR526b, pri-miR655, and *HIF-1α* gene expression in hypoxia and in hypoxia with inhibition: (**A**) Inhibition of pri-miR526b and pri-miR655 gene expression in hypoxia. (**B**) Inhibition of *HIF-1α* gene expression in hypoxia. Data are presented as the mean ± SEM of quadruplicate replicates; * $p < 0.01$ and ** $p < 0.001$.

2.6.2. Inhibition of ROS/SO Production

To examine hypoxia-enhanced ROS and SO production in MCF7, MCF7-miR526b, and MCF7-miR655 cell lines. We tested if CEL, ONO, and WM could significantly block hypoxia-enhanced ROS production in all cell lines. Fluorescence images for MCF7 and MCF7-miR526b ROS-positive cells treated with various inhibitors are presented in Figure 8A,C,E,G and Figure 8B,D,F,H, respectively. Quantification for ROS-positive cell fluorescence is presented in Figure 8I. The fold difference for ROS production before and after inhibitor treatments was very prominent for MCF7-miR526b (between 3.9–11-fold) compared to MCF7 (between 1.8–3.4-fold) (Figure 8I). Both MCF7 and MCF7-miR526b cells show significantly reduced ROS-positive cells with inhibitor treatments. However, MCF7-miR526b cells in hypoxia show a sharper decrease in ROS production after inhibitor treatments than MCF7 cells, as denoted by the fold differences. We observed a very similar phenomenon with SO production. SO production in MCF7 and MCF7-miR526b cell lines in hypoxia were inhibited by CEL, ONO, and WM. Fluorescence images for MCF7 and MCF7-miR526b SO-positive cells treated with various inhibitors are presented in Figure 8K,M,O,Q and Figure 8L,N,P,R, respectively. Quantification for SO cell fluorescence is presented in Figure 8S.

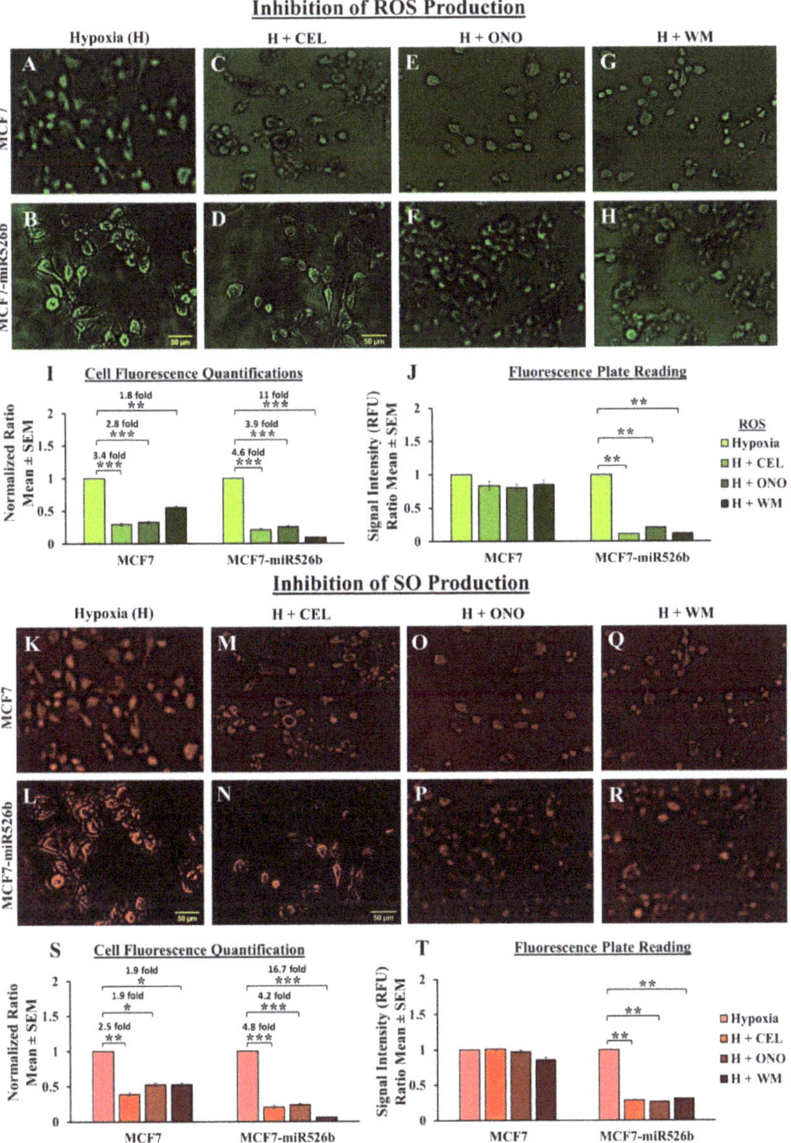

Figure 8. Inhibition of ROS and SO production: Representative MCF7 and MCF7-miR526b cells in (**A**,**B**) hypoxia, (**C**,**D**) hypoxia with CEL, (**E**,**F**) hypoxia with ONO, and (**G**,**H**) hypoxia with WM under the green filter for total ROS detection. (**I**) Quantification ratios for MCF7 and MCF7-miR526b cells positive for ROS. (**J**) Fluorescence microplate assay to quantify total ROS production. Fluorescent MCF7 and MCF7-miR526b cells in (**K**,**L**) hypoxia, (**M**, **N**) hypoxia with CEL, (**O**,**P**) hypoxia with ONO, and

(**Q**,**R**) hypoxia with WM under the red filter for total SO detection. (**S**) Quantification ratios of MCF7 and MCF7-miR526b cells positive for SO. (**T**) Total SO production measured with fluorescence microplate assay. For all pictures, the scale bar represents 50 μM. Quantitative data are presented as the mean of three biological replicates ± SEM. * $p < 0.05$, ** $p < 0.01$ and *** $p < 0.001$.

Similarly, these inhibitors blocked hypoxia-enhanced SO production in MCF7-miR526b cells (4.2–16.7-fold) compared to MCF7 (1.9–2.5-fold) (Figure 8S). We also measured inhibition of total fluorescence emission by cells treated with inhibitors. There was a marginal decrease in total fluorescence by MCF7 cells with treatments, whereas for MCF7-miR526b inhibitors could significantly abrogate hypoxia induced ROS and SO production. Total fluorescence emission measurement for ROS is presented in Figure 8J and for SO is presented in Figure 8T. Fluorescence microplate assays strongly imply that hypoxia enhances miRNAs' promotion of ROS and SO production in miRNA-high cells. This stimulation was significantly inhibited by COX-2, EP4, and PI3K/Akt inhibitors indicate a miRNA-specific function.

2.6.3. Inhibition of Cell Migration

Hypoxia promotes migration of MCF7, MCF7-miR526b, and MCF7-miR655 cell lines. In hypoxia, MCF7 cells migrated marginally (images in Figure 9D–F, quantification in Figure 9AE) compared to the control MCF7 normoxia cells (Figure 9A–C). However, miRNA-high cells migrated significantly and sealed the wound (images in Figure 9S–U, quantification in 9AF) compared to the standard MCF7-miR526b normoxia cells (Figure 9P–R). In MCF7 cells, CEL (Figure 9G–I), ONO (Figure 9J–L), and WM (Figure 9M–O) marginally reduced the wound sizes at 24 h. Quantitative data for MCF7 are presented in Figure 9AE. Hypoxia enhanced cell migration of MCF7-miR526b cells at both 8 h, and 24 h. This migration was inhibited in the presence of CEL (Figure 9V–X); ONO (Figure 9Y–AA); or WM (Figure 9AB–AD). Quantitative data are presented in Figure 9AF. Similarly, it was found that MCF7-miR655 cells in hypoxia had significantly smaller wound sizes at both 8 and 24 h (Figure S7A–C), while hypoxic cells treated with CEL (Figure S7D–F), ONO (Figure S7G–I), or WM (Figure S7J–L) showed marginally smaller wound sizes at 8 h and significantly smaller wound sizes at 24 h. Quantification for MCF7-miR655 is presented in Figure S7M. In both miRNA-high cells, while COX-2 inhibitor and EP4 antagonist could partially block cell migration; irreversible PI3K/Akt inhibitor completely blocked cell migration. These results indicate that hypoxia enhances migration of very significantly in miRNA-high cells via COX-2/EP4/PI3K/Akt pathways, which was evidently absent in miRNA low MCF7 cells. Thus, enhancement of cell migration in hypoxia is due to miRNA.

2.6.4. Inhibition of Vascular Mimicry

We observed inhibition of hypoxia-enhanced tube formation by MCF7-miR526b and MCF7-miR655 cells in the presence of CEL, ONO, and WM. Images are presented in Figure S6N–P,R–T,V–X . Data are presented in Figure S6G–L. This indicates that hypoxia promotes miRNA-induced vascular mimicry phenotypes in miRNA-high cells following the same COX-2 and miRNA-induced angiogenesis and lymphangiogenesis pathways [3,33,34].

2.7. Linking COX-2, EP4, and PI3K/Akt Pathways with Hypoxia and miRNAs

We have shown that miR655 overexpression promotes *COX-2* expression in the ER-positive breast cancer cell line MCF7 (Figure S2E) [14] and here we showed that in hypoxia *COX-2* mRNA expression is enhanced (Figure 6B). We have also shown that *NFκB1* is upregulated in miRNA-high cell lines, and is significantly increased under hypoxia (Figure 6A). COX-2 stimulates the production of PGE2, which activates PGE2 receptor EP4 and consequently activates the PI3K/Akt pathway [32–35]. Moreover, we have shown that the overexpression of miR526b and miR655 upregulates VEGF expression and downregulates *PTEN* [3], a negative regulator of PI3K/Akt and HIF-1α. The absence of *PTEN* results in the upregulation of *HIF-1α* and *VEGFA*. In this study, we found that *VHL*, a tumor suppressor gene and negative regulator of HIF-1α, is downregulated in miRNA-high cells in hypoxic conditions, which leads to upregulation of *HIF-1α*. Hypoxia-enhanced functions could be abrogated in the presence of a COX-2 inhibitor, EP4 antagonist, or PI3K/Akt inhibitor. All of these proposed pathways are illustrated in Figure 10.

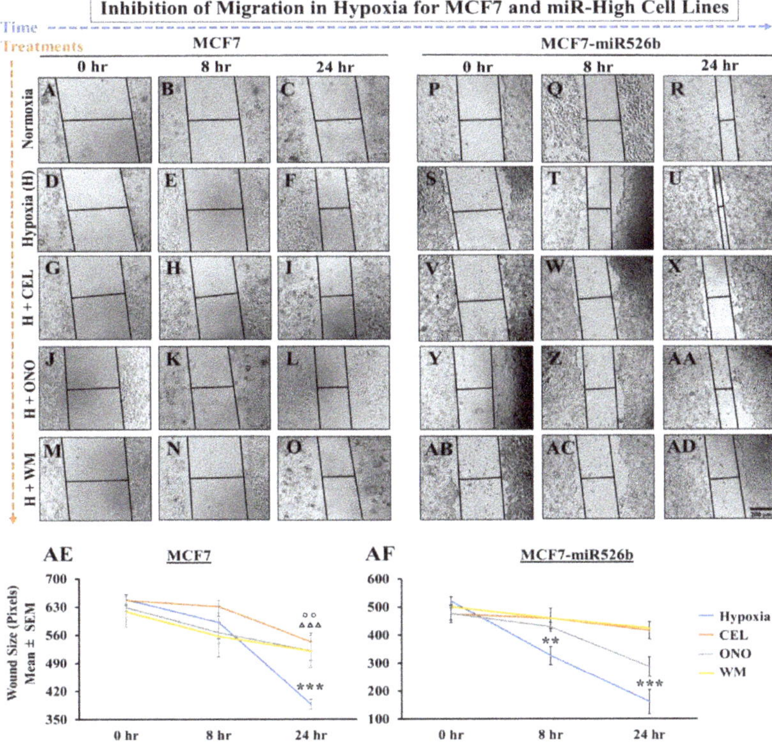

Figure 9. Inhibition of cell migration: Migration of cells recorded at 0, 8 and 24 h. (**A–C**) Representative images of MCF7 cells in normoxia, (**D–F**) in hypoxia, (**G–I**) hypoxia with CEL, (**J–L**) hypoxia with ONO and (**M–O**) hypoxia with WM. (**P–R**) Representative images for MCF7-miR526b in normoxia, (**S–U**) in hypoxia, (**V–X**) hypoxia with CEL, (**Y–AA**) hypoxia with ONO, and (**AB–AD**) hypoxia with WM. (**AE**) Quantitative data for the inhibition of MCF7 cell migration with inhibitors. (**AF**) Quantitative data of inhibition of hypoxia-enhanced migration of MCF7-miR526b. Data are presented as the mean ± SEM of quadruplicate biological replicates; ** $p < 0.01$ and *** $p < 0.001$. * = Hypoxia, ο = Celecoxib, $^{\Delta}$ = ONO-AE3-208.

2.8. Bioinformatics Analysis and Regulation of PTEN and NFκB1 by miRNA

2.8.1. PTEN Regulation

We previously showed that both miR526b and miR655 regulate *PTEN* [3], and that *PTEN* downregulates HIF-1α. Although *PTEN* is not a direct target of miR526b or miR655, both miRNA modulate transcription factors that regulate *PTEN* expression. Using Targetscan via miRBase, we identified that the total number of target genes for miR526b and miR655 was 4133 and 3264, respectively [36]. From the Enrichr database, we found a total of 31 transcription factors (TFs) that regulate *PTEN* gene expression, four of which target human *PTEN*. Transcription factors *ZFX*, *SALL2*, and *SALL4* positively upregulate *PTEN*, while *SREBF* downregulates *PTEN*. Bioinformatics analysis further shows that *SALL2* is a target of miR526b and *SALL4* is directly targeted by miR655 and partially targeted by miR526b (Figure 11A), thus, we decided to measure *SALL4* expression and found an anti-correlation effect with miRNA expression. Here we observed that in both MCF7-miR526b and MCF7-miR655 cell lines, *SALL4* is significantly downregulated compared to MCF7 cells (Figure 11C), indicating an anti-correlation effect between miRNA and *SALL4*. We suggest this to be a plausible explanation for why *PTEN* is significantly downregulated with miRNA upregulation in MCF7 cells.

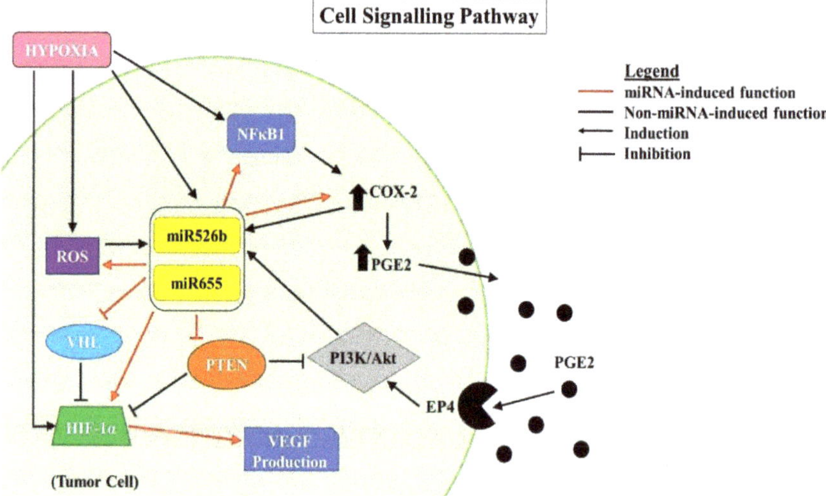

Figure 10. Linking COX-2, EP4, and PI3K/Akt pathways with hypoxia, miR526b, and miR655: red lines indicate functions induced by miR526b and miR655, while black lines indicate functions that are not directly induced by miRNAs. Arrows indicate induction, while T-shaped lines indicate inhibition.

2.8.2. NFκB1 Regulation

Here we show that in hypoxia, miRNA-high cell lines have significant upregulation of *NFκB1* gene expression. We identified a total of 39 transcription factors (TFs) that are associated with the *NFκB1* gene, six of which were identified by the Enrichr database as TFs regulating human *NFκB1*. Amongst these six TFs, two transcription factors, *ZNF207* and *NR2C2*, negatively regulate *NFκB1*. Bioinformatics analysis shows that both *ZNF207* and *NR2C2* are the common target of both miR526b and miR655 (Figure 11B). These two TFs are significantly downregulated in miRNA-high cells compared to miRNA-low MCF7 cells (Figure 11C), indicating an anti-correlation effect between miRNAs and these TFs. The absence of these negative regulators explains why *NFκB1* expression is upregulated in miRNA-high cells. Luciferase reporter assays are needed in the future to validate that *SALL4*, *ZNF207*, and *NR2C2* are targets of miR526b and miR655.

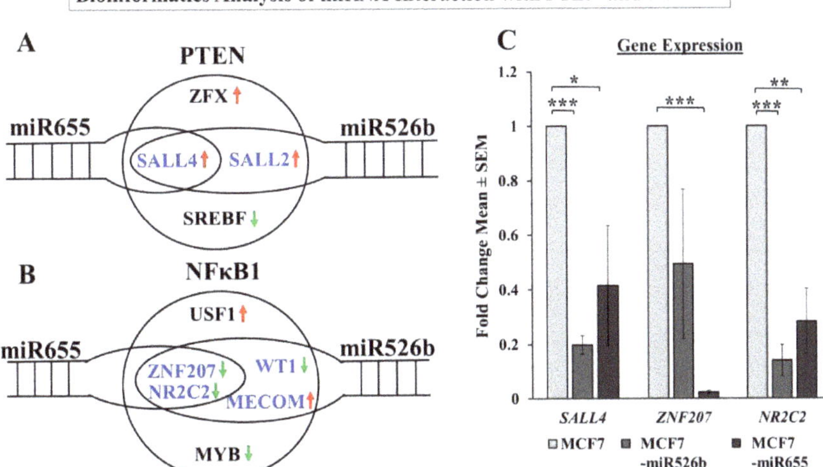

Figure 11. Bioinformatics analysis of the regulation of *PTEN* and *NFκB1* by miRNA-high cells: (**A**) regulation of *PTEN* by miR655 and miR526b through transcription factors *SALL2* and *SALL4*. (**B**) Regulation of *NFκB1* by miR655 and miR526b through transcription factors *MECOM*, *NR2C2*, *WT1*, and *ZNF207*. Red arrows indicate that the transcription factor upregulates gene expression, green arrows indicate that the transcription factor inhibits gene expression. (**C**) Gene expression of transcription factors targeted by both miRNAs (*SALL4*, *ZNF207*, and *NR2C2*) in MCF7, MCF7-miR526b, and MCF7-miR655 cell lines. Data are presented as the mean ± SEM of quadruplicate biological replicates. * $p < 0.05$, ** $p < 0.001$ and *** $p < 0.0001$.

2.9. miR526b and miR655 Expression Significantly Correlates with HIF-1α Expression in Human Breast Tumors

2.9.1. Ontario Tumor Bank Sample Demography

To further validate the relationship between miR526b, miR655, and *HIF-1α* expression, we tested our hypothesis on human breast cancer tissues. We collected 96 tumor tissue and 20 non-cancerous control tissue samples from the Ontario Tumor Bank and extracted total RNA, synthesized cDNA and measured gene and miRNA expressions using Taqman gene and miRNA expression assays. Demographic data of the samples are shown in Table 1. In the tumor sample set, 96.88% are female samples, 25% were considered tobacco smokers, 29.17% were considered social or occasional alcohol consumers, and 3.13% were categorized as regular drinkers. In the data set, 38.85% of tumor samples are ER-positive, and 63.54% are HER2- negative. PR-positive and PR-negative status are almost similar at 32.29% and 31.25%, respectively, and 10.42% are triple-negative breast cancer samples. In this data set, we have seven stage I tumor samples (7.29%), 45 stage II samples (46.87%), 39 stage III samples (40.63%), and five stage IV tumor samples (5.21). Control tissues are histopathologically normal with all females and 5% and 25% had smoking and alcohol habits, respectively.

Table 1. Demography of human benign and malignant tissue samples from Ontario Tumor Bank: this table illustrates tobacco exposures, alcohol consumption, hormone receptor status (ER, PR, HER2), and tumor stage status of the benign and malignant human tissue samples used in this study. Samples were age-matched; the majority of samples are from female patients, only three samples are male. Hormone receptor status of weak and intermediate was considered neither negative nor positive. Age and pack year were presented as mean ± SD.

Subjects		Benign *N* = 20 (%)	Malignant *N* = 96 (%)
Sex	Female Male	20 (100) 0 (0)	93 (96.88) 3 (3.13)
Age Distribution (Years)	Range	52–87	27–92
Age (years)	Mean ± SD	66 ± 11	63 ± 17
Smoking	Smokers	1 (5)	24 (25)
	Pack Year (PY) ± SD	40	27 ± 19
Alcohol Consumption	Social or Occasional Drinker	5 (25)	28 (29.17)
	Regular Drinker	0 (0)	3 (3.13)
ER Status	Positive Negative Unknown	N/A N/A N/A	37 (38.58) 19 (19.79) 6 (6.25)
PR Status	Positive Negative Unknown	N/A N/A N/A	31 (32.29) 30 (31.25) 6 (6.25)
HER2 Status	Positive Negative Unknown	N/A N/A N/A	21 (21.88) 61 (63.54) 14 (14.58)
ER, PR, HER2 Status	Negative	N/A	10 (10.42)
Tumor Stage			
I *		N/A	7 (7.29)
II		N/A	45 (46.87)
III		N/A	39 (40.63)
IV		N/A	5 (5.21)

N/A: Not Applicable. * Stage 0 samples (n = 2) were compiled with stage I samples and considered as stage I (n = 5).

2.9.2. HIF-1α and miRNA Expression in Breast Tumor and Control Tissues

Here, we report that tumor samples showed significant upregulation of *HIF-1α* expression compared to the control tissues (Figure 12A). We also estimated that in stratified samples, *HIF-1α* expression was very significantly high in ER-positive, PR-positive, and HER2-negative breast tumors compared to the control tissues (Figure 12A). We did not find an association of *HIF-1α* expression with triple-negative breast cancer, which could be due to the fact that we had only a few triple-negative breast tumor tissues. In addition, *HIF-1α* expression was very significantly increased in stage I and II tumors, significantly increased in stage III, but only marginally high in stage IV tumors compared to the control tissues (Figure 12B). However, we have only a few stage I and stage IV tumor samples, thus, the observed association of *HIF-1α* expression is specific to stage II and stage III and the observed findings need to be validated with a larger sample set.

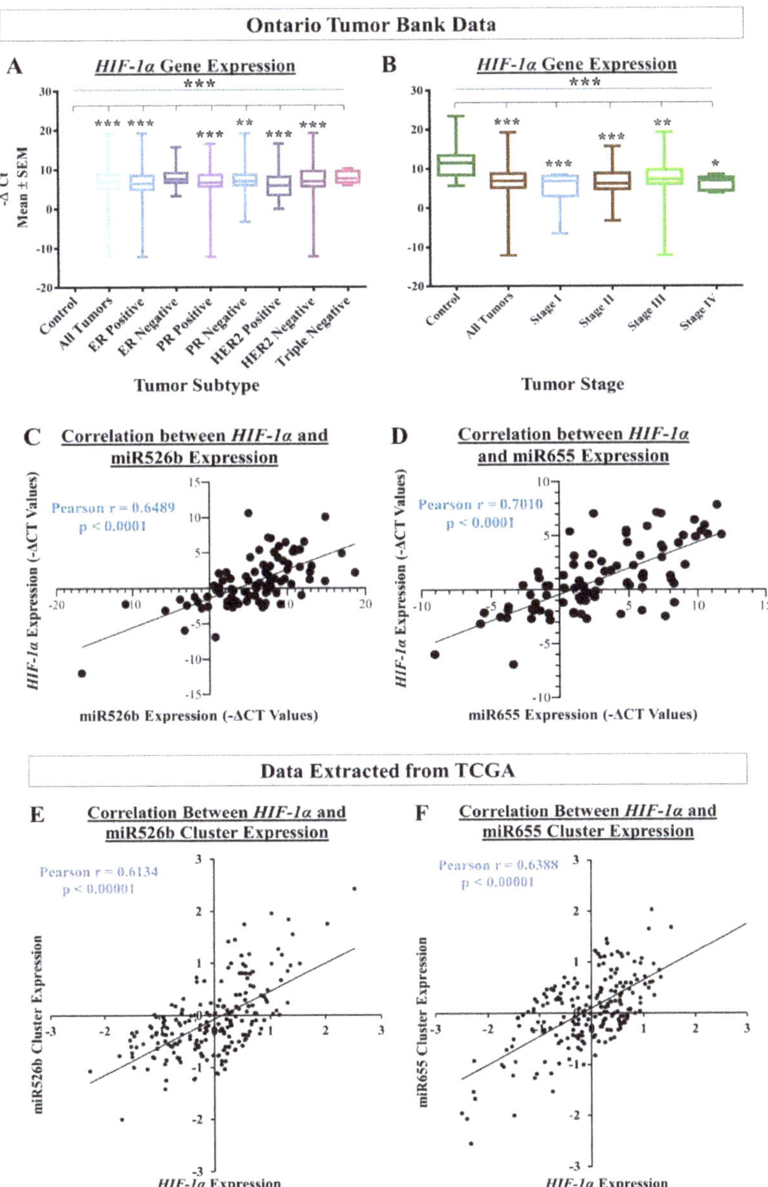

Figure 12. Human tumor and control tissue data: (**A**) Delta CT (ΔCT) of *HIF-1α* expression in control tumor tissues, all tumor tissues, ER-positive/negative, PR-positive/negative, HER2-positive/negative, and triple-negative tumor samples. (**B**) *HIF-1α* expression in stage I, II, III, and IV tumors. (**C**) Correlation between *HIF-1α* and miR526b expression in tumor samples. (**D**) Correlation between *HIF-1α* and miR655 expression in tumor samples. In figures A and B, the *Y*-axis represents -ΔCT, as smaller ΔCT values indicate higher expression. (**E**) Correlation between *HIF-1α* and miR526b cluster expression in tumor samples. (**F**) Correlation between *HIF-1α* and miR655 cluster expression in tumor samples. * $p < 0.05$, ** $p < 0.01$ and *** $p < 0.001$.

We established that in the same tumor sample set, the expressions of both miRNAs are significantly high in tumor compared to control tissues, and both miRNA expressions are associated with poor patient survival [13,14]. In our previous studies, we have shown miR526b and miR655 expression to be proportionally higher in the ER-positive, PR-positive, and HER2-negative samples [3,13,14]. Here in this study, we wanted to investigate a possible link between miRNA expressions with hypoxia in breast cancer. To find any correlation between miRNA expression and *HIF-1α* expression in tumor tissues, we conducted a Pearson correlation coefficient analysis. We observed a very significant positive correlation between miR526b expression and *HIF-1α* expression (Figure 12C), and between miR655 expression and *HIF-1α* expression (Figure 12D). For miR526b and *HIF-1α*, the R-value is 0.6489, and for miR655 and *HIF-1α*, the R-value is 0.7010, showing a strong positive correlation. Due to few samples in stratified tumor subtype categories, we did not perform a correlation analysis between miRNA and *HIF-1α* in each tumor subtype and stage. This should be investigated in future studies.

2.9.3. Data Extracted from the cBioPortal Database Via TCGA

To further validate the link between miRNA expressions with hypoxia in breast cancer, we used human breast tumor gene and miRNA expression data available in the TCGA database. We used the cBioPortal database within TCGA to extract breast cancer specific gene and miRNA expression data [37,38]. In total, we used compiled breast cancer tumor tissue data from 16 breast cancer studies included in cBioPortal. Here, we compared the *HIF-1α* mRNA expression to the mean miRNA cluster expression of either miR526b or miR655. While conducting correlation of *HIF-1α* and miRNA clusters, we excluded samples that did not have data for either miRNA or *HIF-1α* expression. As a result, we had 200 samples for the miR526b cluster analysis and 202 samples for miR655 cluster analysis, which had data for both miRNA clusters and *HIF-1α*. miR526b's miRNA cluster contains 20 miRNAs, of which only two miRNAs, miR516a-1, and miR516a-2, had available expression data. We took the mean expression data of both of these miRNAs and presented this as miR526b cluster expression. The miR655 miRNA cluster also contains 20 miRNAs, nine of which (miR154, miR369, miR381, miR382, miR409, miR410, miR487b, miR539, and miR889) had available expression data. We took the mean of all nine miRNAs expression data and presented this as miR655 miRNA cluster expression.

With Pearson correlation coefficient analysis, miR526b cluster expression showed a very significant correlation with *HIF-1α* expression, with an R-value of 0.6134 and $p < 0.00001$ (Figure 12E). Similarly, the average expression of miR655 cluster was also very significantly correlated with *HIF-1α* expression, with an R-value of 0.6388 and $p < 0.00001$ (Figure 12F). These data, compiled from 16 different studies, shows strong implications of miR526b/miR655 expression correlated to *HIF-1α* expression in breast cancer. These results further strengthen the notion that both miRNAs collaborate in hypoxia to promote aggressive breast cancer.

3. Discussion

The tumor microenvironment plays a major role in tumor growth and metastasis. An aggressively growing tumor goes through a phase of hypoxia, in which the center of the tumor mass is deprived of oxygen. In order to survive, tumor cells release growth factors and chemokines, which in turn promote angiogenesis, thus allowing the tumor to bypass apoptosis [19,30]. Hypoxia promotes angiogenesis, EMT, and oxidative stress in the tumor microenvironment [30]. Our previous studies have demonstrated the roles of miR526b and miR655 as oncogenic miRNAs, promoting aggressive breast cancer phenotypes such as cell migration, invasion, tumor associated angiogenesis, cancer stem cell induction, oxidative stress, tumor growth, and metastasis [3,13,14,27]. Involvement of miRNAs to change and modulate the tumor microenvironment to promote breast cancer metastasis is a growing field of research. Thus, in this article, we tested the interaction and change of functions in two oncogenic miRNAs, miR526b, and miR655, in hypoxia.

Both hypoxia and miRNAs have been associated with the promotion of cancer in various studies, and one has been shown to regulate the other. For instance, Bandara et al. have shown that hypoxia-enhanced miRNAs play an important role in the hypoxic adaptation of cancer cells, and have demonstrated that hypoxia is also a regulator of miRNA biogenesis [39]. Here we observed that hypoxia enhances miR526b and miR655 expression in ER-positive breast cancer cells. Another study by Bhandari et al. also shows hypoxia-enhanced miRNA dysregulation in various cancers, and identified miR133a-3p as a hypoxia-modulated miRNA [40]. Hypoxia-induced miR590-5p was shown to stimulate matrix metalloprotease activity and stimulate cell migration and invasion [41]. Conversely, Krutilina et al. discovered that miR-18a directly targets *HIF-1α*, and downregulates hypoxic gene expression [12] and in colon cancer miR22 was shown to inhibit hypoxia [42].

HIF-1α is a transcription factor that acts as a marker for hypoxia in cells. In this study, we observed that aggressive miR526b/miR655-overexpressing cell lines (MCF7-COX2, MCF7-miR526b, MCF7-miR655, SKBR3-miR526b) produce high HIF-1α in normoxia, while poorly metastatic miRNA-low cell lines show a significantly lesser amount of HIF-1α. These results show that even under normoxic conditions, miRNA-high cell lines are naturally high in hypoxia marker expression, indicating that these miRNAs may be involved in hypoxia in breast cancer. This is supported by Kulshreshtha et al., showing miRNA directly regulates *HIF-1α* gene expression in various cancers [43]. We also observed that in hypoxia, there is a very significant increase in HIF-1α mRNA and protein expression in miRNA-high cell lines, in particular in ER-positive cells (MCF7-miR526b, MCF7-miR655) compared to miR-low MCF7 cells, indicating that this could be an ER-specific phenomena. Additionally, $CoCl_2$ treatment enhanced miR526b and miR655 expression in MCF7 cells as well, thus, an increase in HIF-1α expression in MCF7 cells could be due to miRNA expression upregulation. Furthermore, the expression of HIF-1α is partly controlled by a tumor suppressor pVHL, which tags HIF-1α and sends it for degradation under normoxic conditions [44]. We found that the VHL gene was significantly downregulated in hypoxic conditions in miRNA-high cells, hence, HIF-1α expression enhanced. This established a strong link between miRNA and hypoxia.

Hypoxic conditions are the master regulators of oxidative stress, causing ROS production, DNA damage, promoting inflammation [45,46], and oxidative stress induces inflammatory miRNA production [39,47]. In our previous study, we have shown that miR526b and miR655 directly upregulate oxidative stress in breast cancer [27]. Here, we observed that in MCF7-miR526b and MCF7-miR655 cells, ROS and SO production is further stimulated by hypoxia. The increase in ROS and SO production is greater in the miRNA-high cell lines than the increase in MCF7 cells in hypoxia. TXNRD1 is an enzyme that regulates the production of ROS and SO and overexpression of this enzyme is an indicative marker for oxidative stress. We found that hypoxia enhanced *TXNRD1* expression in all breast cancer cell lines, but the most significant increase was found in MCF7-miR655 cells. The marginal increase in oxidative stress in MCF7 cells after $CoCl_2$ treatment could be a combined effect of hypoxia and miRNA-overexpression in this cell due to hypoxia. This suggests hypoxia and miR526b and miR655 collaborate to enhance oxidative stress in breast cancer.

Hypoxia can completely reprogram tumor cells to induce EMT, and stimulate vasculogenesis to enhance cell migration [48]. We have shown that miR526b and miR655 overexpression in breast cancer cells promotes EMT, cell migration, as well as *VEGFA* upregulation [3,13,14]. Here, we identified that MCF7-miR526b and MCF7-miR655 had higher levels of the mesenchymal markers (*VIM*, *TWIST1*, *SNAIL*) and lower levels of the epithelial marker *CDH1* expression in hypoxia compared to normoxia. Mesenchymal cells are highly migratory, thus, a scratch-wound migration assay was performed and found that the scratch wound closes faster in miRNA-high cells in hypoxia compared to MCF7 cells in hypoxia. We also observed that miRNA-high cell lines show vascular mimicry and promote tube formation in hypoxia. All these phenotypes support that hypoxia enhances functions of miR526b and miR655 to promote breast cancer cell aggressiveness.

In the past research, we have identified that in ER-positive MCF7 breast cancer cells, *COX-2* overexpression significantly upregulates the expression of miR526b and miR655. miR526b and miR655 are known to upregulate *COX-2* and *EP4* expression, and we proposed that miRNA could regulate *COX-2*/*EP4* expression through the NFκB pathway [13,14]. COX-2 activity produces PGE2, which in turn binds to the EP4 receptor. EP4 activation induces PI3K/Akt signaling, which regulates angiogenesis during embryogenesis and in breast cancer metastasis [33,35,49,50]. We have also shown that COX-2, EP4, and PI3K/Akt inhibition could abrogate miRNA-induced angiogenesis in vitro [3]. A link between miRNA regulating HIF-1α expression via PI3K/Akt signaling was shown in other tumor models as well [51]. In the current study, we show that hypoxic conditions enhance *COX-2/EP4* and *NFκB1* expression in ER-positive breast cancer cells, and both COX-2 inhibitor Celecoxib (CEL) and EP4 antagonist ONO-AE3-208 (ONO) significantly abrogate miRNA expression. Therefore, we attempted to block the cancer-promoting phenotypes enhanced by hypoxic conditions in miRNA-high cells. Our findings show that MCF7-miR526b and MCF7-miR655 cell migration, oxidative stress, and vascular mimicry was inhibited by the application of a COX-2 inhibitor, EP4 antagonist, and an irreversible PI3K/Akt inhibitor Wortmannin (WM). The hypoxia-enhanced functions of miRNA-high cells were inhibited to a greater extent than that of miRNA-low MCF7 cells. These results strongly suggest that, in hypoxia, COX-2/EP/PI3k/Akt signaling pathways regulate miRNA functions. However, this does not show the effect of miRNA knockdown or inhibition of miRNA expression in aggressive cell lines. While we have shown in previous studies that the knockdown of miR526b and miR655 in aggressive breast cancer cell lines reduces aggressive breast cancer phenotypes [13,14], here we were unable to test the direct effects of miRNA knockdown in hypoxia. In the future, it would be interesting to investigate the effects of miR526b and miR655 knockdown on hypoxia in breast cancer.

We previously validated that both miR526b and miR655 target *CPEB2*, which is a tumor suppressor gene and strongly correlated with p53 expression in breast cancer [15]. We have previously shown that *PTEN* expression is downregulated in miRNA-overexpressed MCF7 cell lines [3]. *PTEN* is also a tumor suppressor that downregulates the expression of *HIF-1α* and regulates the PI3K/Akt pathway. In the absence of *PTEN*, *HIF-1α* is able to act as a transcription factor for VEGFA, increasing angiogenesis, as well as activating other pathways that promote aggressive cancer phenotypes [9]. *NFκB1* is a transcription factor frequently activated in tumors that is involved in growth, progression, and resistance to chemotherapy. Various alarmin receptors are activated by *HIF-1α*, which in turn strongly activates NFκB and pro-inflammatory pathways, furthering the progression of the malignant phenotype [52]. Here we showed that in the hypoxic conditions, *NFκB1* is upregulated most significantly in miRNA-high MCF7 cells, suggesting that miRNA induces *NFκB1* expression in hypoxic conditions. To establish miRNA-signaling pathways, we examined miRNA target genes list.

Additionally, a bioinformatics approach was taken to determine the direct connection between miR526b and miR655 with *NFκB1* and *PTEN*. A number of transcription factors regulating *NFκB1* and *PTEN* were identified as direct or indirect targets of miRNA. *SALL2* and *SALL4* are positive regulators of *PTEN* and can regulate tumor metastasis [53,54]. In our analysis, *SALL4* expression was significantly downregulated in miRNA-high cell lines compared to MCF7. We identified *ZNF207* and *NR2C2* as transcription factors that are negative regulators of *NFκB1* are significantly downregulated in miRNA-high cells. It was shown that the ZNF207-HER2 fusion protein is oncogenic in gastric cancer [55], and NR2C2 was shown to prevent MCF7 cell proliferation in an ER dependent manner [56]. In our study in miRNA-high cells, both *ZNF207* and *NR2C2* are downregulated, and thus, *NFκB1* is upregulated. Although we were unable to conduct a true miRNA target validation using a luciferase reporter assay, our overall findings finally establish the link between miRNAs, NFκB1, COX-2, EP4, PI3K/Akt, PTEN, and HIF-1α signaling pathways.

To assess the translational impact of the discoveries, we tested the relation between miRNA and *HIF-1α* expression using human breast cancer tissue and non-cancerous control tissues. We found that there is a significant increase in *HIF-1α* gene expression in tumor tissues compared to the control tissues. In particular, ER-positive, PR-positive, and HER2-negative human breast tumor samples showed the highest expression of *HIF-1α*. We also recorded that stage II and stage III tumors showed the highest expression of *HIF-1α*, indicating hypoxia enhances miRNA-induced aggressive breast cancer phenotypes at progressive disease states. In this same set of tumors, we previously published that miR526b and miR655 expression was high in tumors and high expression of both miRNAs were associated with poor-patient survival [13,14]. In tumor tissues, we also recorded a strong correlation between miR526b and *HIF-1α* and between miR655 and *HIF-1α*, which suggests that *HIF-1α* and miRNAs strongly interact to enhance breast cancer progression. These in situ data further confirmed the aggressive breast cancer phenotypes recorded in ER/PR-positive, HER2-negative, and miRNA-overexpressing MCF7-miR526b and MCF7-miR655 cell lines under hypoxia in the present article. To test the correlation between miRNA-cluster expressions with *HIF-1α* expression within tumors in an independent data set, we also extracted data from cBioPortal, which includes data from 16 different breast cancer studies. These independent data sets results also showed a strong and positive correlation between miR526b and miR655 cluster with HIF-1α expression, further strengthening our findings. Here, we discovered a novel collaboration between hypoxia and miR526b/miR655 in breast cancer metastasis. It would be interesting to investigate in the future if these two miRNAs can serve as breast cancer biomarkers, specifically in ER-positive breast cancer, which is the most common type of breast cancer incidence in Canada.

4. Materials and Methods

The overall in vitro methodologies followed in this article are presented in Figure 13. We used the Mind the Graph Platform to create the graphical images.

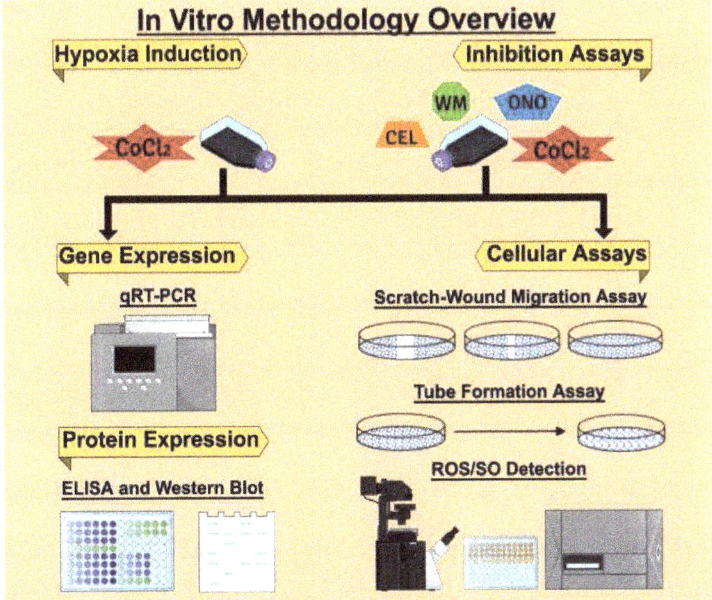

Figure 13. In vitro methodologies overview.

4.1. Ethics Statements

The experiments were conducted at the Department of Biology in Brandon University, following the regulations of Brandon University Research Ethics (#21986, approved on 21 April 2017) and Biohazard Committee (#2017-BIO-02, approved on 13 September 2017).

4.2. Cell Culture

MCF7, T47D, SKBR3, Hs578T, and MDA-MB-231 breast cancer cell lines were purchased from the American Culture Type Collection (ATCC, Rockville, MD, USA). Stable miRNA-overexpression MCF7-miR526b, MCF7-miR655, and SKBR3-miR526b cell lines and COX-2-overexpressing MCF7-COX2 cell line were created by transfecting MCF7 and SKBR3 cells with respective miRNA or COX-2 overexpression plasmids. Transfected cell lines were sustained with Geneticin (Gibco, Mississauga, ON, Canada) following protocols as previously described [13,14,32]. MCF7, MCF7-miR526b, MCF7-miR655, and SKBR3-miR526b cell lines were grown in Roswell Park Memorial Institute (RPMI) 1640 medium (Gibco, Mississauga, ON, Canada), supplemented with 10% fetal bovine serum (FBS) and 1% Penstrep. T47D, SKBR3, MCF7-COX2, Hs578T, and MDA-MB-231 cell lines were grown in Dulbecco's Modified Eagle Medium (DMEM) (Gibco, Mississauga, ON, Canada) supplemented with 10% FBS and 1% Penstrep. All cell lines were maintained in a humidified incubator at 37 °C and 5% CO_2. MCF10A mammary epithelial cells were grown and maintained in Dr. Lala's laboratory at the University of Western Ontario following ATCC protocol. An aliquot of cDNA samples was then transferred to Dr. Majumder laboratory at Brandon University.

4.3. Drugs and Chemicals

Celecoxib (COX-2 inhibitor, CEL) was purchased from Pfizer (Groton, CT, USA). ONO-AE3-208 (selective EP4 antagonist, EP4A, ONO) was a gift from ONO Pharmaceuticals (Osaka, Japan). Wortmannin (irreversible PI3K/Akt inhibitor, WM) was purchased from Sigma-Aldrich (Saint Louis, MO, USA). $CoCl_2$ was purchased from Fisher Scientific (Mississauga, ON, Canada). For all treatments in vitro, hypoxia ($CoCl_2$) served as the positive control and normoxia (sterile water, the solvent of $CoCl_2$) served as the negative control.

4.4. Hypoxia Induction In Vitro with $CoCl_2$ Treatment

Concentrations of $CoCl_2$ were chosen based on other publications tested with breast cancer cells [29,30]. Once 70% confluent, cells were serum starved for 12 h. and $CoCl_2$ was administered at a concentration of either 50 μM or 150 μM. 24 h after $CoCl_2$ treatment, cells were harvested for RNA extraction or used for cell functional assays as described below. We observed that 150 μM induced maximum HIF-1α expression (Figure S1), thus, for all treatments in vitro, 150 μM of $CoCl_2$ treatment was considered as hypoxia, and sterile H_2O served as normoxia.

4.5. RNA Extraction, cDNA Synthesis, and Quantitative Real-Time PCR

Using the miRNeasy Mini Kit (Qiagen, Toronto, ON, Canada), total RNA extractions were done from non-treated T47D, SKBR3, MCF7-COX2, SKBR3-miR526b, Hs578T, MDA-MB-231, MCF7, MCF7-miR526b, and MCF7-miR655 cell lines, as well as $CoCl_2$-treated MCF7, MCF7-miR526b, MCF7-miR655, SKBR3-miR526b, and MCF7-COX2 cell lines. The RNA was then reverse transcribed into cDNA using the High-Capacity cDNA Reverse Transcription Kit (Applied Biosystems, Waltham, MA, USA).

For quantitative Real-Time PCR (qRT-PCR), the TaqMan Gene and miRNA Expression Assays were used. Two genes, *Beta-actin* (Hs01060665_g1) and *RPL5* (Hs03044958_g1) were used as endogenous control genes and RNU44 (assay ID #001094) was considered as an internal control miRNA. The expressions of pri-miR526b (Hs03296227_pri), pri-miR655 (Hs03304873_pri), hsa-miR-526b-5p (assay ID #002382), hsa-miR-655-3p (assay ID #001612), *HIF-1α* (Hs00153153_m1), *VEGFA* (Hs00900055_m1), *COX-2* (Hs00153133_m1), *EP4* (Hs00964382_g1), *VHL* (Hs03046964_s1), *NFκB1* (Hs00231653_m1), *TWIST1* (Hs04989912_s1), *VIM* (Hs00185584_m1), *SNAIL* (Hs00195591_m1), *CDH1* (Hs00170423_m1), *TXNRD1* (Hs00917067_m1), *SALL4* (Hs01010838_g1), *ZNF207* (Hs01045973_m1), and *NR2C2* (Hs00991824_m1) were quantified using relative gene expression analysis. A relative fold change of gene expression was used using the comparative threshold cycle (ΔCt) followed by fold change using the $2^{-\Delta\Delta Ct}$ method [3,13,14,32].

4.6. Enzyme-Linked Immunosorbent Assay (ELISA) Analysis of HIF-1α

HIF-1α protein quantification was carried out using the ab171577-HIF1a Human SimpleStep ELISA Kit (Abcam, Toronto, ON, Canada). This assay is specific to the HIF-1α protein and does not cross-react with HIF-1α homologues, such as HIF-2α (EPAS-1). Three different passages of MCF7, MCF7-miR526b, MCF7-miR655, SKBR3-miR526b, and MCF7-COX2 cell lines were seeded into a six-well plate and grown to 80% confluence. Cells were treated with 150 μM $CoCl_2$ for 24 h. Three experimental replicates were performed for each condition for each passage. The ELISA kit provided standards and was prepared following the manufacturer's instructions. The cells were washed with PBS then solubilized with 1X Cell Extraction Buffer PTR. The cell lysate was then centrifuged, and the supernatant (total protein) was collected. In a 96-well plate, 50 μL of each of the sample protein and prepared standards were added to the wells. Additionally, 50 μL of the HIF-1α antibody cocktail was added, and the plate was incubated on a plate shaker. Next, the wells were washed with 1X Wash Buffer PT, and 100 μL of TMB Substrate was added to each well and incubated. Finally, 100 μL of Stop Solution was added to each well and mixed gently. Microplate readings were then recorded with OD at 450 nm to measure HIF-1α protein levels. Data was collected using the SoftMax Pro 6 Microplate Data Acquisition and Analysis software (Molecular Devices, San Jose, CA, USA). Calculations were performed following the manufacturer's instructions. In all cases, negative control data were subtracted from experimental data for normalization. Provided samples were used to generate a standard curve for protein quantification.

4.7. Western Blot Analysis

Cells were treated with M-PER® Mammalian Protein Extraction Reagent (Thermo Scientific, Rockford, IL, USA), HALT Protease Inhibitor Cocktail (Thermo Scientific), and Phosphatase Inhibitor Cocktail (Thermo Scientific) to extract total protein. Approximately, 15–20 μg of total protein were electrophoresed per well on an 8–10% SDS-polyacrylamide gel and transferred onto Immobilon-FL PVDF membranes (Millipore, Billerica, MA, USA). Blots were incubated with the HIF-1α primary antibody (H1alpha 67): sc-53546 (Santa Cruz Biotechnology, CA, USA) at 1:500 dilution and monoclonal GAPDH antibody (MAB374, from Millipore, Billerica, MA, USA) at 1:10,000 dilutions overnight. After blocking with primary antibodies, membranes were washed and then probed with a mixture of IRDye polyclonal secondary antibodies (LI-COR Biosciences, Lincoln, NE, USA). Images were read with an Odyssey infrared imaging system (LI-COR Biosciences, Lincoln, NE, USA).

4.8. Fluorescence Microplate Assay

Three different passages of MCF7, MCF7-miR526b, and MCF7-miR655 cell lines were seeded in a 96-well plate. Once the cells were grown to 70% confluency, they were treated with either 50 μM or 150 μM of $CoCl_2$ for 24 h. ROS and SO levels were then detected using the ROS-ID Total ROS/SO detection kit (Enzo Life Sciences, Farmingdale, NY, USA), following the manufacturer's protocol. First, the cells were washed with PBS to wash off cell culture media, and ROS/SO detection dyes were added to quantify ROS/SO production. One hour following the addition of detection dyes, microplate readings were done using the standard Fluorescein filter (Excitation/Emission: 485/535 nm) and Rhodamine filter (Excitation/Emission: 550/625 nm). Data was collected using the SoftMax Pro 6 Microplate Data Acquisition and Analysis software (Molecular Devices, San Jose, CA, USA). Concentrations of the ROS and SO produced by cells were determined based on the manufacturer's instructions as we published earlier [27]. ROS/SO production was quantitatively shown as a ratio of hypoxia emissions (emissions from hypoxic cells) to negative control emissions (emissions from normoxic cells). The same process was used for the fluorescence microplate assay with the use of inhibitors, except cells were treated with 150 μM of $CoCl_2$ and supplemented with either 20 μM Celecoxib, 50 μM ONO-AE3-208, or 10 μM Wortmannin for 24 h. To measure the effect of inhibitors, hypoxia treatment was considered as control.

4.9. Fluorescence Microscopy Assay

After total fluorescence emission measurement, we used the same ROS/SO detection kit to determine the total number of cells showing fluorescence and producing ROS and SO following the manufacturer's protocol. MCF7, MCF7-miR526b, and MCF7-miR655 cell lines with or without $CoCl_2$ were seeded in 96 well plates and grown until 70% confluent, then the cells were washed with PBS, and next, ROS/SO detection dyes were added. After 15 min of incubation, images were captured with a Nikon Ds-Ri1 microscopy camera and data were analyzed using the NIS Elements Advanced Research software (Nikon, Melville, NY, USA). The fluorescent cells in each experiment were quantified using the ImageJ software (National Institute of Health, Bethesda, MD, USA) as previously described [27]. For each condition, the negative control (normoxia) was used as a threshold for quantification of hypoxia effect. Total ROS/SO production was presented as quantification ratios, which were calculated by dividing all quantifications by negative control quantifications (i.e., the number of fluorescence-positive cells in the sample divided by the number of fluorescence-positive cells in the control), then dividing the resulting number by the total number of cells present in the well. The same process was used for the fluorescence microscopy assay with the use of inhibitors, except cells were treated with 150 μM of $CoCl_2$ and supplemented with either 20 μM Celecoxib, 50 μM ONO-AE3-208, or 10 μM Wortmannin. To measure the effect of inhibitors, hypoxia treatment was considered as control.

4.10. Scratch-Wound Migration Assay

Cells were grown in RPMI complete (serum-supplemented) media until 90% confluent, then harvested and resuspended in complete RPMI, after which 300 μL of suspended cells (approximately 20,000 cells/mL) was added to a six-well cell culture plate and maintained until 90% confluency. A sterile 2 μL pipette tip was used to scratch the surface of each well, after which the cells were washed with PBS to remove detached cells. The treatment conditions were then applied to the wells. For the hypoxia-mediated migration assay, sterile H_2O was used as the negative control (normoxia), and 150 μM of $CoCl_2$ treatment was considered as hypoxia. A total of 2 mL of the respective conditions (treatments in RPMI basal media) were added to each well. The migratory progress and wound size were captured using a Nikon Ds-Ri1 microscope camera at 0, 16, 24, and 48 h time points. To ensure that we were taking pictures of the same wound-healing site over time, each well was separated into four quadrants manually with a marker pen, and a wound/scratch was made once per coordinate. Additionally, the microscope's coordinate system was used for double validation to ensure photos were taken in the same field of view. We have captured five pictures per quadrant to ensure that the entire wound was captured. Thus, per well, we took at least 20 pictures. NIH ImageJ software was used to measure the width of the scratch wound in pixels and mean data of 20 pictures were considered as data for a single experimental replicate. We used three experimental wells or replicates and three biological replicates per condition [3]. The same process was used for the migration assay with the use of inhibitors, to determine the roles of COX-2, EP4 receptor, and the PI3K/Akt signaling pathways, except the cells were seeded in 24-well plates and treated with 150 μM $CoCl_2$ for 24 h, then supplemented with either 20 μM Celecoxib, 50 μM ONO-AE3-208, or 10 μM Wortmannin for another 24 h. All quantifications were done after 24 h of inhibitor treatment (which is equivalent to 48 h of $CoCl_2$ treatment), as we found an increase in cell death and difficulty in quantification after 24 h.

4.11. Tube Formation Assay

Tube formation assays were carried out as previously described in a 24-well plate [3]. Diluted Matrigel media was prepared as a 1:1 ratio of growth factor reduced Matrigel (BD Biosciences, Bedford, Massachusetts, USA) and basal RPMI media. 200 μL of diluted Matrigel was added to each well in a 24-well plate and incubated for 1 h at 37 °C to allow the Matrigel to crosslink and form the extracellular matrix. Next, 200 μL cells were then seeded into the Matrigel coated plate with a density of approximately 20,000 cells per well. MCF7, MCF7-miR655 and MCF7-miR526b cell lines were resuspended in either RPMI complete media; $CoCl_2$ (150 μM)-RPMI media; or $CoCl_2$-RPMI media along with either 20 μM Celecoxib (COX-2 inhibitor) or 50 μM ONO-AE3-208 (EP4 antagonist), or 10 μM Wortmannin (Irreversible PI3K/Akt pathway inhibitor) as inhibitory conditions. Each condition was tested twice (experimental replicates) and repeated three times (biological replicates). Tube formation was measured at 24 and 48 h, and images were obtained using a Nikon inverted microscope.

Quantification of tubes and branching points was carried out using NIH ImageJ software (NIH, Bethesda, MD, USA).

4.12. Bioinformatics Analysis

miRbase [36] and Enrichr [57] were two online databases used for conducting bioinformatics analysis in this study. miRbase is a miRNA database, which provides predicted miRNA target genes along with miRNA cluster information. The complete target gene list for miR526b and miR655 was downloaded using TargerScanVert release 7.1 [58] in miRbase for five prime mature sequences hsa-mir-526b and hsa-mir-655. The Enrichr database uses enrichment analysis to identify transcription factors regulating genes. All transcription factors associated with *PTEN* and *NFκB1* were downloaded. The two lists generated from miRbase and the Enrichr database were then cross-examined to determine shared target genes and transcription factors.

4.13. Human Breast Cancer Tissue Samples

Frozen human breast tissue samples were obtained from the Ontario Tumour Bank after ethical approval by Ontario Cancer Research Ethics Board (Tec # 010-11), then following approval by the Ethics Review Board of the Tumor bank and collected at the University of Western Ontario at Dr. Lala's laboratory. Qiagen miRNeasy mini kit was used to extract mRNA or miRNA from tissue samples, followed by cDNA synthesis using cDNA Reverse Transcription Kit (Applied Biosystems, USA). An aliquot of all the cDNA samples were transferred to Dr. Majumder's laboratory at Brandon University following the Material Transfer Agreement (MTA) between Brandon University and the University of Western Ontario. All further experiments were conducted at the Department of Biology at Brandon University following Brandon University Ethics and Biohazard protocols.

4.14. In Silico Analysis of cBioPortal Data via TCGA

miR526b and miR655 cluster information was extracted from the miRbase miRNA database [36]. We identified that there are 20 miRNAs within each miRNA cluster for miR526b and miR655. Next, we used the cBioPortal database within TCGA, which includes data from 16 breast cancer studies to extract miR526b and miR655 cluster miRNA expression, along with *HIF-1α* mRNA expression, which were both presented as z-scores [37,38]. For the miR526b miRNA cluster, the cBioPortal database contained miRNA expression data for miR516a-1 and miR516a-2. As for miR655s miRNA cluster, nine miRNA had expression in the cBioPortal database (miR154, miR369, miR381, miR382, miR409, miR410, miR487b, miR539, and miR889). The mean of available miRNAs z-score within each cluster was considered and compared to the *HIF-1α* z-score to determine a correlation between miR526b and miR655 miRNA clusters and *HIF-1α*.

4.15. Statistical Analysis

Statistical calculations were performed using GraphPad Prism (GraphPad Software, Version 8.4.3., San Diego, CA, USA). All parametric data were analyzed with one-way analysis of variance (ANOVA) by Tukey–Kramer or Dunnett post-hoc comparisons. The student's t-test was used when comparing the means of two datasets, and Pearson's correlation coefficient was employed to assess statistical correlations. Statistically relevant differences between means were accepted at $p < 0.05$.

5. Conclusions

Although the roles of miR526b, miR655, and hypoxia are independently studied in various tumor metastasis models, this is the first time an association between miR526b, miR655, and hypoxia to promote metastatic breast cancer phenotypes has been established. These findings further strengthen the roles of these two miRNAs as master regulators of the tumor microenvironment in promoting breast cancer. In addition, these miRNAs can serve as possible therapeutic targets in ER/PR-positive and HER2-negative miR526b/miR655-high breast cancer.

Supplementary Materials: The following are available online at http://www.mdpi.com/2072-6694/12/8/2008/s1, Figure S1: $CoCl_2$ dose-response assay; Figure S2: miRNA, pri-miRNA and COX-2 expression in various cell lines; Figure S3: Western blot analysis to measure endogenous HIF-1α expression in various breast cancer cell lines; Figure S4: MCF7-miR526b and MCF7-miR655 cell densities increase due to hypoxia; Figure S5: MCF7-miR655 cell migration in normoxia and hypoxia; Figure S6: Hypoxia-enhanced vascular mimicry and inhibition of tube formation with COX-2-I, EP4A and PI3K/Akt-I; Figure S7: Inhibition of hypoxia-enhanced migration for MCF7-miR655 cells

Author Contributions: Concept, project design, funding acquisition and supervision: M.M.; Experiments: E.G., B.S., M.C., R.O., and R.F.; Data Analysis: M.M., E.G., B.S., R.O., R.F., and S.M.; Bioinformatics: S.M., R.F., and R.O.; Figures and Image Data Processing: E.G., B.S., R.O., R.F., and S.M.; Manuscript writing: E.G., B.S., M.C., R.O., R.F., and M.M.; Manuscript editing: E.G., B.S., M.C., R.O., R.F., S.M., and M.M. All authors have read and agreed to the published version of the manuscript.

Funding: This project is funded by the NSERC-Discovery grant to M.M. B.S. and M.C. are recipients of the NSERC-USRA scholarship. R.F. was partly supported by the Canadian Summer Job grant to M.M. S.M. is partly supported by a Brandon University Research Committee (BURC) grant to M.M, and R.O. was supported by a Brandon University Student Union (BUSU)-Work study grant to M.M.

Acknowledgments: The authors of this article would like to sincerely thank Peeyush K. Lala for sharing the drugs (Celecoxib and Wortmannin) and human tissue cDNA and Chidambra D. Halari at Lala lab, for sending us MCF7-COX2 cells and MCF10A cDNA; both at the University of Western Ontario. We also thank Lala lab personnel for helping us with Western blot analysis. We sincerely thank Anuraag Shrivastav at the University of Winnipeg for giving us MCF7 cells. We thank Takayuki Maruyama from Ono Pharmaceutical, Osaka, Japan for sharing the ONO-AE3-208. We thank Kingsley Chukwunonso Ugwuagbo for helping us during RNA extraction and cDNA synthesis. We thank Bernadette Ardelli at Brandon University for giving us access to the Fluorescence Microscope, Microplate Reader, and the Rotor Gene PCR machine in her laboratory. We also want to thank Vincent Chen and his lab at Brandon University for allowing us to use the PCR machine.

Conflicts of Interest: The authors declare no conflict of interest.

Abbreviations

CEL	Celecoxib
CDH1	E-cadherin (Epithelial cadherin)
$CoCl_2$	Cobalt chloride
COX-2	Cyclooxygenase-2
CPEB2	Cytoplasmic polyadenylation element-binding protein 2
ELISA	Enzyme-linked immunosorbent assay
EMT	Epithelial-to-mesenchymal transition
EP4	Prostaglandin E_2 receptor 4
ER	Estrogen receptor
HER2	Human epidermal growth factor receptor 2
HIF-1α	Hypoxia-inducible factor 1-alpha
miRNA/miR	MicroRNA
mRNA	Messenger RNA
NFκB	Nuclear Factor Kappa-light-chain-enhancer of activated B cells
NFκB1	Nuclear factor NF-kappa-B p105 subunit
NR2C2	Testicular receptor 4
ONO	ONO-AE3-208
PGE2	Prostaglandin E_2
PHD	Prolyl hydroxylase domain
PI3K	Phosphoinositide 3-kinases
PR	Progesterone receptor
Pri-miRNA	Primary miRNA
PTEN	Phosphatase and tensin homolog
ROS	Reactive oxygen species
SALL2	Sal-like protein 2

SALL4	Sal-like protein 4
SLC	Stem-like cell
SNAIL	Zinc finger protein SNAI1
SO	Superoxide
TCGA	The Cancer Genome Atlas
TWIST1	TWIST1-related protein 1
TXNRD1	Thioredoxin reductase 1
VEGF	Vascular endothelial growth factor
VEGFA	Vascular endothelial growth factor A
VHL	Von Hippel–Lindau tumor suppressor
VIM	Vimentin
WM	Wortmannin
ZNF207	Zinc finger protein 207

References

1. *Canadian Cancer Statistics 2019*; Canadian Cancer Society: Toronto, ON, Canada, 2019.
2. Muz, B.; De La Puente, P.; Azab, F.; Azab, A.K. The role of hypoxia in cancer progression, angiogenesis, metastasis, and resistance to therapy. *Hypoxia* **2015**, *3*, 83–92. [CrossRef] [PubMed]
3. Hunter, S.; Nault, B.; Ugwuagbo, K.C.; Maiti, S.; Majumder, M. Mir526b and Mir655 Promote Tumour Associated Angiogenesis and Lymphangiogenesis in Breast Cancer. *Cancers* **2019**, *11*, 938. [CrossRef] [PubMed]
4. Ahn, G.-O.; Seita, J.; Hong, B.-J.; Kim, Y.-E.; Bok, S.; Lee, C.-J.; Kim, K.S.; Lee, J.; Leeper, N.J.; Cooke, J.P.; et al. Transcriptional activation of hypoxia-inducible factor-1 (HIF-1) in myeloid cells promotes angiogenesis through VEGF and S100A8. *Proc. Natl. Acad. Sci. USA* **2014**, *111*, 2698–2703. [CrossRef]
5. Vaupel, P. The Role of Hypoxia-Induced Factors in Tumor Progression. *Oncologist* **2004**, *9*, 10–17. [CrossRef] [PubMed]
6. Huang, R.; Jin, X.; Gao, Y.; Yuan, H.; Wang, F.; Cao, X. DZNep inhibits Hif-1α and Wnt signalling molecules to attenuate the proliferation and invasion of BGC-823 gastric cancer cells. *Oncol. Lett.* **2019**, *18*, 4308–4316. [CrossRef] [PubMed]
7. Vengellur, A.; Woods, B.G.; Ryan, H.E.; Johnson, R.S.; Lapres, J.J. Gene Expression Profiling of the Hypoxia Signaling Pathway in Hypoxia-Inducible Factor 1α Null Mouse Embryonic Fibroblasts. *Gene Expr.* **2003**, *11*, 181–197. [CrossRef]
8. Camps, C.; Saini, H.; Mole, D.R.; Choudhry, H.; Reczko, M.; Guerra-Assunção, J.A.; Tian, Y.-M.; Buffa, F.M.; Harris, A.L.; Hatzigeorgiou, A.; et al. Integrated analysis of microRNA and mRNA expression and association with HIF binding reveals the complexity of microRNA expression regulation under hypoxia. *Mol. Cancer* **2014**, *13*, 28. [CrossRef]
9. Maruggi, M.; Layng, F.I.A.L.; Lemos, R.; Garcia, G.; James, B.P.; Sevilla, M.; Soldevilla, F.; Baaten, B.J.; De Jong, P.R.; Koh, M.Y.; et al. Absence of HIF1A Leads to Glycogen Accumulation and an Inflammatory Response That Enables Pancreatic Tumor Growth. *Cancer Res.* **2019**, *79*, 5839–5848. [CrossRef]
10. Del Rey, M.J.; Valín, Á.; Usategui, A.; García-Herrero, C.M.; Sánchez-Aragó, M.; Cuezva, J.M.; Galindo, M.; Bravo, B.; Cañete, J.D.; Blanco, F.J.; et al. Hif-1α Knockdown Reduces Glycolytic Metabolism and Induces Cell Death of Human Synovial Fibroblasts Under Normoxic Conditions. *Sci. Rep.* **2017**, *7*, 3644. [CrossRef]
11. Loh, H.-Y.; Norman, B.P.; Lai, K.-S.; Rahman, N.M.A.N.A.; Alitheen, N.; Osman, M.A. The Regulatory Role of MicroRNAs in Breast Cancer. *Int. J. Mol. Sci.* **2019**, *20*, 4940. [CrossRef]
12. Krutilina, R.; Sun, W.; Sethuraman, A.; Brown, M.; Seagroves, T.N.; Pfeffer, L.M.; Ignatova, T.; Fan, M. MicroRNA-18a inhibits hypoxia-inducible factor 1α activity and lung metastasis in basal breast cancers. *Breast Cancer Res.* **2014**, *16*, 1–16. [CrossRef] [PubMed]
13. Majumder, M.; Landman, E.; Liu, L.; Hess, D.; Lala, P.K. COX-2 Elevates Oncogenic miR-526b in Breast Cancer by EP4 Activation. *Mol. Cancer Res.* **2015**, *13*, 1022–1033. [CrossRef] [PubMed]
14. Majumder, M.; Dunn, L.; Liu, L.; Hasan, A.; Vincent, K.; Brackstone, M.; Hess, D.; Lala, P.K. COX-2 induces oncogenic micro RNA miR655 in human breast cancer. *Sci. Rep.* **2018**, *8*, 327. [CrossRef]

15. Tordjman, J.; Majumder, M.; Amiri, M.; Hasan, A.; Hess, D.A.; Lala, P.K. Tumor suppressor role of cytoplasmic polyadenylation element binding protein 2 (CPEB2) in human mammary epithelial cells. *BMC Cancer* **2019**, *19*, 1–16. [CrossRef] [PubMed]
16. Kircher, M.; Bock, C.; Paulsen, M. Structural conservation versus functional divergence of maternally expressed microRNAs in the Dlk1/Gtl2 imprinting region. *BMC Genom.* **2008**, *9*, 346. [CrossRef] [PubMed]
17. Braschi, B.; Denny, P.; Gray, K.; Jones, T.E.M.; Seal, R.; Tweedie, S.; Yates, B.; Bruford, E. Genenames.org: The HGNC and VGNC resources in 2019. *Nucleic Acids Res.* **2018**, *47*, D786–D792. [CrossRef]
18. Jinesh, G.G.; Flores, E.R.; Brohl, A.S. Chromosome 19 miRNA cluster and CEBPB expression specifically mark and potentially drive triple negative breast cancers. *PLoS ONE* **2018**, *13*, e0206008. [CrossRef]
19. Zuazo-Gaztelu, I.; Casanovas, O. Unraveling the Role of Angiogenesis in Cancer Ecosystems. *Front. Oncol.* **2018**, *8*, 248. [CrossRef]
20. Yang, F.; Shao, C.; Wei, K.; Jing, X.; Qin, Z.; Shi, Y.; Shu, Y.; Shen, H. miR-942 promotes tumor migration, invasion, and angiogenesis by regulating EMT via BARX2 in non-small-cell lung cancer. *J. Cell. Physiol.* **2019**, *234*, 23596–23607. [CrossRef]
21. Azad, T.; Ghahremani, M.; Yang, X. The Role of YAP and TAZ in Angiogenesis and Vascular Mimicry. *Cells* **2019**, *8*, 407. [CrossRef]
22. Hendrix, M.J.; Seftor, E.A.; Kirschmann, D.A.; Seftor, R.E. Molecular biology of breast cancer metastasis Molecular expression of vascular markers by aggressive breast cancer cells. *Breast Cancer Res.* **2000**, *2*, 417–422. [CrossRef]
23. Sun, B.; Fang, Y.; Li, Z.; Chen, Z.; Xiang, J. Role of cellular cytoskeleton in epithelial-mesenchymal transition process during cancer progression. *Biomed. Rep.* **2015**, *3*, 603–610. [CrossRef] [PubMed]
24. Quaranta, V. Cell Migration through Extracellular Matrix. *J. Cell Biol.* **2000**, *149*, 1167–1170. [CrossRef] [PubMed]
25. Nantajit, D.; Lin, N.; Li, J.J. The network of epithelial-mesenchymal transition: Potential new targets for tumor resistance. *J. Cancer Res. Clin. Oncol.* **2014**, *141*, 1697–1713. [CrossRef] [PubMed]
26. Buetler, T.M.; Krauskopf, A.; Rüegg, U. Role of Superoxide as a Signaling Molecule. *Physiology* **2004**, *19*, 120–123. [CrossRef]
27. Shin, B.; Feser, R.; Nault, B.; Hunter, S.; Maiti, S.; Ugwuagbo, K.C.; Majumder, M. miR526b and miR655 Induce Oxidative Stress in Breast Cancer. *Int. J. Mol. Sci.* **2019**, *20*, 4039. [CrossRef]
28. Piret, J.-P.; Mottet, D.; Raes, M.; Michiels, C. CoCl2, a Chemical Inducer of Hypoxia-Inducible Factor-1, and Hypoxia Reduce Apoptotic Cell Death in Hepatoma Cell Line HepG2. *Ann. N. Y. Acad. Sci.* **2002**, *973*, 443–447. [CrossRef]
29. Li, Q.; Ma, R.; Zhang, M. CoCl2 increases the expression of hypoxic markers HIF-1α, VEGF and CXCR4 in breast cancer MCF-7 cells. *Oncol. Lett.* **2017**, *15*, 1119–1124. [CrossRef]
30. Rana, N.K.; Singh, P.; Koch, B. CoCl2 simulated hypoxia induce cell proliferation and alter the expression pattern of hypoxia associated genes involved in angiogenesis and apoptosis. *Biol. Res.* **2019**, *52*, 12. [CrossRef]
31. Wu, D.; Yotnda, P. Induction and Testing of Hypoxia in Cell Culture. *J. Vis. Exp.* **2011**. [CrossRef]
32. Majumder, M.; Xin, X.; Liu, L.; Tutunea-Fatan, E.; Rodriguez-Torres, M.; Vincent, K.; Postovit, L.-M.; Hess, D.; Lala, P.K. COX-2 Induces Breast Cancer Stem Cells via EP4/PI3K/AKT/NOTCH/WNT Axis. *Stem Cells* **2016**, *34*, 2290–2305. [CrossRef]
33. Xin, X.; Majumder, M.; Girish, G.V.; Mohindra, V.; Maruyama, T.; Lala, P.K. Targeting COX-2 and EP4 to control tumor growth, angiogenesis, lymphangiogenesis and metastasis to the lungs and lymph nodes in a breast cancer model. *Lab. Investig.* **2012**, *92*, 1115–1128. [CrossRef] [PubMed]
34. Nandi, P.; Girish, G.V.; Majumder, M.; Xin, X.; Tutunea-Fatan, E.; Lala, P.K. PGE2 promotes breast cancer-associated lymphangiogenesis by activation of EP4 receptor on lymphatic endothelial cells. *BMC Cancer* **2017**, *17*, 11. [CrossRef] [PubMed]
35. Majumder, M.; Nandi, P.; Omar, A.; Ugwuagbo, K.C.; Lala, P.K. EP4 as a Therapeutic Target for Aggressive Human Breast Cancer. *Int. J. Mol. Sci.* **2018**, *19*, 1019. [CrossRef] [PubMed]
36. Kozomara, A.; Birgaoanu, M.; Griffiths-Jones, S. miRBase: From microRNA sequences to function. *Nucleic Acids Res.* **2018**, *47*, D155–D162. [CrossRef]

37. Cerami, E.; Gao, J.; Dogrusoz, U.; Gross, B.E.; Sumer, S.O.; Aksoy, B.A.; Skanderup, A.J.; Byrne, C.J.; Heuer, M.L.; Larsson, E.; et al. The cBio cancer genomics portal: An open platform for exploring multidimensional cancer genomics data. *Cancer Discov.* **2012**, *2*, 401–404. [CrossRef] [PubMed]
38. Gao, J.; Aksoy, B.A.; Dogrusoz, U.; Dresdner, G.; Gross, B.; Sumer, S.O.; Sun, Y.; Skanderup, A.J.; Sinha, R.; Larsson, E.; et al. Integrative Analysis of Complex Cancer Genomics and Clinical Profiles Using the cBioPortal. *Sci. Signal.* **2013**, *6*, pl1. [CrossRef]
39. Bandara, K.V.; Michael, M.Z.; Gleadle, J.M. MicroRNA Biogenesis in Hypoxia. *MicroRNA* **2017**, *6*, 80–96. [CrossRef]
40. Bhandari, V.; Hoey, C.; Liu, L.Y.; LaLonde, E.; Ray, J.; Livingstone, J.; Lesurf, R.; Shiah, Y.-J.; Vujcic, T.; Huang, X.; et al. Molecular landmarks of tumor hypoxia across cancer types. *Nat. Genet.* **2019**, *51*, 308–318. [CrossRef]
41. Kim, C.W.; Oh, E.-T.; Kim, J.M.; Park, J.-S.; Lee, D.H.; Lee, J.-S.; Kim, K.K.; Park, H.J. Hypoxia-induced microRNA-590-5p promotes colorectal cancer progression by modulating matrix metalloproteinase activity. *Cancer Lett.* **2017**, *416*, 31–41. [CrossRef]
42. Yamakuchi, M.; Yagi, S.; Ito, T.; Lowenstein, C.J. MicroRNA-22 Regulates Hypoxia Signaling in Colon Cancer Cells. *PLoS ONE* **2011**, *6*, e20291. [CrossRef] [PubMed]
43. Kulshreshtha, R.; Ferracin, M.; Wojcik, S.E.; Garzon, R.; Alder, H.; Agosto-Perez, F.J.; Davuluri, R.; Liu, C.-G.; Croce, C.M.; Negrini, M.; et al. A MicroRNA Signature of Hypoxia. *Mol. Cell. Biol.* **2006**, *27*, 1859–1867. [CrossRef] [PubMed]
44. Tanimoto, K.; Makino, Y.; Pereira, T.; Poellinger, L. Mechanism of regulation of the hypoxia-inducible factor-1alpha by the von Hippel-Lindau tumor suppressor protein. *EMBO J.* **2000**, *19*, 4298–4309. [CrossRef] [PubMed]
45. McGarry, T.; Biniecka, M.; Veale, D.J.; Fearon, U. Hypoxia, oxidative stress and inflammation. *Free. Radic. Biol. Med.* **2018**, *125*, 15–24. [CrossRef] [PubMed]
46. Coimbra-Costa, D.; Alva, N.; Duran, M.; Carbonell, T.; Rama, R. Oxidative stress and apoptosis after acute respiratory hypoxia and reoxygenation in rat brain. *Redox Biol.* **2017**, *12*, 216–225. [CrossRef] [PubMed]
47. Nallamshetty, S.; Chan, S.Y.; Loscalzo, J. Hypoxia: A master regulator of microRNA biogenesis and activity. *Free Radic. Biol. Med.* **2013**, *64*, 20–30. [CrossRef]
48. Qiu, G.-Z.; Jin, M.-Z.; Dai, J.-X.; Sun, W.; Feng, J.-H.; Jin, W.-L. Reprogramming of the Tumor in the Hypoxic Niche: The Emerging Concept and Associated Therapeutic Strategies. *Trends Pharmacol. Sci.* **2017**, *38*, 669–686. [CrossRef]
49. Ugwuagbo, K.C.; Maiti, S.; Omar, A.; Hunter, S.; Nault, B.; Northam, C.; Majumder, M. Prostaglandin E2 promotes embryonic vascular development and maturation in zebrafish. *Biol. Open* **2019**, *8*, bio039768. [CrossRef]
50. Majumder, M.; Xin, X.; Liu, L.; Girish, G.V.; Lala, P.K. Prostaglandin E2 receptor EP4 as the common target on cancer cells and macrophages to abolish angiogenesis, lymphangiogenesis, metastasis, and stem-like cell functions. *Cancer Sci.* **2014**, *105*, 1142–1151. [CrossRef]
51. Sun, G.; Zhou, Y.; Li, H.; Guo, Y.; Shan, J.; Xia, M.; Li, Y.; Li, S.; Long, D.; Li, Y. Over-expression of microRNA-494 up-regulates hypoxia-inducible factor-1 alpha expression via PI3K/Akt pathway and protects against hypoxia-induced apoptosis. *J. Biomed. Sci.* **2013**, *20*, 1–9. [CrossRef]
52. Tafani, M.; Pucci, B.; Russo, A.; Schito, L.; Pellegrini, L.; Perrone, G.A.; Villanova, L.; Salvatori, L.; Ravenna, L.; Petrangeli, E.; et al. Modulators of HIF1α and NFkB in Cancer Treatment: Is it a Rational Approach for Controlling Malignant Progression? *Front. Pharmacol.* **2013**, *4*, 13. [CrossRef]
53. Gao, C.; Dimitrov, T.; Yong, K.J.; Tatetsu, H.; Jeong, H.-W.; Luo, H.R.; Bradner, J.E.; Tenen, D.G.; Chai, L. Targeting transcription factor SALL4 in acute myeloid leukemia by interrupting its interaction with an epigenetic complex. *Blood* **2013**, *121*, 1413–1421. [CrossRef] [PubMed]
54. Ye, L.; Lin, C.; Wang, X.; Peng, X.; Li, Y.; Wang, M.; Zhao, Z.; Wu, X.; Shi, D.; Xiao, Y.; et al. Epigenetic silencing of SALL2 confers tamoxifen resistance in breast cancer. *EMBO Mol. Med.* **2019**, *11*, e10638. [CrossRef] [PubMed]
55. Yu, D.; Tang, L.; Dong, H.; Dong, Z.; Zhang, L.; Fu, J.; Su, X.; Zhang, T.; Fu, H.; Han, L.; et al. Oncogenic HER2 fusions in gastric cancer. *J. Transl. Med.* **2015**, *13*, 116. [CrossRef] [PubMed]
56. Shyr, C.-R.; Hu, Y.-C.; Kim, E.; Chang, C. Modulation of Estrogen Receptor-mediated Transactivation by Orphan Receptor TR4 in MCF-7 Cells. *J. Biol. Chem.* **2002**, *277*, 14622–14628. [CrossRef]

57. Keenan, A.B.; Torre, D.; Lachmann, A.; Leong, A.K.; Wojciechowicz, M.L.; Utti, V.; Jagodnik, K.M.; Kropiwnicki, E.; Wang, Z.; Ma'Ayan, A. ChEA3: Transcription factor enrichment analysis by orthogonal omics integration. *Nucleic Acids Res.* **2019**, *47*, W212–W224. [CrossRef] [PubMed]
58. Agarwal, V.; Bell, G.W.; Nam, J.-W.; Bartel, B. Predicting effective microRNA target sites in mammalian mRNAs. *eLife* **2015**, *4*, e05005. [CrossRef]

MDPI

Article

The microRNA-210-Stathmin1 Axis Decreases Cell Stiffness to Facilitate the Invasiveness of Colorectal Cancer Stem Cells

Tsai-Tsen Liao [1,2,†], Wei-Chung Cheng [3,4,†], Chih-Yung Yang [5,6], Yin-Quan Chen [7], Shu-Han Su [8], Tzu-Yu Yeh [8], Hsin-Yi Lan [1,9,10], Chih-Chan Lee [9], Hung-Hsin Lin [11,12], Chun-Chi Lin [12,13], Ruey-Hwa Lu [14], Arthur Er-Terg Chiou [15], Jeng-Kai Jiang [12,13,*] and Wei-Lun Hwang [7,9,10,*]

1 Graduate Institute of Medical Science, College of Medicine, Taipei Medical University, Taipei 110, Taiwan; liaotsaitsen@tmu.edu.tw (T.-T.L.); sandylan@tmu.edu.tw (H.-Y.L.)
2 Cell Physiology and Molecular Image Research Center, Wan Fang Hospital, Taipei Medical University, Taipei 116, Taiwan
3 Ph.D. Program for Cancer Molecular Biology and Drug Discovery, China Medical University, Taichung 406, Taiwan; wccheng@mail.cmu.edu.tw
4 Research Center for Cancer Biology, China Medical University, Taichung 406, Taiwan
5 Department of Education and Research, Taipei City Hospital, Taipei 106, Taiwan; A3703@tpech.gov.tw
6 General Education Center, University of Taipei, Taipei 100, Taiwan
7 Cancer Progression Research Center, National Yang Ming Chiao Tung University, Taipei 112, Taiwan; ycchen123@ym.edu.tw
8 Institution of Microbiology and Immunology, National Yang-Ming University, Taipei 112, Taiwan; d49702015@ym.edu.tw (S.-H.S.); g30102009@gm.ym.edu.tw (T.-Y.Y.)
9 Department of Biotechnology and Laboratory Science in Medicine, National Yang Ming Chiao Tung University, Taipei 112, Taiwan; chihchanlee@gmail.com
10 Department of Biotechnology and Laboratory Science in Medicine, National Yang-Ming University, Taipei 112, Taiwan
11 Institute of Clinical Medicine, National Yang Ming Chiao Tung University, Taipei 112, Taiwan; hhhlin7@vghtpe.gov.tw
12 Division of Colon & Rectal Surgery, Department of Surgery, Taipei Veterans General Hospital, Taipei 112, Taiwan; cclin15@vghtpe.gov.tw
13 School of Medicine, National Yang Ming Chiao Tung University, Taipei 112, Taiwan
14 Department of Surgery, Zhongxing Branch, Taipei City Hospital, Taipei 106, Taiwan; DAK23@tpech.gov.tw
15 Institute of Biophotonics, National Yang Ming Chiao Tung University, Taipei 112, Taiwan; aechiou@ym.edu.tw
* Correspondence: jkjiang@vghtpe.gov.tw (J.-K.J.); a85296658@ym.edu.tw (W.-L.H.); Tel.: +886-2-2826-7000 (ext. 65832) (W.-L.H.)
† These authors contributed equally to this work.

Citation: Liao, T.-T.; Cheng, W.-C.; Yang, C.-Y.; Chen, Y.-Q.; Su, S.-H.; Yeh, T.-Y.; Lan, H.-Y.; Lee, C.-C.; Lin, H.-H.; Lin, C.-C.; et al. The microRNA-210-Stathmin1 Axis Decreases Cell Stiffness to Facilitate the Invasiveness of Colorectal Cancer Stem Cells. *Cancers* **2021**, *13*, 1833. https://doi.org/10.3390/cancers13081833

Academic Editor: Paola Tucci

Received: 19 February 2021
Accepted: 9 April 2021
Published: 12 April 2021

Simple Summary: Metastasis of tumor cells is the leading cause of death in cancer patients. Concurrent therapy with surgical removal of primary and metastatic lesions is the main approach for cancer therapy. Currently, therapeutic resistant properties of cancer stem cells (CSCs) are known to drive malignant cancer progression, including metastasis. Our study aimed to identify molecular tools dedicated to the detection and treatment of CSCs. We confirmed that microRNA-210-3p (miR-210) was upregulated in colorectal stem-like cancer cells, which targeted stathmin1 (STMN1), to decrease cell elasticity for increasing mobility. We envision that strategies for softening cellular elasticity will reduce the onset of CSC-orientated metastasis.

Abstract: Cell migration is critical for regional dissemination and distal metastasis of cancer cells, which remain the major causes of poor prognosis and death in patients with colorectal cancer (CRC). Although cytoskeletal dynamics and cellular deformability contribute to the migration of cancer cells and metastasis, the mechanisms governing the migratory ability of cancer stem cells (CSCs), a nongenetic source of tumor heterogeneity, are unclear. Here, we expanded colorectal CSCs (CRCSCs) as colonospheres and showed that CRCSCs exhibited higher cell motility in transwell migration assays and 3D invasion assays and greater deformability in particle tracking microrheology than did their parental CRC cells. Mechanistically, in CRCSCs, microRNA-210-3p (miR-210) targeted stathmin1 (STMN1), which is known for inducing microtubule destabilization, to decrease cell elasticity in order to facilitate cell motility without affecting the epithelial–mesenchymal transition (EMT) status. Clinically, the miR-210-STMN1 axis was activated in CRC patients with liver metastasis and correlated with a worse clinical outcome. This study elucidates a miRNA-oriented mechanism regulating the deformability of CRCSCs beyond the EMT process.

Keywords: colon cancer; cancer stem cells; microRNAs; deformability

1. Introduction

Cytoskeletal components, including microtubules, actins, and intermediate filaments, support the structure of eukaryotic cells with appropriate viscoelasticity to regulate physiological cell morphology [1], division [2–4], and movement [5,6]. During cancer progression, intercellular communication and cell-extracellular matrix (ECM) interactions define the localized and premetastatic tumor microenvironment (TME) for stimulating cancer metastasis [7]. Although cellular viscoelasticity has been studied using microrheology to determine the intracellular elastic and viscous moduli [8–10], little is known about the viscoelasticity of cancer cells during stepwise metastatic progression in distinct TMEs.

Cancer stem cells (CSCs), a nongenetic source of phenotypic heterogeneity in bulky tumors, are responsible for tumor initiation, therapeutic resistance, and distal metastasis [11]. In colorectal cancer (CRC), CD133(+) and ESA(+)CD44(+) CSCs have been identified [12–14]. CD26(+) and Lgr5(+) CSCs are further suggested to contribute to the maintenance of distal metastasis [15,16]. CSC phenotypes can be induced by epithelial-mesenchymal transition (EMT) inducers [17], maintained by inflammatory cytokines/chemokines or defined by Wnt activity [18]. However, the mechanism by which CSCs modulate cellular viscoelasticity to promote local invasion and distal metastasis remains elusive.

MicroRNAs (miRNAs) are endogenously expressed small noncoding RNAs of 18–24 nucleotides in length that modulate gene expression at the posttranscriptional level [19]. Dysregulation of miRNAs contributes to tumor formation and progression [20]. Several miRNAs have been associated with EMT, angiogenesis, ECM remodeling, proliferation, invasion, and apoptosis in liver or lung metastases of CRC [21]. miR-20a-5p promotes CRC invasion and metastasis by downregulating Smad4 [22]. miR-885-5p downregulates CPEB2, a negative regulator of TWIST1, and induces cytoskeletal rearrangement by upregulating Rho family small GTPases [23]. However, the miRNome responsible for cellular viscoelasticity is undefined.

This study reveals an EMT-independent mechanism for motility control and demonstrates that modulation of colorectal cancer stem cell (CRCSC) stiffness through miR-210-3p (miR-210) and its downstream target stathmin1 (STMN1) is essential for CRCSC invasiveness. The miR-210^{High}/STMN^{Low} signature is further associated with liver metastasis of CRC and predicts a worse clinical outcome.

2. Materials and Methods

2.1. Plasmids, shRNA Clones and Synthetic Oligonucleotides

The miR-210 antagomir and scramble control were purchased from RiboBio Co., Guangzhou, China. The miR-210 agomir and agomir control were synthesized by GenePharma, Shanghai, China. The miR-Zip Control and miR-Zip-210 plasmids were purchased from System Bioscience, Palo Alto, CA, USA. The pLenti and pLenti-STMN1-Myc-DDKvectors were obtained from OriGene, Rockville, MD, USA. The pCMVΔR8.9, pDVsVg, and pLKO.1-shLuc vectors and the shRNA against STMN1 were obtained from the National RNAi Core Facility of Taiwan for gene silencing. All clones were verified by direct sequencing.

2.2. Cell Culture

Human embryonic kidney (HEK) 293 cells were cultured in Dulbecco's modified Eagle's medium (Gibco/Life Technologies, New York, NY, USA). Human colorectal carcinoma HT29, HCT15 and Colo205 cells were cultured in RPMI-1640 medium (Gibco/Life Technologies). SW1116 human colorectal carcinoma cells were cultured in L-15 medium (Gibco/Life Technologies). The above media were supplemented with 10% fetal bovine serum (FBS, Gibco/Life Technologies). HEK293, HT29, HCT15 and Colo205 cells were cultured in a humidified incubator at 37 °C with 5% CO_2. SW1116 cells were cultured in a humidified incubator at 37 °C with atmospheric air. HCT15-vec and HCT15-Snail stable clones were generated previously [24]. The authenticity of cell lines was verified by examining their DNA-short tandom repeat (STR) profiles.

2.3. Expansion of Colorectal Cancer Stem Cells

To expand sphere-derived cancer stem cells (SDCSCs), a single-cell suspension was prepared, and cells were cultured in stem cell medium (SCM; DMEM/F12 supplemented with N2 Plus Supplement (Invitrogen, New York, NY, USA), 10 ng mL^{-1} recombinant bFGF (PeproTech Asia, Suzhou, China), 10 ng mL^{-1} EGF (PeproTech Asia) and 1% penicillin–streptomycin (Gibco/Life Technologies) for 20 days to form tumor spheres. TryPLE express (Gibco/Life Technologies) was used to dissociate cells and SDCSCs to prepare single-cell suspensions for experiments. Cells were cultured in a humidified incubator at 37 °C with 5% CO_2.

2.4. Transwell Migration Assay

RPMI medium (600 μL) supplemented with 10% FBS was added to the bottom wells, and 2×10^5 cells suspended in basal RPMI medium were then seeded in the 6.5 mm diameter upper chamber with an 8 μm pore size membrane (Corning, New York, NY, USA) and incubated for 20 h. Cell suspensions in the upper inserts were discarded, and the remaining cells were removed with cotton swabs. Cells adhering to the underside of the membranes were fixed with 4% paraformaldehyde (Sigma-Aldrich, St. Louis, MO, USA) for 15 min and stained with 1% crystal violet reagent (Sigma-Aldrich) for 30 min at room temperature. Images were acquired with an inverted microscope (Eclipse Ts-2, Nikon Instruments Inc., Tokyo, Japan), and the migrated cells in a 10x low-power field (LPF) were counted for quantification.

2.5. Two-and-a-Half Dimentional (2.5D) Time-Lapse Trajectory

For measuring 2.5D cell motility, a mixture of 0.85 mL of 3 mg mL^{-1} PureCor bovine collagen solution (Advance Biomatrix, Carlsbad, CA, USA), 0.3 mL of 5× RPMI basal medium, 6 μL of 1 M NaOH and 0.35 mL of water to a total volume of 1.5 mL was prepared as the collagen solution. Then, 250 μL of the collagen solution was added to wells in a 24-well plate for solidification at 37 °C for 30 min. A total of 3×10^4 cells were suspended in RPMI basal medium and seeded on top of the thick collagen layer for 6 h prior to time-lapse recording using an IX83 inverted microscope (Olympus Corporation, Tokyo, Japan). Images were acquired every 10 min for up to 5 h, and a video was exported using cellSens software (Olympus Corporation).

2.6. Three-Dimensional (3D) Invasion Assay

A total of 1×10^6 cells were suspended in 500 μL of basal RPMI medium and plated in one well of a four-well chambered borosilicate coverglass slide (Lab-tek, New York, NY, USA) overnight for attachment. Then, the supernatant was removed, and the surface of the well was covered with 350 μL of the collagen solution described above for 30 min to allow solidification. Then, 700 μL of basal RPMI medium was added, and the slide was incubated at 37 °C and 5% CO_2 for 72 h. After incubation, cells were fixed with 4% formaldehyde (Sigma-Aldrich) at 4 °C overnight and mounted with Fluoroshield with DAPI (Sigma-Aldrich). Confocal images (49 layers) of each well were acquired at 1.5 μm steps from the bottom to a height of 73.5 μm with an Olympus FV1000 laser confocal microscope (Olympus Corporation).

2.7. Paired Cell Assay

For BrdU labeling, cells were cultured in medium containing 0.5 μM BrdU (Sigma-Aldrich) for 2 weeks to ensure BrdU incorporation in cells. Cells were then synchronized with 40 ng mL^{-1} nocodazole (Sigma-Aldrich) overnight in the presence of 0.5 μM BrdU. For the BrdU chase, cells were washed intensively, trypsinized and seeded on poly-L lysine (Sigma-Aldrich)-coated coverslips placed in wells of 6-well plates in BrdU-free medium and synchronized through sequential exposure to thymidine, nocodazole, and blebbistatin [25] to control cell division and entry into the second mitosis. Briefly, cells were treated with 5 mM thymidine (Sigma-Aldrich) for 24 h and released for 10 h prior to additional thymidine treatment for 24 h (the first round of division). The thymidine was then removed for 6 h prior to nocodazole treatment for 16 h to enrich mitotic-phase cells. The nocodazole was then removed by washing for 15 min prior to 50 μM blebbistatin (Sigma-Aldrich) treatment for up to two hours (the second round of division). Paired cells were observed at this stage. Cells were fixed with 4% paraformaldehyde for 30 min at 4 °C and permeabilized with 0.1% Triton X-100 for 5 min. Then, cells were immersed first in 1 N HCl for 10 min on ice and then in 2 N HCl/1% Triton X-100 for 45 min in a 37 °C incubator to open the DNA helix. Immediately after one acid wash with PBS, borate buffer (0.1 M, pH = 8) was used to buffer cells for 12 min at room temperature. Cells were washed again with PBS three times and incubated overnight with an antibody against BrdU (1:200, 14-5071-82, eBioscience, San Diego, CA, USA) and a fluorescein-conjugated goat anti-mouse antibody (1:200, F2761, Invitrogen). Cells were mounted with Fluoroshield with DAPI (Sigma-Aldrich), and images were acquired with a Leica DM600B fluorescence microscope (Leica Microsystems, Wetzlar, Germany).

2.8. Elasticity Measurements

Video particle tracking microrheology (VPTM) was used to measure the elastic modulus of cells [4,26]. A total of 2×10^6 cells were suspended in basal RPMI medium in a 10-ch dish, and 20 μL of 200 nM fluorescent carboxylated polystyrene particles (F8810, Invitrogen, fluorescence excitation/emission peaks: 580 nm/605 nm) was then injected into the cells with a biolistic particle delivery system (PDS-100; pressure, Bio-Rad, 450 psi, Hercules, CA, USA). Three hours after particle injection, the cells were washed twice with PBS and transferred to 35 mm glass bottom culture dishes (Alpha Plus, Saitama, Japan). After incubation for 4 h, the two-dimensional Brownian motion of intracellular fluorescent beads was recorded with an inverted epifluorescence microscope (Eclipse Ti, Nikon Instruments Inc.), equipped with an oil immersion objective (100×, NA = 1.45), a sCMOS camera (Hamamatsu, Hamamatsu-shi, Japan), and a cell incubation chamber (INUB-GSI-F1, TOKAI HIT, Fujinomiya City, Japan). The two-dimensional projection of the trajectories of the intracellular fluorescent beads was recorded for 10 s at a frame rate of 100 Hz and analyzed via customized MATLAB software (MathWorks, Natic, MA, USA) [27]. From the two-dimensional (2-D) position (x(t), y(t)) of each particle as a function of time, we calculated the ensemble-averaged mean square displacement (MSD, Kenilworth, NJ, USA), the effective creep compliance, and the elastic modulus $G'(\omega)$ [28]. The subcellular locations of injected particles were observed using confocal microscopy (LSM 880, ZEISS, Oberkochen, Germany), and 3D images were generated using ZEISS Zen software.

2.9. Cell Viability, Clonogenicity and Sphere Formation Assays

A total of 1×10^4 cells were suspended in 100 μL of complete RPMI medium and seeded in wells in 96-well plates for 48 h. The medium was discarded, and MTT reagent (Sigma-Aldrich) was added to the cells for 45 min at 37 °C. Mitochondrial MTT crystals were dissolved with DMSO (J.T. Baker, Phillipsburg, NJ, USA), and the optical density values were then read in a microplate reader (SpectraMax 250, Molecular Devices Corp.,

San Jose, CA, USA). To evaluate clonogenicity, 1×10^4 cells were resuspended in complete RPMI medium and seeded in wells of a 6-well plate for 10 days. Colonies were visualized by crystal violet staining prior to counting. For sphere formation assays, 1000 cells were suspended in SCM, and the spheroids were counted after 10 days. Images were acquired with an inverted microscope (Eclipse Ts-2, Nikon Instruments Inc.), and spheroids in a 4× LPF were counted for quantification.

2.10. Lentivirus Production and Transduction

For virus packaging, pCMVΔR8.9, pDVsVg and expression lentivectors (miR-Zip control, miR-Zip-210, pLenti-vector, pLenti-STMN1 Myc-DDK, pLKO.1-shLuc, and shRNA clones) were cotransfected into 293T cells with T-Pro NTR III reagent (T-Pro Biotechnology, Taiwan) overnight according to the manufacturer's protocol. The virus-containing supernatant was harvested 48 h after transfection. Cells were seeded and supplemented with 8 μg/mL polybrene (Sigma-Aldrich) for overnight virus transduction.

2.11. RNA Extraction and Quantitative RT-PCR

Cells were immersed in TRIzol® reagent (Life Technologies). Total RNA was reverse transcribed using a RevertAidTM Reverse transcriptase kit (Fermentas, Waltham, MA, USA), and FAST SYBR Green Master Mix (Applied Biosystems Inc., Foster City, CA, USA) was used for real-time PCR in a StepOne-Plus real-time PCR system (Applied Biosystems Inc.). Cellular mRNA and miRNA expression levels were normalized to the expression levels of GAPDH and U6, respectively. The sequences of the primers used are indicated below. Primer for reverse transcription of miR-210-3p (GTC GTA TCC AGT GCA GGG TCC GAG GTA TTC GCA CTG GAT ACG ACA CAG GC). Primers for qPCR analysis: MiR-210-3p (forward primer: GGG GGG AAT ATA ACA CAG ATG GCC, reverse primer: TGC AGG GTC CGA GGT), GAPDH (forward primer: GGA GTC CAC TGG CGT CTT CA, reverse primer: TGG TTC ACA CCC ATG ACG AA); U6 (forward primer: CGC TTC GGC AGC ACA TAT AC, reverse primer: TTC ACG AAT TTG CGT GTC AT), LRRC2 (forward primer: CTT GGC AGA AGA AGG AGG TG, reverse primer: AGT ATA CAG CCT GGG GGA TG), CDKN2A (forward primer: ACC AGA GGC AGT AAC CAT GC, reverse primer: AAG TTT CCC GAG GTT TCT CA), PARD3: (forward primer: TTT CAG CCT CAT CCA GCA G, reverse primer: TTC CTC CAT CTC CAT GTT CC), RCBTB2 (forward primer: TCG TCA GGC TTG TGT CTT TG, reverse primer: CGT CAC CTA ACC CCA AAC AG), STMN1 (forward primer: TAC ACT GCC TGT CGC TTG TC, reverse primer: AGG GCT GAG AAT CAG CTC AA), CDH1 (forward primer: AGA TGG CCT TAG AGG TGG GT, reverse primer: AGG CTG TGC CTT CCT ACA GA), CDH2 (forward primer: AGC TTC TCA CGG CAT ACA CC, reverse primer: GTG CAT GAA GGA CAG CCT CT), SNAI1 (forward primer: GCT GCC AAT GCT CAT CTG GGA CTC T, reverse primer: TTG AAG GGC TTT CGA GCC TGG AGA T), NANOG (forward primer: CAA CCA GAC CCA GAA CAT CC, reverse primer: TTC CAA AGC AGC CTC CAA G), POU5F1 (forward primer: ACC GAG TGA GAG GCA ACC, reverse primer: TGA GAA AGG AGA CCC AGC AG), LGR5 (forward primer: TGT TGG GAG ATC TGC TTT C, reverse primer: CAG ACG GTT TGA GGA AGA GA), CD44 (forward primer: CCA GAT GGA GAA AGC TCT GA, reverse primer: GTC ATA CTG GGA GGT GTT GG), VIM (forward primer: CAA TGT TAA GAT GGC CCT TG, reverse primer: GGG TAT CAA CCA GAG GGA GT), BMP4 (forward primer: CTC CTG GTC ACC TTT GGC CA, reverse primer: ATT CCA GCC CAC ATC GCT GA), and CDX2 (forward primer: CTG GAG CTG GAG AAG GAG TTT C, reverse primer: ATT TTA ACC TGC CTC TCA GAG AGC).

2.12. Bioinformatic Analysis

The small RNA-seq (smRNA-seq) data of HT29 cells, HCT15 cells and expanded SDCSCs were collected from GSE43793. The smRNA-seq and RNA-seq data for the TCGA COAD dataset were retrieved from established databases: DriverDB [29] and YM500 [30]. In-house pipelines were used to estimate the expression levels of miRNAs (30) as reads per million (RPM) values from fastq files. Gene expression array and microRNA array data of the NCI-60 cell line panel implemented with Affymetrix HG-U133 Plus 2 and Agilent Human microRNA-V2 chip platforms, respectively, were downloaded from the CellMiner database [31]. We used Gene Set Enrichment Analysis (GSEA) (http://www.broadinstitute.org/gsea (accessed on 20 June 2020) to assess degree of association defined signature and expression profiles of CRC patients downloaded from GSE17538. The clinical phenotypes were used for permutation.

2.13. Immunoblotting

Whole-cell lysates were extracted with cell culture lysis buffer (Promega, Madison, WI, USA), and protein concentrations were quantified with a Pierce BCA Protein Assay Kit (Thermo Fisher Scientific, USA) according to the manufacturer's protocol. The transfer membrane was blocked and probed with the following antibodies prepared in 5% BSA (Sigma-Aldrich) overnight at 4 °C: anti-STMN1 (1:1000, 13655S, Cell Signaling, Danvers, MA, USA), anti-FLAG-M2-HRP (1:1000, A8592, Sigma-Aldrich), and anti-β-actin (1:5000, A5441, Sigma-Aldrich). The membrane was then probed at room temperature for 1 h with the corresponding secondary antibodies: bovine anti-rabbit IgG-HRP (1:3000, sc-2370, Santa Cruz Biotechnology, Dallas, TX, USA) and chicken anti-mouse IgG-HRP (1:5000, sc-2954, Santa Cruz Biotechnology). Immunoblots were visualized in an ImageQuant LAS 4000 chemiluminescence detection system (GE Healthcare Bio-Sciences, Piscataway, NJ, USA). All Uncropped blots can be seen in Figure S8.

2.14. Immunohistochemical (IHC) Assay

Sections of tissues (4 μm thick) from microarrays were deparaffinized and rehydrated before staining. Tissue antigens were retrieved by autoclaving in 10 mM (pH 6) citrate buffer for 10 min. Sections were cooled on ice for 30 min before treatment with 3% H_2O_2. Samples were permeabilized with 0.2% Triton X-100 (Sigma) in DPBS and reacted with a diluted primary STMN1 antibody (1:200, 13655S, Cell Signaling). Signals were amplified and detected with a Super Sensitive™ Link-Label IHC Detection System (BioGenex, Fremont, CA, USA) according to the instructions and counterstained with hematoxylin QS (Vector, Burlingame, CA, USA) for 20 s. The H-scores represent the percentage of STMN1 immunoreactivity-positive regions multiplied by the STMN1 staining intensity. Images were acquired with a BX43 light microscope equipped with a DP22 CCD camera (Olympus).

2.15. Preparation of Patient-Derived Xenografts (PDXs)

The experimental animal procedure was approved by the Institutional Animal Care and Use Committee (IACUC) of Taipei Veterans General Hospital (2018-191). The CRC specimens were first rinsed twice with DPBS and immersed in Matrigel (Becton-Dickinson, Franklin Lakes, NJ, USA) at 37 °C. The tumors were then cut into L mm^3 pieces and implanted subcutaneously into 4-week-old female nude mice to establish patient-derived xenografts (PDXs). The mice were sacrificed, and tumors were homogenized in TRIzol® reagent (Life Technologies) and subjected to total RNA isolation.

2.16. Biological Samples

This study was approved by the Institutional Ethics Committee/Institutional Board of Taipei Veterans General Hospital (2016-03-001BC, 2018-11-002C). Two sets of human specimens were used. First, 2 CRC specimens (one primary tumor and one unpaired liver metastatic tumor) were collected to prepare PDXs with informed consents. Second, 11 paraffin-embedded sections from the paired primary and metastatic CRC specimens collected from the tissue biobank were subjected to IHC staining.

2.17. Statistical Analysis

Unless indicated otherwise, Student's *t*-test was used to assess the significance of differences. The Pearson correlation analysis was used to analyze correlations between two factors described by continuous data. The log-rank test was used for survival analysis. The x^2 test was applied for comparisons of dichotomous variables. Two-tailed *p*-values of <0.05 were considered to indicate significant differences.

3. Results

3.1. Small RNA Sequencing (smRNA-seq) Reveals Enhanced miR-210-3p Expression in CRCSCs

In an attempt to discover mechanisms regulating the motility of CSCs, we initiated this study by expanding CRCSCs from two CRC cell lines, HT29 and HCT15, using a serum-free cultivation platform, because stem-like cancer populations were enriched as cancer spheroids [32]. We found that the expanded HT29- and HCT15-CRCSCs grew as suspended colonospheres (Figure 1a) and showed increased expression of stemness genes, including NANOG, POU5F1, LGR5, CD44, and SNAI1 (Figure S1a).

The resultant spheroids are referred to as sphere-derived cancer stem cells (SDCSCs) hereafter. Both HT29- and HCT15-SDCSCs exhibited higher transwell migration capacity (Figure 1b) and enhanced three-dimensional (3D) vertical invasiveness (Figure 1c) than their parental cells, and cell viability was not affected (Figure 1d). The top 500 upregulated gene signature analyzed in HT29-SDCSCs (GSE14773) was positively associated with the expression profiles of recurrent (Figure S1b, upper) and late stage (Figure S1b, lower) CRC patients deposited at GSE17538, suggesting the expanded SDCSCs are migrating CRCSCs.

Next, we sought to identify the primary microRNA(s) (miRNAs) responsible for CRCSC motility, because miRNA deregulation is critically involved in cancer progression [33]. Global miRNA expression patterns of HT29-SDCSCs, HCT15-SDCSCs, and their parental cells (GSE43793) were explored by small RNA sequencing (smRNA-seq). The miRNAs with log counts per million (logCPM) values of >1 and fold changes of ≥ 2 were selected for examination of their clinical relevance. The Venn diagram shows the nine differentially expressed miRNAs in HT29-SDCSCs and HCT15-SDCSCs (Figure 1e); four miRNAs were significantly upregulated and three were downregulated in both SDCSC datasets (Figure 1f). To explore the clinical relevance of these seven dysregulated miRNAs in CRC, we evaluated these miRNAs according to the patient survival data in the TCGA COAD dataset ($n = 425$ patients) and found that miR-210-3p (miR-210) was the only miRNA both enriched in SDCSCs and associated with poor overall survival (Figure 1g). The increased expression of miR-210 in HT29-SDCSCs (Figure 1h, left panel) and HCT15-SDCSCs (Figure 1h, right panel) was confirmed. The enhanced expression of miR-210 in stage IV tumor-derived Colo205 CRC cells (Figure S1c) and liver metastatic PDX specimens (Figure S1d) implies roles of miR-210 in CRCSC metastasis.

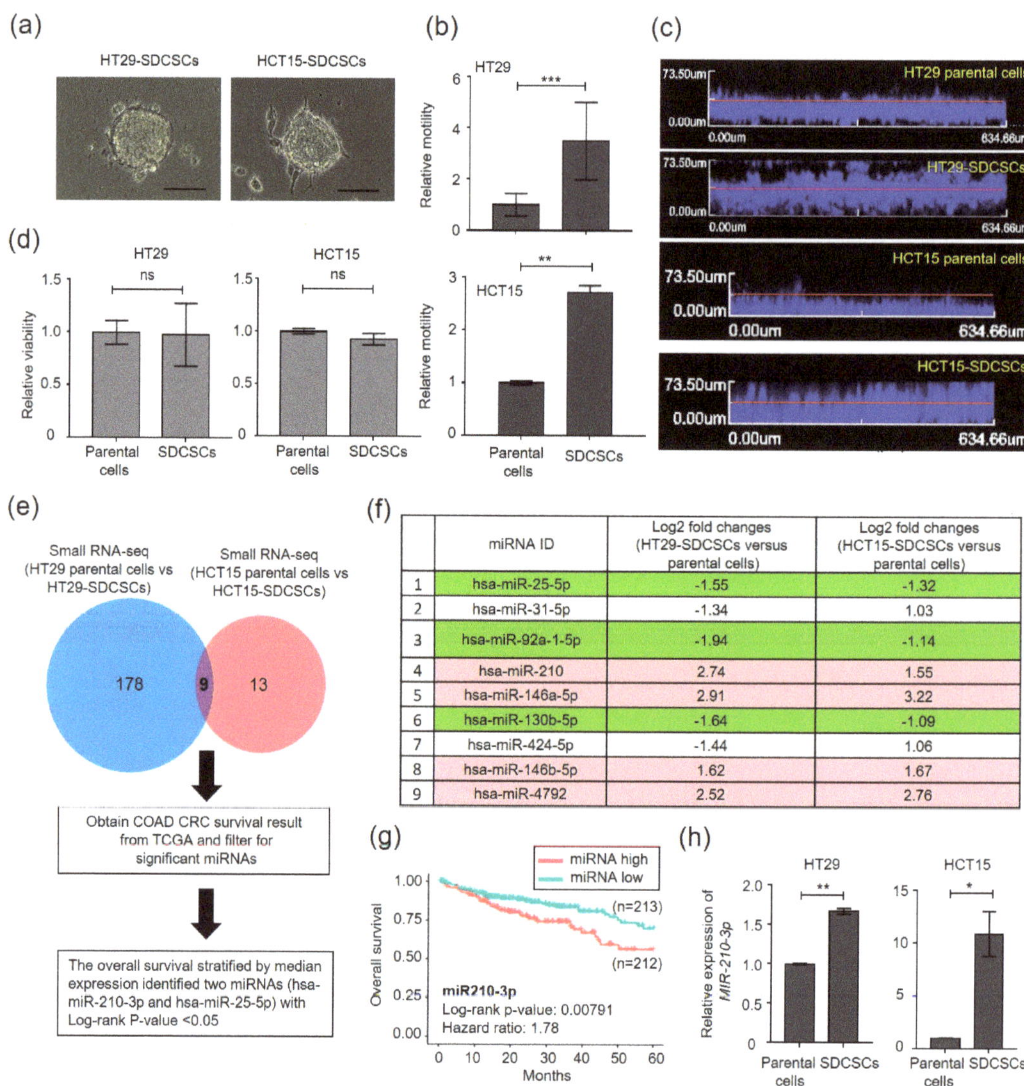

	miRNA ID	Log2 fold changes (HT29-SDCSCs versus parental cells)	Log2 fold changes (HCT15-SDCSCs versus parental cells)
1	hsa-miR-25-5p	-1.55	-1.32
2	hsa-miR-31-5p	-1.34	1.03
3	hsa-miR-92a-1-5p	-1.94	-1.14
4	hsa-miR-210	2.74	1.55
5	hsa-miR-146a-5p	2.91	3.22
6	hsa-miR-130b-5p	-1.64	-1.09
7	hsa-miR-424-5p	-1.44	1.06
8	hsa-miR-146b-5p	1.62	1.67
9	hsa-miR-4792	2.52	2.76

Figure 1. Increased expression of miR-210-3p in SDCSCs. (**a**) Representative images of SDCSCs from two CRC cell lines. SDCSCs, sphere-derived cancer stem cells. Scale bar, 100 µm. (**b**) Histograms showing the relative transwell migration ability of cells. (**c**) Representative images of vertical invasion of cells. (**d**) Relative viability of cells as assessed by an MTT assay. ns, nonsignificant. (**e**) Flow charts for identifying differentially expressed miRNAs in the CRCSC miRNome associated with poor patient outcome in the TCGA-COAD dataset. The numbers of miRNAs with the same differential regulation patterns in HT29- and HCT15-SDCSCs are indicated in Venn diagrams. TCGA-COAD, The Cancer Genome Atlas Colon Adenocarcinoma. (**f**) A table illustrating the relative miRNA expression levels in SDCSCs. Red, miRNAs upregulated in both SDCSC lines. Green, miRNAs downregulated in both SDCSC lines. (**g**) Kaplan–Meier analysis of overall survival in patients in a TCGA-COAD dataset (n = 425). The median miR-210-3p expression level was used for patient stratification. miRNA high, CRC patients with high miR-210-3p expression. miRNA low, patients with low miR-210-3p expression. The p-value was estimated by the log-rank test. (**h**) RT-qPCR validation of miR-210-3p expression in the indicated cells. Unless otherwise stated, all data in bar charts are expressed as the mean ± SD values. * $p < 0.05$; ** $p < 0.01$; *** $p < 0.001$ (Student's t-test).

3.2. MiR-210 Is Required for the Migration and Invasiveness of CRCSCs

We investigated the functional roles of miR-210 in SDCSCs. In human cancers, accumulated evidence suggests that defects in asymmetric cell division (ACD) and increased symmetric cell division (SCD) of somatic stem cells expand stem cell pools and fuel tumor growth [34,35]. As stem cells tend to retain the mother strand (old) DNA template in one daughter stem cell and segregate newly synthesized DNA strands to differentiating daughter cells [36], we investigated the segregation of mother strand DNA in SDCSCs by pulse-chase BrdU labeling and paired cell assays (Figure S2a). Upon knockdown of miR-210 in HT29 SDCSCs with a specific antagomir (Figure S2b), predominantly symmetrical BrdU segregation (i.e., SCD) was observed (Figure S2c,d), indicating that silencing miR-210 did not promote early differentiation of SDCSCs. The observation of the reduced sphere-forming capacity of HT29-SDCSCs with stable miR-210 knockdown (Zip-210) (Figure 2a,b) but not the corresponding HCT15-SDCSCs (Zip-210) (Figure 2a,c) also suggests limited effects of miR-210 on the self-renewal of SDCSCs. The expression of strmness markers (CD44, NANOG, and POU5F1) and differentiation markers (BMP4 and CDX2) were not altered upon silencing miR-210 in both HT29- and HCT15-SDCSCs (Figure S2e). In contrast, knockdown of miR-210 markedly reduced the transwell migration capacity (Figure 2d,e) and 3D vertical invasiveness (Figure 2f,g) of both HCT15- and HT29-SDCSCs without affecting cell viability (Figure 2h).

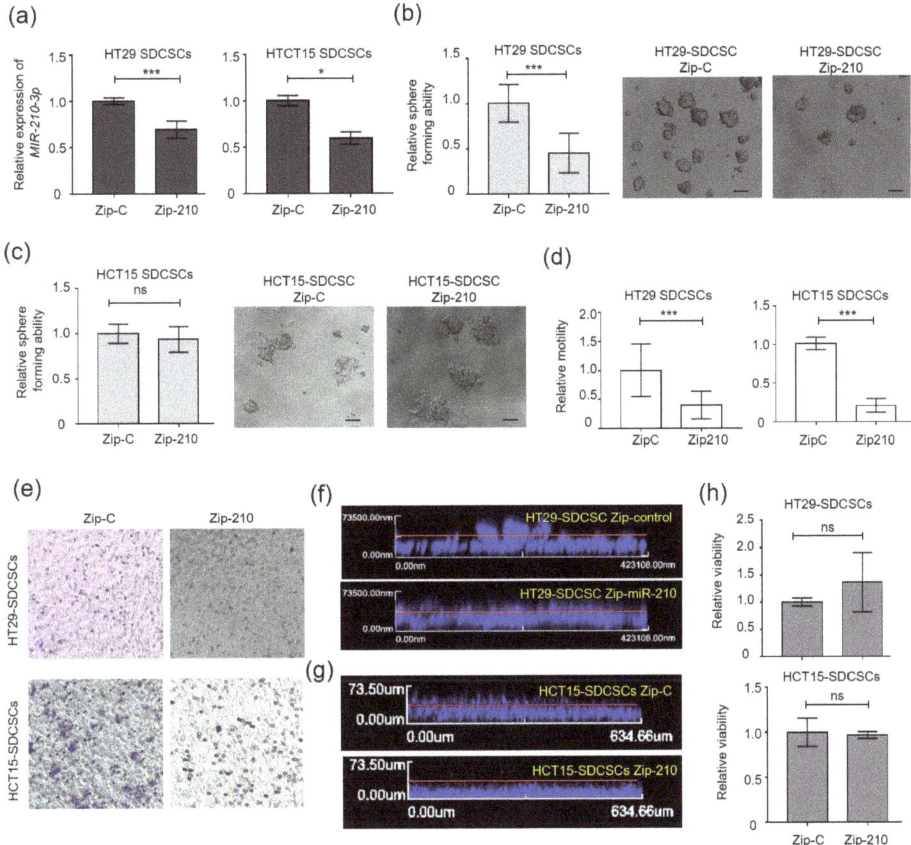

Figure 2. Silencing miR-210-3p expression inhibits the migratory and invasive capabilities of SDCSCs. (**a**) RT-qPCR validation of miR-210-3p expression in SDCSCs. Zip-C, cells receiving the scramble control; Zip-210, miR-210-3p-knockdown

cells receiving a shRNA targeting miR-210-3p. (**b**) Left: sphere-forming capacity of HT29-SDCSCs receiving the scramble control (Zip-C) or shRNA targeting miR-210-3p (Zip-210). Right: representative pictures of sphere formation in miR-210-3p-knockdown SDCSCs. Scale bar, 50 μm. (**c**) Left: sphere-forming capacity of control HCT15-SDCSCs (Zip-C) and miR-210-3p-knockdown SDCSCs (Zip-210). Right: representative pictures of sphere formation in miR-210-3p-knockdown SDCSCs. Scale bar, 50 μm. (**d**) Histograms showing the relative transwell migration ability of cells. (**e**) Representative images showing migrated SDCSCs. (**f**) Representative images of vertical invasion of control HT29-SDCSCs and miR-210-3p-silenced HT29-SDCSCs. (**g**) Representative images of vertical invasion of control HCT15-SDCSCs and miR-210-3p-silenced HCT15-SDCSCs. (**h**) Relative viability of cells as assessed by an MTT assay. ns, nonsignificant. All data in bar charts are expressed as the mean ± SD values. * $p < 0.05$; *** $p < 0.001$ (Student's t-test).

3.3. MiR-210 Suppresses Stathmin1 in CRCSCs

We hypothesized that miR-210 targets a critical cytoskeletal regulator mediating cell motility and invasiveness. To identify the miR-210 target(s) responsible for these functions, we identified the overlapped downregulated genes in HT29-SDCSCs (GSE14773) with the miR-210 targets predicted by SVmicro [37] and miRtar [38] software (Figure 3a). We focused on genes associated with cell motility or tumor suppression annotated in DAVID (https://david.ncifcrf.gov/home.jsp (accessed on 4 July 2014)) and examined the expression of five putative candidates identified (LRRC2, CDKN2A, PARD3, RCBTB, and STMN1) (Figure 3b). The expression of LRRC2, RCBTB2, and STMN1 was reduced in HT29- and HCT15-SDCSCs (Figure 3c). Stathmin 1 (STMN1) expression was found to be decreased in HCT15 (Figure 3d) and HT29 (Figure 3e) cells receiving miR-210 agomiRs. The decreased expression of STMN1 protein in HT29 cells receiving miR-210 agomiRs was confirmed (Figure 3f). As STMN1 is a known miR-210 target [39], a negative association between STMN1 and miR-210 levels was observed in the NCI-60 panel (Figure 3g). Decreased expression of STMN1 at the protein level was also observed in HT29- and HCT15-SDCSCs (Figure 3h). Additionally, knockdown of miR-210 expression restored the protein expression of STMN1 in HT29- and HCT15-SDCSCs, confirming the existence of the miR-210-STMN1 axis in CRCSCs (Figure 3i).

3.4. Stathmin1 Expression Attenuates the Motility of CRCSCs

STMN1 is a phosphoprotein regulated by extracellular signals and can bind to α/β-tubulin to modulate microtubule dynamics [40,41]. We next evaluated whether STMN1 is involved in regulating the motility CRCSCs. We found that restoration of STMN1 expression in HT29- and HCT15-SDCSCs (Figure 4a) did not alter the viability (Figure 4b), sphere-forming capacity (Figure 4c), or clonogenicity (Figure 4d,e) of these cells. Although STMN1 was shown to regulate EMT [42], restoration of STMN1 expression decreased the transwell migration ability (Figure 4f) and invasiveness (Figure 4g,h) of both HT29- and HCT15-SDCSCs without affecting a complete EMT program (Figure S3a–c).

In an attempt to verify the participation of the miR-210-STMN1 axis in mediating SDCSC motility, we silenced STMN1 expression in miR-210 knockdown HCT15-SDCSCs. The reduced expression of STMN1 was first confirmed (Figure 5a). We found that silencing STMN1 expression had no effects on the viability (Figure 5b), sphere-forming capacity (Figure 5c), or clonogenicity (Figure 5d,e) of miR-210-knockdown HCT15-SDCSCs. Moreover, enhanced transwell migration potential (Figure 5f) and invasiveness (Figure 5g) were noted in miR-210 knockdown HCT15-SDCSCs, but no effects on the expression of E-cadherin (CDH1), N-cadherin (CDH2) and Vimentin (VIM) were observed (Figure 5h).

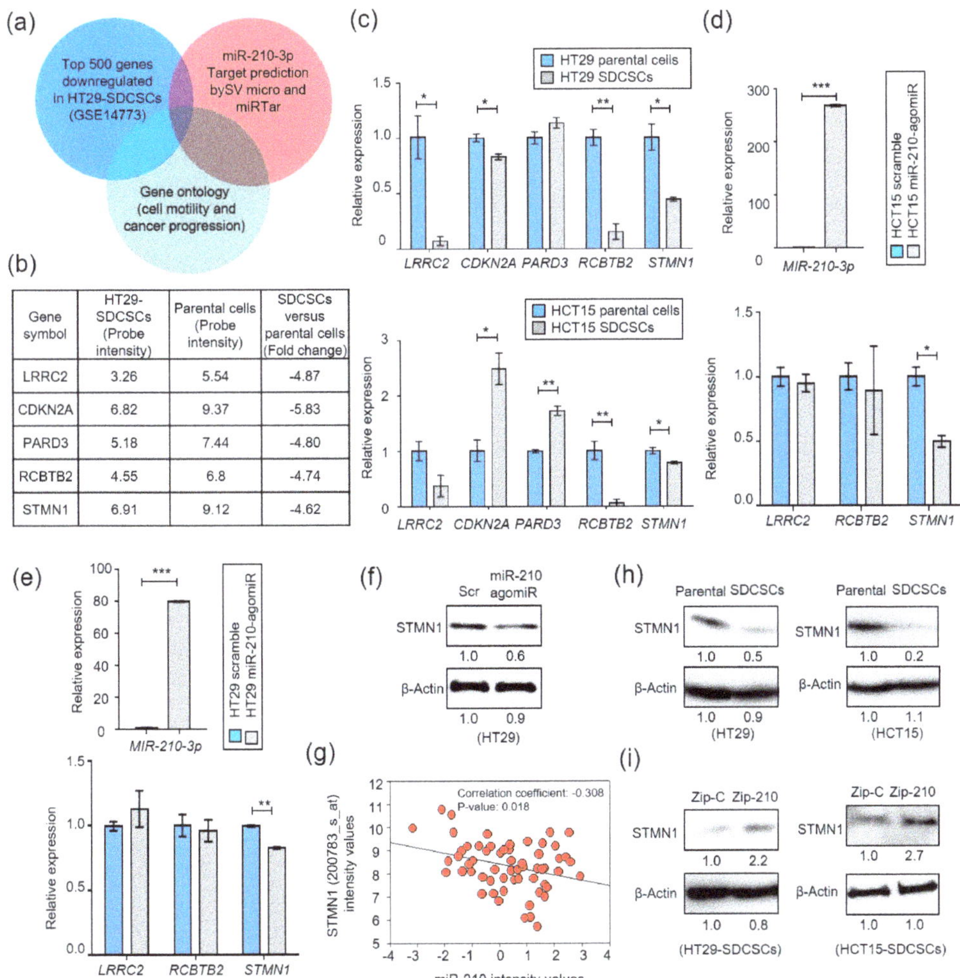

Gene symbol	HT29-SDCSCs (Probe intensity)	Parental cells (Probe intensity)	SDCSCs versus parental cells (Fold change)
LRRC2	3.26	5.54	-4.87
CDKN2A	6.82	9.37	-5.83
PARD3	5.18	7.44	-4.80
RCBTB2	4.55	6.8	-4.74
STMN1	6.91	9.12	-4.62

Figure 3. Identification of the miR-210-3p-STMN1 axis in SDCSCs. (**a**) Strategy for identifying miR-210 targets. The putative targets were obtained from the set of differentially expressed genes in SDCSCs versus parental HT29 cells (GSE14773) and subjected to target prediction with SVmicro. Migration-, tumor suppressor- and stemness-related genes were selected. (**b**) A table illustrating the expression levels of five candidate targets of miR-210-3p from microarray analysis. (**c**) RT-qPCR examining the expression of five candidate targets of miR-210-3p in SDCSCs and their parental cells. (**d**,**e**) RT-qPCR examining the expression of miR-210-3p or LRRC2, RCBTB2 and STMN1 in HCT15 cells (**d**) and HT29 cells transfected with control agomiR (Scramble, 100 nM) or miR-210-3p agomiR (miR-210-AgomiR, 100 nM). (**f**) Western blot showing the expression of STMN1 in HT29 cells transfected with control agomiR (Scramble, Scr, 100 nM) or miR-210-3p agomiR (miR-210-AgomiR, 100nM). (**g**) Analysis of the NCI-60 panel revealed an inverse correlation between STMN1 (200783_s_at) and miR-210-3p expression. The pvalue of the Pearson correlation was assessed, and the correlation coefficient is reported. (**h**) Immunoblots showing STMN1 expression. (**i**) Western blots showing the expression of STMN1 in control SDCSCs (Zip-C) and miR-210-3p-silenced SDCSCs (Zip-210). Data in bar charts are expressed as the mean ± SD values. * $p < 0.05$; ** $p < 0.01$; *** $p < 0.001$ (Student's t-test).

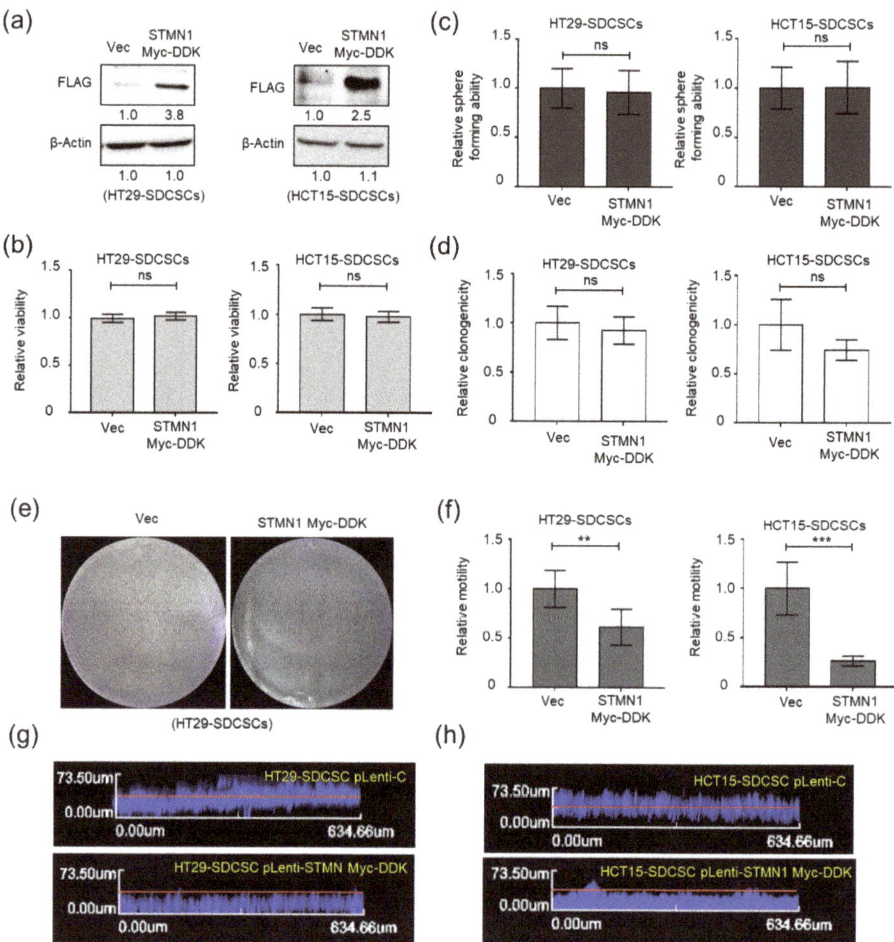

Figure 4. Restoration of STMN1 expression abolishes the migratory and invasive abilities of SDCSCs. (**a**) Western blots showing the expression of STMN1 in control SDCSCs (Vec) and SDCSCs ectopically expressing STMN1. An anti-FLAG antibody was used to detect the expression of exogenous STMN1 with a DKK-Myc tag. (**b**) Relative viability of cells as assessed by an MTT assay. ns, nonsignificant. (**c**) The sphere-forming capacity of control SDCSCs (Vec) and STMN-expressing SDCSCs (STMN1 DKK-Myc). ns, nonsignificant. (**d**) The colony formation of control SDCSCs (Vec) and STMN-expressing SDCSCs (STMN1 DKK-Myc). ns, nonsignificant. (**e**) Representative images showing the colonies generated. (**f**) Histograms showing the relative transwell migration ability of cells. (**g**,**h**) Representative images of vertical invasion of vector control HT29-SDCSCs (pLenti-C) and STMN-expressing SDCSCs (pLenti-STMN1 Myc-DDK). Data in bar charts are expressed as the mean ± SD values. ** $p < 0.01$; *** $p < 0.001$ (Student's t-test).

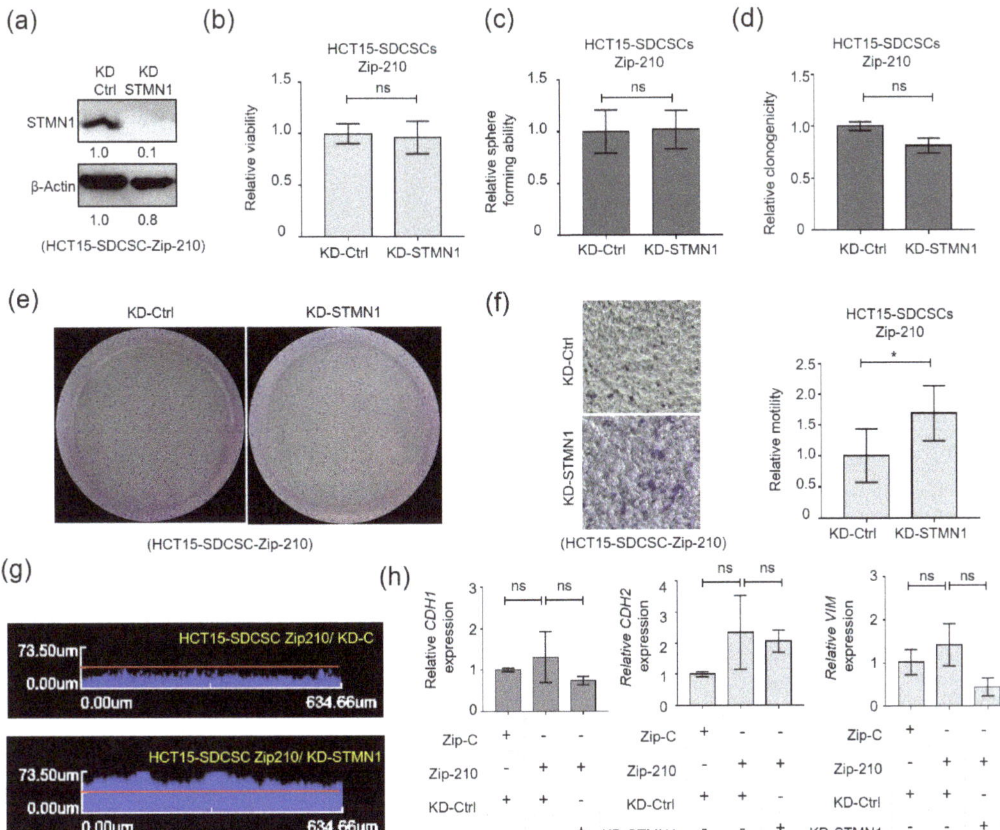

Figure 5. Silencing STMN1 expression restored the attenuated migratory and invasive abilities of miR-210-3p-knockdown SDCSCs. (**a**) Western blots showing the expression of STMN1 in miR-210-3p-knockdown HCT15-SDCSCs. (**b**) Relative viability of cells as assessed by an MTT assay. ns, nonsignificant. (**c**) The sphere-forming capacity of miR-210-3p-silenced HCT15-SDCSCs receiving control shRNA (KD-Ctrl) and shRNA against STMN1 (KD-STMN1). ns, nonsignificant. (**d**) The colony formation ability of miR-210-3p-silenced HCT15-SDCSCs receiving control shRNA (KD-Ctrl) and shRNA against STMN1 (KD-STMN1). ns, nonsignificant. (**e**) Representative images showing colonies generated from the indicated cells. (**f**) Left: representative images of migrated cells. Right: histograms showing the relative transwell migration ability of cells. (**g**) Representative images of vertical invasion of the indicated cells. (**h**) RT-qPCR validation of the expression of an epithelial cell marker (E-cadherin, CDH1) and mesenchymal marker (N-cadherin, CDH2) in control HCT15-SDCSCs, miR-210-3p-silenced HCT15-SDCSCs receiving shRNA control (KD-Ctrl) and miR-210-3p-silenced HCT15-SDCSCs receiving shRNA against STMN1 (KD-STMN1). ns, nonsignificant. Data in bar charts are expressed as the mean ± SD values. * $p < 0.05$ (Student's *t*-test).

3.5. The miR-210-STMN1 Axis Determines the Stiffness of CRCSCs

As SDCSCs must change their shapes while migrating through 8-μm pores in the transwell migration assay and through the collagen matrix in the 3D invasion assay, and the deformability of cancer cells is associated with their metastatic competence [43], we next examined the deformability of SDCSCs. To this end, we performed elasticity measurements by monitoring the time-lapse trajectory of injected fluorescent carboxylated polystyrene beads in dissociated single CRC cells (Figure 6a).

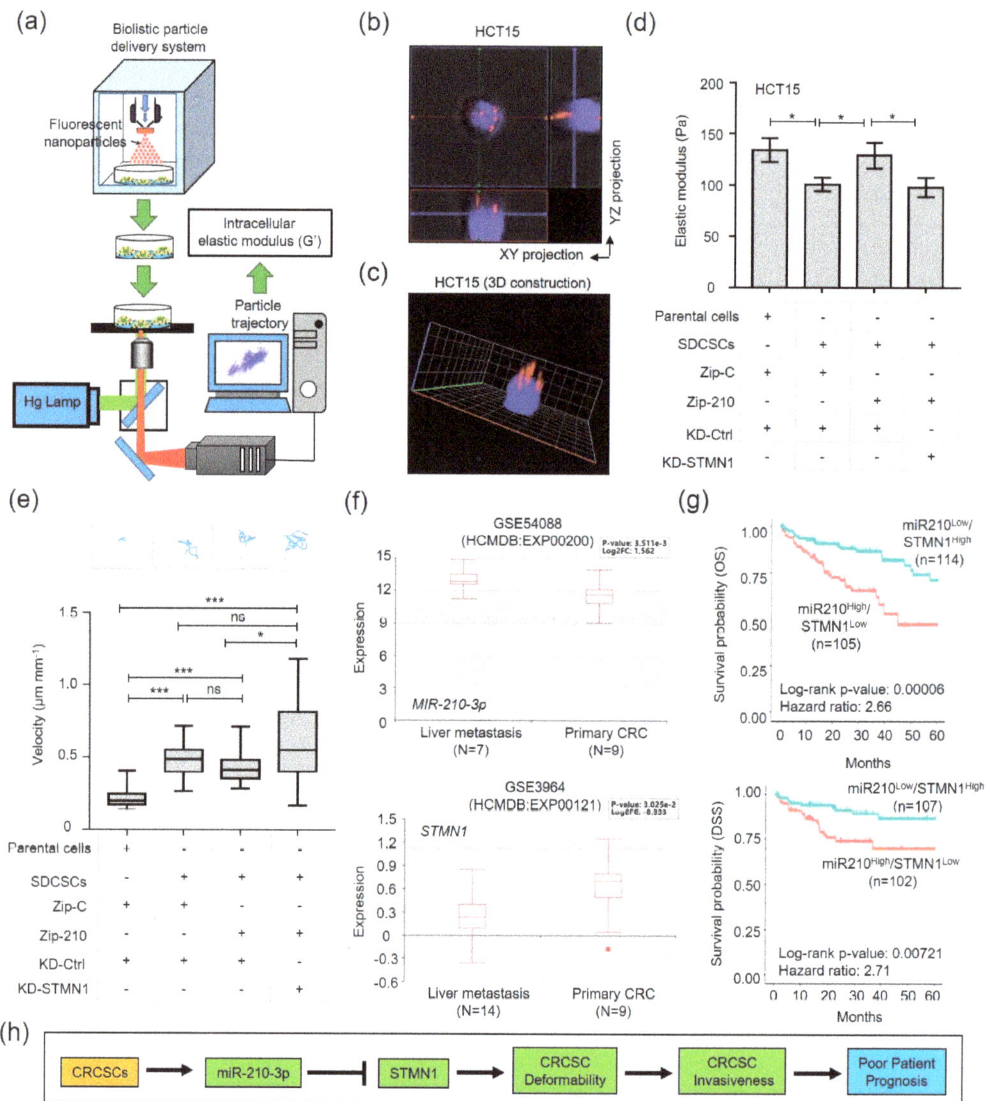

Figure 6. The miR-210-3p-STMN1 axis determines the elasticity of SDCSCs and correlates with liver metastasis in CRC patients. (**a**) A representative image illustrating the biolistic particle delivery system and measurement of the intracellular elastic modulus. (**b**) Representative images showing the distribution of injected fluorescent nanoparticles in HCT15 cells by confocal imaging. The XY and YZ projections are shown. Blue, nuclear staining with Hoechst 33342; red, fluorescent nanoparticles. (**c**) A three-dimensional (3D) constructed image from (**b**). (**d**) Histograms showing the elastic modulus of the indicated HCT15 cells at 100 Hz. Zip-C, scramble control; Zip-210, shRNA against miR-210-3p; KD-Ctrl, scramble control; KD-STMN1, shRNA targeting STMN1. Results are expressed as the mean ± SEM values. (**e**) Upper: trajectory of the indicated cells in the 2.5D assay. Lower: the velocity values in the indicated cells. The box plots show the sample maximum (upper end of the whisker), upper quartile (top edge of the box), median (band in the box), lower quartile (bottom edge of the box), and sample minimum (lower end of the whisker) values. * $p < 0.05$; *** $p < 0.001$ (Student's t-test). (**f**) The bar charts showing the expression of miR-210-3p (upper panel) and STMN1 (lower panel) in GSE54088 and GSE3964 datasets retrieved from the Human Cancer Metastasis Database (HCMDB), respectively. The box plots show the sample maximum

(upper end of the whisker), upper quartile (top edge of the box), median (band in the box), lower quartile (bottom edge of the box), and sample minimum (lower end of the whisker) values. (**g**) Kaplan–Meier analysis of the overall survival (upper panel) and disease-specific survival (lower panel) of patients in a TCGA-COAD dataset. The median expression levels of miR-210-3p and STMN1 were used for patient stratification. OS, overall survival; DSS, disease-specific survival. The *p*-values were estimated by the log-rank test. Hazard ratios are reported. (**h**) A schematic summarizing the observations in this study.

The intracellular fluorescent particles were mainly distributed in the cytoplasm (Figure 6b,c). The enhanced movement of the intracellular carboxylated polystyrene beads in HCT15-SDCSCs (Video S1) suggested the reduced intracellular elasticity (i.e., enhanced deformability) of SDCSCs (Figure 6d, bars 1–2). Furthermore, knockdown of miR-210 (Zip-210) in HCT15-SDCSCs restored the elastic modulus (stiffness), and silencing STMN1 expression reversed the reduction in elasticity in miR-210 knockdown HCT15-SDCSCs (Figure 6d, bars 3–4). In contrast, interference with the miR-210-STMN1 axis had limited impact on the 2.5D horizontal movement of HCT15-SDCSCs on the collagen gel (Figure 6e). Collectively, these results indicated that the miR-210-STMN1 axis determined the deformability of SDCSCs to facilitate their motility.

3.6. The miR-210-STMN1 Axis Promotes the Stiffness of CRC Cells

To validate impacts of the miR-210-STMN1 axis on parental CRC cells, we ectopically express STMN1 in HT29 parental cells receiving miR-210 agomiRs. The expression of miR-210 (Figure S4a) and Myc-DDK-tagged STMN1 (Figure S4b) were conformed. It was found that the miR-210-STMN1 axis had limited impacts on the viability (Figure S4c) and clonogenicity (Figure S4d,e) of HT29 parental cells. On the contrary, the enhanced motility (Figure S4f) and reduced elasticity (Figure S4g) of HT29 cells receiving miR-210 agomiR could be reverted upon expressing STMN1 without modulating the EMT markers (CDH1. CDH2 and VIM) (Figure S4h). Consistently, we found that expression of STMN1 in parental HT29 cells (Figure S5a) had no effect on cell viability (Figure S5b), clonogenicity (Figure S5c,d) or EMT marker expression (Figure S5e). However, the transwell migration ability was reduced (Figure S5f) and the cellular elasticity was elevated (Figure S5g) upon STMN1 expression. This results indicated the miR-210-STMN1 axis identified in CRCSCs also contributes to deformability and motility of CRC cells.

3.7. The miR-210^{High}/STMN1^{Low} Expression Signature Is Associated with Liver Metastasis and Predicts a Poor Clinical Outcome in CRC Patients

As enhanced cellular deformability benefits local dissemination in the extracellular matrix (ECM), along with intravasation and extravasation, the miR-210^{High}/STMN1^{Low} expression signature may predict distal metastasis in CRC patients. To this end, we examined the expression of miR-210-STMN1 axis components in primary and paired liver metastatic CRC specimens from our collection and databases in the public domain. Increased expression of miR-210 and decreased expression of STMN1 were observed in liver metastatic CRC samples from GSE54088 and GSE3964 datasets, respectively (Figure 6f). Additionally, decreased STMN expression in liver metastastic CRC patients was verified in paired, paraffin-embedded tissues by IHC staining (Figure S6a,b). However, the expression of STMN1 were not changed in lymph node metastatic (Figure S6c) or lung metastatic CRC specimens (Figure S6d) comparing to liver metastasis of CRC. As miR-210^{High}/STMN1^{Low} expression signature was associated with CRC metastasis that contributes to poor patient outcomes, we verified the clinical significance of the miR-210^{High}/STMN1^{Low} expression pattern. In analysis of the TCGA data sets, the miR-210^{High}/STMN1^{Low} expression signature predicted worse CRC patient survival (Figure 6g). Taken together, our results suggested the elevated expression of miR-210 attenuated STMN1 expression to engender deformability of CRCSCs for facilitating invasiveness, resulting in poor prognosis of CRC patients (Figure 6h).

4. Discussion

STMN1, also called oncoprotein 18 (Op18), metablastin (p19) and prosolin, is identified as a cytosolic microtubule-destabilizing phosphoprotein [44]. Unphosphorylated STMN1 promotes microtubule depolymerization by sequestering soluble tubulin and promotes microtubule catastrophe [45,46]. STMN1 contains four serine phosphorylation sites (Ser16, Ser25, Ser38, and Ser63), and the microtubule-destabilizing ability of STMN1 is regulated by its phosphorylation [47,48]. Phosphorylation of STMN1 in early mitosis abolishes its microtubule-destabilizing ability, allowing the formation of mitotic spindles, and it becomes dephosphorylated when cells exit mitosis and undergo cytokinesis [40]. Overexpression or inhibition of STMN1 expression in K562 cells resulted in accumulation of mitotic cells that were arrested in early and late mitotic stages, respectively [40,49], suggesting a threshold level of STMN1 is required for mitosis progression. Aside from cell cycle regulation, roles of STMN1 in hematopoiesis have been addressed in leukemia cells. STMN1 is abundant in acute leukemia blasts [50] and its expression was decreased when inducing differentiation by exposing an acute promyelocytic leukemia cell line HL60 to Me_2SO or exposing erythroleukemia cells K562 to hemin [51]. Inhibition of STMN1 promoted higher megakaryocytic differentiation and polyploidization of phorbol ester-induced K562 cells [52]. On the contrary, overexpression of STMN1 in human primary CD34(+) cells reduced the megakaryocyte maturation and platelet production [53]. The megaloblastic anemia and thrombocytosis phenotypes observed in aged Stmn1 knockout mice further support STMN1′s roles in hematopoiesis [54]. Additionally, aged Stmn1 deficient mice also developed a progressive axonopathy [55]. Under social defeat stress, Stmn1 deficient mice showed anxious hyperactivity, impaired object recognition, and decreased levels of social investigating behaviors [56]. Thus, pleiotropic roles of STMN1 are highlighted.

In cancers, STMN1 expression correlates with a malignant phenotypes and has been suggested as a therapeutic target [57]. Silencing STMN1 expression inhibited the metastatic ability of a CRC cell line HCT-116 [58]. STMN1 expression was associated with aggressive phenotypes in breast cancer [59]. The oncogenic Stathmin1 is also regulated by a tumor suppressor miRNA-223 in gastric cancer [60] and liver cancer [61] or a tumor suppressor miR-34a in prostate cancer [62]. Overexpression of the somatic STMN1 Q18E mutation identified in esophageal adenocarcinoma promoted the malignant transformation of 3T3 fibroblast cells [63] and chromosomal instability in K562 cells [64]. A S31Y STMN1 missense mutation was noted in colorectal cancer patients analyzed with TumorPortal (http://www.tumorportal.org (accessed on 4 May 2020)) without functional annotation [65]. Nevertheless, D'Andrea et al. showed that Stmn1 knockout mice showed no impact on the onset of the p53-dependent nor RAS-driven tumorigenesis in bladder and fibrosarcomas or skin carcinomas in mice, respectively [66], suggesting cellular context may contribute to diverse functions of STMN1 during oncogenesis.

As local inactivation of STMN1 at the leading edge of the migrating Xenopus A6 cells potentiated localized microtubule growth, STMN1 may function as a negative regulator in cell movement [67]. Consistently, tumor suppressive roles of STMN1 were identified in prostate cancer cells [42]. Williams et al. showed that the highly invasive, EMT-like prostate cancer cells isolated from undifferentiated adenocarcinoma exhibited low STMN1 expression. Inhibition of STMN1 expression in a prostate cancer cell line DU-145, a standard prostate cancer cell line used for CSC enrichment [68], accelerated the metastatic process by initiating an EMT program via activation of p38 and cooperation of TGF-β signaling [42]. In this study, we identified an increased expression of an oncomiR miR-210 (Figure 1h) in both HT29- and HCT15-CRCSCs characterized previously [24] and showed the miR-210 mediated STMN1 suppression in CRCSCs (Figure 3i). The tumor suppressive roles of STMN1 were demonstrated by observing the reduced invasiveness of STMN1-restored CRCSCs (Figure 4g,h) and decreased motility of STMN1-overexpressed HT29 cells (Figure S5f), a CRC cell line exhibits higher stem-like properties [24]. The EMT program was found to be disconnected from the miR-210-STMN1 activated invasiveness of both CRCSCs (Figure 5h) and HT29 cells (Figure S4h). Here, we unraveled metastatic inhibition effects of STMN1 in our CRC cell models. Reduced STMN1 expression was also observed in paired, liver metastatic CRC specimens (Figure S6a,b). Taken together, these findings indicate that STMN1 tends to function as a metastatic suppressor in stem-like tumor cells and suggest that understanding the stemness profiles and numbers of stem-like cells in cancer patients are crucial for utilizing STMN1 as a therapeutic target.

According to the present results and our previous findings about CRCSCs, we propose a model in which CRCSCs trigger different signaling pathways to maintain cancer stemness and subsequent metastasis: CRCSCs are Snail-dominant cells that undergo EMT [24]. In CRCSCs, Snail suppresses E-cadherin, leading to EMT and cellular disaggregation. Decreased E-cadherin expression results in nuclear translocation of β-catenin and activation of the Wnt pathway, which induces miR-146a expression in CRCSCs. During serum-induced differentiation, mIR-146a could be segregated non-randomly into CD44(+), Snail(+) daughter colorectal stem cells to initiate a feedforward β-catenin/TCF signaling to maintain stem cell pools without promoting CRCSC migration by targeting *NUMB* [35]. Here, we identified one additional oncomiR, namely miR-210, that suppressed *STMN1* expression to facilitate invasiveness of CRCSCs. The ectopic Snail expression was not found to activate the miR-210-STMN1 axis in CRC cells (Figure S7a,b), indicating the miR-210-STMN1 axis was disconnected from an EMT program. Our findings suggest that Snail-dominant CRCSCs uncouple cancer cell division mode and deformability by utilizing distinct miRNAs for maintaining aggressive CSC phenotypes. The sequential activation of miR-210 and miR-146 and the collaboration of these miRNAs with other coding and noncoding genes in the TME require further investigation.

Our study has some limitations. First, our findings mainly rely on CRC cell line-derived cell models, primary cells or CSCs isolated from different tumor types may help to delineate dual roles of STMN1 under diverse context of cells or tissues. Second, the molecular mediators driving dual roles of STMN1 and STMN1-driven metastasis need further exploration.

5. Conclusions

The significance of our study is double-edged. Scientifically, the miR-210-STMN1 axis determines the invasiveness of CRCSCs without affecting cancer stemness. As the low STMN1 expression is essential for migrating of CRCSCs, using STMN1 as a therapeutic target might accelerate metastasis of CRCSCs. Clinically, this study proposes a miR-210^{High}/$STMN1^{Low}$ expression pattern as a potential indicator for monitoring the liver metastasis longitudinally along with therapeutic regimes.

Supplementary Materials: The following are available online at https://www.mdpi.com/article/10.3390/cancers13081833/s1, Figure S1: Characterization of sphere-derived cancer stem cells (SDCSCs), Figure S2: Silencing miR-210-3p expression does not affect the symmetric cell division (SCD) of HT29-SDCSCs, Figure S3: Restoration of STMN1 does not consistently alter the expression of epithelial and mesenchymal markers in HT29- and HCT15-SDCSCs, Figure S4: Ectopically expression of STMN1 reverses the miR-210-regulated trans well migration ability and elasticity of HT29 parental cells, Figure S5: Overexpression of STMN1 inhibits the transwell migration potential and promotes intracellular elasticity of HT29 parental cells, Figure S6:Decreased expression of STMN1 in liver metastatic CRC specimens, Figure S7: Examining the expression of miR-210-3p-STMN1 axis components in Snail-expressing HCT15 cells, Figure S8: Uncropped blots, Video S1: A video showing the Brownian motion of intracellular fluorescent beads in HCT15 cells and HCT15-SDCSCs.

Author Contributions: Conceptualization, T.-T.L., W.-C.C., C.-Y.Y., J.-K.J. and W.-L.H.; methodology, T.-T.L., W.-C.C., C.-Y.Y., Y.-Q.C. and A.E.-T.C.; validation, S.-H.S., T.-Y.Y., H.-Y.L., C.-C.L. (Chih-Chan Lee) and W.-L.H.; formal analysis, T.-T.L., W.-C.C., Y.-Q.C., S.-H.S., T.-Y.Y., H.-Y.L. and C.-C.L.(Chih-Chan Lee); investigation, J.-K.J. and W.-L.H.; resources, H.-H.L., C.-C.L. (Chun-Chi Lin), R.-H.L., A.E.-T.C. and J.-K.J.; data curation, T.-T.L., H.-Y.L., C.-C.L. (Chih-Chan Lee) and W.-L.H.; writing—original draft preparation, W.-L.H.; writing—review and editing, W.-L.H. and J.-K.J.; visualization, S.-H.S., T.-Y.Y. and W.-L.H.; supervision, J.-K.J. and W.-L.H.; project administration, J.-K.J. and W.-

L.H.; funding acquisition, T.-T.L., W.-C.C., C.-Y.Y., J.-K.J., Y.-Q.C. and W.-L.H. All authors have read and agreed to the published version of the manuscript.

Funding: Ministry of Science and Technology (109-2636-B-038-001 and 110-2636-B-038-005 to T.-T.L.; 109-2314-B-075 -033 to J.-K.J.; 107-2221-E-010-009-MY2 and 109-2221-E-010-018 to Y.-Q.C.; 108-2622-E-039-005-CC2, 109-2628-E-039-001-MY3 and MOST 109-2327-B-039-002 to W.-C.C.; 109-2320-B-010-021 to W.-L.H.). Taipei Medical University (TMU108-AE1-B25 to T.-T.L.). China Medical University (CMU107-S-24, CMU109-MF-61 and CMU109-MF-93 to W.-C.C.). Taipei Veterans General Hospital (V108C-164 to J.-K.J.). Taipei City Hospital (TPCH-109-01 to C.-Y.Y.). Yen Tjing Ling Medical Foundation (CI-109-11 to W.-L.H.). Ministry of Health and Welfare, Center of Excellence for Cancer Research (MOHW107-TDU-B-211-114019, 109 CRC-T208 and 110 CRC-T208 to W.-L.H.).

Institutional Review Board Statement: This study was approved by the Institutional Ethics Committee/Institutional Board of Taipei Veterans General Hospital (2016-03-001BC, 2018-11-002C).

Informed Consent Statement: Informed consent was obtained from all subjects involved in the study.

Data Availability Statement: The datasets used and analyzed during the current study are available from the corresponding author on reasonable request.

Acknowledgments: We would like to dedicate this paper to the memory of Hsei-Wei Wang (Institute of Microbiology and Immunology, National Yang-Ming University), who passed away during the period of this research. Wang initiated this study in 2014, and this paper could not have been completed without his enduring devotion to cancer genomics. We appreciate the assistance of the Biobank at Taipei Veterans General Hospital for providing human specimens. This work was supported by Taipei City Hospital, Yen Tjing Ling Medical Foundation, in Taiwan and by the Cancer Progression Research Center, National Yang-Ming University, from The Featured Areas Research Center Program within the framework of the Higher Education Sprout Project by the Ministry of Education (MOE). We thank Springer Nature Author Services for editing assistance.

Conflicts of Interest: The authors declare no conflict of interest.

References

1. Chiang, M.Y.; Yangben, Y.; Lin, N.J.; Zhong, J.L.; Yang, L. Relationships among cell morphology, intrinsic cell stiffness and cell substrate interactions. *Biomaterials* **2013**, *34*, 9754–9762. [CrossRef]
2. Matzke, R.; Jacobson, K.; Radmacher, M. Direct, high-resolution measurement of furrow stiffening during division of adherent cells. *Nat. Cell Biol.* **2001**, *3*, 607–610. [CrossRef] [PubMed]
3. Shimamoto, Y.; Maeda, Y.T.; Ishiwata, S.; Libchaber, A.J.; Kapoor, T.M. Insights into the micromechanical properties of the metaphase spindle. *Cell* **2011**, *145*, 1062–1074. [CrossRef]
4. Chen, Y.Q.; Kuo, C.Y.; Wei, M.T.; Wu, K.; Su, P.T.; Huang, C.S.; Chiou, A.E. Intracellular viscoelasticity of HeLa cells during cell division studied by video particle-tracking microrheology. *J. Biomed. Opt.* **2014**, *19*, 011008. [CrossRef] [PubMed]
5. Yu, H.W.; Chen, Y.Q.; Huang, C.M.; Liu, C.Y.; Chiou, A.; Wang, Y.K.; Tang, M.J.; Kuo, J.C. beta-PIX controls intracellular viscoelasticity to regulate lung cancer cell migration. *J. Cell. Mol. Med.* **2015**, *19*, 934–947. [CrossRef]
6. Kole, T.P.; Tseng, Y.; Jiang, I.; Katz, J.L.; Wirtz, D. Intracellular mechanics of migrating fibroblasts. *Mol. Biol. Cell* **2005**, *16*, 328–338. [CrossRef]
7. Winkler, J.; Abisoye-Ogunniyan, A.; Metcalf, K.J.; Werb, Z. Concepts of extracellular matrix remodelling in tumour progression and metastasis. *Nat. Commun.* **2020**, *11*, 5120. [CrossRef] [PubMed]
8. Mizuno, D.; Tardin, C.; Schmidt, C.F.; Mackintosh, F.C. Nonequilibrium mechanics of active cytoskeletal networks. *Science* **2007**, *315*, 370–373. [CrossRef] [PubMed]
9. Wirtz, D. Particle-tracking microrheology of living cells: Principles and applications. *Annu. Rev. Biophys.* **2009**, *38*, 301–326. [CrossRef]
10. Copley, A.L.; King, R.G.; Chien, S.; Usami, S.; Skalak, R.; Huang, C.R. Microscopic observations of viscoelasticity of human blood in steady and oscillatory shear. *Biorheology* **1975**, *12*, 257–263. [CrossRef]
11. Visvader, J.E.; Lindeman, G.J. Cancer stem cells: Current status and evolving complexities. *Cell Stem Cell* **2012**, *10*, 717–728. [CrossRef] [PubMed]
12. O'Brien, C.A.; Pollett, A.; Gallinger, S.; Dick, J.E. A human colon cancer cell capable of initiating tumour growth in immunodeficient mice. *Nature* **2006**, *445*, 106–110. [CrossRef]
13. Ricci-Vitiani, L.; Lombardi, D.G.; Pilozzi, E.; Biffoni, M.; Todaro, M.; Peschle, C.; De Maria, R. Identification and expansion of human colon-cancer-initiating cells. *Nature* **2006**, *445*, 111–115. [CrossRef] [PubMed]
14. Dylla, S.J.; Beviglia, L.; Park, I.K.; Chartier, C.; Raval, J.; Ngan, L.; Pickell, K.; Aguilar, J.; Lazetic, S.; Smith-Berdan, S.; et al. Colorectal cancer stem cells are enriched in xenogeneic tumors following chemotherapy. *PLoS ONE* **2008**, *3*, e2428. [CrossRef]

15. Pang, R.; Law, W.L.; Chu, A.C.; Poon, J.T.; Lam, C.S.; Chow, A.K.; Ng, L.; Cheung, L.W.; Lan, X.R.; Lan, H.Y.; et al. A subpopulation of CD26+ cancer stem cells with metastatic capacity in human colorectal cancer. *Cell Stem Cell* **2010**, *6*, 603–615. [CrossRef]
16. de Sousa e Melo, F.; Kurtova, A.V.; Harnoss, J.M.; Kljavin, N.; Hoeck, J.D.; Hung, J.; Anderson, J.E.; Storm, E.E.; Modrusan, Z.; Koeppen, H.; et al. A distinct role for Lgr5+ stem cells in primary and metastatic colon cancer. *Nature* **2017**, *543*, 676–680. [CrossRef]
17. Mani, S.A.; Guo, W.; Liao, M.J.; Eaton, E.N.; Ayyanan, A.; Zhou, A.Y.; Brooks, M.; Reinhard, F.; Zhang, C.C.; Shipitsin, M.; et al. The epithelial-mesenchymal transition generates cells with properties of stem cells. *Cell* **2008**, *133*, 704–715. [CrossRef]
18. Vermeulen, L.; De Sousa, E.M.F.; van der Heijden, M.; Cameron, K.; de Jong, J.H.; Borovski, T.; Tuynman, J.B.; Todaro, M.; Merz, C.; Rodermond, H.; et al. Wnt activity defines colon cancer stem cells and is regulated by the microenvironment. *Nat. Cell Biol.* **2010**, *12*, 468–476. [CrossRef]
19. Bartel, D.P. MicroRNAs: Target recognition and regulatory functions. *Cell* **2009**, *136*, 215–233. [CrossRef]
20. Croce, C.M. Causes and consequences of microRNA dysregulation in cancer. *Nat. Rev. Genet.* **2009**, *10*, 704–714. [CrossRef] [PubMed]
21. Huang, S.; Tan, X.; Huang, Z.; Chen, Z.; Lin, P.; Fu, S.W. microRNA biomarkers in colorectal cancer liver metastasis. *J. Cancer* **2018**, *9*, 3867–3873. [CrossRef] [PubMed]
22. Cheng, D.; Zhao, S.; Tang, H.; Zhang, D.; Sun, H.; Yu, F.; Jiang, W.; Yue, B.; Wang, J.; Zhang, M.; et al. MicroRNA-20a-5p promotes colorectal cancer invasion and metastasis by downregulating Smad4. *Oncotarget* **2016**, *7*, 45199–45213. [CrossRef]
23. Lam, C.S.; Ng, L.; Chow, A.K.; Wan, T.M.; Yau, S.; Cheng, N.S.; Wong, S.K.M.; Man, J.H.W.; Lo, O.S.H.; Foo, D.C.C.; et al. Identification of microRNA 885-5p as a novel regulator of tumor metastasis by targeting CPEB2 in colorectal cancer. *Oncotarget* **2017**, *8*, 26858–26870. [CrossRef]
24. Hwang, W.L.; Yang, M.H.; Tsai, M.L.; Lan, H.Y.; Su, S.H.; Chang, S.C.; Teng, H.W.; Yang, S.H.; Lan, Y.T.; Chiou, S.H.; et al. SNAIL regulates interleukin-8 expression, stem cell-like activity, and tumorigenicity of human colorectal carcinoma cells. *Gastroenterology* **2011**, *141*, 279–291.e5. [CrossRef]
25. Matsui, Y.; Nakayama, Y.; Okamoto, M.; Fukumoto, Y.; Yamaguchi, N. Enrichment of cell populations in metaphase, anaphase, and telophase by synchronization using nocodazole and blebbistatin: A novel method suitable for examining dynamic changes in proteins during mitotic progression. *Eur. J. Cell Biol.* **2012**, *91*, 413–419. [CrossRef]
26. Wu, P.H.; Hale, C.M.; Chen, W.C.; Lee, J.S.; Tseng, Y.; Wirtz, D. High-throughput ballistic injection nanorheology to measure cell mechanics. *Nat. Protoc.* **2012**, *7*, 155–170. [CrossRef]
27. del Alamo, J.C.; Norwich, G.N.; Li, Y.S.; Lasheras, J.C.; Chien, S. Anisotropic rheology and directional mechanotransduction in vascular endothelial cells. *Proc. Natl. Acad. Sci. USA* **2008**, *105*, 15411–15416. [CrossRef]
28. Baker, E.L.; Bonnecaze, R.T.; Zaman, M.H. Extracellular matrix stiffness and architecture govern intracellular rheology in cancer. *Biophys. J.* **2009**, *97*, 1013–1021. [CrossRef]
29. Liu, S.H.; Shen, P.C.; Chen, C.Y.; Hsu, A.N.; Cho, Y.C.; Lai, Y.L.; Chen, F.H.; Li, C.Y.; Wang, S.C.; Chen, M.; et al. DriverDBv3: A multi-omics database for cancer driver gene research. *Nucleic Acids Res.* **2020**, *48*, D863–D870. [CrossRef]
30. Chung, I.F.; Chang, S.J.; Chen, C.Y.; Liu, S.H.; Li, C.Y.; Chan, C.H.; Shih, C.C.; Cheng, W.C. YM500v3: A database for small RNA sequencing in human cancer research. *Nucleic Acids Res.* **2017**, *45*, D925–D931. [CrossRef]
31. Reinhold, W.C.; Sunshine, M.; Liu, H.; Varma, S.; Kohn, K.W.; Morris, J.; Doroshow, J.; Pommier, Y. CellMiner: A web-based suite of genomic and pharmacologic tools to explore transcript and drug patterns in the NCI-60 cell line set. *Cancer Res.* **2012**, *72*, 3499–3511. [CrossRef] [PubMed]
32. Lee, J.; Kotliarova, S.; Kotliarov, Y.; Li, A.; Su, Q.; Donin, N.M.; Pastorino, S.; Purow, B.W.; Christopher, N.; Zhang, W.; et al. Tumor stem cells derived from glioblastomas cultured in bFGF and EGF more closely mirror the phenotype and genotype of primary tumors than do serum-cultured cell lines. *Cancer Cell* **2006**, *9*, 391–403. [CrossRef]
33. Peng, Y.; Croce, C.M. The role of MicroRNAs in human cancer. *Signal Transduct Target Ther.* **2016**, *1*, 15004. [CrossRef]
34. Sugiarto, S.; Persson, A.I.; Munoz, E.G.; Waldhuber, M.; Lamagna, C.; Andor, N.; Hanecker, P.; Ayers-Ringler, J.; Phillips, J.; Siu, J.; et al. Asymmetry-defective oligodendrocyte progenitors are glioma precursors. *Cancer Cell* **2011**, *20*, 328–340. [CrossRef]
35. Hwang, W.L.; Jiang, J.K.; Yang, S.H.; Huang, T.S.; Lan, H.Y.; Teng, H.W.; Yang, C.Y.; Tsai, Y.P.; Lin, C.H.; Wang, H.W.; et al. MicroRNA-146a directs the symmetric division of Snail-dominant colorectal cancer stem cells. *Nat. Cell Biol.* **2014**, *16*, 268–280. [CrossRef]
36. Rando, T.A. The immortal strand hypothesis: Segregation and reconstruction. *Cell* **2007**, *129*, 1239–1243. [CrossRef]
37. Liu, H.; Yue, D.; Chen, Y.; Gao, S.J.; Huang, Y. Improving performance of mammalian microRNA target prediction. *BMC Bioinform.* **2010**, *11*, 476. [CrossRef]
38. Hsu, J.B.; Chiu, C.M.; Hsu, S.D.; Huang, W.Y.; Chien, C.H.; Lee, T.Y.; Huang, H.D. miRTar: An integrated system for identifying miRNA-target interactions in human. *BMC Bioinform.* **2011**, *12*, 300. [CrossRef]
39. Kiga, K.; Mimuro, H.; Suzuki, M.; Shinozaki-Ushiku, A.; Kobayashi, T.; Sanada, T.; Kim, M.; Ogawa, M.; Iwasaki, Y.W.; Kayo, H.; et al. Epigenetic silencing of miR-210 increases the proliferation of gastric epithelium during chronic Helicobacter pylori infection. *Nat. Commun.* **2014**, *5*, 4497. [CrossRef] [PubMed]
40. Marklund, U.; Larsson, N.; Gradin, H.M.; Brattsand, G.; Gullberg, M. Oncoprotein 18 is a phosphorylation-responsive regulator of microtubule dynamics. *EMBO J.* **1996**, *15*, 5290–5298. [CrossRef]
41. Cassimeris, L. The oncoprotein 18/stathmin family of microtubule destabilizers. *Curr. Opin. Cell Biol.* **2002**, *14*, 18–24. [CrossRef]

42. Williams, K.; Ghosh, R.; Giridhar, P.V.; Gu, G.; Case, T.; Belcher, S.M.; Kasper, S. Inhibition of stathmin1 accelerates the metastatic process. *Cancer Res.* **2012**, *72*, 5407–5417. [CrossRef] [PubMed]
43. Suresh, S. Biomechanics and biophysics of cancer cells. *Acta Biomater.* **2007**, *3*, 413–438. [CrossRef]
44. Belmont, L.D.; Mitchison, T.J. Identification of a protein that interacts with tubulin dimers and increases the catastrophe rate of microtubules. *Cell* **1996**, *84*, 623–631. [CrossRef]
45. Jourdain, L.; Curmi, P.; Sobel, A.; Pantaloni, D.; Carlier, M.F. Stathmin: A tubulin-sequestering protein which forms a ternary T2S complex with two tubulin molecules. *Biochemistry* **1997**, *36*, 10817–10821. [CrossRef]
46. Curmi, P.A.; Andersen, S.S.; Lachkar, S.; Gavet, O.; Karsenti, E.; Knossow, M.; Sobel, A. The stathmin/tubulin interaction in vitro. *J. Biol. Chem.* **1997**, *272*, 25029–25036. [CrossRef] [PubMed]
47. Larsson, N.; Marklund, U.; Gradin, H.M.; Brattsand, G.; Gullberg, M. Control of microtubule dynamics by oncoprotein 18: Dissection of the regulatory role of multisite phosphorylation during mitosis. *Mol. Cell Biol.* **1997**, *17*, 5530–5539. [CrossRef] [PubMed]
48. Wittmann, T.; Bokoch, G.M.; Waterman-Storer, C.M. Regulation of microtubule destabilizing activity of Op18/stathmin downstream of Rac1. *J. Biol. Chem.* **2004**, *279*, 6196–6203. [CrossRef] [PubMed]
49. Iancu, C.; Mistry, S.J.; Arkin, S.; Wallenstein, S.; Atweh, G.F. Effects of stathmin inhibition on the mitotic spindle. *J. Cell Sci.* **2001**, *114 Pt 5*, 909–916.
50. Machado-Neto, J.A.; de Melo Campos, P.; Favaro, P.; Lazarini, M.; Lorand-Metze, I.; Costa, F.F.; Saad, S.T.O.; Traina, F. Stathmin 1 is involved in the highly proliferative phenotype of high-risk myelodysplastic syndromes and acute leukemia cells. *Leuk. Res.* **2014**, *38*, 251–257. [CrossRef] [PubMed]
51. Luo, X.N.; Arcasoy, M.O.; Brickner, H.E.; Mistry, S.; Schechter, A.D.; Atweh, G.F. Regulated expression of p18, a major phosphoprotein of leukemic cells. *J. Biol. Chem.* **1991**, *266*, 21004–21010. [CrossRef]
52. Rubin, C.I.; French, D.L.; Atweh, G.F. Stathmin expression and megakaryocyte differentiation: A potential role in polyploidy. *Exp. Hematol.* **2003**, *31*, 389–397. [CrossRef]
53. Iancu-Rubin, C.; Gajzer, D.; Tripodi, J.; Najfeld, V.; Gordon, R.E.; Hoffman, R.; Atweh, G.F. Down-regulation of stathmin expression is required for megakaryocyte maturation and platelet production. *Blood* **2011**, *117*, 4580–4589. [CrossRef]
54. Ramlogan-Steel, C.A.; Steel, J.C.; Fathallah, H.; Iancu-Rubin, C.; Soleimani, M.; Dong, Z.; Atweh, G.F. The role of Stathmin, a regulator of mitosis, in hematopoiesis. *Blood* **2012**, *120*, 3453. [CrossRef]
55. Liedtke, W.; Leman, E.E.; Fyffe, R.E.; Raine, C.S.; Schubart, U.K. Stathmin-deficient mice develop an age-dependent axonopathy of the central and peripheral nervous systems. *Am. J. Pathol.* **2002**, *160*, 469–480. [CrossRef]
56. Nguyen, T.B.; Prabhu, V.V.; Piao, Y.H.; Oh, Y.E.; Zahra, R.F.; Chung, Y.C. Effects of Stathmin 1 Gene Knockout on Behaviors and Dopaminergic Markers in Mice Exposed to Social Defeat Stress. *Brain Sci.* **2019**, *9*, 215. [CrossRef]
57. Biaoxue, R.; Xiguang, C.; Hua, L.; Shuanying, Y. Stathmin-dependent molecular targeting therapy for malignant tumor: The latest 5 years' discoveries and developments. *J. Transl. Med.* **2016**, *14*, 279. [CrossRef]
58. Wu, W.; Tan, X.F.; Tan, H.T.; Lim, T.K.; Chung, M.C. Unbiased proteomic and transcript analyses reveal that stathmin-1 silencing inhibits colorectal cancer metastasis and sensitizes to 5-fluorouracil treatment. *Mol. Cancer Res.* **2014**, *12*, 1717–1728. [CrossRef]
59. Obayashi, S.; Horiguchi, J.; Higuchi, T.; Katayama, A.; Handa, T.; Altan, B.; Bai, T.; Bao, P.; Bao, H.; Yokobori, T.; et al. Stathmin1 expression is associated with aggressive phenotypes and cancer stem cell marker expression in breast cancer patients. *Int. J. Oncol.* **2017**, *51*, 781–790. [CrossRef]
60. Kang, W.; Tong, J.H.; Chan, A.W.; Lung, R.W.; Chau, S.L.; Wong, Q.W.; Wong, N.; Yu, J.; Cheng, A.S.; To, K.F. Stathmin1 plays oncogenic role and is a target of microRNA-223 in gastric cancer. *PLoS ONE* **2012**, *7*, e33919. [CrossRef]
61. Wong, Q.W.; Lung, R.W.; Law, P.T.; Lai, P.B.; Chan, K.Y.; To, K.F.; Wong, N. MicroRNA-223 is commonly repressed in hepatocellular carcinoma and potentiates expression of Stathmin1. *Gastroenterology* **2008**, *135*, 257–269. [CrossRef]
62. Chakravarthi, B.; Chandrashekar, D.S.; Agarwal, S.; Balasubramanya, S.A.H.; Pathi, S.S.; Goswami, M.T.; Jing, X.; Wang, R.; Mehra, R.; Asangani, I.A.; et al. miR-34a Regulates Expression of the Stathmin-1 Oncoprotein and Prostate Cancer Progression. *Mol. Cancer Res.* **2018**, *16*, 1125–1137. [CrossRef]
63. Misek, D.E.; Chang, C.L.; Kuick, R.; Hinderer, R.; Giordano, T.J.; Beer, D.G.; Hanash, S.M. Transforming properties of a Q18–>E mutation of the microtubule regulator Op18. *Cancer Cell* **2002**, *2*, 217–228. [CrossRef]
64. Holmfeldt, P.; Sellin, M.E.; Gullberg, M. Upregulated Op18/stathmin activity causes chromosomal instability through a mechanism that evades the spindle assembly checkpoint. *Exp. Cell Res.* **2010**, *316*, 2017–2026. [CrossRef]
65. Lawrence, M.S.; Stojanov, P.; Mermel, C.H.; Robinson, J.T.; Garraway, L.A.; Golub, T.R.; Meyerson, M.; Gabriel, S.B.; Lander, E.S.; Getz, G. Discovery and saturation analysis of cancer genes across 21 tumour types. *Nature* **2014**, *505*, 495–501. [CrossRef] [PubMed]
66. D'Andrea, S.; Berton, S.; Segatto, I.; Fabris, L.; Canzonieri, V.; Colombatti, A.; Vecchione, A.; Belletti, B.; Baldassarre, G. Stathmin is dispensable for tumor onset in mice. *PLoS ONE* **2012**, *7*, e45561. [CrossRef] [PubMed]
67. Niethammer, P.; Bastiaens, P.; Karsenti, E. Stathmin-tubulin interaction gradients in motile and mitotic cells. *Science* **2004**, *303*, 1862–1866. [CrossRef]
68. Wang, L.; Huang, X.; Zheng, X.; Wang, X.; Li, S.; Zhang, L.; Yang, Z.; Xia, Z. Enrichment of prostate cancer stem-like cells from human prostate cancer cell lines by culture in serum-free medium and chemoradiotherapy. *Int. J. Biol. Sci.* **2013**, *9*, 472–479. [CrossRef] [PubMed]

MDPI

Review

Roles of microRNAs in Regulating Cancer Stemness in Head and Neck Cancers

Melysa Fitriana [1,2,†], Wei-Lun Hwang [3,4,5,†], Pak-Yue Chan [6], Tai-Yuan Hsueh [6] and Tsai-Tsen Liao [1,7,*]

1 Graduate Institute of Medical Sciences, College of Medicine, Taipei Medical University, Taipei 11031, Taiwan; melysa.fitriana@mail.ugm.ac.id
2 Otorhinolaryngology Head and Neck Surgery Department, Faculty of Medicine, Public Health and Nursing, Universitas Gadjah Mada, Yogyakarta 55281, Indonesia
3 Department of Biotechnology and Laboratory Science in Medicine, National Yang Ming Chiao Tung University, Taipei 11221, Taiwan; a85296658@ym.edu.tw
4 Department of Biotechnology and Laboratory Science in Medicine, National Yang-Ming University, Taipei 11221, Taiwan
5 Cancer Progression Center of Excellence, National Yang Ming Chiao Tung University, Taipei 11221, Taiwan
6 School of Medicine, Taipei Medical University, Taipei 11031, Taiwan; b101108138@tmu.edu.tw (P.-Y.C.); b101108149@tmu.edu.tw (T.-Y.H.)
7 Cell Physiology and Molecular Image Research Center, Wan Fang Hospital, Taipei Medical University, Taipei 11696, Taiwan
* Correspondence: liaotsaitsen@tmu.edu.tw; Tel.: +886-2736-1661 (ext. 3435)
† These two authors contributed equally to this work.

Simple Summary: Head and neck squamous cell carcinomas (HNSCCs) are highly heterogeneous human malignancies associated with genetic and environmental factors. In HNSCCs, cancer stem cells (CSCs) provide the plasticity for cancer cell progression, metastasis, therapeutic resistance, and recurrence. During carcinogenesis, microRNAs (miRNAs) play important roles in regulating the maintenance and acquisition of cancer stem cell features. Therefore, in this review, we summarize the roles of miRNAs in regulating the cancer stemness of HNSCCs to provide potential therapeutic applications.

Abstract: Head and neck squamous cell carcinomas (HNSCCs) are epithelial malignancies with 5-year overall survival rates of approximately 40–50%. Emerging evidence indicates that a small population of cells in HNSCC patients, named cancer stem cells (CSCs), play vital roles in the processes of tumor initiation, progression, metastasis, immune evasion, chemo-/radioresistance, and recurrence. The acquisition of stem-like properties of cancer cells further provides cellular plasticity for stress adaptation and contributes to therapeutic resistance, resulting in a worse clinical outcome. Thus, targeting cancer stemness is fundamental for cancer treatment. MicroRNAs (miRNAs) are known to regulate stem cell features in the development and tissue regeneration through a miRNA–target interactive network. In HNSCCs, miRNAs act as tumor suppressors and/or oncogenes to modulate cancer stemness and therapeutic efficacy by regulating the CSC-specific tumor microenvironment (TME) and signaling pathways, such as epithelial-to-mesenchymal transition (EMT), Wnt/β-catenin signaling, and epidermal growth factor receptor (EGFR) or insulin-like growth factor 1 receptor (IGF1R) signaling pathways. Owing to a deeper understanding of disease-relevant miRNAs and advances in in vivo delivery systems, the administration of miRNA-based therapeutics is feasible and safe in humans, with encouraging efficacy results in early-phase clinical trials. In this review, we summarize the present findings to better understand the mechanical actions of miRNAs in maintaining CSCs and acquiring the stem-like features of cancer cells during HNSCC pathogenesis.

Keywords: microRNA; cancer stem cell; stemness; head and neck squamous cell carcinoma

Citation: Fitriana, M.; Hwang, W.-L.; Chan, P.-Y.; Hsueh, T.-Y.; Liao, T.-T. Roles of microRNAs in Regulating Cancer Stemness in Head and Neck Cancers. *Cancers* **2021**, *13*, 1742. https://doi.org/10.3390/cancers13071742

Academic Editor: Paola Tucci

Received: 31 January 2021
Accepted: 2 April 2021
Published: 6 April 2021

1. Introduction

Cancer is responsible for about 30% of all premature deaths from non-communicable diseases (NCDs) in adults aged approximately 30–69 years [1]. In 2018, there were 18.1 million people who suffered from cancer worldwide, and 9.6 million of them died from cancer (around one in six deaths globally). In addition, 354,864 (2% of all cancer sites) of new cases were of the lip and oral cavity, accounting for 177,384 (1.9% of all cancer sites) deaths [1,2]. Head and neck squamous cell carcinomas (HNSCCs) are epithelial malignancies located in the oral cavity, nasal cavity, pharynx (nasopharynx, oropharynx, and hypopharynx), and larynx [3,4]. HNSCC subtypes include oral SCCs (OSCCs), laryngeal SCCs (LSCCs), nasopharyngeal carcinomas (NPCs), and oropharyngeal SCCs (OPSCCs) [5,6]. It should be noted that with tongue squamous cell carcinomas, an OSCC includes the anterior two-thirds of the tongue (anterior to the circumvallate papillae) and an OPSCC consists of the base (or posterior one-third) of the tongue [7]. The incidence of HNSCC continues to rise and is anticipated to increase by 30% with 1.08 million new cases annually by 2030 [2,8,9]. Risk factors for the malignant incidence of HNSCCs include tobacco use, alcohol abuse, and human papillomavirus (HPV) infection [3,6]. Signs and symptoms can manifest as a lesion in the nose, mouth, or throat; a lump or neck mass; and ear discomfort; and functional abnormalities such as difficulty swallowing and/or chewing are often found in the later stages of these diseases [3]. The tumor, node, and metastasis (TNM) staging system is used for clinical staging and as a basis for treatment choice [10]. Therapeutic approaches include surgery, radiotherapy, chemoradiotherapy, a combination of surgery with radiotherapy or chemotherapy, and a combination of surgery with adjuvant chemotherapy and radiotherapy. Chemoradiotherapy may be taken as adjuvant therapy in advanced stages [11]. Unfortunately, despite several treatment options being available, outcomes of HNSCC treatment remain poor, patients generally develop resistance, and, as a result, the five-year overall survival rates of HNSCC patients are approximately 40–50% [6,12]. Advanced approaches have been developed by applying immunotherapy or combined immunotherapy treatment to treat resistant and recurrent cases [13].

The major obstacle in cancer therapy is tumor heterogeneity. Cancer stem cells (CSCs) are small populations of cancer cells and are well-known for their association with cancer resistance, relapse, tumorigenesis, and poor clinical outcomes in HNSCCs, which has promoted the development of novel and effective therapeutic protocols for better clinical outcomes [5,14]. Therefore, targeting CSCs has become an attractive approach for potential strategies to treat HNSCCs [15,16]. The abnormal activation of signaling cassettes, genetic and epigenetic modification, and microRNA (miR or miRNA) regulation are central regulators of CSC malignancy [17–19]. miRNAs work as hub regulators to modulate cell functions by binding to multiple 3′-untranslated regions (3′-UTRs) of target messenger RNAs (mRNAs) and cause the translational inhibition and/or degradation of transcripts [19–21]. Therefore, in this review, we address the roles of miRNAs in regulating the cancer stemness of HNSCCs.

2. CSCs and Cancer Stemness

CSCs are a minor population of cells within tumor tissues with a tumor-initiating capacity [22] and stem-like features, including self-renewal [23,24] and asymmetric cell division (ACD) [25]. Under chemotherapy, the cycling rates of CSCs slow and they enter the G0 phase in order to survive, accounting for therapeutic heterogeneity [26–28]. Cancer patients with higher stem cell signatures present poorly differentiated histological properties and are associated with a worse clinical outcome [29]. CSCs are heterogeneous populations. In colorectal cancer (CRC) tissues, prominin-1 (CD133) is the first molecular marker used to isolate colorectal cancer stem cells (CRCSCs) [30,31]. The epithelial-specific antigen (ESA)(+)/CD44 molecule and Indian blood group (CD44)(+) CRCSCs are associated with tumor recurrence after chemotherapy [32]. Dipeptidyl peptidase 4 (CD26)(+) CRCSCs enriched from CD133(+)/CD44(+) cells drive tumor metastasis [33]. Leucine-rich repeat-containing G protein-coupled receptor 5 (Lgr5)(+) CRCSCs are considered to be responsible for liver metastasis [34]. CD24 molecule (CD24) and activated leukocyte cell adhesion molecule (CD166) surface antigens are often combined with CD44 or CD133 for the identification and separation of CRCSCs [35]. In HNSCC, CSCs are grouped in accordance with the expression of surface markers such as CD44+ and aldehyde dehydrogenase 1+ (ALDH1+). CD44 mediates the adhesion, migration, and metastasis of CSCs [36], while ALDH1 ameliorate oxidative stress under therapeutic regimens such as platinum, taxanes, and oxazaphosphorine [37,38].

Despite the fact that the origins of CSCs have been linked to genetic mutation, epigenetic alterations, and unrestrained signaling pathways for the normal stem cells and progenitor cells [39,40], CSC properties would be induced or maintained by inflammatory mediators. Inflammatory cytokines and chemokines secreted by CSCs, including interleukin (IL)-1, IL-4, IL-6, and IL-8, sustain CSC niches in an autocrine manner [41–44]. Besides, the expression of IL-8 promotes the migratory and tube-forming capacities of endothelial cells [44]. IL-6 is also involved in cancer metastasis [45]. IL-6 activates Janus kinase 1 (JAK1) and phosphorylates programmed death–ligand 1(PD-L1) and promotes PD-L1 protein stability [46]. CSCs also enhance PD-L1 expression to escape immune surveillance, thereby enriching the CSC subpopulation [47–49]. In addition to secretory proteins, CSCs create an immunosuppressive, pro-tumoral microenvironment by releasing CSC exosomes for cancer progression [50,51].

To target CSCs, researchers have focused on deciphering how cancer cells acquire stemness properties. The major mechanisms involve the expression of genes associated with early development and aberrant intracellular signaling activation. Activation of stemness regulators sustains the stemness properties of HNSCCs, including the MYC proto-oncogene, bHLH transcription factor (MYC), sex-determining region Y-box 2 (SOX2), Nanog homeobox (NANOG), Krüppel-like factor 4 (KLF4), octamer-binding transcription factor 4 (OCT4), high-mobility group AT-hook 2 (HMGA2), cytokines, and epithelial-to-mesenchymal transition (EMT) transcription factors (EMT-TFs) [52–58]. On the other hand, abnormal signaling activation in Notch, Wnt(wingless)/β-catenin, transforming growth factor-β (TGF-β), Janus-activated kinase/signal transducer and activator of transcription (JAK/STAT), nuclear factor-κB (NF-κB), and the sonic hedgehog (SHH) pathway maintains cancer stemness [59–64]. Therefore, the rationale for identifying combinatorial therapeutic strategies combating CSC is intriguing.

3. miRNAs

miRNAs are non-coding (nc) RNA components with approximately 21–23 nucleotides that bind to and repress complementary mRNA targets [21,65]. Previously, ncRNAs were only considered to be evolutionary junk, but emerging evidence has indicated that miRNAs have important cellular functional roles and act as post-translational regulators [21,65–67]. miRNAs control around 30% of human genes, and about half of those genes are tumor associated or situated in vulnerable loci [68–70]; other studies have even suggested that miRNAs can regulate the expression of more than 60% of human genes [71,72]. miRNA expression is modulated by several mechanisms, such as transcriptional control, epigenetic modulation, and post-transcriptional regulation [67,73,74]. On the other hand, the biogenesis of miRNAs can mainly be divided into six steps: (1) RNA polymerase II transcribes miRNA genes into primary (pri)-miRNAs in the nucleus [75], (2) intermediate precursor (pre)-miRNA is released by pri-miRNA after being processed by the Drosha/DiGeorge syndrome critical region 8 (DGCR8) complex [75,76], (3) pre-miRNA bonds to exportin-5 (Exp5)/ras-related nuclear protein (Ran)-guanosine 5′-triphosphate (GTP) complex and is transferred to the cytoplasm [77], (4) the Dicer/(HIV-1 transactivating response (TAR)) RNA-binding protein (TRBP)/PACT complex turns pre-miRNA into double-stranded (ds) RNA in the cytoplasm [78–80], (5) the miRNA duplex is released into single strands by helicase [81], and (6) the miRNA-induced silencing complex (RISC) is bound to the 3′-UTRs of target mRNAs via the seed region of miRNA and subsequently triggers inhibition of the translation or degradation of target mRNAs [82]. The seed region of an miRNA (also known as the seed sequence) is a short, conserved sequence at nucleotides 2–8 at the 5′ end of the miRNA [83–85]. Therefore, the miRNA target prediction tools rely on an algorithm with the thermodynamics-based modeling of RNA, i.e., RNA duplex interactions with comparative sequence analysis to evaluate the seed region matching to the mRNAs [86].

The squamous epithelium covering the oral mucosa and skin depends on epithelial stem cells for tissue renewal [87]. In the oral mucosa, the basal cell layer harbors the self-renewing stem cells and their immediate descendants, the transient amplifying progenitor cells, to produce expanded terminally differentiating cells [88]. The terminally differentiating cells then leave the basal layer and form the outer layers to maintain the oral mucosa integrity. Therefore, stem cells and the proper controls between the phase transition of stem cells and differentiating cells are critical to maintaining tissue homeostasis. Evidence has shown that miRNA expression patterns control the epithelium stem cells' characteristics. For example, Peng et al. indicated that the *miR-103/107* family is highly expressed in the stem-cell-enriched limbal epithelium. The *miR-103/107* family regulates and integrates these stem cell characteristics, thereby sustaining tissue maintenance and regeneration [89]. Moreover, studies have also indicated that the epidermal-specific deletion of enzymes responsible for miRNA maturation, such as DICER, Drosha, and DGCR8, severely impairs the homeostasis and morphogenesis of the epidermis [90–93]. These results indicate that miRNA expression is critical for the proper development of the epidermis and oral mucosa. Moreover, emerging evidence also highlights the importance of miRNAs in regulating carcinogenesis and CSCs. Better characterizations of miRNAs in regulating the stemness features of CSCs will contribute to better cancer treatment strategies (Figure 1).

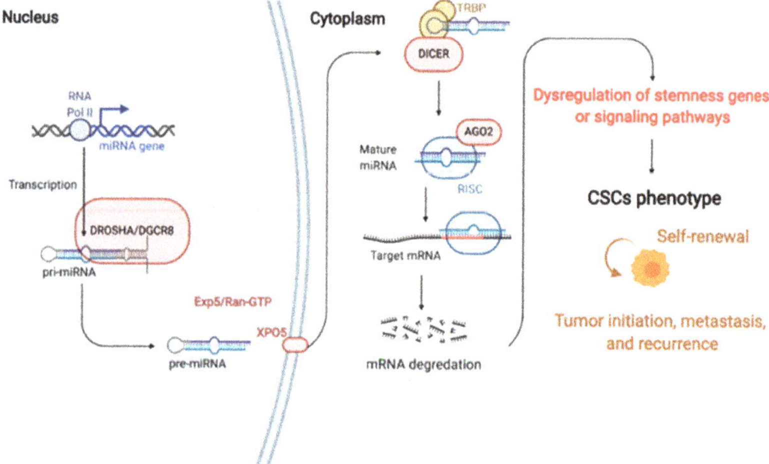

Figure 1. The generations and roles of miRNAs in cancer stem cell (CSC) regulation. In miRNA generation, miRNA genes are commonly transcribed by RNA Pol II in the nucleus. This transcription, which is cleaved by Drosha, produces a multiprotein complex with the DGCR8 protein. Cleavage induces pre-miRNA binding to the nuclear export factor EXO5 and then transported from the nucleus to the cytoplasm. In the cytoplasm, Dicer (another RNase III), which forms a complex with the double-stranded RNA-binding protein TRBP, cuts out the hairpin and produces an RNA duplex with approximately 22 nucleotides. The RNA duplex is dissociated to a single strand via AGO2 mediation, incorporated into the RISC, and binds to 3′UTR of the target mRNA to suppress the gene expression. Therefore, miRNAs can regulate cancer stemness properties and phenotypes by targeting the critical genes that control the activation of signaling pathways, transcriptional factors, and secreted factors. RNA Pol II, RNA polymerase II; pri-miRNAs, primary miRNAs; pre-miRNAs, precursor miRNAs; EXO5, exportin 5; AGO2, argonaute 2; RISC, miRNA-induced silencing complex; mRNA, messenger RNA; miRNA, microRNA (created with Biorender.com (accessed on 30 January 2021)).

4. miRNAs in Regulating Cancer Stemness

CSCs maintain and acquire stemness features through complex mechanisms, including abnormal activation of oncogenes, cytokines, signaling pathways, and EMT-TFs, as mentioned in Section 2 above. Studies indicated that miRNAs that regulate cancer stemness mainly depend on post-translational regulation to modulate activation of those stemness-related factors. Several studies have proven that abnormal miRNA expressions can act as oncogenes, tumor suppressors, or dual-role regulators [94,95]. All of these data highlight the potential for targeting miRNAs to eradicate CSCs, and researchers are working on anti-miRNA drugs and are searching for diagnostic miRNAs [96–98]. miRNAs have been applied as biomarkers to determine cancer prognoses and diagnoses due to their stability [99–101]. Xia et al. indicated that various tumor mutational burden levels had different miRNA expression patterns in HNSCC patients [102], and correlations between miRNA prognostic values as applied to HNSCCs have generated significant interest among researchers [103–105].

4.1. miRNAs as Oncomirs

As oncomirs, miRNAs can act as oncogenic miRNAs that promote biological processes such as proliferation, migration, angiogenesis, invasion, EMT, and stemness [106–111]. Oncomirs regulate cancer stemness through targeting their downstream targets which results in activation of stemness-related factors and signaling pathways. Therefore, oncomirs were shown to enhance tumor initiation and progression by modifying CSC properties such as self-renewal, tumorigenesis, drug resistance, and signaling pathways in cancer [112–115].

Several mechanisms for the oncogenicity of HNSCCs can be affected by miRNA presence. For example, *miR-125a* enhances the proliferation, migration, invasion, and stemness maintenance in cancer cells via suppressing *p53* expression [116]. The overexpression of p53 makes cell viability significantly decrease and induces cell cycle arrest at the G0/G1 phase [116]. *miR-134* suppresses E-cadherin expression and promotes OSCC cell progression through targeting programmed cell death 7 (*PDCD7*) [117]. E-cadherin can suppress cancer stemness by regulating the expressions of pluripotent genes (*MYC, NESTIN, POU5F1*, and *SOX2*) via the activation of Wnt/β-catenin signaling [118]. On the other hand, by suppressing the expression of the WW domain-containing oxidoreductase (*WWOX*) gene, *miR-134* can trigger oncogenicity and metastasis in HNSCCs [119]. WWOX is a tumor stemness suppressor that reduces the self-renewal ability of CSCs, differentiation potential, in vivo tumorigenic capability, and multidrug resistance [120]. Consistently, the downregulation of WWOX was indicated to induce EMT, enhance stemness, and increase chemoresistance in breast cancer [121]. *miR-1246* confers tumorigenicity and affects cancer stemness in OSCC through suppressing cyclin-G2 (*CCNG2*) [122] CCNG2 has been shown to suppress EMT by disrupting Wnt/β-catenin signaling [123], which has been proven to be involved in the migration and invasion of OSCCs [124].

Protocadherins are cell–cell adhesion molecules. The loss of protocadherins may contribute to cancer development not only by altering cell–cell adhesion but also by enhancing proliferation and EMT via activating the Wnt signaling pathway [125,126]. With LSCC, Giefing et al. showed that protocadherin 17 (*PCDH-17*) acts as a tumor suppressor gene [127]. Inhibition of *miR-196b* can suppress cell proliferation, migration, and invasion abilities but promote apoptosis by targeting *PCDH-17* in LSCC cells [128]. Moreover, LSCC patients with low expression of *miR-196b* and high expression of *PCDH-17* were shown to have an increase in the 5-year survival rates [128]. *miR-19a* and *miR-424* inhibit the TGF-β type III receptor (*TGFBR3*), also known as β-glycan, which results in promoting the EMT of tongue squamous carcinoma cells [108]. Other studies have also indicated that *miR-19a* promotes migration and EMT in gastric cancer, CRC, and lung cancer [129–131].

Mitogen-activated protein kinase (MAPK) signaling cascades are critical signal pathways related to EMT, which promotes cancer cell progression and metastasis in CSCs [132]. *miR-106A-5p* facilitates a malignant phenotype by acting as an autophagic suppressor through targeting BTG anti-proliferation factor 3 (*BTG3*) and activates autophagy-regulating

MAPK signaling in NPC [133]. MYC target 1 (*MYCT1*), a direct target gene of MYC, is a novel candidate tumor suppressor gene cloned from LSCC [134,135]. MYCT1 protein suppresses *miR-629-3p* expression by reducing specificity protein 1 (*SP1*) expression. SP1 is also a TF for *miR-629-3p*, and its suppression enhances the expression of *miR-629-3p*'s downstream target, epithelial splicing regulatory protein 2 (*ESRP2*). Taken together, MYCT1 protein suppresses the EMT of laryngeal cancer via the SP1/*miR-629-3p*/ESRP2 pathway [136] Previous studies have shown that oral CSCs switch from expressing the CD44-variant form (CD44v) to expressing the standard form (CD44s) during the acquisition of cisplatin resistance, which results in EMT induction [137] During the process, CD44s induces *miR-629-3p* expression, which inhibits apoptotic cell death under cisplatin treatment conditions and promotes cell migration in HNSCCs [138]. Therefore, *miR-629-3p* serves as a therapeutic target to reverse chemotherapy resistance. Altogether, the miRNAs as oncomirs that regulate the stemness process of HNSCC are summarized in Table 1.

Table 1. miRNAs as oncomirs that are involved in the stemness of head and neck squamous cell carcinomas (HNSCCs).

Oncomirs	Target(s)	Molecular Mechanism	Action Mode	Refs
miR-125a	*p53*	*miR-125a* enhances cell proliferation, migration, invasion, and stemness maintenance through suppression of *p53*.	Gene expression	[116]
miR-134	*PDCD7*	*miR-134* reduces E-cadherin expression by suppressing *PDCD7*. E-cadherin inhibition enhances expressions of pluripotent genes.	Gene expression	[117,118]
miR-134	*WWOX*	*miR-134* suppresses *WWOX*, a tumor stemness suppressor.	Suppressor inhibition	[119,120]
miR-1246	*CCNG2*	*miR-1246* promotes cancer stemness and tumorigenicity by suppressing *CCNG2*.	Suppressor inhibition	[122,123]
miR-196b	*PCDH-17*	*miR-196b* promotes cell proliferation, migration, and invasion abilities by inhibiting *PCDH-17*.	Suppressor inhibition	[127,128]
miR-19a, *miR-424*	*TGFBR3*	*miR-19a* and *miR-424* promote EMT by suppressing *TGFBR3*.	Signal transduction	[108]
miR-106A-5p	*BTG3*	*miR-106A-5p* inhibits autophagy and activates MAPK signaling by targeting *BTG3*.	Signal transduction	[133]
miR-629-3p	*ESRP2*	*miR-629-3p* enhances EMT via targeting *ESRP2*.	EMT process	[136]

EMT, epithelial-to-mesenchymal transition; MAPK, mitogen-activated protein kinase.

4.2. miRNAs as Tumor Suppressors

In contrast, tumor suppressor miRNAs were found to suppress activation of stemness factors, thereby decreasing CSC populations and tumor progression. Studies indicated that expressions of tumor suppressor miRNAs were commonly reduced in tumor samples. Conversely, their corresponding oncogenic downstream targets were activated, thereby activating stemness factors and enhancing the ability of cancer cells to acquire stemness features.

4.2.1. miRNAs in HNSCC

The miRNA *let-7* family controls normal cellular development and differentiation, and a reduction in *let-7* contributes to carcinogenesis via the upregulation of oncogenic downstream targets and stemness properties [99]. Therefore, members of the *let-7* family are considered to be tumor suppressors for various cancers [139]. Ten members of the human *let-7* family have been identified, i.e., *let-7a*, *let-7b*, *let-7c*, *let-7d*, *let-7e*, *let-7f*, *let-7g*, *let-7i*, *miR-98*, and *miR-202*, which share the same seed region sequence [140,141]. Expressions of *let-7* family members decrease in HNSCCs patients, and among them, *let-7i* has been shown to most significantly suppress the expression of the chromatin modifier AT-rich interacting domain 3B (*ARID3B*). By suppressing *let-7i* expression, cells enhance ARID3B expression and acquire stemness features by activating embryonic SC (ESC)-specific genes such as *POU5F1*, *NANOG*, and *SOX2* via histone modifications [142]. The study also indicated that the EMT factor twist family bHLH transcription factor 1 (Twist1) cooperates with B lymphoma Mo-MLV insertion region 1 (BMI1), suppresses *let-7i* expression, and contributes to stem-like properties, thus enabling mesenchymal movements [143]. In OSCC of the tongue, ALDH1+ cells with cancer stemness characteristics show decreased expression of *let-7a* and high expressions of *NANOG* and *POU5F1*. *let-7a* overexpression in ALDH1+ cells further inhibited tumor formation and metastasis in vivo, suggesting that the *let-7a* gene plays an important role in modulating tumorigenesis stemness of HNSCC cells [144].

Moreover, radioresistance poses a major challenge in HNSCC treatment, in which CSCs are relatively radioresistant owing to different intrinsic and extrinsic factors [145]. Evidence has indicated that miRNAs might regulate not only stemness properties but also radiotherapy response. For example, *let-7c* contributes to oral cancer stemness and radio/chemoresistance through suppressing *CXCL8* (IL-8) [146]. Similarly, *CXCL8* was identified as a direct target of *miR-203*, and the reduction in *miR-203* promoted radioresistance by activating IL-8/AKT serine/threonine kinase 1 (AKT) signaling in NPC cells [147]. The low expression of *miR-203* was also showed to enhance EMT and result in intrinsic radioresistance of HNSCC, which could enable identification and treatment modification of radioresistant tumors [148]. *miR-520b* attenuates cell mobility via EMT suppression and suppressed spheroid cell formation, as well as reduced expressions of multiple stemness regulators (Nestin, Twist1, NANOG, OCT4) through targeting suppression of *CD44* in HNSCC cells [149]. Moreover, *miR-520b* also sensitized cells to therapeutic drugs and irradiation through targeting *CD44* [149]. CD44 is an adhesion molecule expressed in CSCs, which interacts with a glutamate–cystine transporter and controls the intracellular level of reduced glutathione (GSH). Therefore, CSCs with high CD44 expression show an enhanced capacity for GSH synthesis, resulting in higher reactive oxygen species (ROS) defense and radiotherapy resistance [150,151]. Therefore, *miR-520b* suppresses *CD44* and not only inhibits cancer stemness and multiple malignant properties but also sensitizes cells to chemoradiotherapy [149].

miR-101 acts as a potent tumor suppressor, and its downregulation is associated with oral carcinomas [152]. In HNSCCs, low expressions of *miR-101* upregulate the oncogene Zeste homolog 2 (*EZH2*), which subsequently downregulates another tumor suppressor gene *rap1GAP*, which promotes HNSCC progression. EZH2 is a histone methyltransferase that belongs to the polycomb repressive complex 2 (PRC2) family that facilitates the trimethylation of H3K27 on the *rap1GAP* promoter to suppress its activation [153,154]. EZH2 can regulate cancer stemness by mediating the NOTCH1 activator and signaling to promote the initiation and growth of SCs [155]. EZH2 was shown to promote cell migration, invasion, and metastasis, and EMT, thereby enhancing cellular plasticity for oral tongue squamous cell carcinomas [156,157]. The *miR-29* family is also significantly downregulated in HNSCC patients [158]. Moreover, *miR-29b* suppresses DNA methyltransferase 3 beta (*DNMT3B*), resulting in inhibited EMT and promoted invasiveness of HNSCC cell lines through restoring E-cadherin expression by the demethylation of the promoter region [159].

miR-204-5p is a tumor suppressor in HNSCCs, which inhibits tumor growth, metastasis, and stemness by suppressing the signal transducer and activator of transcription 3 (STAT3) signaling and EMT via targeting *SNAI2, SUZ12, HDAC1*, and *JAK2* [160]. *STAT3* is a critical regulator of CSCs because of its relationship with EMT as one of the major proposed mechanisms for generating CSCs. It also plays a critical role in the angiogenesis and regulation of the tumor microenvironment (TME), which provides signals for differentiation or proliferation, especially through its involvement in the inflammatory NF-κB pathway [161]. *miR-124* was observed to target *STAT3* to repress tumor growth and metastasis in NPCs [162]. *miR-365-3p* targets the ETS homologous factor (*EHF*), a keratin 16 (KRT16) transcription factor, thereby suppressing KRT16 expression. The decrease in KRT16 further enhances the lysosomal degradation of β5-integrin and c-Met, leading to inhibition of their downstream Src/STAT3 signaling. In OSCC cells, *miR-365-*

3p decreases migration, invasion, metastasis, cancer stemness, and chemoresistance via inhibiting Src/STAT3 signaling [163].

4.2.2. miRNAs in OSCC

In OSCC, *let-7d* was shown to function as a negative regulator of EMT and exhibited chemoresistant properties and silencing of enhanced mesenchymal, stem-like, and chemoresistant traits through suppressing *TWIST1* and Snail family transcriptional repressor 1 (*SNAI1*) expression [164]. *miR-98* acts as a tumor suppressor, which reduces tumor cell growth and metastasis through targeting the insulin-like growth factor 1 receptor (*IGF1R*) in OSCCs [165]. IGF1R is critical in the human embryonic niche for self-renewal and SC expansion and regulates SC maintenance in normal tissue processes [166,167]. Moreover, the IGF1R pathway is critical for EMT induction/maintenance and the expansion of cancer stem-like cells [167–170]. In HNSCCs, Leong et al. indicated increased epidermal growth factor (EGF) receptor (EGFR) and IGF1R expressions and phosphorylation, which increased the activation of downstream pathways in ALDH1+ cells compared to ALDH- cells. Importantly, treatment with EGFR and IGF1R inhibitors reduced SC fractions, implying that the IGF1R is critical for maintaining HNSCC CSCs [171].

miR-139-5p overexpression inhibits OSCC cell proliferation, in vitro mobility of OSCC, and the expression of WNT-responsive *MYC*, *CCND1*, and *BCL2* through suppressing CXC chemokine receptor 4 (*CXCR4*) [172]. MYC-related signaling regulates CSC chemotherapeutic resistance and CRC organoids [173]. In addition, *miR-139* triggers the apoptosis of an oral cancer cell line, Tca8113 cells, through the Akt signaling pathway [174]. Another study suggested that *miR-139-5p* suppresses the tumorigenesis process and OSCC cell mobility by targeting homeobox (HOX)-A9 (*HOXA9*) [175]. HOX genes can encode master regulatory TFs that regulate SCs during development in various cancers; *HOX4* and *HOXA9* were observed to upregulate expression of the SC marker ALDH1 and increase SC self-renewal [176]. Similarly, *miR-495* was observed to suppress EMT, proliferation, migration, and invasion and promote the apoptosis of CSCs by inhibiting the *HOXC6*-mediated TGF-β signaling pathway in OSCCs [177]. Other studies have also indicated that *miR-495* significantly inhibits cell proliferation, migration, invasion, and EMT through the *miR-495*/IGF1/AKT signaling axis or by targeting *NOTCH1* in OSCCs [178,179].

The *miR-34* family contains three members, *miR-34a*, *miR-34b*, and *miR-34c*, clustered on two different chromosomal loci on chromosomes 1p36.22 (*Mir34a*) and 11q23.1 (*Mir34b/c*) [180,181]. In OSCCs and OPSCCs, *miR-34a* is described as a regulator of SCs [182]. *miR-34a* was observed to be downregulated in HNSCC tumors and cell lines [183]. Sun et al. observed that CSC enrichment by a spheroid culture showed significant downregulation of *miR-34a* expression. Furthermore, the restoration of *miR-34a* significantly inhibited EMT formation of the CSC phenotype and functionally reduced clonogenic and invasive capacities in HNSCC cell lines [184]. During the EMT process, cancer cells acquire the ability for tumor metastasis, invasion, drug resistance, and recurrence, which are associated with CSC functions. Gregory et al. first indicated that *miR-205* and the *miR-200* family (*miR-200a*, *miR-200b*, *miR-200c*, *miR-141*, and *miR-429*) suppressed the EMT by targeting zinc finger E-box binding homeobox 1 (*ZEB1*) and Smad-interacting protein 1 (*SIP1*, also known as *ZEB2*) in breast cancer [185]. Similarly, the *miR-200* family was indicated to enhance EMT through a reciprocal feedback loop between the *miR-200* family and ZEB1 in HNSCCs [186,187]. Recent emerging evidence has indicated that the EMT process might not simply be divided into a dichotomous system but may actually be an EMT spectrum. The epithelial/mesenchymal (E/M) hybrid status provides plasticity for cells with mixed E and M characteristics [188]. Lu et al. devised a unique model of miRNA-based coupled chimeric modules to elucidate the core regulatory network that underlies the hybrid E/M status. In that model system, two double-negative feedback loops of *miR-34*/SNAI1 and the *miR-200*/ZEB mutually regulate the E and M phenotypes and the hybrid phenotype. *miR-200*/ZEB was indicated to act as the decision-making module for cancer cells to undergo partial or complete EMT [189,190].

miR-22 inhibits phosphatidylinositol 3-kinase (PI3K)/Akt/NF-κB signaling via downregulating activators such as *S100A8*, platelet-derived growth factor (*PDGF*), and vascular endothelial growth factor (*VEGF*), which implies a tumor suppressor role of *miR-22* in tongue squamous cell carcinoma [191]. PI3K is well known as a regulator for stemness-related signaling, including RAS/mitogen-activated protein kinase (MAPK) [192,193], NF-κB [194,195], Wnt/β-catenin [196,197], and TGF-β [198–202]. The NF-κB pathway maintains stemness by regulating many tumor-promoting inflammation-related cytokines, like tumor necrosis factor (TNF)-α [203], IL-1 [204], IL-6 [205,206], monocyte chemoattractant protein 1 (MCP1) [207], cytochrome oxidase subunit 2 (COX2) [203], and inducible nitric oxide synthase (iNOS) [203,208]. Simultaneously, the NF-κB pathway downregulates the expression of matrix metalloproteinases (MMPs) to increase tumor cell invasion [209]. *miR-22* also targets the expression of node-like receptor (NLR) family pyrin domain-containing 3 (*NLRP3*) and suppresses OSCC cell growth, migration, and invasion [210]. The NLRP3 inflammasome was associated with the carcinogenesis and CSC self-renewal activation in HNSCC patients with upregulated expression of BMI1, ALDH1, and CD44 [211]. The overexpression of *miR-22* results in reduced cell viability and an increase in the OSCC cell apoptotic rate by targeting the Wnt/β-catenin signaling pathway [212]. Qiu et al. indicated that the downregulation of *miR-22* would result in the upregulation of CD147 in tongue squamous cell carcinomas [213]. CD147 is also known as an extracellular MMP inducer, which promotes tumor initiation and progression through NF-κB signaling and also mediates the TGF-β1-induced EMT in HNSCC cells [214,215]. Therefore, CD147 might be a potential prognostic and treatment biomarker for HNSCCs.

4.2.3. miRNAs in LSCC

In LSCC, *miR-98* was shown significantly reduced in both clinical specimens and cell lines, and *miR-98* directly targeted HMGA2-POSTN signaling and then suppressed cell migration, metastasis, invasion, and EMT-TFs of SNAI1 and Twist1, as well as SC-like features [216]. Moreover, *miR-101* inhibited tumorigenesis progression by regulating the Wnt/β-catenin signaling pathway by directly targeting cyclin-dependent kinase 8 (*CDK8*) in LSCC [217]. *CDK8* plays an important role in regulating biological processes at the transcription level in the Wnt/β-catenin signaling pathway, and it is considered a CRC oncogene [218,219].

4.2.4. miRNAs in NPC

miR-139-5p inhibits the proliferation, invasion, and migration of human NPC cells by modulating EMT [220]. EMT enhances cancer cell motility and dissemination, which led to the concept of migrating CSCs as the basis of metastasis [221]. Findings have demonstrated a direct molecular link between EMT and stemness, where EMT activators such as Twist1 can co-induce EMT and stemness properties, thereby linking the EMT and CSC concepts [188]. EMT plays an important role in tumor metastasis and recurrence, and thus it is closely related to CSC functions [222,223]. Moreover, *miR-139-5p* reduces cisplatin resistance in NPC cells [220].

miR-488-3p activates the p53 pathway through suppressing zinc finger and BTB domain-containing protein 2 (*ZBTB2*), a reader of unmethylated DNA that regulates embryonic stem cell differentiation, thereby inhibiting proliferation and inducing apoptosis in esophageal SCCs [224,225]. p53 is able to suppress *CD44*, which is a CSC marker and suppresses cellular plasticity [226]. In NPCs, *miR-372* promotes radiosensitivity by activating the p53 signaling pathway via the inhibition of PDZ-binding kinase (*PBK*) [227]. Moreover, p53 represses EMT by mediating *miR-200c* expression, which causes the inhibition the translation of *ZEB1* and *BMI1* [228]. By downregulating *ATF3* expression, *miRNA-488* suppresses cell invasion and EMT in tongue squamous cell carcinoma cells [229]. Taken together, miRNAs as tumor suppressors that regulate the stemness process are summarized in Table 2.

Table 2. miRNAs as tumor suppressors that are involved in the stemness of head and neck squamous cell carcinomas (HNSCCs).

Suppressors miRNAs	Target(s)	Molecular Mechanism	Action Mode	Refs
		HNSCC		
let-7i	*ARID3B*	*let-7i* inhibition enhances *ARID3B* expression and activates the expression of *POU5F1, NANOG*, and *SOX2*.	Gene expression	[142]
let-7a	*NANOG, POU5F1*	Upregulation of *let-7a* in ALDH1$^+$ cell suppresses tumor formation and metastasis.	Gene expression	[144]
let-7c	*CXCL8*	*let-7c* inhibition enhances stemness and radio-/chemoresistance by suppressing *CXCL8*.	Signal transduction	[146]
miR-203	*CXCL8*	*miR-203* reduction promotes EMT and activates IL-8/AKT signaling to trigger radioresistance.	EMT process/ Signal transduction	[147]
miR-520b	*CD44*	*miR-520b* inhibits EMT and the expression of stemness regulators and sensitizes cells to chemoradiotherapy through suppression of *CD44*.	EMT process/ Gene expression	[149]
miR-101	*EZH2*	*miR-101* inhibits *EZH2* and suppresses metastasis and EMT.	EMT process/ Signal transduction	[153,156,157]
miR-101	*CDK8*	*miR-101* inhibits *CDK8* expression and subsequently suppresses Wnt/β-catenin signaling and tumorigenesis.	Signal transduction	[217]
miR-29b	*DNMT3B*	*miR-29* suppresses *DNMT3B*, resulting in inhibition of EMT.	EMT process	[159]
miR-204-5p	*SNAI2, SUZ12, HDAC1*, and *JAK2*	*miR-204-5p* inhibits stemness by suppressing STAT3 signaling and EMT via targeting *SNAI2, SUZ12, HDAC1*, and *JAK2*.	EMT process/ Signal transduction	[160]
miR-124	*STAT3*	*miR-124* inhibits tumor growth and metastasis by suppressing *STAT3*.	Signal transduction	[162]
miR-365-3p	*EHF*	*miR-365-3p* decreases metastasis, stemness, and chemoresistance by suppressing EHF, which inhibits Src/STAT3 signaling.	Signal transduction	[163]
		OSCC		
let-7d	*TWIST1, SNAI1*	*let-7d* suppresses EMT.	EMT process	[164]
miR-98	*IGF1R*	*miR-98* reduces self-renewal by suppressing *IGF1R*.	Signal transduction	[165]
miR-139-5p	*CXCR4*	*miR-139-5p* inhibits cell proliferation and the expression of WNT-responsive *MYC, CCND1*, and *BCL2* via inhibiting *CXCR4*.	Signal transduction	[172]
miR-139-5p	*HOXA9*	*miR-139-5p* inhibits *HOXA9*. *HOXA9* can increase stem cell self-renewal.	Gene expression/ Signal transduction	[175,176]
miR-495	HOXC6-mediated TGF-β signaling pathway	*miR-495* inhibits the HOXC6-mediated TGF-β signaling pathway and then suppresses EMT, proliferation, migration, and invasion.	Signal transduction	[177]
miR-495	IGF1/AKT signaling axis and *NOTCH1*	*miR-495* inhibits cell proliferation migration, invasion, and EMT through targeting the IGF1/AKT signaling axis or *NOTCH1*.	Signal transduction	[178,179]
miR-34a	EMT	*miR-34a* inhibits EMT formation.	EMT process	[184]
miR-200 family	*ZEB1/2*	The *miR-200* family suppresses EMT by targeting *ZEB1/2*.	EMT process	[186,187]

Table 2. *Cont.*

Suppressors miRNAs	Target(s)	Molecular Mechanism	Action Mode	Refs
miR-22	KAT6B	*miR-22* inhibits activators of PI3K/Akt/NF-κB signaling.	Signal transduction	[191]
miR-22	*NLRP3*	*miR-22* inhibits *NLRP3* which suppresses expressions of BMI1, ALDH1, and CD44.	Signal transduction	[210,211]
miR-22	*CD147*	*miR-22* inhibits *CD147*, which suppresses tumor initiation and progression through NF-κB signaling and mediates TGF-β1-induced EMT.	EMT process/ Signal transduction	[213–215]
		LSCC		
miR-98	HMGA2-POSTN signaling	*miR-98* inhibits HMGA2-POSTN signaling, which suppresses metastasis and EMT-TFs.	EMT process/ Signal transduction	[216]
miR-101	*CDK8*	*miR-101* inhibits *CDK8* expression, which suppresses Wnt/β-catenin signaling and tumorigenesis.	Signal transduction	[217]
		NPC		
miR-139-5p	EMT	*miR-139-5p* suppresses EMT and inhibits proliferation, invasion, migration, and cisplatin resistance.	EMT process	[220]
miR-488-3p	*ZBTB2*	*miR-488-3p* activates the p53 pathway by suppressing *ZBTB2* to inhibit proliferation and induce apoptosis.	Signal transduction	[224,225]
miR-372	*PBK*	*miR-372* activates the p53 signaling pathway via repressing *PBK* to promotes radiosensitivity.	Signal transduction	[227]

EMT, epithelial-to-mesenchymal transition; TFs, transcription factors; OSCC, oral squamous cell carcinoma; LSCC, laryngeal squamous cell carcinoma; NPC, nasopharyngeal carcinoma; IL-8, interleukin 8; EHF, ETS homologous factor; STAT3, signal transducer and activator of transcription 3; IGF1, insulin-like growth factor 1; AKT, AKT serine/threonine kinase 1; TGF-β, transforming growth factor-β; HMGA2, high-mobility group AT-hook 2; NF-κB, nuclear factor-κB.

4.3. miRNAs as Pleiotropic Functions

Some miRNAs play dual roles in oncogenes and tumor suppression, depending on the specific cell/tissue context. This reflects the complexity of the miRNA–target regulatory network. For example, *miR-107* was observed to antagonize and degrade *let-7*. *miR-107* suppressed *let-7* expression and activated downstream oncoprotein expressions such as HMGA2 and RAS and enhanced the tumorigenic and metastatic potential of cancer cells [230,231]. In HNSCCs, a *miR-107* increment was found in patients with lymph node metastasis, suggesting an oncogenic role for *miR-107* [232]; however, *miR-107* was indicated to suppress the proliferation, invasion, and colony formation of cells in LSCCs via inhibiting the voltage-gated calcium channel subunit α2δ1 (α2δ1) (encoded by *CACNA2D1*) [233]. In non-small-cell lung cancer (NSCLC), α2δ1 also enhances radioresistance in cancer stem-like cells by enhancing the efficiency of DNA damage repair [234]. Those results indicate the pleiotropic functions of *miR107* in HNSCCs.

miRNAs also mediated the regulation of cytokines/chemokines and the TME that modulates the CSC signaling pathway and sustains the CSC niche for acquiring and maintaining CSC features. For example, downregulation of *miR-9*, *miR-542-3p*, and *miR-34a*, and significant upregulation of *miR-21* were shown in CD44-positive CSCs with increased IL-6 and IL-8 expressions via targeting of the CD44v6/NANOG/phosphatase and tensin homolog (PTEN) axis in oral cancer [235]. *miR-9* acts as a tumor suppressor microRNA in HNSCC, and its role seems to be mediated through CXCR4 suppression [236]. Studies have indicated that *miR-9* overexpression results in decreased cellular proliferation and inhibited colony formation in soft agars when targeting *CXCR4* in HNSCC cells [236]. Conversely, another study has also indicated that *miR-9* was expressed at high levels in patients with recurrent HNSCCs [237]. Similar results were also shown for breast cancer with *miR-9* acting as a tumor suppressor in breast cancer proliferation in the early stage of breast cancer, while, with a higher malignancy, *miR-9* plays an opposite role in the metastatic process [238]. Thus, *miR-9* was suggested to have dual roles in carcinogenesis. Taken together, miRNAs with pleiotropic functions in HNSCCs are summarized in Table 3.

Table 3. miRNAs with pleiotropic functions that are involved in the stemness of head and neck squamous cell carcinomas (HNSCCs).

Pleiotropic miRNAs	Target(s)	Molecular Mechanism	Action Mode	Refs
miR-107	*let-7*	*miR-107* suppresses *let-7* expression and activates downstream oncoprotein expressions for enhancing tumorigenic and metastasis.	Gene expression/ Signal transduction	[230,231]
miR-107	*CACNA2D1*	*miR-107* suppresses proliferation, invasion, and colony formation of LSCC cells via inhibiting *CACNA2D1*.	Signal transduction	[233]
miR-9	CD44v6/NANOG/PTEN axis	*miR-9* inhibits the CD44v6/NANOG/PTEN axis for suppressing IL-6 and IL-8 signaling.	Signal transduction	[235]
miR-9	*CXCR4*	*miR-9* decreases proliferation and colony formation by targeting *CXCR4*.	Signal transduction	[236]

LSCC, laryngeal squamous cell carcinoma; NANOG, Nanog homeobox; PTEN, phosphatase and tensin homolog.

5. Conclusions

miRNAs can function as cancer suppressors or oncogenes, or even exhibit dual roles during cancer development, depending on the different cancer types or tumorigenesis stage. miRNAs are critical to tumor initiation, progression, metastasis, EMT, and chemoresistance via regulating CSC functions. miRNAs regulate important EMT-TFs and signaling pathways and modulate the TME to sustain and enhance cancer stemness. Therefore, targeting CSCs through miRNA manipulation provides a therapeutic opportunity for managing metastatic diseases. Moreover, with an understanding of miRNAs during tumorigenesis, we can take advantage of miRNA stability and use it as a diagnostic marker for primary diagnoses and patient follow-ups. We can also monitor miRNA changes to predict therapeutic responses as a non-invasive detection method. Recent studies have indicated that exosomal miRNAs can be better sources of biomarkers due to their advantages in terms of their quantity, quality, and stability [239]. Ludwig et al. indicated that *miR-205-5p* was exclusively expressed in HPV(+) exosomes, whereas *miR-1972* was only detected in HPV(−) exosomes. These miRNAs emerge as potential discriminating HPV-associated biomarkers [240] Intriguingly, human papillomavirus 16 (HPV16) infection has been indicated to enhance CSC properties, including ALDH1 activity, migration/invasion, and CSC-related factor expression, and enhances tumor growth OSCC cells [241]. Whether tumor-derived exosomes (TEX)-miRNAs are also involved in regulating the recipient cell stemness is unclear. In contrast to the extensive studies for cellular miRNAs in regulating cancer stemness, TEX-miRNA knowledge is relatively limited.

Moreover, Huang et al. indicated that only 5.63% of miRNAs were detected in both cells and TEX, which implies that cells can selectively pack certain miRNAs into exosomes in OSCC cells [242]. Meanwhile, exosomes can be released by various cell types, such as cancer-associated fibroblasts (CAFs) [243], dendritic cells [244], B cells [245], T cells [246], and tumor cells [247]. For example, Li et al. indicated that *miR-34a-5p* was significantly reduced in CAF-derived exosomes in OSCC patients. CAF transfers *miR-34a-5p*-devoid exosomes to OSCC cells and results in promoting the proliferation and motility of OSCC cells by upregulating the downstream target AXL (encoding AXL receptor tyrosine kinase). Therefore, the *miR-34a-5p*/AXL axis promotes the proliferation, metastasis, and EMT of oral cancer cells through the AKT/glycogen synthase kinase (GSK)-3β (GSK-3β)/β-catenin/SNAI1 signaling cascade [248]. Consistently, the cellular *miR-34a* significantly inhibited EMT formation of the CSC phenotype in HNSCC cell lines [184]. Hence, the sources and biological functions of exosomal miRNAs warrant further research before using them for screening and surveillance.

Among the tumor suppressor microRNAs of HNSCCs, *miR-34* is the only one that has been used in a clinical trial applied to treat primary liver cancer, small-cell lung carcinomas, lymphomas, multiple myelomas, and renal cell carcinomas. In 2013, the first microRNA-associated therapeutic drug was tested in a clinical trial (NCT01829971), MRX34, a special amphoteric lipid nanoparticle filled with *miR-34* mimics. Although this phase I study provided a dose-dependent modulation of relevant target genes that provide a proof-of-concept for MRX34 application for cancer therapy, severe adverse events were reported in five patients in terms of experiencing serious immune responses [249,250]. Hence, leading up to the MRX34 phase 2 clinical trials, NCT02862145 for melanomas has been withdrawn [251]. Other clinical trials have mainly focused on observational studies to explore the prognostic value of miRNAs in HNSCCs, for example, the prognostic value of *miR-29b* in the tissue, blood, and saliva in oral squamous cell carcinomas (NCT02009852). The *miR-29* family has been used to investigate Twist1-mediated cancer metastasis in HNSCCs (NCT01927354) (Table 4). Further research is warranted to determine the molecular functions and mechanisms of cellular or exosomal miRNAs, as well as their potential as miRNA-based diagnostics and therapeutics for HNSCCs.

Table 4. List of miRNA-related clinical trials in head and neck squamous cell carcinomas (HNSCCs).

miRNAs	Clinical Trials	Trial ID
miR-29b	Observational study to explore the prognostic value of *miR-29b* in tissue, blood, and saliva in OSCC	NCT02009852
miR-29 family	Observational study to investigate the role of *miR-29* family in Twist1-mediated cancer metastasis in HNSCC	NCT01927354

OSCC, oral squamous cell carcinoma.

Author Contributions: M.F., W.-L.H., and T.-T.L. wrote and revised the paper with assistance from P.-Y.C. and T.-Y.H. All authors have read and agreed to the published version of the manuscript.

Funding: This work was supported by the Yen Tjing Ling Medical Foundation in Taiwan and the Cancer Progression Research Center, National Yang Ming Chiao Tung University, from the Featured Areas Research Center Program within the framework of the Higher Education Sprout Project by the Ministry of Education (MOE). We appreciate the funding from the Ministry of Science and Technology (109-2320-B-010-021 to W.-L.H.; MOST 109-2636-B-038-001 and MOST 110-2636-B-038-005 to T.-T.L.) and Taipei Medical University (TMU108-AE1-B25 to T.-T.L), the Yen Tjing Ling Medical Foundation (CI-109-11 to W.-L.H.), and the Ministry of Health and Welfare, Center of Excellence for Cancer Research (MOHW107-TDU-B-211-114019, 109 CRC-T208, and 110 CRC-T208 to W.-L.H.).

Institutional Review Board Statement: Not applicable.

Informed Consent Statement: Not applicable.

Conflicts of Interest: The authors declare no conflict of interest.

References

1. WHO. *WHO Report on Cancer: Setting Priorities, Investing Wisely and Providing Care for All*; World Health Organization: Geneva, Switzerland, 2020.
2. Bray, F.; Ferlay, J.; Soerjomataram, I.; Siegel, R.L.; Torre, L.A.; Jemal, A. Global cancer statistics 2018: GLOBOCAN estimates of incidence and mortality worldwide for 36 cancers in 185 countries. *CA Cancer J. Clin.* **2018**, *68*, 394–424. [CrossRef]
3. Simon, S. *Cancer Facts & Figures 2019*; American Cancer Society: Atlanta, GA, USA, 2019; pp. 1–71.

4. Dale, O.T.; Pring, M.; Davies, A.; Leary, S.; Ingarfield, K.; Toms, S.; Waterboer, T.; Pawlita, M.; Ness, A.R.; Thomas, S.J. Squamous cell carcinoma of the nasal cavity: A descriptive analysis of cases from the head and neck 5000 study. *Clin. Otolaryngol.* **2019**, *44*, 961–967. [CrossRef]
5. Siegel, R.L.; Miller, K.D.; Jemal, A. Cancer statistics, 2019. *CA Cancer J. Clin.* **2019**, *69*, 7–34. [CrossRef] [PubMed]
6. Argirion, I.; Zarins, K.R.; Defever, K.; Suwanrungruang, K.; Chang, J.T.; Pongnikorn, D.; Chitapanarux, I.; Sriplung, H.; Vatanasapt, P.; Rozek, L.S. Temporal Changes in Head and Neck Cancer Incidence in Thailand Suggest Changing Oropharyngeal Epidemiology in the Region. *J. Glob. Oncol.* **2019**, *5*, 1–11. [CrossRef] [PubMed]
7. Chi, A.C.; Day, T.A.; Neville, B.W. Oral cavity and oropharyngeal squamous cell carcinoma-an update. *CA Cancer J. Clin.* **2015**, *65*, 401–421. [CrossRef] [PubMed]
8. Ferlay, J.; Colombet, M.; Soerjomataram, I.; Mathers, C.; Parkin, D.M.; Pineros, M.; Znaor, A.; Bray, F. Estimating the global cancer incidence and mortality in 2018: GLOBOCAN sources and methods. *Int. J. Cancer* **2019**, *144*, 1941–1953. [CrossRef] [PubMed]
9. Johnson, D.E.; Burtness, B.; Leemans, C.R.; Lui, V.W.Y.; Bauman, J.E.; Grandis, J.R. Head and neck squamous cell carcinoma. *Nat. Rev. Dis. Primers* **2020**, *6*, 1–22. [CrossRef]
10. Sethi, S.; Lu, M.; Kapke, A.; Benninger, M.S.; Worsham, M.J. Patient and tumor factors at diagnosis in a multi-ethnic primary head and neck squamous cell carcinoma cohort. *J. Surg. Oncol.* **2008**, *99*, 104–108. [CrossRef]
11. Machiels, J.-P.; Leemans, C.R.; Golusinski, W.; Grau, C.; Licitra, L.; Gregoire, V. Squamous cell carcinoma of the oral cavity, larynx, oropharynx and hypopharynx: EHNS–ESMO–ESTRO Clinical Practice Guidelines for diagnosis, treatment and follow-up. *Ann. Oncol.* **2020**, *31*, 1462–1475. [CrossRef] [PubMed]
12. Lu, L.; Wu, Y.; Feng, M.; Xue, X.; Fan, Y. A novel seven-miRNA prognostic model to predict overall survival in head and neck squamous cell carcinoma patients. *Mol. Med. Rep.* **2019**, *20*, 4340–4348. [CrossRef]
13. Cohen, E.E.W.; Bell, R.B.; Bifulco, C.B.; Burtness, B.; Gillison, M.L.; Harrington, K.J.; Le, Q.-T.; Lee, N.Y.; Leidner, R.; Lewis, R.L.; et al. The Society for Immunotherapy of Cancer consensus statement on immunotherapy for the treatment of squamous cell carcinoma of the head and neck (HNSCC). *J. Immunother. Cancer* **2019**, *7*, 184. [CrossRef] [PubMed]
14. Lv, X.; Zheng, X.; Yu, J.; Ma, H.; Hua, C.; Gao, R. EGFR enhances the stemness and progression of oral cancer through inhibiting autophagic degradation of SOX2. *Cancer Med.* **2019**, *9*, 1131–1140. [CrossRef] [PubMed]
15. Lee, Y.; Shin, J.H.; Longmire, M.; Wang, H.; Kohrt, H.E.; Chang, H.Y.; Sunwoo, J.B. CD44+ Cells in Head and Neck Squamous Cell Carcinoma Suppress T-Cell–Mediated Immunity by Selective Constitutive and Inducible Expression of PD-L1. *Clin. Cancer Res.* **2016**, *22*, 3571–3581. [CrossRef]
16. Chen, D.; Wu, M.; Li, Y.; Chang, I.; Yuan, Q.; Ekimyan-Salvo, M.; Deng, P.; Yu, B.; Yu, Y.; Dong, J.; et al. Targeting BMI1 + Cancer Stem Cells Overcomes Chemoresistance and Inhibits Metastases in Squamous Cell Carcinoma. *Cell Stem Cell* **2017**, *20*, 621–634.e6. [CrossRef]
17. Giudice, F.S.; Jr, D.S.P.; Nör, J.E.; Squarize, C.H.; Castilho, R.M. Inhibition of Histone Deacetylase Impacts Cancer Stem Cells and Induces Epithelial-Mesenchyme Transition of Head and Neck Cancer. *PLoS ONE* **2013**, *8*, e58672. [CrossRef]
18. Stransky, N.; Egloff, A.M.; Tward, A.D.; Kostic, A.D.; Cibulskis, K.; Sivachenko, A.; Kryukov, G.V.; Lawrence, M.S.; Sougnez, C.; McKenna, A.; et al. The Mutational Landscape of Head and Neck Squamous Cell Carcinoma. *Science* **2011**, *333*, 1157–1160. [CrossRef] [PubMed]
19. Huang, Y.; Liu, Z.; Zhong, L.; Wen, Y.; Ye, Q.; Cao, D.; Li, P.; Liu, Y. Construction of an 11-microRNA-based signature and a prognostic nomogram to predict the overall survival of head and neck squamous cell carcinoma patients. *BMC Genom.* **2020**, *21*, 1–11. [CrossRef] [PubMed]
20. Yata, K.; Beder, L.B.; Tamagawa, S.; Hotomi, M.; Hirohashi, Y.; Grenman, R.; Yamanaka, N. MicroRNA expression profiles of cancer stem cells in head and neck squamous cell carcinoma. *Int. J. Oncol.* **2015**, *47*, 1249–1256. [CrossRef]
21. Baek, D.; Villén, J.; Shin, C.; Camargo, F.D.; Gygi, S.P.; Bartel, D.P. The impact of microRNAs on protein output. *Nat. Cell Biol.* **2008**, *455*, 64–71. [CrossRef]
22. Ishizawa, K.; Rasheed, Z.A.; Karisch, R.; Wang, Q.; Kowalski, J.; Susky, E.; Pereira, K.; Karamboulas, C.; Moghal, N.; Rajeshkumar, N.; et al. Tumor-Initiating Cells Are Rare in Many Human Tumors. *Cell Stem Cell* **2010**, *7*, 279–282. [CrossRef]
23. O'Brien, C.A.; Kreso, A.; Jamieson, C.H. Cancer Stem Cells and Self-renewal. *Clin. Cancer Res.* **2010**, *16*, 3113–3120. [CrossRef]
24. Chen, R.; Nishimura, M.C.; Bumbaca, S.M.; Kharbanda, S.; Forrest, W.F.; Kasman, I.M.; Greve, J.M.; Soriano, R.H.; Gilmour, L.L.; Rivers, C.S.; et al. A Hierarchy of Self-Renewing Tumor-Initiating Cell Types in Glioblastoma. *Cancer Cell* **2010**, *17*, 362–375. [CrossRef] [PubMed]
25. Yoo, Y.D.; Kwon, Y.T. Molecular mechanisms controlling asymmetric and symmetric self-renewal of cancer stem cells. *J. Anal. Sci. Technol.* **2015**, *6*, 1–6. [CrossRef] [PubMed]
26. Chen, W.; Dong, J.; Haiech, J.; Kilhoffer, M.-C.; Zeniou, M. Cancer Stem Cell Quiescence and Plasticity as Major Challenges in Cancer Therapy. *Stem Cells Int.* **2016**, *2016*, 1–16. [CrossRef]
27. Cheung, T.H.T.; Rando, T.A. Molecular regulation of stem cell quiescence. *Nat. Rev. Mol. Cell Biol.* **2013**, *14*, 329–340. [CrossRef]
28. Yang, L.; Shi, P.; Zhao, G.; Xu, J.; Peng, W.; Zhang, J.; Zhang, G.; Wang, X.; Dong, Z.; Chen, F.; et al. Targeting cancer stem cell pathways for cancer therapy. *Signal Transduct. Target. Ther.* **2020**, *5*, 1–35. [CrossRef] [PubMed]
29. Merlos-Suárez, A.; Barriga, F.M.; Jung, P.; Iglesias, M.; Céspedes, M.V.; Rossell, D.; Sevillano, M.; Hernando-Momblona, X.; da Silva-Diz, V.; Muñoz, P.; et al. The Intestinal Stem Cell Signature Identifies Colorectal Cancer Stem Cells and Predicts Disease Relapse. *Cell Stem Cell* **2011**, *8*, 511–524. [CrossRef] [PubMed]

30. O'Brien, C.A.; Pollett, A.; Gallinger, S.; Dick, J.E. A human colon cancer cell capable of initiating tumour growth in immunodeficient mice. *Nat. Cell Biol.* **2006**, *445*, 106–110. [CrossRef]
31. Ricci-Vitiani, L.; Lombardi, D.G.; Pilozzi, E.; Biffoni, M.; Todaro, M.; Peschle, C.; De Maria, R. Identification and expansion of human colon-cancer-initiating cells. *Nature* **2007**, *445*, 111–115. [CrossRef]
32. Dylla, S.J.; Beviglia, L.; Park, I.-K.; Chartier, C.; Raval, J.; Ngan, L.; Pickell, K.; Aguilar, J.; Lazetic, S.; Smith-Berdan, S.; et al. Colorectal Cancer Stem Cells Are Enriched in Xenogeneic Tumors Following Chemotherapy. *PLoS ONE* **2008**, *3*, e2428. [CrossRef]
33. Pang, R.; Law, W.L.; Chu, A.C.; Poon, J.T.; Lam, C.S.; Chow, A.K.; Ng, L.; Cheung, L.W.; Lan, X.R.; Lan, H.Y.; et al. A Subpopulation of CD26+ Cancer Stem Cells with Metastatic Capacity in Human Colorectal Cancer. *Cell Stem Cell* **2010**, *6*, 603–615. [CrossRef]
34. De Sousa e Melo, F.; Kurtova, A.V.; Harnoss, J.M.; Kljavin, N.; Hoeck, J.D.; Hung, J.; Anderson, J.E.; Storm, E.E.; Modrusan, Z.; Koeppen, H.; et al. A distinct role for Lgr5+ stem cells in primary and metastatic colon cancer. *Nature* **2017**, *543*, 676–680. [CrossRef] [PubMed]
35. Todaro, M.; Francipane, M.G.; Medema, J.P.; Stassi, G. Colon Cancer Stem Cells: Promise of Targeted Therapy. *Gastroenterology* **2010**, *138*, 2151–2162. [CrossRef] [PubMed]
36. Allegra, E.; Trapasso, S. Role of CD44 as a marker of cancer stem cells in head and neck cancer. *Biol. Targets Ther.* **2012**, *6*, 379–383. [CrossRef]
37. Landen, C.N.; Goodman, B.; Katre, A.A.; Steg, A.D.; Nick, A.M.; Stone, R.L.; Miller, L.D.; Mejia, P.V.; Jennings, N.B.; Gershenson, D.M.; et al. Targeting Aldehyde Dehydrogenase Cancer Stem Cells in Ovarian Cancer. *Mol. Cancer Ther.* **2010**, *9*, 3186–3199. [CrossRef] [PubMed]
38. Singh, S.; Brocker, C.; Koppaka, V.; Chen, Y.; Jackson, B.C.; Matsumoto, A.; Thompson, D.C.; Vasiliou, V. Aldehyde dehydrogenases in cellular responses to oxidative/electrophilicstress. *Free Radic. Biol. Med.* **2013**, *56*, 89–101. [CrossRef] [PubMed]
39. Pardal, R.; Clarke, M.F.; Morrison, S.J. Applying the principles of stem-cell biology to cancer. *Nat. Rev. Cancer* **2003**, *3*, 895–902. [CrossRef]
40. Reya, T.; Morrison, S.J.; Clarke, M.F.; Weissman, I.L. Stem cells, cancer, and cancer stem cells. *Nature* **2001**, *414*, 105–111. [CrossRef]
41. Korkaya, H.; Liu, S.; Wicha, M.S. Regulation of Cancer Stem Cells by Cytokine Networks: Attacking Cancer's Inflammatory Roots: Figure 1. *Clin. Cancer Res.* **2011**, *17*, 6125–6129. [CrossRef]
42. Todaro, M.; Alea, M.P.; Di Stefano, A.B.; Cammareri, P.; Vermeulen, L.; Iovino, F.; Tripodo, C.; Russo, A.; Gulotta, G.; Medema, J.P.; et al. Colon Cancer Stem Cells Dictate Tumor Growth and Resist Cell Death by Production of Interleukin-4. *Cell Stem Cell* **2007**, *1*, 389–402. [CrossRef]
43. Samanta, D.; Gilkes, D.M.; Chaturvedi, P.; Xiang, L.; Semenza, G.L. Hypoxia-inducible factors are required for chemotherapy resistance of breast cancer stem cells. *Proc. Natl. Acad. Sci. USA* **2014**, *111*, E5429–E5438. [CrossRef]
44. Hwang, W.; Yang, M.; Tsai, M.; Lan, H.; Su, S.; Chang, S.; Teng, H.; Yang, S.; Lan, Y.; Chiou, S. SNAIL Regulates Interleukin-8 Expression, Stem Cell–Like Activity, and Tumorigenicity of Human Colorectal Carcinoma Cells. *Gastroenterology* **2011**, *141*, 279–291.e5. [CrossRef] [PubMed]
45. Korkaya, H.; Kim, G.-I.; Davis, A.; Malik, F.; Henry, N.L.; Ithimakin, S.; Quraishi, A.A.; Tawakkol, N.; D'Angelo, R.; Paulson, A.K.; et al. Activation of an IL6 Inflammatory Loop Mediates Trastuzumab Resistance in HER2+ Breast Cancer by Expanding the Cancer Stem Cell Population. *Mol. Cell* **2012**, *47*, 570–584. [CrossRef] [PubMed]
46. Chan, L.-C.; Li, C.-W.; Xia, W.; Hsu, J.-M.; Lee, H.-H.; Cha, J.-H.; Wang, H.-L.; Yang, W.-H.; Yen, E.-Y.; Chang, W.-C.; et al. IL-6/JAK1 pathway drives PD-L1 Y112 phosphorylation to promote cancer immune evasion. *J. Clin. Investig.* **2019**, *129*, 3324–3338. [CrossRef] [PubMed]
47. Reiman, J.M.; Knutson, K.L.; Radisky, D.C. Immune Promotion of Epithelial-mesenchymal Transition and Generation of Breast Cancer Stem Cells: Figure 1. *Cancer Res.* **2010**, *70*, 3005–3008. [CrossRef]
48. Hsu, J.-M.; Xia, W.; Hsu, Y.-H.; Chan, L.-C.; Yu, W.-H.; Cha, J.-H.; Chen, C.-T.; Liao, H.-W.; Kuo, C.-W.; Khoo, K.-H.; et al. STT3-dependent PD-L1 accumulation on cancer stem cells promotes immune evasion. *Nat. Commun.* **2018**, *9*, 1–17. [CrossRef]
49. Liao, T.-T.; Lin, C.-C.; Jiang, J.-K.; Yang, S.-H.; Teng, H.-W.; Yang, M.-H. Harnessing stemness and PD-L1 expression by AT-rich interaction domain-containing protein 3B in colorectal cancer. *Theranostics* **2020**, *10*, 6095–6112. [CrossRef]
50. Zhang, X.; Shi, H.; Yuan, X.; Jiang, P.; Qian, H.; Xu, W. Tumor-derived exosomes induce N2 polarization of neutrophils to promote gastric cancer cell migration. *Mol. Cancer* **2018**, *17*, 1–16. [CrossRef]
51. Hwang, W.-L.; Lan, H.-Y.; Cheng, W.-C.; Huang, S.-C.; Yang, M.-H. Tumor stem-like cell-derived exosomal RNAs prime neutrophils for facilitating tumorigenesis of colon cancer. *J. Hematol. Oncol.* **2019**, *12*, 1–17. [CrossRef]
52. Takahashi, K.; Yamanaka, S. Induction of Pluripotent Stem Cells from Mouse Embryonic and Adult Fibroblast Cultures by Defined Factors. *Cell* **2006**, *126*, 663–676. [CrossRef]
53. Maherali, N.; Sridharan, R.; Xie, W.; Utikal, J.; Eminli, S.; Arnold, K.; Stadtfeld, M.; Yachechko, R.; Tchieu, J.; Jaenisch, R.; et al. Directly Reprogrammed Fibroblasts Show Global Epigenetic Remodeling and Widespread Tissue Contribution. *Cell Stem Cell* **2007**, *1*, 55–70. [CrossRef] [PubMed]
54. Okita, K.; Ichisaka, T.; Yamanaka, S. Generation of germline-competent induced pluripotent stem cells. *Nat. Cell Biol.* **2007**, *448*, 313–317. [CrossRef] [PubMed]
55. Li, W.; Wang, Z.; Zha, L.; Kong, D.; Liao, G.; Li, H. HMGA2 regulates epithelial-mesenchymal transition and the acquisition of tumor stem cell properties through TWIST1 in gastric cancer. *Oncol. Rep.* **2016**, *37*, 185–192. [CrossRef] [PubMed]

56. Lee, S.H.; Hong, H.S.; Liu, Z.X.; Kim, R.H.; Kang, M.K.; Park, N.-H.; Shin, K.-H. TNFα enhances cancer stem cell-like phenotype via Notch-Hes1 activation in oral squamous cell carcinoma cells. *Biochem. Biophys. Res. Commun.* **2012**, *424*, 58–64. [CrossRef] [PubMed]
57. Kim, S.-Y.; Kang, J.W.; Song, X.; Kim, B.K.; Yoo, Y.D.; Kwon, Y.T.; Lee, Y.J. Role of the IL-6-JAK1-STAT3-Oct-4 pathway in the conversion of non-stem cancer cells into cancer stem-like cells. *Cell. Signal.* **2013**, *25*, 961–969. [CrossRef] [PubMed]
58. Liao, T.-T.; Yang, M.-H. Revisiting epithelial-mesenchymal transition in cancer metastasis: The connection between epithelial plasticity and stemness. *Mol. Oncol.* **2017**, *11*, 792–804. [CrossRef]
59. Venkatesh, V.; Nataraj, R.; Thangaraj, G.S.; Karthikeyan, M.; Gnanasekaran, A.; Kaginelli, S.B.; Kuppanna, G.; Kallappa, C.G.; Basalingappa, K.M. Targeting Notch signalling pathway of cancer stem cells. *Stem Cell Investig.* **2018**, *5*, 5. [CrossRef] [PubMed]
60. Pelullo, M.; Zema, S.; Nardozza, F.; Checquolo, S.; Screpanti, I.; Bellavia, D. Wnt, Notch, and TGF-β Pathways Impinge on Hedgehog Signaling Complexity: An Open Window on Cancer. *Front. Genet.* **2019**, *10*, 711. [CrossRef]
61. Vermeulen, L.; Melo, F.D.S.E.; Van Der Heijden, M.; Cameron, K.; De Jong, J.H.; Borovski, T.; Tuynman, J.B.; Todaro, M.; Merz, C.; Rodermond, H.; et al. Wnt activity defines colon cancer stem cells and is regulated by the microenvironment. *Nat. Cell Biol.* **2010**, *12*, 468–476. [CrossRef]
62. Yeung, J.; Esposito, M.T.; Gandillet, A.; Zeisig, B.B.; Griessinger, E.; Bonnet, D.; So, C.W.E. β-Catenin Mediates the Establishment and Drug Resistance of MLL Leukemic Stem Cells. *Cancer Cell* **2010**, *18*, 606–618. [CrossRef]
63. Dierks, C.; Beigi, R.; Guo, G.-R.; Zirlik, K.; Stegert, M.R.; Manley, P.; Trussell, C.; Schmitt-Graeff, A.; Landwerlin, K.; Veelken, H.; et al. Expansion of Bcr-Abl-Positive Leukemic Stem Cells Is Dependent on Hedgehog Pathway Activation. *Cancer Cell* **2008**, *14*, 238–249. [CrossRef]
64. Peacock, C.D.; Wang, Q.; Gesell, G.S.; Corcoran-Schwartz, I.M.; Jones, E.; Kim, J.; Devereux, W.L.; Rhodes, J.T.; Huff, C.A.; Beachy, P.A.; et al. Hedgehog signaling maintains a tumor stem cell compartment in multiple myeloma. *Proc. Natl. Acad. Sci. USA* **2007**, *104*, 4048–4053. [CrossRef] [PubMed]
65. A Broderick, J.; Zamore, P.D. MicroRNA therapeutics. *Gene Ther.* **2011**, *18*, 1104–1110. [CrossRef] [PubMed]
66. Wright, M.W.; Bruford, E.A. Naming 'junk': Human non-protein coding RNA (ncRNA) gene nomenclature. *Hum. Genom.* **2011**, *5*, 1–9. [CrossRef]
67. Wang, Y.; Medvid, R.; Melton, C.; Jaenisch, R.; Blelloch, R. DGCR8 is essential for microRNA biogenesis and silencing of embryonic stem cell self-renewal. *Nat. Genet.* **2007**, *39*, 380–385. [CrossRef] [PubMed]
68. Lewis, B.P.; Burge, C.B.; Bartel, D.P. Conserved Seed Pairing, Often Flanked by Adenosines, Indicates that Thousands of Human Genes are MicroRNA Targets. *Cell* **2005**, *120*, 15–20. [CrossRef]
69. Calin, G.A.; Sevignani, C.; Dumitru, C.D.; Hyslop, T.; Noch, E.; Yendamuri, S.; Shimizu, M.; Rattan, S.; Bullrich, F.; Negrini, M.; et al. Human microRNA genes are frequently located at fragile sites and genomic regions involved in cancers. *Proc. Natl. Acad. Sci. USA* **2004**, *101*, 2999–3004. [CrossRef] [PubMed]
70. Garzon, R.; Fabbri, M.; Cimmino, A.; Calin, G.A.; Croce, C.M. MicroRNA expression and function in cancer. *Trends Mol. Med.* **2006**, *12*, 580–587. [CrossRef]
71. Zhang, F.; Wang, D. The Pattern of microRNA Binding Site Distribution. *Genes* **2017**, *8*, 296. [CrossRef] [PubMed]
72. Friedman, R.C.; Farh, K.K.-H.; Burge, C.B.; Bartel, D.P. Most mammalian mRNAs are conserved targets of microRNAs. *Genome Res.* **2008**, *19*, 92–105. [CrossRef]
73. Lujambio, A.; Ropero, S.; Ballestar, E.; Fraga, M.F.; Cerrato, C.; Setién, F.; Casado, S.; Suarez-Gauthier, A.; Sanchez-Cespedes, M.; Gitt, A.; et al. Genetic Unmasking of an Epigenetically Silenced microRNA in Human Cancer Cells. *Cancer Res.* **2007**, *67*, 1424–1429. [CrossRef]
74. Obernosterer, G.; Leuschner, P.J.; Alenius, M.; Martinez, J. Post-transcriptional regulation of microRNA expression. *RNA* **2006**, *12*, 1161–1167. [CrossRef]
75. Lee, Y.; Kim, M.; Han, J.; Yeom, K.-H.; Lee, S.; Baek, S.H.; Kim, V.N. MicroRNA genes are transcribed by RNA polymerase II. *EMBO J.* **2004**, *23*, 4051–4060. [CrossRef] [PubMed]
76. Lee, Y.; Ahn, C.; Han, J.; Choi, H.; Kim, J.; Yim, J.; Lee, J.; Provost, P.; Rådmark, O.; Kim, S.; et al. The nuclear RNase III Drosha initiates microRNA processing. *Nat. Cell Biol.* **2003**, *425*, 415–419. [CrossRef]
77. Yi, R.; Qin, Y.; Macara, I.G.; Cullen, B.R. Exportin-5 mediates the nuclear export of pre-microRNAs and short hairpin RNAs. *Genes Dev.* **2003**, *17*, 3011–3016. [CrossRef] [PubMed]
78. Zhang, H.; A Kolb, F.; Jaskiewicz, L.; Westhof, E.; Filipowicz, W. Single Processing Center Models for Human Dicer and Bacterial RNase III. *Cell* **2004**, *118*, 57–68. [CrossRef]
79. Chendrimada, T.P.; Gregory, R.I.; Kumaraswamy, E.; Norman, J.; Cooch, N.; Nishikura, K.; Shiekhattar, R. TRBP recruits the Dicer complex to Ago2 for microRNA processing and gene silencing. *Nat. Cell Biol.* **2005**, *436*, 740–744. [CrossRef] [PubMed]
80. Lee, Y.; Hur, I.; Park, S.-Y.; Kim, Y.-K.; Suh, M.R.; Kim, V.N. The role of PACT in the RNA silencing pathway. *EMBO J.* **2006**, *25*, 522–532. [CrossRef]
81. Fukuda, T.; Yamagata, K.; Fujiyama, S.; Matsumoto, T.; Koshida, I.; Yoshimura, K.; Mihara, M.; Naitou, M.; Endoh, H.; Nakamura, T.; et al. DEAD-box RNA helicase subunits of the Drosha complex are required for processing of rRNA and a subset of microRNAs. *Nat. Cell Biol.* **2007**, *9*, 604–611. [CrossRef]
82. Gregory, R.I.; Chendrimada, T.P.; Cooch, N.; Shiekhattar, R. Human RISC Couples MicroRNA Biogenesis and Posttranscriptional Gene Silencing. *Cell* **2005**, *123*, 631–640. [CrossRef]

83. Lee, R.C.; Feinbaum, R.L.; Ambros, V. The C. elegans heterochronic gene lin-4 encodes small RNAs with antisense complementarity to lin-14. *Cell* **1993**, *75*, 843–854. [CrossRef]
84. Wightman, B.; Ha, I.; Ruvkun, G. Posttranscriptional regulation of the heterochronic gene lin-14 by lin-4 mediates temporal pattern formation in C. elegans. *Cell* **1993**, *75*, 855–862. [CrossRef]
85. Wightman, B.; Burglin, T.R.; Gatto, J.; Arasu, P.; Ruvkun, G. Negative regulatory sequences in the lin-14 3′-untranslated region are necessary to generate a temporal switch during Caenorhabditis elegans development. *Genes Dev.* **1991**, *5*, 1813–1824. [CrossRef] [PubMed]
86. Lewis, B.P.; Shih, I.-H.; Jones-Rhoades, M.W.; Bartel, D.P.; Burge, C.B. Prediction of Mammalian MicroRNA Targets. *Cell* **2003**, *115*, 787–798. [CrossRef]
87. Iglesias-Bartolome, R.; Callejas-Valera, J.L.; Gutkind, J.S. Control of the epithelial stem cell epigenome: The shaping of epithelial stem cell identity. *Curr. Opin. Cell Biol.* **2013**, *25*, 162–169. [CrossRef]
88. Papagerakis, S.; Pannone, G.; Zheng, L.; About, I.; Taqi, N.; Nguyen, N.P.; Matossian, M.; McAlpin, B.; Santoro, A.; McHugh, J.; et al. Oral epithelial stem cells—Implications in normal development and cancer metastasis. *Exp. Cell Res.* **2014**, *325*, 111–129. [CrossRef]
89. Peng, H.; Park, J.K.; Katsnelson, J.; Kaplan, N.; Yang, W.; Getsios, S.; Lavker, R.M. microRNA-103/107 Family Regulates Multiple Epithelial Stem Cell Characteristics. *Stem Cells* **2015**, *33*, 1642–1656. [CrossRef]
90. Andl, T.; Murchison, E.P.; Liu, F.; Zhang, Y.; Yunta-Gonzalez, M.; Tobias, J.W.; Andl, C.D.; Seykora, J.T.; Hannon, G.J.; Millar, S.E. The miRNA-Processing Enzyme Dicer Is Essential for the Morphogenesis and Maintenance of Hair Follicles. *Curr. Biol.* **2006**, *16*, 1041–1049. [CrossRef]
91. Yi, R.; Pasolli, H.A.; Landthaler, M.; Hafner, M.; Ojo, T.; Sheridan, R.; Sander, C.; O'Carroll, D.; Stoffel, M.; Tuschl, T.; et al. DGCR8-dependent microRNA biogenesis is essential for skin development. *Proc. Natl. Acad. Sci. USA* **2008**, *106*, 498–502. [CrossRef]
92. Yi, R.; O'Carroll, D.; A Pasolli, H.; Zhang, Z.; Dietrich, F.S.; Tarakhovsky, A.; Fuchs, E. Morphogenesis in skin is governed by discrete sets of differentially expressed microRNAs. *Nat. Genet.* **2006**, *38*, 356–362. [CrossRef]
93. Otsuka-Tanaka, Y.; Oommen, S.; Kawasaki, M.; Kawasaki, K.; Imam, N.; Jalani-Ghazani, F.; Hindges, R.; Sharpe, P.T.; Ohazama, A. Oral Lining Mucosa Development Depends on Mesenchymal microRNAs. *J. Dent. Res.* **2012**, *92*, 229–234. [CrossRef]
94. Calin, G.A.; Croce, C.M. MicroRNA Signatures in Human Cancers. *Nat. Rev. Cancer* **2006**, *6*, 857–866. [CrossRef]
95. Khan, A.Q.; Ahmed, E.I.; Elareer, N.R.; Junejo, K.; Steinhoff, M.; Uddin, S. Role of miRNA-Regulated Cancer Stem Cells in the Pathogenesis of Human Malignancies. *Cells* **2019**, *8*, 840. [CrossRef] [PubMed]
96. Esau, C.C. Inhibition of microRNA with antisense oligonucleotides. *Methods* **2008**, *44*, 55–60. [CrossRef] [PubMed]
97. Mercatelli, N.; Coppola, V.; Bonci, D.; Miele, F.; Costantini, A.; Guadagnoli, M.; Bonanno, E.; Muto, G.; Frajese, G.V.; De Maria, R.; et al. The Inhibition of the Highly Expressed Mir-221 and Mir-222 Impairs the Growth of Prostate Carcinoma Xenografts in Mice. *PLoS ONE* **2008**, *3*, e4029. [CrossRef] [PubMed]
98. Van Roosbroeck, K.; Fanini, F.; Setoyama, T.; Ivan, C.; Rodriguez-Aguayo, C.; Fuentes-Mattei, E.; Xiao, L.; Vannini, I.; Redis, R.S.; D'Abundo, L.; et al. Combining Anti-Mir-155 with Chemotherapy for the Treatment of Lung Cancers. *Clin. Cancer Res.* **2016**, *23*, 2891–2904. [CrossRef] [PubMed]
99. Chirshev, E.; Oberg, K.C.; Ioffe, Y.J.; Unternaehrer, J.J. Let-7as biomarker, prognostic indicator, and therapy for precision medicine in cancer. *Clin. Transl. Med.* **2019**, *8*, 24. [CrossRef]
100. Zhao, Q.; Zheng, X.; Guo, H.; Xue, X.; Zhang, Y.; Niu, M.; Cui, J.; Liu, H.; Luo, H.; Yang, D.; et al. Serum Exosomal miR-941 as a promising Oncogenic Biomarker for Laryngeal Squamous Cell Carcinoma. *J. Cancer* **2020**, *11*, 5329–5344. [CrossRef]
101. Troschel, F.M.; Böhly, N.; Borrmann, K.; Braun, T.; Schwickert, A.; Kiesel, L.; Eich, H.T.; Götte, M.; Greve, B. miR-142-3p attenuates breast cancer stem cell characteristics and decreases radioresistance in vitro. *Tumor Biol.* **2018**, *40*, 1010428318791887. [CrossRef]
102. Xia, Y.; Wang, Q.; Huang, X.; Yin, X.; Song, J.; Ke, Z.; Duan, X. miRNA-Based Feature Classifier Is Associated with Tumor Mutational Burden in Head and Neck Squamous Cell Carcinoma. *BioMed Res. Int.* **2020**, *2020*, 1686480. [CrossRef]
103. Avilés-Jurado, F.X.; Muñoz, C.; Meler, C.; Flores, J.C.; Gumà, J.; Benaiges, E.; Mora, J.; Camacho, M.; León, X.; Vilaseca, I.; et al. Circulating microRNAs modulating glycolysis as non-invasive prognostic biomarkers of HNSCC. *Eur. Arch. Oto-Rhino-Laryngol.* **2020**, 1–10. [CrossRef] [PubMed]
104. Petronacci, C.C.; García, A.G.; Iruegas, E.P.; Mundiña, B.R.; Pouso, A.L.; Sayáns, M.P. Identification of Prognosis Associated microRNAs in HNSCC Subtypes Based on TCGA Dataset. *Medicina* **2020**, *56*, 535. [CrossRef] [PubMed]
105. Ganci, F.; Sacconi, A.; Manciocco, V.; Sperduti, I.; Battaglia, P.; Covello, R.; Muti, P.; Strano, S.; Spriano, G.; Fontemaggi, G.; et al. MicroRNA expression as predictor of local recurrence risk in oral squamous cell carcinoma. *Head Neck* **2016**, *38*, E189–E197. [CrossRef] [PubMed]
106. Islam, M.; Datta, J.; Lang, J.C.; Teknos, T.N. Down regulation of RhoC by microRNA-138 results in de-activation of FAK, Src and Erk1/2 signaling pathway in head and neck squamous cell carcinoma. *Oral Oncol.* **2014**, *50*, 448–456. [CrossRef] [PubMed]
107. Geng, J.; Liu, Y.; Jin, Y.; Tai, J.; Zhang, J.; Xiao, X.; Chu, P.; Yu, Y.; Wang, S.C.; Lu, J.; et al. MicroRNA-365a-3p promotes tumor growth and metastasis in laryngeal squamous cell carcinoma. *Oncol. Rep.* **2016**, *35*, 2017–2026. [CrossRef]
108. Li, D.; Liu, K.; Li, Z.; Wang, J.; Wang, X. miR-19a and miR-424 target TGFBR3 to promote epithelial-to-mesenchymal transition and migration of tongue squamous cell carcinoma cells. *Cell Adhes. Migr.* **2017**, *12*, 236–246. [CrossRef] [PubMed]

109. Chen, S.; Zhang, J.; Sun, L.; Li, X.; Bai, J.; Zhang, H.; Li, T. miR-611 promotes the proliferation, migration and invasion of tongue squamous cell carcinoma cells by targeting FOXN3. *Oral Dis.* **2019**, *25*, 1906–1918. [CrossRef]
110. Yao, X.; Xie, L.; Zeng, Y. MiR-9 Promotes Angiogenesis via Targeting on Sphingosine-1- Phosphate Receptor 1. *Front. Cell Dev. Biol.* **2020**, *8*, 755. [CrossRef]
111. Lages, E. MicroRNAs: Molecular features and role in cancer. *Front. Biosci.* **2012**, *17*, 2508–2540. [CrossRef]
112. Xu, W.; Ji, J.; Xu, Y.; Liu, Y.; Shi, L.; Liu, Y.; Lu, X.; Zhao, Y.; Luo, F.; Wang, B.; et al. MicroRNA-191, by promoting the EMT and increasing CSC-like properties, is involved in neoplastic and metastatic properties of transformed human bronchial epithelial cells. *Mol. Carcinog.* **2014**, *54*, E148–E161. [CrossRef]
113. Ceppi, P.; E Peter, M. MicroRNAs regulate both epithelial-to-mesenchymal transition and cancer stem cells. *Oncogene* **2013**, *33*, 269–278. [CrossRef] [PubMed]
114. Kim, J.; Yao, F.; Xiao, Z.; Sun, Y.; Ma, L. MicroRNAs and metastasis: Small RNAs play big roles. *Cancer Metastasis Rev.* **2017**, *37*, 5–15. [CrossRef] [PubMed]
115. Song, S.J.; Ito, K.; Ala, U.; Kats, L.; Webster, K.; Sun, S.M.; Jongen-Lavrencic, M.; Manova-Todorova, K.; Teruya-Feldstein, J.; Avigan, D.E.; et al. The Oncogenic MicroRNA miR-22 Targets the TET2 Tumor Suppressor to Promote Hematopoietic Stem Cell Self-Renewal and Transformation. *Cell Stem Cell* **2013**, *13*, 87–101. [CrossRef] [PubMed]
116. Chen, J.; Ouyang, H.; An, X.; Liu, S. miR-125a is upregulated in cancer stem-like cells derived from TW01 and is responsible for maintaining stemness by inhibiting p53. *Oncol. Lett.* **2018**, *17*, 87–94. [CrossRef] [PubMed]
117. Peng, S.-Y.; Tu, H.-F.; Yang, C.-C.; Wu, C.-H.; Liu, C.-J.; Chang, K.-W.; Lin, S.-C. miR-134targetsPDCD7to reduce E-cadherin expression and enhance oral cancer progression. *Int. J. Cancer* **2018**, *143*, 2892–2904. [CrossRef]
118. Farmakovskaya, M.; Khromova, N.; Rybko, V.; Dugina, V.; Kopnin, B.; Kopnin, P. E-Cadherin repression increases amount of cancer stem cells in human A549 lung adenocarcinoma and stimulates tumor growth. *Cell Cycle* **2016**, *15*, 1084–1092. [CrossRef]
119. Liu, C.-J.; Shen, W.G.; Peng, S.-Y.; Cheng, H.-W.; Kao, S.-Y.; Lin, S.-C.; Chang, K.-W. miR-134induces oncogenicity and metastasis in head and neck carcinoma through targetingWWOXgene. *Int. J. Cancer* **2013**, *134*, 811–821. [CrossRef]
120. Yan, H.C.; Xu, J.; Fang, L.S.; Qiu, Y.Y.; Lin, X.M.; Huang, H.X.; Han, Q.Y. Ectopic expression of the WWOX gene suppresses stemness of human ovarian cancer stem cells. *Oncol. Lett.* **2015**, *9*, 1614–1620. [CrossRef]
121. Li, J.; Liu, J.; Li, P.; Zhou, C.; Liu, P. The downregulation of WWOX induces epithelial–mesenchymal transition and enhances stemness and chemoresistance in breast cancer. *Exp. Biol. Med.* **2018**, *243*, 1066–1073. [CrossRef]
122. Lin, S.-S.; Peng, C.-Y.; Liao, Y.-W.; Chou, M.-Y.; Hsieh, P.-L.; Yu, C.-C. miR-1246 Targets CCNG2 to Enhance Cancer Stemness and Chemoresistance in Oral Carcinomas. *Cancers* **2018**, *10*, 272. [CrossRef]
123. Bernaudo, S.; Salem, M.; Qi, X.; Zhou, W.; Zhang, C.; Yang, W.; Rosman, D.; Deng, Z.; Ye, G.; Yang, B.B.; et al. Cyclin G2 inhibits epithelial-to-mesenchymal transition by disrupting Wnt/β-catenin signaling. *Oncogene* **2016**, *35*, 4816–4827. [CrossRef] [PubMed]
124. Iwai, S.; Yonekawa, A.; Harada, C.; Hamada, M.; Katagiri, W.; Nakazawa, M.; Yura, Y. Involvement of the Wnt-β-catenin pathway in invasion and migration of oral squamous carcinoma cells. *Int. J. Oncol.* **2010**, *37*, 1095–1103. [CrossRef] [PubMed]
125. Mah, K.M.; Weiner, J.A. Regulation of Wnt signaling by protocadherins. *Semin. Cell Dev. Biol.* **2017**, *69*, 158–171. [CrossRef] [PubMed]
126. DiMeo, T.A.; Anderson, K.; Phadke, P.; Feng, C.; Perou, C.M.; Naber, S.; Kuperwasser, C. A Novel Lung Metastasis Signature Links Wnt Signaling with Cancer Cell Self-Renewal and Epithelial-Mesenchymal Transition in Basal-like Breast Cancer. *Cancer Res.* **2009**, *69*, 5364–5373. [CrossRef] [PubMed]
127. Giefing, M.; Zemke, N.; Brauze, D.; Kostrzewska-Poczekaj, M.; Luczak, M.; Szaumkessel, M.; Pelinska, K.; Kiwerska, K.; Tönnies, H.; Grenman, R.; et al. High resolution ArrayCGH and expression profiling identifies PTPRD and PCDH17/PCH68 as tumor suppressor gene candidates in laryngeal squamous cell carcinoma. *Genes Chromosom. Cancer* **2010**, *50*, 154–166. [CrossRef] [PubMed]
128. Luo, M.; Sun, G.; Sun, J.-W. MiR-196b affects the progression and prognosis of human LSCC through targeting PCDH-17. *Auris Nasus Larynx* **2019**, *46*, 583–592. [CrossRef] [PubMed]
129. Lu, W.; Xu, Z.; Zhang, M.; Zuo, Y. MiR-19a promotes epithelial-mesenchymal transition through PI3K/AKT pathway in gastric cancer. *Int. J. Clin. Exp. Pathol.* **2014**, *7*, 7286–7296. [CrossRef]
130. Huang, L.; Wang, X.; Wen, C.; Yang, X.; Song, M.; Chen, J.; Wang, C.; Zhang, B.; Wang, L.; Iwamoto, A.; et al. Hsa-miR-19a is associated with lymph metastasis and mediates the TNF-α induced epithelial-to-mesenchymal transition in colorectal cancer. *Sci. Rep.* **2015**, *5*, 13350. [CrossRef]
131. Li, J.; Yang, S.; Yan, W.; Yang, J.; Qin, Y.-J.; Lin, X.-L.; Xie, R.-Y.; Wang, S.-C.; Jin, W.; Gao, F.; et al. MicroRNA-19 triggers epithelial–mesenchymal transition of lung cancer cells accompanied by growth inhibition. *Lab. Investig.* **2015**, *95*, 1056–1070. [CrossRef]
132. Xu, M.; Wang, S.; Wang, Y.; Wu, H.; Frank, J.A.; Zhang, Z.; Luo, J. Role of p38γ MAPK in regulation of EMT and cancer stem cells. *Biochim. Biophys. Acta Mol. Basis Dis.* **2018**, *1864*, 3605–3617. [CrossRef]
133. Zhu, Q.; Zhang, Q.; Gu, M.; Zhang, K.; Xia, T.; Zhang, S.; Chen, W.; Yin, H.; Yao, H.; Fan, Y.; et al. MIR106A-5p upregulation suppresses autophagy and accelerates malignant phenotype in nasopharyngeal carcinoma. *Autophagy* **2020**, *10*, 1–17. [CrossRef]
134. Fu, S.; Fu, Y.; Chen, F.; Hu, Y.; Quan, B.; Zhang, J. Overexpression of MYCT1 Inhibits Proliferation and Induces Apoptosis in Human Acute Myeloid Leukemia HL-60 and KG-1a Cells in vitro and in vivo. *Front. Pharmacol.* **2018**, *9*, 1045. [CrossRef] [PubMed]

135. Fu, S.; Guo, Y.; Chen, H.; Xu, Z.-M.; Qiu, G.-B.; Zhong, M.; Sun, K.-L.; Fu, W.-N. MYCT1-TV, A Novel MYCT1 Transcript, Is Regulated by c-Myc and May Participate in Laryngeal Carcinogenesis. *PLoS ONE* **2011**, *6*, e25648. [CrossRef] [PubMed]
136. Yue, P.-J.; Sun, Y.-Y.; Li, Y.-H.; Xu, Z.-M.; Fu, W.-N. MYCT1 inhibits the EMT and migration of laryngeal cancer cells via the SP1/miR-629-3p/ESRP2 pathway. *Cell. Signal.* **2020**, *74*, 109709. [CrossRef] [PubMed]
137. Miyazaki, H.; Takahashi, R.-U.; Prieto-Vila, M.; Kawamura, Y.; Kondo, S.; Shirota, T.; Ochiya, T. CD44 exerts a functional role during EMT induction in cisplatin-resistant head and neck cancer cells. *Oncotarget* **2018**, *9*, 10029–10041. [CrossRef]
138. Chikuda, J.; Otsuka, K.; Shimomura, I.; Ito, K.; Miyazaki, H.; Takahashi, R.-U.; Nagasaki, M.; Mukudai, Y.; Ochiya, T.; Shimane, T.; et al. CD44s Induces miR-629-3p Expression in Association with Cisplatin Resistance in Head and Neck Cancer Cells. *Cancers* **2020**, *12*, 856. [CrossRef]
139. Büssing, I.; Slack, F.J.; Grosshans, H. let-7 microRNAs in development, stem cells and cancer. *Trends Mol. Med.* **2008**, *14*, 400–409. [CrossRef] [PubMed]
140. Roush, S.; Slack, F.J. The let-7 family of microRNAs. *Trends Cell Biol.* **2008**, *18*, 505–516. [CrossRef]
141. Wang, X.; Cao, L.; Wang, Y.; Wang, X.; Liu, N.; You, Y. Regulation of let-7 and its target oncogenes (Review). *Oncol. Lett.* **2012**, *3*, 955–960. [CrossRef]
142. Liao, T.-T.; Hsu, W.-H.; Ho, C.-H.; Hwang, W.-L.; Lan, H.-Y.; Lo, T.; Chang, C.-C.; Tai, S.-K.; Yang, M.-H. let-7 Modulates Chromatin Configuration and Target Gene Repression through Regulation of the ARID3B Complex. *Cell Rep.* **2016**, *14*, 520–533. [CrossRef] [PubMed]
143. Yang, W.-H.; Lan, H.-Y.; Huang, C.-H.; Tai, S.-K.; Tzeng, C.-H.; Kao, S.-Y.; Wu, K.-J.; Hung, M.-C.; Yang, M.-H. RAC1 activation mediates Twist1-induced cancer cell migration. *Nat. Cell Biol.* **2012**, *14*, 366–374. [CrossRef] [PubMed]
144. Yu, C.-C.; Chen, Y.-W.; Chiou, G.-Y.; Tsai, L.-L.; Huang, P.-I.; Chang, C.-Y.; Tseng, L.-M.; Chiou, S.-H.; Yen, S.-H.; Chou, M.-Y.; et al. MicroRNA let-7a represses chemoresistance and tumourigenicity in head and neck cancer via stem-like properties ablation. *Oral Oncol.* **2011**, *47*, 202–210. [CrossRef] [PubMed]
145. Hittelman, W.N.; Liao, Y.; Wang, L.; Milas, L. Are cancer stem cells radioresistant? *Future Oncol.* **2010**, *6*, 1563–1576. [CrossRef] [PubMed]
146. Peng, C.-Y.; Wang, T.-Y.; Lee, S.-S.; Hsieh, P.-L.; Liao, Y.-W.; Tsai, L.-L.; Fang, C.-Y.; Yu, C.-C.; Hsieh, C.-S. Let-7c restores radiosensitivity and chemosensitivity and impairs stemness in oral cancer cells through inhibiting interleukin-8. *J. Oral Pathol. Med.* **2018**, *47*, 590–597. [CrossRef]
147. Qu, J.-Q.; Yi, H.-M.; Ye, X.; Zhu, J.-F.; Li, L.-N.; Xiao, T.; Yuan, L.; Li, J.-Y.; Wang, Y.-Y.; Feng, J.; et al. MiRNA-203 Reduces Nasopharyngeal Carcinoma Radioresistance by Targeting IL8/AKT Signaling. *Mol. Cancer Ther.* **2015**, *14*, 2653–2664. [CrossRef]
148. De Jong, M.C.; Hoeve, J.J.T.; Grénman, R.; Wessels, L.F.; Kerkhoven, R.M.; Riele, H.T.; Brekel, M.W.M.V.D.; Verheij, M.; Begg, A.C. Pretreatment microRNA Expression Impacting on Epithelial-to-Mesenchymal Transition Predicts Intrinsic Radiosensitivity in Head and Neck Cancer Cell Lines and Patients. *Clin. Cancer Res.* **2015**, *21*, 5630–5638. [CrossRef]
149. Lu, Y.-C.; Cheng, A.-J.; Lee, L.-Y.; You, G.-R.; Li, Y.-L.; Chen, H.-Y.; Chang, J.T. MiR-520b as a novel molecular target for suppressing stemness phenotype of head-neck cancer by inhibiting CD44. *Sci. Rep.* **2017**, *7*, 2042. [CrossRef]
150. Diehn, M.; Cho, R.W.; Lobo, N.A.; Kalisky, T.; Dorie, M.J.; Kulp, A.N.; Qian, D.; Lam, J.S.; Ailles, L.E.; Wong, M.; et al. Association of reactive oxygen species levels and radioresistance in cancer stem cells. *Nat. Cell Biol.* **2009**, *458*, 780–783. [CrossRef]
151. Ishimoto, T.; Nagano, O.; Yae, T.; Tamada, M.; Motohara, T.; Oshima, H.; Oshima, M.; Ikeda, T.; Asaba, R.; Yagi, H.; et al. CD44 Variant Regulates Redox Status in Cancer Cells by Stabilizing the xCT Subunit of System xc− and Thereby Promotes Tumor Growth. *Cancer Cell* **2011**, *19*, 387–400. [CrossRef]
152. Dioguardi, M.; Caloro, G.A.; Laino, L.; Alovisi, M.; Sovereto, D.; Crincoli, V.; Aiuto, R.; Coccia, E.; Troiano, G.; Muzio, L.L. Circulating miR-21 as a Potential Biomarker for the Diagnosis of Oral Cancer: A Systematic Review with Meta-Analysis. *Cancers* **2020**, *12*, 936. [CrossRef]
153. Banerjee, R.; Mani, R.-S.; Russo, N.; Scanlon, C.S.; Tsodikov, A.; Jing, X.; Cao, Q.; Palanisamy, N.; Metwally, T.; Inglehart, R.C.; et al. The tumor suppressor gene rap1GAP is silenced by miR-101-mediated EZH2 overexpression in invasive squamous cell carcinoma. *Oncogene* **2011**, *30*, 4339–4349. [CrossRef]
154. Simon, J.A.; Lange, C.A. Roles of the EZH2 histone methyltransferase in cancer epigenetics. *Mutat. Res. Mol. Mech. Mutagen.* **2008**, *647*, 21–29. [CrossRef]
155. Gonzalez, M.E.; Moore, H.M.; Li, X.; Toy, K.A.; Huang, W.; Sabel, M.S.; Kidwell, K.M.; Kleer, C.G. EZH2 expands breast stem cells through activation of NOTCH1 signaling. *Proc. Natl. Acad. Sci. USA* **2014**, *111*, 3098–3103. [CrossRef]
156. Wang, C.; Liu, X.; Chen, Z.; Huang, H.; Jin, Y.; Kolokythas, A.; Wang, A.; Dai, Y.; Wong, D.T.W.; Zhou, X. Polycomb group protein EZH2-mediated E-cadherin repression promotes metastasis of oral tongue squamous cell carcinoma. *Mol. Carcinog.* **2011**, *52*, 229–236. [CrossRef]
157. Luo, H.; Jiang, Y.; Ma, S.; Chang, H.; Yi, C.; Cao, H.; Gao, Y.; Guo, H.; Hou, J.; Yan, J.; et al. EZH2 promotes invasion and metastasis of laryngeal squamous cells carcinoma via epithelial-mesenchymal transition through H3K27me3. *Biochem. Biophys. Res. Commun.* **2016**, *479*, 253–259. [CrossRef] [PubMed]
158. Kinoshita, T.; Nohata, N.; Hanazawa, T.; Kikkawa, N.; Yamamoto, N.; Yoshino, H.; Itesako, T.; Enokida, H.; Nakagawa, M.; Okamoto, Y.; et al. Tumour-suppressive microRNA-29s inhibit cancer cell migration and invasion by targeting laminin–integrin signalling in head and neck squamous cell carcinoma. *Br. J. Cancer* **2013**, *109*, 2636–2645. [CrossRef] [PubMed]

159. Chen, L.-H.; Hsu, W.-L.; Tseng, Y.-J.; Liu, D.-W.; Weng, C.-F. Involvement of DNMT 3B promotes epithelial-mesenchymal transition and gene expression profile of invasive head and neck squamous cell carcinomas cell lines. *BMC Cancer* **2016**, *16*, 431. [CrossRef]
160. Zhuang, Z.; Yu, P.; Xie, N.; Wu, Y.; Liu, H.; Zhang, M.; Tao, Y.; Wang, W.; Yin, H.; Zou, B.; et al. MicroRNA-204-5p is a tumor suppressor and potential therapeutic target in head and neck squamous cell carcinoma. *Theranostics* **2020**, *10*, 1433–1453. [CrossRef] [PubMed]
161. Galoczova, M.; Coates, P.; Vojtesek, B. STAT3, stem cells, cancer stem cells and p63. *Cell. Mol. Biol. Lett.* **2018**, *23*, 1–20. [CrossRef] [PubMed]
162. Li, X.; Zhao, Z.; Li, M.; Liu, M.; Bahena, A.; Zhang, Y.; Zhang, Y.; Nambiar, C.; Liu, G. Sulforaphane promotes apoptosis, and inhibits proliferation and self-renewal of nasopharyngeal cancer cells by targeting STAT signal through miRNA-124-3p. *Biomed. Pharmacother.* **2018**, *103*, 473–481. [CrossRef]
163. Huang, W.-C.; Jang, T.-H.; Tung, S.-L.; Yen, T.-C.; Chan, S.-H.; Wang, L.-H. A novel miR-365-3p/EHF/keratin 16 axis promotes oral squamous cell carcinoma metastasis, cancer stemness and drug resistance via enhancing β5-integrin/c-met signaling pathway. *J. Exp. Clin. Cancer Res.* **2019**, *38*, 1–17. [CrossRef] [PubMed]
164. Yu, C.-C.; Chang, C.-J.; Hsu, C.-C.; Tsai, L.-L.; Lu, S.-W.; Huang, H.-S.; Wang, J.-J.; Chou, M.-Y.; Hu, F.-W. Let-7d functions as novel regulator of epithelial-mesenchymal transition and chemoresistant property in oral cancer. *Oncol. Rep.* **2011**, *26*, 1003–1010. [CrossRef] [PubMed]
165. Du, Y.; Li, Y.; Lv, H.; Zhou, S.; Sun, Z.; Wang, M. miR-98 suppresses tumor cell growth and metastasis by targeting IGF1R in oral squamous cell carcinoma. *Int. J. Clin. Exp. Pathol.* **2015**, *8*, 12252–12259. [PubMed]
166. Bendall, S.C.; Stewart, M.H.; Menendez, P.; George, D.; Vijayaragavan, K.; Werbowetski-Ogilvie, T.; Ramos-Mejia, V.; Rouleau, A.; Yang, J.; Bossé, M.; et al. IGF and FGF cooperatively establish the regulatory stem cell niche of pluripotent human cells in vitro. *Nat. Cell Biol.* **2007**, *448*, 1015–1021. [CrossRef] [PubMed]
167. Farabaugh, S.M.; Boone, D.N.; Lee, A.V. Role of IGF1R in Breast Cancer Subtypes, Stemness, and Lineage Differentiation. *Front. Endocrinol.* **2015**, *6*, 59. [CrossRef] [PubMed]
168. Xu, C.; Xie, D.; Yu, S.-C.; Yang, X.-J.; He, L.-R.; Yang, J.; Ping, Y.-F.; Wang, B.; Yang, L.; Xu, S.-L.; et al. β-Catenin/POU5F1/SOX2 Transcription Factor Complex Mediates IGF-I Receptor Signaling and Predicts Poor Prognosis in Lung Adenocarcinoma. *Cancer Res.* **2013**, *73*, 3181–3189. [CrossRef]
169. Malaguarnera, R.; Belfiore, A. The Emerging Role of Insulin and Insulin-Like Growth Factor Signaling in Cancer Stem Cells. *Front. Endocrinol.* **2014**, *5*, 10. [CrossRef]
170. Wang, X.-H.; Wu, H.-Y.; Gao, J.; Wang, X.-H.; Gao, T.-H.; Zhang, S.-F. IGF1R facilitates epithelial-mesenchymal transition and cancer stem cell properties in neuroblastoma via the STAT3/AKT axis. *Cancer Manag. Res.* **2019**, *11*, 5459–5472. [CrossRef]
171. Leong, H.S.; Chong, F.T.; Sew, P.H.; Lau, D.P.; Wong, B.H.; Teh, B.-T.; Tan, D.S.; Iyer, N.G. Targeting Cancer Stem Cell Plasticity Through Modulation of Epidermal Growth Factor and Insulin-Like Growth Factor Receptor Signaling in Head and Neck Squamous Cell Cancer. *Stem Cells Transl. Med.* **2014**, *3*, 1055–1065. [CrossRef]
172. Jiang, Q.; Cao, Y.; Qiu, Y.; Li, C.; Liu, L.; Xu, G. Progression of squamous cell carcinoma is regulated by miR-139-5p/CXCR4. *Front. Biosci.* **2020**, *25*, 1732–1745.
173. Elbadawy, M.; Usui, T.; Yamawaki, H.; Sasaki, K. Emerging Roles of C-Myc in Cancer Stem Cell-Related Signaling and Resistance to Cancer Chemotherapy: A Potential Therapeutic Target against Colorectal Cancer. *Int. J. Mol. Sci.* **2019**, *20*, 2340. [CrossRef] [PubMed]
174. Ren, Y.; Zhu, H.; Chi, C.; Yang, F.; Xu, X. MiRNA-139 regulates oral cancer Tca8113 cells apoptosis through Akt signaling pathway. *Int. J. Clin. Exp. Pathol.* **2015**, *8*, 4588–4594. [PubMed]
175. Wang, K.; Jin, J.; Ma, T.; Zhai, H. MiR-139-5p inhibits the tumorigenesis and progression of oral squamous carcinoma cells by targetingHOXA9. *J. Cell. Mol. Med.* **2017**, *21*, 3730–3740. [CrossRef] [PubMed]
176. Bhatlekar, S.; Viswanathan, V.; Fields, J.Z.; Boman, B.M. Overexpression of HOXA4 and HOXA9 genes promotes self-renewal and contributes to colon cancer stem cell overpopulation. *J. Cell. Physiol.* **2018**, *233*, 727–735. [CrossRef]
177. You, X.; Zhou, Z.; Chen, W.; Wei, X.; Zhou, H.; Luo, W. MicroRNA-495 confers inhibitory effects on cancer stem cells in oral squamous cell carcinoma through the HOXC6-mediated TGF-β signaling pathway. *Stem Cell Res. Ther.* **2020**, *11*, 1–14. [CrossRef] [PubMed]
178. Wang, Y.; Jia, L.; Wang, B.; Diao, S.; Jia, R.; Shang, J. MiR-495/IGF-1/AKT Signaling as a Novel Axis Is Involved in the Epithelial-to-Mesenchymal Transition of Oral Squamous Cell Carcinoma. *J. Oral Maxillofac. Surg.* **2019**, *77*, 1009–1021. [CrossRef] [PubMed]
179. Lv, L.; Wang, Q.; Yang, Y.; Ji, H. MicroRNA-495 targets Notch1 to prohibit cell proliferation and invasion in oral squamous cell carcinoma. *Mol. Med. Rep.* **2018**, *19*, 693–702. [CrossRef]
180. Zhang, L.; Liao, Y.; Tang, L. MicroRNA-34 family: A potential tumor suppressor and therapeutic candidate in cancer. *J. Exp. Clin. Cancer Res.* **2019**, *38*, 1–13. [CrossRef]
181. Kim, J.S.; Kim, E.J.; Lee, S.; Tan, X.; Liu, X.; Park, S.; Kang, K.; Yoon, J.-S.; Ko, Y.H.; Kurie, J.M.; et al. MiR-34a and miR-34b/c have distinct effects on the suppression of lung adenocarcinomas. *Exp. Mol. Med.* **2019**, *51*, 1–10. [CrossRef]
182. Brito, B.D.L.; Lourenço, S.V.; Damascena, A.S.; Kowalski, L.P.; Soares, F.A.; Coutinho-Camillo, C.M. Expression of stem cell-regulating miRNAs in oral cavity and oropharynx squamous cell carcinoma. *J. Oral Pathol. Med.* **2016**, *45*, 647–654. [CrossRef]

183. Kumar, B.; Yadav, A.; Lang, J.; Teknos, T.N.; Kumar, P. Dysregulation of MicroRNA-34a Expression in Head and Neck Squamous Cell Carcinoma Promotes Tumor Growth and Tumor Angiogenesis. *PLoS ONE* **2012**, *7*, e37601. [CrossRef]
184. Sun, Z.; Hu, W.; Xu, J.; Kaufmann, A.M.; Albers, A.E. MicroRNA-34a regulates epithelial-mesenchymal transition and cancer stem cell phenotype of head and neck squamous cell carcinoma in vitro. *Int. J. Oncol.* **2015**, *47*, 1339–1350. [CrossRef]
185. Gregory, P.A.; Bert, A.G.; Paterson, E.L.; Barry, S.C.; Tsykin, A.; Farshid, G.; Vadas, M.A.; Khew-Goodall, Y.; Goodall, G.J. The miR-200 family and miR-205 regulate epithelial to mesenchymal transition by targeting ZEB1 and SIP1. *Nat. Cell Biol.* **2008**, *10*, 593–601. [CrossRef] [PubMed]
186. Arunkumar, G.; Rao, A.K.D.M.; Manikandan, M.; Rao, H.P.S.; Subbiah, S.; Ilangovan, R.; Murugan, A.K.; Munirajan, A.K. Dysregulation of miR-200 family microRNAs and epithelial-mesenchymal transition markers in oral squamous cell carcinoma. *Oncol. Lett.* **2017**, *15*, 649–657. [CrossRef] [PubMed]
187. Tamagawa, S.; Beder, L.B.; Hotomi, M.; Gunduz, M.; Yata, K.; Grenman, R.; Yamanaka, N. Role of miR-200c/miR-141 in the regulation of epithelial-mesenchymal transition and migration in head and neck squamous cell carcinoma. *Int. J. Mol. Med.* **2014**, *33*, 879–886. [CrossRef] [PubMed]
188. Liao, T.-T.; Yang, M.-H. Hybrid Epithelial/Mesenchymal State in Cancer Metastasis: Clinical Significance and Regulatory Mechanisms. *Cells* **2020**, *9*, 623. [CrossRef]
189. He, P.; Qiu, K.; Jia, Y. Modeling of mesenchymal hybrid epithelial state and phenotypic transitions in EMT and MET processes of cancer cells. *Sci. Rep.* **2018**, *8*, 14323. [CrossRef] [PubMed]
190. Lu, M.; Jolly, M.K.; Levine, H.; Onuchic, J.N.; Ben-Jacob, E. MicroRNA-based regulation of epithelial-hybrid-mesenchymal fate determination. *Proc. Natl. Acad. Sci. USA* **2013**, *110*, 18144–18149. [CrossRef]
191. Gu, Y.; Liu, H.; Kong, F.; Ye, J.; Jia, X.; Zhang, Z.; Li, N.; Yin, J.; Zheng, G.; He, Z. miR-22/KAT6B axis is a chemotherapeutic determiner via regulation of PI3k-Akt-NF-kB pathway in tongue squamous cell carcinoma. *J. Exp. Clin. Cancer Res.* **2018**, *37*, 164. [CrossRef]
192. Mendoza, M.C.; Er, E.E.; Blenis, J. The Ras-ERK and PI3K-mTOR pathways: Cross-talk and compensation. *Trends Biochem. Sci.* **2011**, *36*, 320–328. [CrossRef]
193. Castellano, E.; Molina-Arcas, M.; Krygowska, A.A.; East, P.; Warne, P.; Nicol, A.; Downward, J. RAS signalling through PI3-Kinase controls cell migration via modulation of Reelin expression. *Nat. Commun.* **2016**, *7*, 11245. [CrossRef]
194. Park, S.; Zhao, D.; Hatanpaa, K.J.; Mickey, B.E.; Saha, D.; Boothman, D.A.; Story, M.D.; Wong, E.T.; Burma, S.; Georgescu, M.-M.; et al. RIP1 Activates PI3K-Akt via a Dual Mechanism Involving NF-κB–Mediated Inhibition of the mTOR-S6K-IRS1 Negative Feedback Loop and Down-regulation of PTEN. *Cancer Res.* **2009**, *69*, 4107–4111. [CrossRef]
195. Oeckinghaus, A.; Hayden, M.S.; Ghosh, S. Crosstalk in NF-κB signaling pathways. *Nat. Immunol.* **2011**, *12*, 695–708. [CrossRef]
196. Inoki, K.; Ouyang, H.; Zhu, T.; Lindvall, C.; Wang, Y.; Zhang, X.; Yang, Q.; Bennett, C.; Harada, Y.; Stankunas, K.; et al. TSC2 Integrates Wnt and Energy Signals via a Coordinated Phosphorylation by AMPK and GSK3 to Regulate Cell Growth. *Cell* **2006**, *126*, 955–968. [CrossRef]
197. Fang, D.; Hawke, D.; Zheng, Y.; Xia, Y.; Meisenhelder, J.; Nika, H.; Mills, G.B.; Kobayashi, R.; Hunter, T.; Lu, Z. Phosphorylation of β-Catenin by AKT Promotes β-Catenin Transcriptional Activity. *J. Biol. Chem.* **2007**, *282*, 11221–11229. [CrossRef] [PubMed]
198. Zhang, L.; Zhou, F.; Drabsch, Y.; Gao, R.; Snaar-Jagalska, B.E.; Mickanin, C.; Huang, H.; Sheppard, K.-A.; Porter, J.A.; Lu, C.X.; et al. USP4 is regulated by AKT phosphorylation and directly deubiquitylates TGF-β type I receptor. *Nat. Cell Biol.* **2012**, *14*, 717–726. [CrossRef] [PubMed]
199. Zhang, L.; Zhou, F.; Dijke, P.T. Signaling interplay between transforming growth factor-β receptor and PI3K/AKT pathways in cancer. *Trends Biochem. Sci.* **2013**, *38*, 612–620. [CrossRef]
200. Budi, E.H.; Muthusamy, B.P.; Derynck, R. The insulin response integrates increased TGF-β signaling through Akt-induced enhancement of cell surface delivery of TGF-β receptors. *Sci. Signal.* **2015**, *8*, ra96. [CrossRef] [PubMed]
201. Hamidi, A.; Song, J.; Thakur, N.; Itoh, S.; Marcusson, A.; Bergh, A.; Heldin, C.-H.; Landström, M. TGF-β promotes PI3K-AKT signaling and prostate cancer cell migration through the TRAF6-mediated ubiquitylation of p85α. *Sci. Signal.* **2017**, *10*, eaal4186. [CrossRef] [PubMed]
202. Madsen, R.R. PI3K in stemness regulation: From development to cancer. *Biochem. Soc. Trans.* **2020**, *48*, 301–315. [CrossRef] [PubMed]
203. Natarajan, K.; Abraham, P.; Kota, R.; Isaac, B. NF-κB-iNOS-COX2-TNF α inflammatory signaling pathway plays an important role in methotrexate induced small intestinal injury in rats. *Food Chem. Toxicol.* **2018**, *118*, 766–783. [CrossRef] [PubMed]
204. Yamamoto, K.; Arakawa, T.; Ueda, N.; Yamamoto, S. Transcriptional Roles of Nuclear Factor κB and Nuclear Factor-Interleukin-6 in the Tumor Necrosis Factor α-Dependent Induction of Cyclooxygenase-2 in MC3T3-E1 Cells. *J. Biol. Chem.* **1995**, *270*, 31315–31320. [CrossRef] [PubMed]
205. Yang, H.; Qi, H.; Ren, J.; Cui, J.; Li, Z.; Waldum, H.L.; Cui, G. Involvement of NF-κB/IL-6 Pathway in the Processing of Colorectal Carcinogenesis in Colitis Mice. *Int. J. Inflamm.* **2014**, *2014*, 1–7. [CrossRef]
206. Greten, F.R.; Arkan, M.C.; Bollrath, J.; Hsu, L.-C.; Goode, J.; Miething, C.; Göktuna, S.I.; Neuenhahn, M.; Fierer, J.; Paxian, S.; et al. NF-κB Is a Negative Regulator of IL-1β Secretion as Revealed by Genetic and Pharmacological Inhibition of IKKβ. *Cell* **2007**, *130*, 918–931. [CrossRef] [PubMed]
207. Rovin, B.H.; Dickerson, J.A.; Tan, L.C.; Hebert, C.A. Activation of nuclear factor-κB correlates with MCP-1 expression by human mesangial cells. *Kidney Int.* **1995**, *48*, 1263–1271. [CrossRef] [PubMed]

208. Yamamoto, Y.; Gaynor, R.B. IκB kinases: Key regulators of the NF-κB pathway. *Trends Biochem. Sci.* **2004**, *29*, 72–79. [CrossRef]
209. Rinkenbaugh, A.L.; Baldwin, A.S. The NF-κB Pathway and Cancer Stem Cells. *Cells* **2016**, *5*, 16. [CrossRef]
210. Feng, X.; Luo, Q.; Wang, H.; Zhang, H.; Chen, F. MicroRNA-22 suppresses cell proliferation, migration and invasion in oral squamous cell carcinoma by targeting NLRP3. *J. Cell. Physiol.* **2018**, *233*, 6705–6713. [CrossRef]
211. Huang, C.-F.; Chen, L.; Li, Y.-C.; Wu, L.; Yu, G.-T.; Zhang, W.-F.; Sun, Z.-J. NLRP3 inflammasome activation promotes inflammation-induced carcinogenesis in head and neck squamous cell carcinoma. *J. Exp. Clin. Cancer Res.* **2017**, *36*, 1–13. [CrossRef]
212. Zhang, C.; Hao, Y.; Sun, Y.; Liu, P. Quercetin suppresses the tumorigenesis of oral squamous cell carcinoma by regulating microRNA-22/WNT1/β-catenin axis. *J. Pharmacol. Sci.* **2019**, *140*, 128–136. [CrossRef]
213. Qiu, K.; Huang, Z.; Huang, Z.; He, Z.; You, S. miR-22 regulates cell invasion, migration and proliferation in vitro through inhibiting CD147 expression in tongue squamous cell carcinoma. *Arch. Oral Biol.* **2016**, *66*, 92–97. [CrossRef] [PubMed]
214. Yu, B.; Zhang, Y.; Wu, K.; Wang, L.; Jiang, Y.; Chen, W.; Yan, M. CD147 promotes progression of head and neck squamous cell carcinoma via NF-kappa B signaling. *J. Cell. Mol. Med.* **2019**, *23*, 954–966. [CrossRef] [PubMed]
215. Suzuki, S.; Toyoma, S.; Tsuji, T.; Kawasaki, Y.; Yamada, T. CD147 mediates transforming growth factor-β1-induced epithelial-mesenchymal transition and cell invasion in squamous cell carcinoma of the tongue. *Exp. Ther. Med.* **2019**, *17*, 2855–2860. [CrossRef]
216. Zhu, M.; Zhang, C.; Chen, D.; Chen, S.; Zheng, H. MicroRNA-98-HMGA2-POSTN signal pathway reverses epithelial-to-mesenchymal transition in laryngeal squamous cell carcinoma. *Biomed. Pharmacother.* **2019**, *117*, 108998. [CrossRef] [PubMed]
217. Li, M.; Tian, L.; Ren, H.; Chen, X.; Wang, Y.; Ge, J.; Wu, S.; Sun, Y.; Liu, M.; Xiao, H. MicroRNA-101 is a potential prognostic indicator of laryngeal squamous cell carcinoma and modulates CDK8. *J. Transl. Med.* **2015**, *13*, 271. [CrossRef] [PubMed]
218. Firestein, R.; Bass, A.J.; Kim, S.Y.; Dunn, I.F.; Silver, S.J.; Guney, I.; Freed, E.; Ligon, A.H.; Vena, N.; Ogino, S.; et al. CDK8 is a colorectal cancer oncogene that regulates β-catenin activity. *Nature* **2008**, *455*, 547–551. [CrossRef] [PubMed]
219. He, S.-B.; Yuan, Y.; Wang, L.; Yu, M.-J.; Zhu, Y.-B.; Zhu, X.-G. Effects of cyclin-dependent kinase 8 specific siRNA on the proliferation and apoptosis of colon cancer cells. *J. Exp. Clin. Cancer Res.* **2011**, *30*, 109. [CrossRef]
220. Shao, Q.; Zhang, P.; Ma, Y.; Lu, Z.; Meng, J.; Li, H.; Wang, X.; Chen, D.; Zhang, M.; Han, Y.; et al. MicroRNA-139-5p affects cisplatin sensitivity in human nasopharyngeal carcinoma cells by regulating the epithelial-to-mesenchymal transition. *Gene* **2018**, *652*, 48–58. [CrossRef]
221. Brabletz, T.; Jung, A.; Spaderna, S.; Hlubek, F.; Kirchner, T. Migrating cancer stem cells—an integrated concept of malignant tumour progression. *Nat. Rev. Cancer* **2005**, *5*, 744–749. [CrossRef] [PubMed]
222. Mani, S.A.; Guo, W.; Liao, M.J.; Eaton, E.N.; Ayyanan, A.; Zhou, A.Y.; Brooks, M.; Reinhard, F.; Zhang, C.C.; Shipitsin, M.; et al. The Epithelial-Mesenchymal Transition Generates Cells with Properties of Stem Cells. *Cell* **2008**, *133*, 704–715. [CrossRef]
223. Polyak, K.; Weinberg, R.A. Transitions between epithelial and mesenchymal states: Acquisition of malignant and stem cell traits. *Nat. Rev. Cancer* **2009**, *9*, 265–273. [CrossRef] [PubMed]
224. Karemaker, I.D.; Vermeulen, M. ZBTB 2 reads unmethylated CpG island promoters and regulates embryonic stem cell differentiation. *EMBO Rep.* **2018**, *19*, e44993. [CrossRef]
225. Yang, Y.; Li, H.; He, Z.; Xie, D.; Ni, J.; Lin, X. MicroRNA-488-3p inhibits proliferation and induces apoptosis by targeting ZBTB2 in esophageal squamous cell carcinoma. *J. Cell. Biochem.* **2019**, *120*, 18702–18713. [CrossRef] [PubMed]
226. Godar, S.; Ince, T.A.; Bell, G.W.; Feldser, D.; Donaher, J.L.; Bergh, J.; Liu, A.; Miu, K.; Watnick, R.S.; Reinhardt, F.; et al. Growth-Inhibitory and Tumor- Suppressive Functions of p53 Depend on Its Repression of CD44 Expression. *Cell* **2008**, *134*, 62–73. [CrossRef]
227. Wang, Z.; Mao, J.-W.; Liu, G.-Y.; Wang, F.-G.; Ju, Z.-S.; Zhou, N.; Wang, R.-Y. MicroRNA-372 enhances radiosensitivity while inhibiting cell invasion and metastasis in nasopharyngeal carcinoma through activating the PBK-dependent p53 signaling pathway. *Cancer Med.* **2019**, *8*, 712–728. [CrossRef] [PubMed]
228. Chang, C.-J.; Chao, C.-H.; Xia, W.; Yang, J.-Y.; Xiong, Y.; Li, C.-W.; Yu, W.-H.; Rehman, S.K.; Hsu, J.L.; Lee, H.-H.; et al. p53 regulates epithelial–mesenchymal transition and stem cell properties through modulating miRNAs. *Nat. Cell Biol.* **2011**, *13*, 317–323. [CrossRef]
229. Shi, B.; Yan, W.; Liu, G.; Guo, Y. MicroRNA-488 inhibits tongue squamous carcinoma cell invasion and EMT by directly targeting ATF3. *Cell. Mol. Biol. Lett.* **2018**, *23*, 28. [CrossRef]
230. Su, J.-L.; Chen, P.-S.; Johansson, G.; Kuo, M.-L. Function and Regulation of Let-7 Family microRNAs. *MicroRNA* **2012**, *1*, 34–39. [CrossRef]
231. Chen, P.-S.; Su, J.-L.; Cha, S.-T.; Tarn, W.-Y.; Wang, M.-Y.; Hsu, H.-C.; Lin, M.-T.; Chu, C.-Y.; Hua, K.-T.; Chen, C.-N.; et al. miR-107 promotes tumor progression by targeting the let-7 microRNA in mice and humans. *J. Clin. Investig.* **2011**, *121*, 3442–3455. [CrossRef]
232. González-Arriagada, W.A.; Olivero, P.; Rodríguez, B.; Lozano-Burgos, C.; De Oliveira, C.E.; Coletta, R.D. Clinicopathological significance of miR-26, miR-107, miR-125b, and miR-203 in head and neck carcinomas. *Oral Dis.* **2018**, *24*, 930–939. [CrossRef]
233. Huang, C.; Wang, Z.; Zhang, K.; Dong, Y.; Zhang, A.; Lu, C.; Liu, L. MicroRNA-107 inhibits proliferation and invasion of laryngeal squamous cell carcinoma cells by targeting CACNA2D1 in vitro. *Anti-Cancer Drugs* **2020**, *31*, 260–271. [CrossRef]
234. Sui, X.; Geng, J.-H.; Li, Y.-H.; Zhu, G.-Y.; Wang, W.-H. Calcium channel α2δ1 subunit (CACNA2D1) enhances radioresistance in cancer stem-like cells in non-small cell lung cancer cell lines. *Cancer Manag. Res.* **2018**, *10*, 5009–5018. [CrossRef]

235. Patel, S.; Rawal, R. Role of miRNA dynamics and cytokine profile in governing CD44v6/Nanog/PTEN axis in oral cancer: Modulating the master regulators. *Tumor Biol.* **2016**, *37*, 14565–14575. [CrossRef]
236. Hersi, H.M.; Raulf, N.; Gaken, J.; Folarin, N.; Tavassoli, M. Micro RNA-9 inhibits growth and invasion of head and neck cancer cells and is a predictive biomarker of response to plerixafor, an inhibitor of its target CXCR 4. *Mol. Oncol.* **2018**, *12*, 2023–2041. [CrossRef]
237. Citron, F.; Armenia, J.; Franchin, G.; Polesel, J.; Talamini, R.; D'Andrea, S.; Sulfaro, S.; Croce, C.M.; Klement, W.; Otasek, D.; et al. An Integrated Approach Identifies Mediators of Local Recurrence in Head and Neck Squamous Carcinoma. *Clin. Cancer Res.* **2017**, *23*, 3769–3780. [CrossRef] [PubMed]
238. Li, X.; Zeng, Z.; Wang, J.; Wu, Y.; Chen, W.; Zheng, L.; Xi, T.; Wang, A.; Lu, Y. MicroRNA-9 and breast cancer. *Biomed. Pharmacother.* **2020**, *122*, 109687. [CrossRef]
239. Kamal, N.N.S.B.N.M.; Shahidan, W.N.S. Non-Exosomal and Exosomal Circulatory Micrornas: Which Are More Valid as Biomarkers? *Front. Pharmacol.* **2020**, *10*, 1500. [CrossRef] [PubMed]
240. Ludwig, S.; Sharma, P.; Wise, P.; Sposto, R.; Hollingshead, D.; Lamb, J.; Lang, S.; Fabbri, M.; Whiteside, T.L. mRNA and miRNA Profiles of Exosomes from Cultured Tumor Cells Reveal Biomarkers Specific for HPV16-Positive and HPV16-Negative Head and Neck Cancer. *Int. J. Mol. Sci.* **2020**, *21*, 8570. [CrossRef] [PubMed]
241. Lee, S.H.; Lee, C.-R.; Rigas, N.K.; Kim, R.H.; Kang, M.K.; Park, N.-H.; Shin, K.-H. Human papillomavirus 16 (HPV16) enhances tumor growth and cancer stemness of HPV-negative oral/oropharyngeal squamous cell carcinoma cells via miR-181 regulation. *Papillomavirus Res.* **2015**, *1*, 116–125. [CrossRef]
242. Huang, Q.; Yang, J.; Zheng, J.; Hsueh, C.; Guo, Y.; Zhou, L. Characterization of selective exosomal microRNA expression profile derived from laryngeal squamous cell carcinoma detected by next generation sequencing. *Oncol. Rep.* **2018**, *40*, 2584–2594. [CrossRef]
243. Yang, X.; Li, Y.; Zou, L.; Zhu, Z. Role of Exosomes in Crosstalk between Cancer-Associated Fibroblasts and Cancer Cells. *Front. Oncol.* **2019**, *9*, 356. [CrossRef]
244. Morelli, A.E.; Larregina, A.T.; Shufesky, W.J.; Sullivan, M.L.G.; Stolz, D.B.; Papworth, G.D.; Zahorchak, A.F.; Logar, A.J.; Wang, Z.; Watkins, S.C.; et al. Endocytosis, intracellular sorting, and processing of exosomes by dendritic cells. *Blood* **2004**, *104*, 3257–3266. [CrossRef]
245. Benjamins, J.A.; Nedelkoska, L.; Touil, H.; Stemmer, P.M.; Carruthers, N.J.; Jena, B.P.; Naik, A.R.; Bar-Or, A.; Lisak, R.P. Exosome-enriched fractions from MS B cells induce oligodendrocyte death. *Neurol. Neuroimmunol. Neuroinflamm.* **2019**, *6*, e550. [CrossRef] [PubMed]
246. Wang, X.; Shen, H.; He, Q.; Tian, W.; Xia, A.; Lu, X.-J. Exosomes derived from exhausted CD8+ T cells impaired the anticancer function of normal CD8+ T cells. *J. Med. Genet.* **2019**, *56*, 29–31. [CrossRef] [PubMed]
247. Whiteside, T.L. Tumor-Derived Exosomes and Their Role in Cancer Progression. *Int. Rev. Cytol.* **2016**, *74*, 103–141. [CrossRef]
248. Li, Y.-Y.; Tao, Y.-W.; Gao, S.; Li, P.; Zheng, J.-M.; Zhang, S.-E.; Liang, J.; Zhang, Y. Cancer-associated fibroblasts contribute to oral cancer cells proliferation and metastasis via exosome-mediated paracrine miR-34a-5p. *EBioMedicine* **2018**, *36*, 209–220. [CrossRef]
249. Beg, M.S.; Brenner, A.J.; Sachdev, J.; Borad, M.; Kang, Y.-K.; Stoudemire, J.; Smith, S.; Bader, A.G.; Kim, S.; Hong, D.S. Phase I study of MRX34, a liposomal miR-34a mimic, administered twice weekly in patients with advanced solid tumors. *Investig. New Drugs* **2017**, *35*, 180–188. [CrossRef]
250. Hong, D.S.; Kang, Y.-K.; Borad, M.; Sachdev, J.; Ejadi, S.; Lim, H.Y.; Brenner, A.J.; Park, K.; Lee, J.-L.; Kim, T.-Y.; et al. Phase 1 study of MRX34, a liposomal miR-34a mimic, in patients with advanced solid tumours. *Br. J. Cancer* **2020**, *122*, 1630–1637. [CrossRef] [PubMed]
251. Calin, G.A. miRNAs in oncogenesis and tumor suppression. In *International Review of Cell and Molecular Biology, MiRNAs in Differentiation and Development*, 1st ed.; Galluzzi, L., Vitale, I., Eds.; Academic Press: Cambridge, MA, USA, 2017; Volume 333.

 cancers

Review

Current Implications of microRNAs in Genome Stability and Stress Responses of Ovarian Cancer

Arkadiusz Gajek, Patrycja Gralewska, Agnieszka Marczak and Aneta Rogalska *

Department of Medical Biophysics, Faculty of Biology and Environmental Protection, Institute of Biophysics, University of Lodz, Pomorska 141/143, 90-236 Lodz, Poland; arkadiusz.gajek@biol.uni.lodz.pl (A.G.); patrycja.gralewska@edu.uni.lodz.pl (P.G.); agnieszka.marczak@biol.uni.lodz.pl (A.M.)
* Correspondence: aneta.rogalska@biol.uni.lodz.pl; Tel.: +48-42-635-44-77

Simple Summary: Ovarian cancer is the leading cause of death from gynecological malignancies. Recent studies have focused on ovarian cancer-associated microRNAs that play strong regulatory roles in various cellular processes. While miRNAs have been shown to participate in regulation of tumorigenesis and drug responses through modulating the DNA damage response (DDR), little is known about their potential influence on sensitivity to chemotherapy. The main objective of this review is to summarize recent findings on the utility of miRNAs as ovarian cancer biomarkers and their regulation of DDR or modified replication stress response proteins.

Abstract: Genomic alterations and aberrant DNA damage signaling are hallmarks of ovarian cancer (OC), the leading cause of mortality among gynecological cancers worldwide. Owing to the lack of specific symptoms and late-stage diagnosis, survival chances of patients are significantly reduced. Poly (ADP-ribose) polymerase (PARP) inhibitors and replication stress response inhibitors present attractive therapeutic strategies for OC. Recent research has focused on ovarian cancer-associated microRNAs (miRNAs) that play significant regulatory roles in various cellular processes. While miRNAs have been shown to participate in regulation of tumorigenesis and drug responses through modulating the DNA damage response (DDR), little is known about their potential influence on sensitivity to chemotherapy. The main objective of this review is to summarize recent findings on the utility of miRNAs as cancer biomarkers, in particular, ovarian cancer, and their regulation of DDR or modified replication stress response proteins. We further discuss the suppressive and promotional effects of various miRNAs on ovarian cancer and their participation in cell cycle disturbance, response to DNA damage, and therapeutic functions in multiple cancer types, with particular focus on ovarian cancer. Improved understanding of the mechanisms by which miRNAs regulate drug resistance should facilitate the development of effective combination therapies for ovarian cancer.

Keywords: microRNA; ovarian cancer; PARP; replication stress; targeted therapy

Citation: Gajek, A.; Gralewska, P.; Marczak, A.; Rogalska, A. Current Implications of microRNAs in Genome Stability and Stress Responses of Ovarian Cancer. *Cancers* **2021**, *13*, 2690. https://doi.org/10.3390/cancers13112690

Academic Editor: Paola Tucci

Received: 13 April 2021
Accepted: 26 May 2021
Published: 29 May 2021

Publisher's Note: MDPI stays neutral with regard to jurisdictional claims in published maps and institutional affiliations.

1. Introduction

Ovarian cancer is the leading cause of death from gynecological malignancies. The high fatality rate is linked to the complexity of the disease and consequent difficulty in making an accurate diagnosis. At the initial stages of disease progression, patients present with non-specific symptoms [1]. The majority of cases are diagnosed at the third or fourth stage of clinical advancement following the spread of disease to other organs. At present, standard treatment for ovarian cancer involves total removal of tumor mass and any tissue that may pose risk of spread. In the case of highly advanced tumors, radical surgery is not possible [2]. Debulking surgery is performed, followed by adjuvant treatment with drugs containing platinum compounds or taxane-based chemotherapy, which has shown success in improving the survival rates of patients in the fourth stage of the disease. However, the five-year survival rate for advanced-stage cases remains below 30% [3]. Ovarian cancer cells often carry BRCA1 or BRCA2 (BRCA1/2), germline or somatic mutations in ataxia telangiectasia and Rad3-related protein (ATR) or checkpoint kinase 1 (CHK1) genes of the homologous recombination (HR) pathway [4]. Poly (ADP ribose) polymerase inhibitors (PARPis) have been identified as the most promising targeted therapy for ovarian cancer. In the United States, the FDA approved olaparib for maintenance treatment of patients with BRCA^{MUT} advanced epithelial ovarian cancer (EOC) showing complete or partial response to first-line platinum-based chemotherapy in 2018 [5].

Elucidation of molecular alterations in serous ovarian carcinoma cells is necessary to identify novel targets for early detection and treatment. Recent studies have focused on ovarian cancer-associated microRNAs (miRNAs) that play strong regulatory roles in various cellular processes. Initial findings support a potential correlation between miRNAs and cancer development. Even low-level disruption in expression patterns of individual miRNAs can lead to significant pathological changes, such as neoplasia. Alterations in miRNA expression are widely reported in multiple cancer types, especially ovarian cancer resistant to chemotherapy [6]. Knowledge of the specific associations between miRNAs and DNA damage response (DDR) or DNA repair should aid in expanding applications of miRNAs in cancer therapy. Cells detect DNA damage and coordinate an appropriate response involving activation of repair pathways, such as nucleotide excision repair (NER) and HR. If damage is too excessive for repair, an apoptotic response is initiated through activating death receptors or triggering intrinsic apoptosis.

In this review, earlier findings on the direct effects of miRNAs on sensitivity of ovarian cancer cells to replication stress response (RSR) inhibitors are described. The therapeutic potential of the miRNAs regulating DDR/DNA repair is discussed, along with the mechanisms by which miRNAs affect sensitivity to PARP, ATR, and CHK1 inhibitor therapy. The identification of mediator miRNAs that improve response to treatment with checkpoint inhibitors would increase the proportion of patients benefiting from therapy.

2. Participation of miRNAs in Pathogenesis and Development of Neoplastic Diseases

MiRNAs are a class of small, endogenous, and noncoding RNAs that post-transcriptionally regulate gene expression. MiRNA is transcripted in the nucleus, usually by RNA polymerase II, to produce the primary miRNA (pri-miRNA). Pri-miRNA is identified and cleaved by the Drosha, an RNase III enzyme, and its cofactor DGCR8 (Pasha), which form a hairpin precursor miRNA (pre-mRNA). The hairpin precursor is exported out of the nucleus by Exportin 5, where Dicer (RNase III enzyme) cleaves double stranded mi-RNA and creates miRNA as transient 21–24 nucleotide duplex miRNA. The strand of mature sequence is then transported onto Argonaute (Ago) and is loaded into a protein complex called RISC. MiRNA recognizes their target sequences based on complementarity to the 3′untranslated region (3′UTR) of mRNA transcripts, leading to translational inhibition and/or mRNA degradation [7,8], Figure 1.

Almost half the miRNA genes are located in fragile sites of the genome where chromosome fragments are lost or rearranged with high frequency. Mutations in these areas are often linked to cancer development, implicating microRNAs in the formation and progression of neoplasms [7–9]. Expression of miRNA genes located near these regions is commonly disrupted. One example is miR-15a and miR-16-1 genes located on the long arm of chromosome 13 in region 14.2 where deletions are frequent. Reduced levels or complete absence of miR-15a and miR-16-1 are detected in many patients with B-cell chronic lymphocytic leukemia, ovarian and prostate cancer, mantle cell lymphoma, and multiple myeloma [10–12]. MiRNAs are also secreted from both normal and cancer cells in exosomes, small vesicles that play a key role in cell-to-cell communication in the body [13]. In addition, variable environmental factors, such as low pH and hypoxia (characteristic of most solid tumors, including ovarian cancer), affect miRNA expression and promote exosome secretion. Several indications support the potential utility of miRNAs circulating in the bloodstream as biomarkers of cancer. One hypothesis is that miRNAs appear in the bloodstream through two mechanisms, one associated with tissue damage, as demonstrated in earlier studies (e.g., miR-208 is observed in serum after cardiac muscle damage) [14], and the second related to so-called microbubbles (exosomes) involved in tumor-associated immunosuppression, metastasis, and angiogenesis that are derived directly from the cytoplasmic membrane and reflect the antigenic composition of parent cells [15]. Secreted miRNAs can also be an additional source of information about defects in the DNA repair system, including those related to replication stress (miR-200c, miR-214 [16], miR-185-5p [17], miR-126, miR-17, miR-92a [18], and miR-34a [19]). In 2008, Taylor et al. first demonstrated that eight exosomal miRNAs (miR-21, miR-141, miR-200a, miR-200b, miR-200c, miR-203, miR-205, and miR-214) were elevated in the serum of ovarian cancer patients, even in the case of patients with early stages of the disease. Very importantly, the miRNA profiles observed in exosomes were similar to those in the originating tumor cells. Circulating miRNA profiles accurately reflect the tumor profiles, which make them potential biomarkers and relevant for ovarian cancer diagnosis prognosis and therapeutics [16,20,21]. Since that time, other studies confirmed other circulating miRNA profiles in plasma samples of ovarian cancer as possible biomarkers of which some miRNA were significantly increased, e.g., hsa-miR-106a-5p, hsa-let-7d-5p, hsa-miR-93-5p [17], miR-1274a, miR-625-3p, and miR-720 [18], and others significantly decreased, e.g., hsa-miR-122-5p, hsa-miR-185-5p and hsa-miR-99b-5p [17], miR-106b, miR-126, miR-150, miR-17, miR-20a, and miR-92a [18]. Maeda et al. recently described the potential role of serum miR-34a in early diagnosis of ovarian cancer and for histological subtyping of EOC [19]. In the case of ovarian cancer patients, the elevated level of miRNA (in comparison to healthy) was reported not only in serum exosomes, but also, e.g., in ascites: miR-21, miR-23b, or miR-29a [22], or in urine: miR-30-5p [23]. It is considered that these exosomes are responsible for inducing more aggressive disease, so it confirms that they also might serve as a promising diagnostic and therapeutic targets [21,22]. The correlation between the increased levels of miR-200b and miR-200c with the main marker of ovarian cancer-CA125, commonly used in diagnosis [24], was also observed. Additionally, studies performed by Kapetanakis et al. [25] demonstrated that miR-200b was able to predict the sensitivity to treatment in much more sensitive manner than CA125. After primary treatment (surgery and chemotherapy) of the group of 33 ovarian cancer patients, CA125 very quickly (even after 1 month after treatment) returned to a normal level in almost all patients, whereas there was a difference in the level of miR-200b between individuals. The patients with a negative miR-200b variation had a longer progression-free survival (PFS), than those patients with a positive variation. Increased levels of specific mRNAs characteristic of certain cancer types (breast, lung, ovary, prostate, pancreas, liver, and colon cancer and chronic myeloid leukemia) are often associated with tumor invasiveness or metastasis. The miRNA molecules known to inhibit the processes of migration and invasion of neoplastic cells include miR-149 (breast cancer), miR-138 (ovarian and kidney cancer), miR-126 (lung and stomach cancer), and miR-206 (melanoma and cervical cancer), among others. Additionally, miR-373 and miR-520c are associated with the invasive and metastatic ability of cancer cells. These molecules directly inhibit expression of the CD44 surface receptor responsible for binding hyaluronan (the main component of the extracellular matrix), an intermediary for several stimulatory processes, such as migration and proliferation. Moreover, miRNAs are involved in epithelial-to-mesenchymal transition (EMT), a necessary step in metastasis. Increased levels of specific miRNAs are additionally associated with the occurrence of epigenetic abnormalities in cancer cells [26,27] Molecular functions of miRNAs in ovarian cancer acting as oncogenes or suppressors are presented in Table 1.

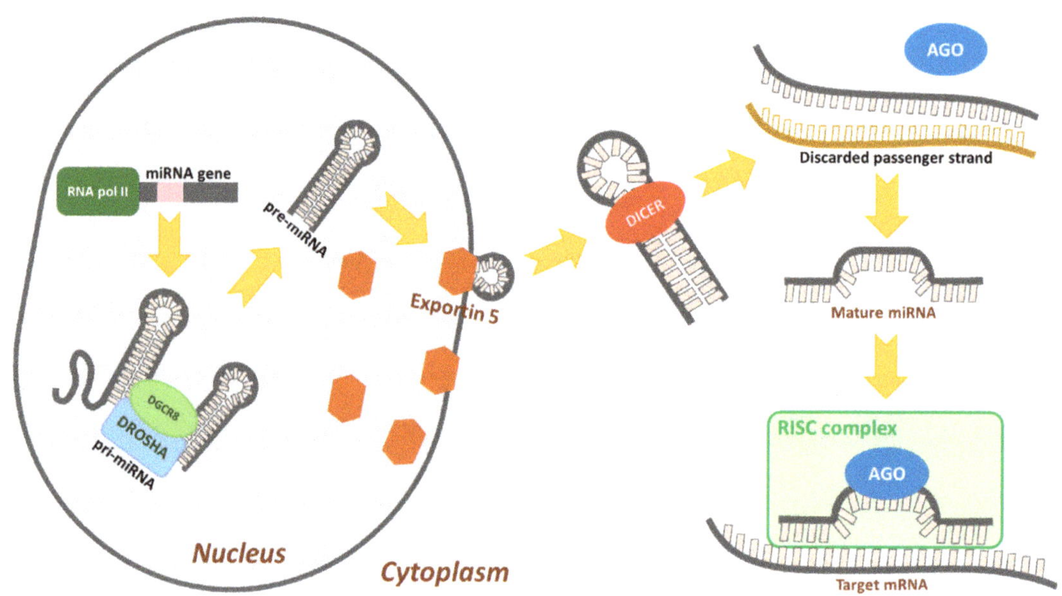

Figure 1. The main molecular steps involved in miRNA formation and organization.

Table 1. Molecular functions of miRNAs in ovarian cancer.

MiRNAs as Oncogenes	Ref	MiRNAs as Suppressors	Ref
miR-138	[26,27]	miR-16	[28,29]
miR-200 a, a-3p, b, c	[30–33]	miR-10a, 10b	[34]
miR-141	[30,33]	miR-29	[35,36]
miR-429	[30,33]	miR-let-7	[37–40]
miR-205	[41]	miR-31, 31-5p	[42,43]
miR-126-3p	[44]	miR-506-3p	[45]
miR-183	[46]	miR-424-5p	[47]
miR-760	[48,49]	miR-503-5p	[47]
miR-151	[50]	miR-199a-5p	[51]
miR-21-5p	[52]	miR-34	[53]
miR-106a	[54]	miR-340-5p	[55]
miR-195	[54]	miR-138	[56]
miR-222	[57,58]	miR-509-3	[59]
miR-221	[57,58,60]	miR-335-5p	[61]
miR-520b	[62]	miR-383	[63]
miR-10b	[64]	miR-185	[65]
miR-21	[66]	miR-126	[67]
miR-17-92	[66]	miR-708	[68]
miR-622	[69]	miR-200c	[18,70,71]
miR-424-5p	[72]		

3. Aberrant Expression Profiles of miRNAs in Ovarian Cancer

Recent studies revealed differences in the miRNA expression profiles in tissues from patients with ovarian cancer and healthy individuals. Increased miRNA levels were not detected in patients with benign ovarian disease [73]. Further comparison of expression levels of miRNAs in ovarian cancer revealed distinct roles of different miRNAs. Expression of miR-200a, miR-200b, and miR-200c was significantly higher than that in normal tissues, whereas mir-199a, miR-140, miR-145, and miR-125b1 displayed low expression in ovarian cancer tissues [74]. MiR-10a and miR-10b suppressed proliferation of granulosa cells in the ovary. The miR-10 family suppressed expression of several key genes in the transforming growth factor beta (TGF-β) pathway, suggesting a negative feedback loop between the miR-10 family and TGF-β pathway [34]. Numerous studies performed on various tissue types have validated the utility of miRNAs as a prognostic marker of ovarian cancer [75–77].

3.1. MiRNAs as Tumor Suppressors

Suppressor genes, also known as anti-oncogenes, encode proteins that inhibit the processes of cell growth and differentiation and maintain genetic stability of the cell. Mutations in these genes can lead to uncontrolled cell proliferation and, consequently, development of cancer. The effect of miRNAs in the case of their activation or deactivation will lead to an insufficient number of target genes or their overexpression, respectively. Target miRNA transcripts determine whether the miRNA should be considered an oncogene or a tumor suppressor [78,79]. Biological functions of miRNAs depend on the cellular context, tumor molecular subtype, stage of tumor progression, or interactions with therapy [80]. It was observed that miR-200c and miR-141 produce resistance to carboplatin while sensitizing MES-OV/TP cells to paclitaxel. The authors suggest that the effects of these miRNAs on drug sensitivity are cell context dependent [6]. Higher miR-200c levels were also associated with better progression-free survival in stage I epithelial ovarian cancer [81]. Liu et al. found that miR-200b and miR-200c increased cisplatin sensitivity through downregulation of DNA methyltransferases (DNMT3A/DNMT3B) and the indirect downregulation of DNMT1 by targeting Sp1 transcription factor [26]. Based on 220 ovarian cancer patients' analysis, it was observed that overexpression of miR-200c correlated with poor or good outcome depending on the cellular localization of HuR (RNA binding protein). MiR-200c can act either as a suppressor or enhancer of the aggressive phenotype, depending upon the localization of HuR. Suppressor genes contribute to drug resistance of several types of solid tumors [82]. Tumor suppressor miRNAs prevent tumor development through negative regulation of genes that control cell differentiation or apoptosis. To date, a number of miRNAs have been identified as tumor suppressors [83]. For instance, miR-29 significantly reduces migration of highly metastatic ovarian cancer cells [36]. Expression of miR-29 alone or in combination with cisplatin could effectively reduce tumorigenicity of CP70 ovarian cancer cells [35].

One of the most well-characterized tumor suppressors of the miRNA family in ovarian cancer is Let-7, which belongs to a family of highly homologous members. Ten mature subtypes of the human let-7 family have been identified to date, whereby mature let-7a and let-7f are processed from precursor sequences (let-7a-1, let-7a-2, let-7a-3; let-7f-1, and let-7f-2) [38–40]. Let-7 inhibits cell proliferation and increases apoptosis by inhibiting expression of proto-oncogenes, such as the small GTPase RAS, high mobility group AT-hook 2 (HMGA 2), c-Myc, cell division cycle homolog 25A (CDC25A), cdk 6, and cyclin 2 [39,84]. Overexpression of let-7g miRNA in OVCAR3 and HEY-A8 EOC cells induced cell cycle arrest, slowed progression of EMT, and significantly improved cell response to cis-platinum treatment. Let-7g worked through vimentin and reduced the expression of Snail and Slug (the protein product of snail family transcriptional repressor 2) [39]. Other studies have demonstrated overexpression of miR-16 in ovarian cancer tissues, including SKOV3 and OVCAR3 cell lines, compared with normal ovarian epithelial cells. MiR-16 is reported to exert suppressive effects on cell migration and invasion by inactivation of the Wnt/β-catenin signaling pathway through decreasing expression of matrix metallopeptidase MMP2 and MMP9. Additionally, miR-16 regulates the expression of mesenchymal markers (cadherin 1 and 2, snail 1 and 2, vimentin, and twist family BHLH transcription factor) [28]. MiR-31 is another microRNA with biological significance. This miRNA is expressed at low levels in serous ovarian cancer cells and tissues and induces suppression of cell proliferation, clonogenic potential, and cell migration and invasion [43]. Recent research indicates that miR503HG interacts with and promotes methylation of miR-31-5p that play a role in inhibition of ovarian cancer cell invasion and migration [42]. MiR-506-3p inhibits proliferation and promotes apoptosis via inactivation of the NAD-dependent protein deacetylase sirtuin-1 (SIRT1)/AKT/Forkhead box protein 3a (FOXO3a) signaling pathway [45]. Myotubularin-related protein 6 (MTMR6) has been identified as another functional target of miR-506-3p. Several recent studies indicate that malignant biological behaviors are regulated by the myotubularin (MTM) protein family [85]. Other miRNAs acting as suppressors include miR-424-5p and miR-503-5p that directly target the 3′UTR of KIF23 (kinesin-6, a plus-end-directed motor protein in mitosis) to suppress its expression and inhibit ovarian cancer cell proliferation and migration [47]. Additionally, miR-199a-5p is reported to function as a suppressor of ovarian cancer (HO-8910 and ES-2) cell proliferation and invasion through inhibiting NF-κB1 expression. Notably, expression patterns of matrix metalloproteinases (MMP-2 and MMP-9) are altered in a similar manner as NF-κB1 upon exogenous expression of miR-199a-5p [51]. The anti-oncomiR list includes miRNAs from the miR-34 family that inhibit oncogenes, such as c-MYC and c-MET, or promote mitosis CDKs [53] and miR-340-5p. Deficiency of miR-340-5p promotes expression of serine/threonine-protein kinase B-raf (BRAF), NF-kB and ATP-binding cassette sub-family B member 5, also known as P-glycoprotein (ABCB5), resulting in development of drug resistance [55].

3.2. MiRNAs as Oncogenes

Alterations in expression of several miRNAs are observed in many cancer types [81,86,87]. Mutation in a single allele of proto-oncogenes can trigger transformation into oncogenes. These genes promote cancer development by negatively regulating the tumor genes responsible for cell differentiation or apoptosis [88]. Several miRNAs in tumor cells exhibit oncogenic traits and promote tumorigenesis. Notably, almost all members of the miR-200 family (miR-200a, miR-200b, miR-200c, miR-141, and miR-429) are upregulated in ovarian cancer [89]. Different miRNA types, including miR-182 and the miR-200 family (specifically, miR-200a, miR-200b, and miR-200c), are highly overexpressed in high-grade serous epithelial ovarian cancer (SEOC). The miR-200 family participates in EMT through regulating E-cadherin by inhibiting zinc-finger E-box-binding homeobox 1 (ZEB1) and zinc-finger E-box-binding homeobox 2 (ZEB2) [30] and improves response to paclitaxel (PTX) due to repression of the miR-200c target, ZEB1. The transcription factor, Grainyhead-like 2n (GRHL2), acts as a pivotal gatekeeper of EMT in EOC via miR-200-ZEB1 [31]. The miR-200 family also sensitizes ovarian cancer cells to PTX through downregulation of TUBB3/class III beta-tubulin, a component of microtubules that binds paclitaxel [90]. Moreover, in PTX resistant cells (A2780/1A9, MES-OV, OVCAR-3, ES-2), miR-200b and miR-200c were downregulated and associated with EMT, with increased vimentin, fibronectin1, MMP2, or MMP9 [90]. MiR-200a is reported to enhance sensitivity to PTX-induced reactive oxygen species production. Overexpression of miR-200a-3p markedly promotes proliferation, colony formation, and invasion of ovarian cancer cells. Expression of this miRNA in ovarian cancer tissues is significantly negatively correlated with that of Protocadherin-9, a potential tumor suppressor, in a variety of cancers [32]. Moreover, the miR-200 family plays a major role in regulating EMT and sensitivity to carboplatin and PTX in OVCAR-3 and MES-OV cells. Inhibition of miR-200c and miR-141 resulted in the downregulation of E-cadherin and the upregulation of vimentin and fibronectin [33].

MiR-205 expression is significantly increased with a simultaneous decrease in transcription factor 21 (TCF21, a tumor suppressor gene) in epithelial ovarian carcinoma compared to normal ovarian cells. Thus, miR-205 is regarded as an oncogene in ovarian cancer that plays critical roles in tumor invasion and metastasis [41]. MiRNA-126-3p is also implicated in cancer progression and inflammation. Overexpression of miR-126-3p in OVCAR3 cells is reported to suppress cell proliferation and invasion as well as phosphorylation of serine/threonine-specific protein kinase B (AKT) and extracellular signal-regulated kinases $\frac{1}{2}$ (ERK1/2) [44]. MiR-183 exerts tumor-promoting effects in ovarian cancer by regulating one of the transcription factor proteins, Mothers against decapentaplegic homolog 4 (Smad 4), via the TGF-β/Smad4 pathway. MiR-183 is upregulated in OC tissues and cell lines. Downregulation of miR-183 via cell transfection inhibited proliferation and invasion and induced apoptosis in SKOV3 and OVCAR3 cells [46]. Expression of miR-760 is markedly upregulated in association with an aggressive phenotype of OC and poor prognosis [48,49]. Additionally, miR-151 plays an oncogenic role in carcinogenesis and progression of ovarian cancer by activating AKT/mTOR signaling through effects on the Rho guanine nucleotide dissociation inhibitor (RhoGDIA). MiR-151 activates Ras-related C3 botulinum toxin substrate 1 (Rac1), Cdc42, and Rho GTPase by directly targeting the 3-UTR of RhoGDIA, a metastasis suppressor [50]. Examples of oncomiRs include miR-21-5p, which controls the suppressor gene phosphatase and tensin homolog (PTEN, an inhibitor of the Akt kinase pathway) [52], miR-106a, which regulates the p21 protein level, and miR-195, which controls WEE1 kinase, an inhibitor of cell division [54]. MiR-222 is overexpressed in EOC cases and promotes proliferation through downregulation of target cyclin-dependent kinase inhibitor p27Kip1 [57]. An earlier study reported upregulation of miR-221 in 63 samples of ovarian cancer. A negative correlation between expression of apoptosis protease activator 1 (APAF1) protein and miR-221 in 5 of 63 ovarian cancer tissues and six cell lines was observed, including A2780, OVCAR3, SKOV3, and 3AO5 [60]. An in vitro cell viability assay showed that downregulation of miR-221/222 sensitized A2780/CP cells to cisplatin-induced cytotoxicity [58]. Another identified oncomir shown to promote proliferation of SKOV3, Hey, and OVCAR3 cells is miR-520b, which targets the ring finger protein 216 (RNF216) gene to promote cell growth. The negative correlation between miR-520b and RNF216 may present a new strategy for ovarian cancer [62]. In addition, numerous studies have shown that oncomirs play an important role in the acquisition of the ability to invade and form metastases by cancer cells. Overexpression of miR-10b in ovarian cancer has been reported in association with reduced amounts of transcription factor, HOXD10, in altered cells, leading to an increase in the levels of ras homolog family member C (RhoC) and matrix metallopeptidase 14 (MMP14), which are responsible for metastasis [64].

4. MiRNA Functions in Cancer Based on Regulation of DDR

The DNA damage response is a complex network involving proteins that are activated to facilitate detection of DNA damage and determine the survival or death of cells exposed to stress via stimulation of the signal transduction cascade [91]. Activation of the DDR pathway triggers cell cycle checkpoint activation and dividing alternation, in turn preventing the transfer of damaged DNA to daughter cells. Simultaneously, DNA repair mechanisms are activated. Upon repair of damage, cell cycle and division resume, allowing survival and continuation of function. If repair is not possible due to an excessive number of lesions, cells are eliminated by triggering programmed cell death or cellular aging, irreversible cell cycle arrest, and division processes [92], as presented in Figure 2. DDR modulates miRNA expression in transcriptional and post-transcriptional levels and involves miRNA degradation [66,93,94]. On the other hand, miRNAs may directly modulate the expression of multiple proteins in the DDR pathways.

ATR and Ataxia telangiectasia mutated (ATM) kinases belonging to the phosphatidylinositol 3-kinase-related kinases (PIKK) family play central roles in activation of the DDR pathway [95]. Histone H2AX, one of the first known substrates for ATR and ATM kinases, is expressed in high-grade SOC, mucinous adenocarcinomas, and clear cell carcinomas. Significant changes in the gene and protein levels of H2AX have been reported in OC, supporting its predictive value as a biomarker [96].

4.1. MiRNAs Are Involved in Cell Cycle Disruption

Imbalances in activities of miRNA molecules significantly affect cell cycle regulation, leading to excessive proliferation. Disruption of this process is often associated with direct interactions of miRNAs with key regulatory molecules of signaling pathways underlying proliferation, e.g., PTEN, Myc, Ras, and V-abl Abelson murine leukemia viral oncogene homolog 1 (ABL1), as well as proteins from the Rb pathway, cyclin-CDK complexes, or cell cycle inhibitors from families of inhibitors of CDK4 (INK4) and CDK-interacting protein/kinase inhibitory protein (Cip/Kip) [97]. Examples include miR-21, which is overexpressed in breast, ovary, and liver cancer, and a group of miR-17-92 members that inhibit PTEN phosphatase activity. Suppression of the gene encoding PTEN promotes cell proliferation. Another miRNA that influences the cell cycle is miR-15b, Figure 3. Decreased expression of miR-15b leads to an increase in cellular cyclin E1, resulting in lack of control during the transition from G1 to S phase. Ectopic expression of miR-192/215 induces cell cycle arrest and targets a number of transcripts that regulate G1/S and G2/M checkpoints [98,99]. These miRNAs are transcriptional targets of p53 and also upregulate p53 by downregulating Murine Double Minute gene 2 protein (MDM2) [100].

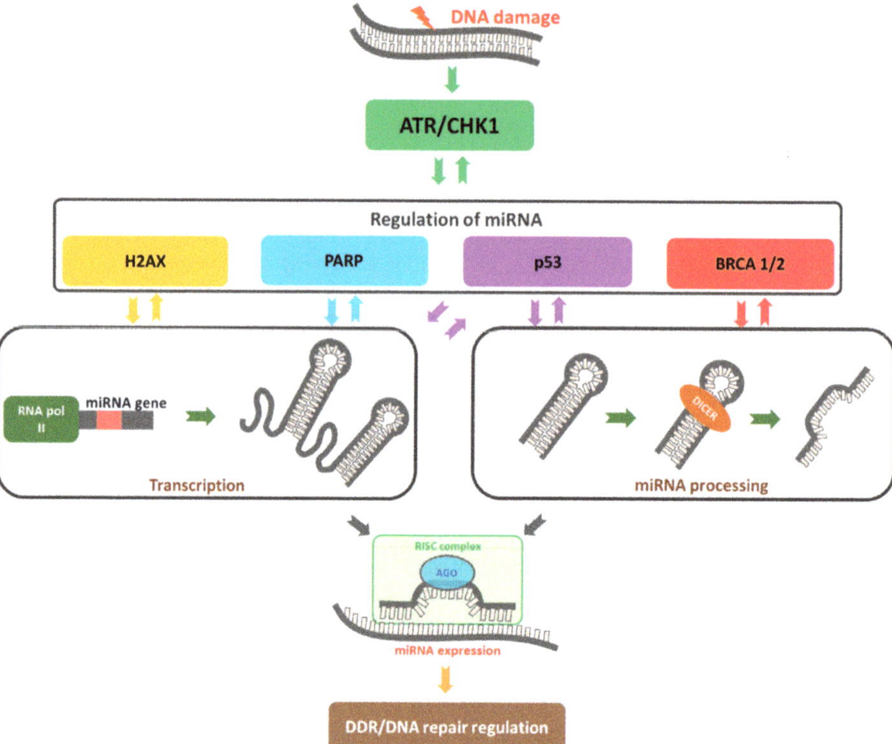

Figure 2. The role of miRNAs as a regulator of DDR. DNA damage triggers the activation of the ATR/CHK1 pathway, which is responsible for substrates phosphorylation (i.e., H2AX, p53, and BRCA 1/2) to repair DNA damage. Single-strand DNA damage is identified and repaired by poly (ADP-ribose) polymerase (PARP) pathway activation, predominantly through base excision repair (BER). With continuous PARP inhibition, ssDNA breaks are converted to dsDNA breaks during DNA replication in which replication forks stall at the point of DNA damage. Over-expressed or down-regulated miRNAs may affect DDR by changing the expression of repair genes.

4.2. Functional miRNAs in Activation of the "Response Track" to DNA Damage and the Role of H2AX Histone

In response to DNA damage, H2AX is phosphorylated by DNA-dependent protein kinase, catalytic subunit (DNA-PKc), which is also a member of the PIKK family. The histone is phosphorylated at serine 139 (known as γH2AX) and initiates attachment of subsequent elements of the signaling pathway [101]. At the same time, histone H2AX is dephosphorylated at tyrosine 142 and constitutively phosphorylated under conditions of no DNA damage [102]. Dephosphorylation promotes direct attachment of the mediator of

DNA damage checkpoint protein 1 (MDC1) protein to γH2AX. Anchoring of MDC1 at the site of damage is a platform for activation of other proteins belonging to the DDR pathway and the MRN (MRE11, Rad50, NBS1)/ATM complex. This enhances local ATM kinase activity and extension of the H2AX phosphorylation region to include nucleosomes adjacent to DNA damage [103,104]. The clusters favor extensive formation of γH2AX, which plays an important role in accumulation and maintenance of components of the DDR pathway, such as MRN, and proteins related to DNA repair, including BRCA1 and p53-binding protein 1 (53BP1). Binding of phosphorylated MDC1 to γH2AX facilitates attachment of E3, RNF8 (E3 ubiquitin-protein ligase), and RNF168 ubiquitin ligases to the lesion site, which promote association of BRCA1 and 53BP1 via ubiquitination of chromatin [91,105–108]. Downregulation of ubiquitin ligase RNF8, which is necessary for γH2AX to recruit DNA repair proteins to DNA damage sites, via miR-214, induces chromosomal instability in ovarian cancer [109], Figure 4. Thus, H2AX histone appears to play a pivotal role as an early indicator protein for DDR. Previous reports showed that miR-24 and miR-138 regulate H2AX via 3′-UTR attachment. Overexpression of miR-138 inhibited homologous recombination and enhanced cellular sensitivity to multiple DNA damage agents (cisplatin, camptothecin, and ionizing radiation) [98]. MiR-138 was recently identified as an effective tumor suppressor in multiple malignancies including ovarian cancer [56]. MiR-24 mediates suppression of H2AX in terminally differentiated blood cells, which renders them hypersensitive to gamma-irradiation, deficient in DSB repair, and susceptible to chromosomal instability [110]. Another study reported that overexpression of miR-24-insensitive CHEK1 does not rescue the DNA repair phenotype induced by miR-24 [111]. Moreover, γH2AX has been shown to regulate miR-3196 gene expression. H3K27 trimethylation in the miR-3196 promoter region regulated via H2AX phosphorylation at Ser139 is a key step in H2AX-mediated apoptosis [112]. Furthermore, Fra-1 transcriptional factor and miR-134 are upregulated in ovarian cancer tissues. MiR-134 enhances H2AX S139 phosphorylation via activation of c-Jun NH2 kinase (JNK) and promotes DNA repair through non-homologous end-joining (NHEJ) [113].

4.3. MiRNAs Contributes to DSB DNA Damage Repair System

The primary function of the DDR pathway is to identify DNA damage and, where possible, initiate repair processes. The majority of DNA damage is repaired by the triggering of catalytic event sequences involving multiple proteins, including base excision repair (BER), NER, mismatch repair (MMR), HR, and NHEJ. Two types of nucleotide excision repair pathways exist. One is active during transcription (transcription coupled repair, TCR), while the other is independent of transcription (Global Genomic Repair, GGR) [114,115]. Activation of a specific mechanism depends on the type of DNA damage. BER, NER, and MMR pathways play key roles in repairing damage such as single DNA strand breaks (SSB), replication errors, insertions, deletions, and adducts [116,117].

Double-strand breaks (DSB) are one of the most dangerous types of DNA damage, and a single unrepaired DSB is sufficient to trigger apoptosis [118]. Two processes are involved in repair of double-strand breaks, specifically, HR and NHEJ. Homologous recombination can occur in the S and G2 phases of the cell cycle. On the other hand, repair of damage by non-homologous recombination is possible at any phase of the cell cycle, including G0 [119–121]. In HR repair, proteins of the MRN complex and BRCA1 C-terminal Interacting Protein (CtIP) play a key role. These proteins are involved in formation of short sections of single-stranded DNA (ssDNA), which initiate repair of damage through homologous recombination. With the aid of BRCA1, BRCA2, and RAD51 proteins, short sections of single-stranded DNA are joined to the undamaged template. In conjunction with the activities of polymerase, nuclease, helicase, and other proteins, damage is repaired. HR is also involved in resumption of replication caused by blockage of replication forks [122,123]. One of the key proteins of the MRN complex is RAD51. In an earlier study, upregulation of miR-210 significantly suppressed expression of RAD51, while upregulation of miR-373 inhibited RAD52 (which recognizes double-strand breaks and adheres to the free ends of the break) [110]. Another study by Moskwa et al. [124] consistently showed that miR-182 downregulates BRCA1 expression. MiR-182 enhances BRCA1 protein levels and protects against irradiation-induced cell death, while its overexpression reduces BRCA1 protein, impairs homologous recombination-mediated repair, and renders cells hypersensitive to irradiation. Subsequently, ability of HR to stimulate DSB repair is significantly decreased [124].

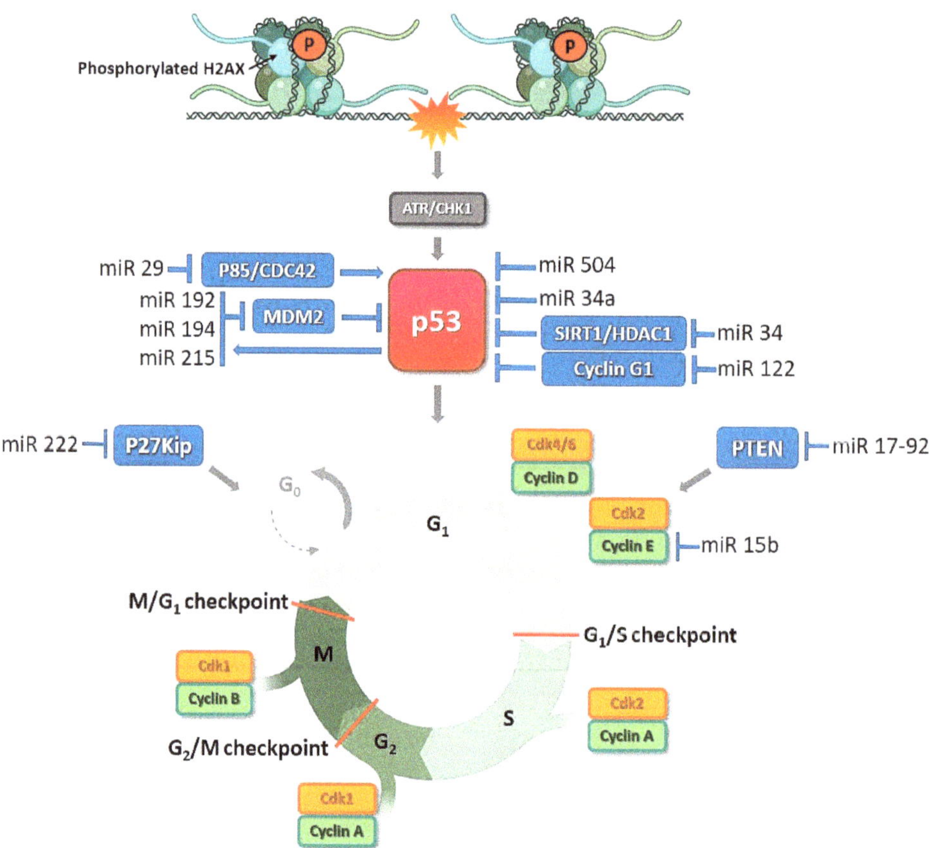

Figure 3. P53-induced miRNAs control cell cycle and cell survival in ovarian cancer. DNA damage stimulates ATR or DNA-PK kinases that activate TP53 to directly induce many kinds of miRNAs that repress cell-cycle regulators or allow to DNA repair. ATR transduces the DDR signal by phosphorylation of the checkpoint kinase CHK1, which results in cell cycle arrest and DNA repair. MiRNAs, e.g., miR-34a, a direct transcriptional target of TP53, participate in the regulation of TP53 activity. Activated TP53 translocates into the nucleus where it induces the transcription of several targets involved in cell cycle control, DNA repair, or apoptosis. DDR-related proteins that are MiRNAs recognizable are shown in blue box.

In the case of NHEJ, DSB are recognized by the heterodimeric Ku70/Ku80 protein complex, which binds DNA-PKc kinase. Subsequently, DNA polymerases and DNA ligase IV, enzymes that process DNA ends, are recruited and activated. In addition, it is possible to repair DNA damage related to joining non-homologous ends. This process, known as alternative NHEJ (alt-NHEJ) or microhomological-mediated end joining (MMEJ), occurs independently of the Ku protein [125,126]. Earlier literature suggests that miR-101 is able to successfully regulate DNA-PKcs and ATM through attaching to their 3′-UTRs. Specifically, upregulation of miR-101 significantly reduced the protein levels of DNA-PKcs and ATM in tumor cells and sensitized them to radiation, both in vitro and in vivo. Thus, miR-101 is a potential option for use in DNA DSB repair gene targeting to optimize the effects of tumor radiotherapy [127].

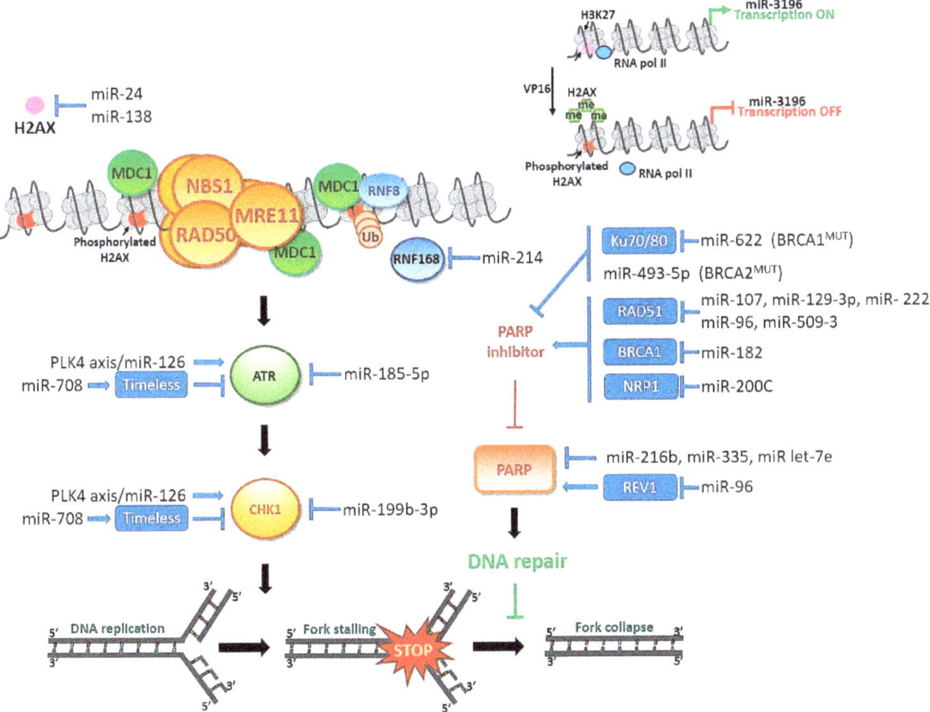

Figure 4. MiRNAs' involvement in the DNA damage response in ovarian cancer. DNA synthesis inhibition or damage induces checkpoint responses. PARP and checkpoint proteins controlled by the ATR–CHK1 pathway prevent fork collapse, replication stress, and genome instability. DDR-related proteins that are MiRNAs recognizable are shown in blue box.

4.4. MiRNAs Modulate Activity of p53, a Key Protein of the DDR Pathway

The p53 protein is a key suppressor of neoplastic transformation that regulates transcription of numerous genes and interacts directly with multiple proteins. p53 is implicated in a number of critical cell processes, including DNA repair, cell cycle, and programmed cell death. Under conditions where the cell is not exposed to stress factors, the p53 protein level is relatively low. This may be attributed to interactions with (MDM2), which blocks transcriptional activity of p53, leading to its ubiquitination-dependent degradation. MDM2 synthesis is regulated by p53, generating a negative feedback loop leading to a decrease in p53 levels after induction. The imbalance between p53 and MDM2 levels is a critical step in p53 activation [128] and occurs when activated ATM and/or ATR kinase phosphorylates the p53 protein at serine 15 and CHK2 at serine 20. ATM also phosphorylates MDM2 in response to DNA damaging agents. As a result of these modifications, interactions of MDM2 with p53 are blocked, leading to the inhibition of MDM2-dependent degradation and, consequently, accumulation of p53. Thus p53 is activated as a result of post-translational modifications, such as phosphorylation, acetylation, methylation, and ubiquitination. The p53 protein serves as a transcriptional factor to regulate expression of target genes, which also occurs through recruitment of coactivators or corepressors. Among these molecules, acetyltransferases are known to play an important role. Enzymes such as CREB-binding protein (CBP), p300, Tip60, human males absent on the first (Hmof), or P300/CBP-associated factor (PCAF) acetylate p53 and histones alter chromatin conformation, increasing the availability of the DNA template for transcription machinery. In response to DNA damage, CBP and/or p300 acetylate p53 at six lysine residues, which present a target for MDM2 ubiquitination, thereby increasing the stability of p53 and binding ability to DNA [129].

Depending on the type and extent of DNA damage, various post-translational modifications of p53 are initiated, which translate into different cellular responses. Thus, p53 serves as the main decision switch for survival or death. Several groups can be distinguished among the genes regulated by p53 in response to DNA damage. One of these categories is negative regulators/inhibitors of the cell cycle, such as p21, 14-3-3σ, and GADD45α, which trigger cell cycle arrest and division, facilitating repair of DNA damage [130]. In response to DNA damage, p53 is involved in the regulation of processes related to cell metabolism and autophagy. In addition, transcription-independent and miRNA-dependent p53 functions have been reported. MiRNAs either directly target the 3′ UTR of p53 or indirectly regulate p53 activity by modulating proteins associated with p53. Among these microRNAs, miR-504 negatively regulates p53 expression through binding two DNA *cis* elements located in the 3′ UTR region [131]. DNA damage promotes the p53-dependent upregulation of miR-192, miR-194, and miR-215. Studies also have revealed the existence of a specific p53 binding site around the miR-194/miR-215 cluster [132].

In addition to direct binding to p53, several miRNAs, including miR-34a, miR-29, and miR-122, indirectly modify p53 activity [133–136]. MiR-34a is a direct transcriptional target of p53 [137–139], whereby p53 upregulates miR-34a expression via binding to specific promoter regions. MiR-34a positively regulates p53-dependent apoptosis through another SIRT1 [133]. MiR-34a expression is low in patients with chromosomal abnormalities involving the tumor protein p53 (TP53) gene locus and is associated with poorer prognosis and shorter survival. Mutations or deletions in the 17p13 region of the TP53 gene locus may indirectly lead to reduced miR-34a expression [140]. Another miRNA family involved in p53 regulation is miR-29. Members of this family directly suppress phosphoinositide 3-kinase subunit (P85a) and cell division control protein 42 homolog (CDC42), both of which negatively regulate p53. As a result, miR-29 positively upregulates the p53 level and induces apoptosis and DNA repair in a p53 dependent manner [134].

5. MiRNAs Associated with DNA Repair Checkpoint Proteins: New Options for Optimizing Ovarian Cancer Therapy

5.1. PARP

PARP is an important protein involved in the repair of single-stranded DNA breaks, seen in Figure 3. PARPis have been shown to selectively kill cells with defective HR pathways as a result of synthetic lethality [141]. However, a large proportion of HR-mutated cancers gain resistance to these therapeutic agents. PARPi sensitivity is modulated through downregulation of critical DNA repair genes as a consequence of alterations in miRNA profiles. PARPi resistance may be promoted by miR-622 that modulates the balance of DNA repair through selective inhibition of expression of NHEJ proteins, such as KU70/80, which maintain genome stabilization after treatment with DNA-damaging agents or PARPi. High expression of miR-622 in BRCA1^{MUT} epithelial ovarian cancer is associated with prediction of poorer disease-free and overall survival [69]. The functional impact of miR-493-5p has been characterized in BRCA2^{MUT} cancer cells. MiR-493-5p induces platinum and PARPi resistance by affecting several pathways, including single-strand annealing (SSA), R-loops, and replication fork stability [142]. In contrast, miR-107, miR-129-3p, and miR-222 increase sensitivity to PARP inhibitors and ionizing radiation by causing a reduction in the DNA damage response via impairing the HR pathway based on targeting of RAD51 [143]. Mi182 exerts similar effects and enhances PARPi sensitivity by downregulating BRCA1 [124]. Moreover, expression of miR-96 is increased in many cancer types. This miRNA enhances sensitivity to platinum agents and PARP via downregulation of the DNA repair proteins REV1 and RAD51 [144]. Another study on a patient-derived xenograft (PDX) model of high-grade serous ovarian carcinoma (HGSOC) revealed an essential role of miR-509-3 in tumor suppression and HR signaling, along with increased sensitivity to PARPi treatment [59].

Furthermore, PARP1 expression could be altered by miR-335 or miR-216-b. MiR-335 plays a dual role as either a tumor promoter or suppressor in a wide variety of cancers. However, expression is reduced in ovarian cancer cells and miR-335 shown to effectively increase sensitivity to cisplatin treatment [61]. MiR-216-b regulates apoptosis and autophagy and directly binds to PARP1 mRNA, leading to inhibition of its expression. Lower expression of miR-216b is reported in cisplatin-resistant ovarian cancer cells [145]. Recent studies have additionally demonstrated a role of Neuropilin 1 (NRP1) in response to ovarian cancer therapies. NRP1 is expressed at high levels in resistant cells (SKOV3) and shown to be upregulated in partially sensitive cells (UWB-BRCA) upon prolonged olaparib treatment, resulting in poor drug response. MiR-200c targets and suppresses NRP1 expression in OC cells resistant to therapy, leading to the restoration of olaparib sensitivity [89].

Platinum-resistant ovarian tumors display low miR-Let7i expression. Conversely, its gain of function results in restoration of drug sensitivity in chemoresistant ovarian cancer cells [146]. Agomir is a type of specially labelled and chemically modified double-stranded microRNA that can regulate the biological functions of target genes by mimicking endogenous microRNAs. Let-7e agomir suppressed the mRNA levels of PARP1 and insulin-like growth factor I (IGF-1) while its downregulation enhanced PARP1 and IGF-1 expression [37]. Specific miRNA expression profiles could therefore serve as biomarkers in ovarian cancer to predict response to PARPi therapy.

5.2. ATR

The ATR protein belonging to the phosphatidylinositol 3-kinase-related (PI3K) family is involved in the signaling of stalled replication forks and maintaining genomic stability during the S phase, along with its partners ATR interacting protein (ATRIP) and replication protein A (RPA) [147]. A broad spectrum of DNA damage, such as single- and double-stranded DNA breaks, cross-links, and adducts, can lead to the activation of ATR [148]. ATR is referred to as the "master of DDR", highlighting the relevance of miRNAs implicated in DDR pathways as novel therapeutic targets for ovarian cancer. MiR-383-5p and miR-185-5p have been shown to be associated with ATR kinase. MiRNA-383-5p is predominantly downregulated and acts as a tumor suppressor in several human cancer types, such as gastric, glioma, medulloblastoma, and testicular embryonal carcinomas. In the mammalian ovary, miR-383 plays a functional role in follicle development [63]. MiR-185 suppresses expression of ATR and activation of its downstream effector, CHK1, which are induced by ionizing radiation. Furthermore, miR-185 is reported to induce G1 cell cycle arrest and apoptosis, inhibiting cancer cell proliferation [149].

A serine/threonine-protein kinase, PLK-4, has been identified as a target of miR-126, which is downregulated in various cancers in correlation with tumor progression and poor prognosis. Earlier experiments showed that PLK-4 knockdown led to a decrease in expression of ATR and CHK1, supporting its interactions with the ATR/CHK1 pathway. Moreover, changes in miR-126 expression led to PLK-4, ATR, and CHK1 dysregulation [67]. Based on these findings, it is proposed that miR-126 inhibits cancer progression via regulation of the cell cycle through inducing alterations in the ATR/CHK1 pathway.

MiR-708 overexpression is associated with suppression of the ATR/CHK1 pathway. Timeless was a direct target of miR-708. Total and phosphorylated ATR and CHK1 levels were decreased in cells overexpressing miR-708 after cisplatin treatment [68], supporting the utility of this miRNA as a potential therapeutic target. Overall, the effects of miRNAs on ATR kinase levels signify their potential application as new therapeutic targets for ovarian cancer.

5.3. CHK1

Checkpoint kinase 1 is a serine/threonine kinase encoded by the CHEK1 gene activated in response to DNA damage and replication stress that is proposed to regulate mitotic progression [150]. ATR and CHK1 share the same signaling pathway. However, in addition to ATR-induced activation, CHK1 can be autophosphorylated and activated independently of ATR [151].

Numerous studies have validated the oncogenic association of miR-424. Decreased expression of miR-424-5p is significantly associated with distant metastasis in high-stage (stage III and IV) ovarian cancers [72]. Moreover, downregulation of miR-424 contributes to the progression of cervical cancer via upregulation of target CHEK1 gene expression and phosphorylation of CHK1 protein, while its overexpression inhibits CHK1 expression [152].

Another miRNA downregulated in serous ovarian tumours is miR-195-5p [153]. In lung tumor tissues, miR-195 expression is low and associated with poor survival outcomes, while overexpression of miR-195 results in suppression of cancer cell growth, migration, and invasion. CHK1 has been identified as a direct target of miR-195. Low expression of miR-195 leads to high expression of CHK1, which is associated with poor prognosis in patients with lung tumors [154].

Expression of miR-330-5p regulates the development of different tumor cell types. In cutaneous malignant melanoma, miR-330 suppresses cell proliferation as well as expression of tyrosinase and protein disulfide-isomerase A3 (PDIA3) [155]. Conversely, its overexpression could promote apoptosis of prostate cancer cells through E2F1-mediated suppression of RAC-alpha serine/threonine-protein kinase (Akt) phosphorylation [156]. In esophageal adenocarcinoma, miR-330 was shown to modulate neoadjuvant chemoradiotherapy sensitivity [157], while in non-small cell lung cancer, its overexpression inhibited NIN1/RPN12 binding protein 1 homolog (NOB1) expression and cancer cell growth [158]. On the other hand, downregulation of miR-330-5p is reported in epithelial ovarian cancer tissues [159]. Moreover, the long non-coding RNA LINC01224 modulates expression of miR-330-5p, resulting in the downregulation of CHEK1 in hepatocellular carcinoma [160]. CHEK1 has also been identified as a direct target of miR-497, whereby expression of CHK1 protein is negatively regulated by miR-497 and upregulated under conditions of downregulation of miR-497 [161]. Other miRNAs responsible for suppressing expression of CHK1 and Wee1 are miR-16 and miR-26a. During genotoxic stress, p53 upregulates miR-16 and miR-26a, in turn attenuating expression of Wee1 and CHK1 [29]. These effects promote accumulation of cells in the G1 phase and, consequently, apoptosis. Additionally, miR-199b-3p overexpression in ovarian cancer suppresses E-box binding homeobox (ZEB)1 and CHK1. Moreover, E-cadherin and EMT expression were increased, which led to the conclusion that miRNA-199b-3p may suppress the progression of ovarian cancer via the CHK1/E-cadherin/EMT signaling pathway [162].

The collective findings highlight the significance of CHK1 as a key pharmacological target. Inhibition of CHK1 protein induces sensitization of cancer cells to genotoxic therapy and is recognized as beneficial in the treatment of ovarian cancer [163]. Thus, downregulation of CHK1 through targeted miRNAs may present an effective novel therapeutic strategy.

6. Conclusions

Since early detection tools are lacking, ovarian cancer is often diagnosed at late stages, which substantially contributes to the high mortality rates. MiRNAs are implicated in regulating almost every aspect of the DDR, DNA repair, and cell cycle arrest (Figures 2–5). MiRNAs may be an alternative method to identify DDR defects in patient therapy. Previously, a miRNA-score was developed that was associated with genome instability and predicted the outcome of ovarian cancer based on mutations in caretaker genes. The authors described 10 miRNAs. Six of them had higher expression than the median value across the dataset and were associated with a high frequency of mutation (miR-151, miR-301b, miR-505, miR-324, miR-502, and miR-421). The other four (let-7a, miR-320, miR-146a, and miR-193a) had lower expression associated with a lower frequency of mutation in the cancer genome [164].

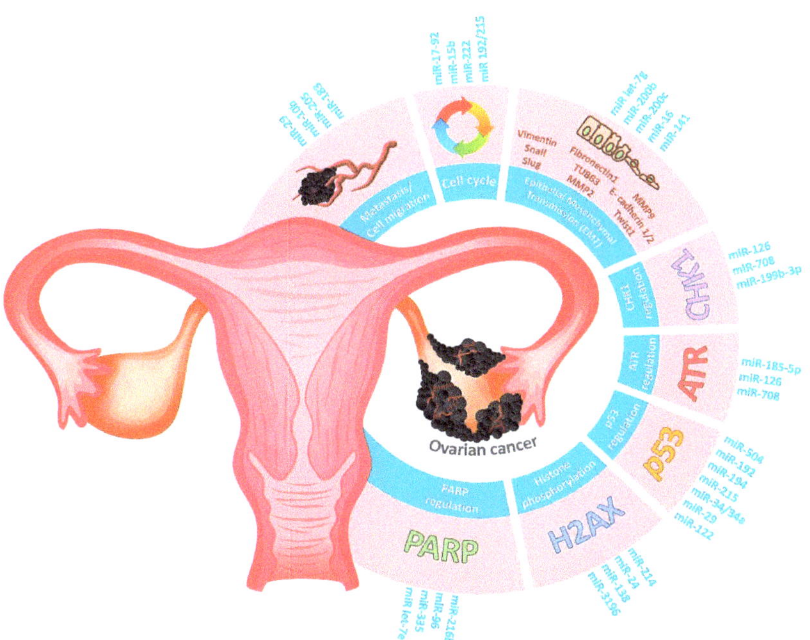

Figure 5. Schematic diagram representing several hallmarks of cancer contributing to pathogenesis of ovarian cancer. RNA markers involved in DDR signaling (PARP1, histone H2AX, P53, ATR, CHK1 regulation), cell cycle disturbance, metastasis, or epithelial–mesenchymal transmission are highlighted in blue.

Improved understanding of the critical roles of miRNAs in DDR and chemotherapy may therefore provide novel insights with a view to expanding their application as potential tools, biomarkers, or sensitizers in cancer treatment. Promising for increasing the effectiveness of ovarian cancer treatment is the combined therapy with miRNA and chemotherapeutic agents. The role of miRNA in modulating the ovarian cancer cells' sensitivity to chemotherapeutic agents in multidrug-resistance has been confirmed. It has been revealed that, e.g., decreased resistance to paclitaxel was associated with the upregulation of miR-29b, let-7i, miR-199a, miR-200a, miR-200c, and miR-215, while decreased resistance to platinum agents is related to the upregulation of miR-149, miR-155, miR152, miR-199a, miR200b, miR- 200c, miR-30d, miR-34c, miR-363, miR-497, miR-506, miR-9, and let-7i and to the downregulation of miR-23a and miR-603 [165]. Recent studies demonstrated also that miR-200c significantly enhanced the anticancer efficacy of olaparib in drug-resistant OC cells, which gives hope for optimizing the clinical use of PARPi [89]. Further research is warranted to clarify the correlations among miRNAs, DDR, and ovarian cancer. Continued advancements in miRNA research should allow clarification of the mechanisms' underlying cancer development, individualization of treatment, and improvement in prognosis for patients with ovarian cancer.

Author Contributions: A.G. and P.G. contributed equally to this work. P.G., A.G., A.M., and A.R. (in order of contribution) wrote this review. All authors have read and agreed to the published version of the manuscript.

Funding: This research was funded by the National Science Centre, Poland (Project grant number: Sonata Bis 2019/34/E/NZ7/00056).

Data Availability Statement: The data presented in this study are available on request from the corresponding author.

Conflicts of Interest: The authors declare no conflict of interest.

References

1. Ahmed, N.; Kadife, E.; Raza, A.; Short, M.; Jubinsky, P.T.; Kannourakis, G. Ovarian Cancer, Cancer Stem Cells and Current Treatment Strategies: A Potential Role of Magmas in the Current Treatment Methods. *Cells* **2020**, *9*, 719. [CrossRef] [PubMed]
2. Islam, M.J.; Roshid, B.; Pervin, S.; Kabir, S.; Chigurupati, S.; Hasan, M.N. A 35 Year Old Bangladeshi Lady with Hereditary Mucinous Ovarian Cancer, Complicated with Omental Metastasis. *Mymensingh Med. J. MMJ* **2019**, *28*, 484–489.
3. Davidson, B.; Tropé, C.G. Ovarian Cancer: Diagnostic, Biological and Prognostic Aspects. *Women's Health* **2014**, *10*, 519–533. [CrossRef]
4. Bonadio, R.R.D.C.C.; Fogace, R.N.; Miranda, V.C.; Diz, M.D.P.E. Homologous recombination deficiency in ovarian cancer: A review of its epidemiology and management. *Clinics* **2018**, *73*, e450s. [CrossRef]
5. Pujade-Lauraine, E.; Ledermann, J.A.; Selle, F.; Gebski, V.; Penson, R.T.; Oza, A.M.; Korach, J.; Huzarski, T.; Poveda, A.; Pignata, S.; et al. Olaparib tablets as maintenance therapy in patients with platinum-sensitive, relapsed ovarian cancer and a BRCA1/2 mutation (SOLO2/ENGOT-Ov21): A double-blind, randomised, placebo-controlled, phase 3 trial. *Lancet Oncol.* **2017**, *18*, 1274–1284. [CrossRef]
6. Xu, S.; Fu, G.-B.; Tao, Z.; Ouyang, J.; Kong, F.; Jiang, B.-H.; Wan, X.; Chen, K. MiR-497 decreases cisplatin resistance in ovarian cancer cells by targeting mTOR/P70S6K1. *Oncotarget* **2015**, *6*, 26457–26471. [CrossRef]
7. Zhou, L.; Liu, F.; Wang, X.; Ouyang, G. The roles of microRNAs in the regulation of tumor metastasis. *Cell Biosci.* **2015**, *5*, 32. [CrossRef]
8. Si, W.; Shen, J.; Zheng, H.; Fan, W. The role and mechanisms of action of microRNAs in cancer drug resistance. *Clin. Epigenet.* **2019**, *11*, 25. [CrossRef]
9. Han, W.; Cui, H.; Liang, J.; Su, X. Role of MicroRNA-30c in cancer progression. *J. Cancer* **2020**, *11*, 2593–2601. [CrossRef]
10. Hanlon, K.; Rudin, C.E.; Harries, L.W. Investigating the Targets of MIR-15a and MIR-16-1 in Patients with Chronic Lymphocytic Leukemia (CLL). *PLoS ONE* **2009**, *4*, e7169. [CrossRef]
11. Acunzo, M.; Croce, C.M. Downregulation of miR-15a and miR-16-1 at 13q14 in Chronic Lymphocytic Leukemia. *Clin. Chem.* **2016**, *62*, 655–656. [CrossRef] [PubMed]
12. Rampazzo, E.; Bojnik, E.; Trentin, L.; Bonaldi, L.; Del Bianco, P.; Frezzato, F.; Visentin, A.; Facco, M.; Semenzato, G.; De Rossi, A. Role of miR-15a/miR-16-1 and the TP53 axis in regulating telomerase expression in chronic lymphocytic leukemia. *Haematologica* **2017**, *102*, e253–e256. [CrossRef] [PubMed]
13. Hunter, M.P.; Ismail, N.; Zhang, X.; Aguda, B.D.; Lee, E.J.; Yu, L.; Xiao, T.; Schafer, J.; Lee, M.-L.T.; Schmittgen, T.D.; et al. Detection of microRNA Expression in Human Peripheral Blood Microvesicles. *PLoS ONE* **2008**, *3*, e3694. [CrossRef] [PubMed]
14. Ji, X.; Takahashi, R.; Hiura, Y.; Hirokawa, G.; Fukushima, Y.; Iwai, N. Plasma miR-208 as a Biomarker of Myocardial Injury. *Clin. Chem.* **2009**, *55*, 1944–1949. [CrossRef]
15. Valadi, H.; Ekström, K.; Bossios, A.; Sjöstrand, M.; Lee, J.J.; Lötvall, J. Exosome-mediated transfer of mRNAs and microRNAs is a novel mechanism of genetic exchange between cells. *Nat. Cell Biol.* **2007**, *9*, 654–659. [CrossRef]
16. Taylor, D.D.; Gercel-Taylor, C. MicroRNA signatures of tumor-derived exosomes as diagnostic biomarkers of ovarian cancer. *Gynecol. Oncol.* **2008**, *110*, 13–21. [CrossRef] [PubMed]
17. Zhang, H.; Xu, S.; Liu, X. MicroRNA profiling of plasma exosomes from patients with ovarian cancer using high-throughput sequencing. *Oncol. Lett.* **2019**, *17*, 5601–5607. [CrossRef]
18. Shapira, I.; Oswald, M.; Lovecchio, J.; Khalili, H.; Menzin, A.; Whyte, J.; Dos Santos, L.; Liang, S.; Bhuiya, T.; Keogh, M.; et al. Circulating biomarkers for detection of ovarian cancer and predicting cancer outcomes. *Br. J. Cancer* **2014**, *110*, 976–983. [CrossRef]
19. Maeda, K.; Sasaki, H.; Ueda, S.; Miyamoto, S.; Terada, S.; Konishi, H.; Kogata, Y.; Ashihara, K.; Fujiwara, S.; Tanaka, Y.; et al. Serum exosomal microRNA-34a as a potential biomarker in epithelial ovarian cancer. *J. Ovarian Res.* **2020**, *13*, 47. [CrossRef]
20. Nakamura, K.; Sawada, K.; Yoshimura, A.; Kinose, Y.; Nakatsuka, E.; Kimura, T. Clinical relevance of circulating cell-free microRNAs in ovarian cancer. *Mol. Cancer* **2016**, *15*, 48. [CrossRef]
21. Staicu, C.E.; Predescu, D.-V.; Rusu, C.M.; Radu, B.M.; Cretoiu, D.; Suciu, N.; Crețoiu, S.M.; Voinea, S.-C. Role of microRNAs as Clinical Cancer Biomarkers for Ovarian Cancer: A Short Overview. *Cells* **2020**, *9*, 169. [CrossRef]
22. Vaksman, O.; Tropé, C.; Davidson, B.; Reich, R. Exosome-derived miRNAs and ovarian carcinoma progression. *Carcinogenesis* **2014**, *35*, 2113–2120. [CrossRef]
23. Zhou, J.; Gong, G.; Tan, H.; Dai, F.; Zhu, X.; Chen, Y.; Wang, J.; Liu, Y.; Chen, P.; Wu, X.; et al. Urinary microRNA-30a-5p is a potential biomarker for ovarian serous adenocarcinoma. *Oncol. Rep.* **2015**, *33*, 2915–2923. [CrossRef]
24. Meng, X.; Müller, V.; Milde-Langosch, K.; Trillsch, F.; Pantel, K.; Schwarzenbach, H. Diagnostic and prognostic relevance of circulating exosomal miR-373, miR-200a, miR-200b and miR-200c in patients with epithelial ovarian cancer. *Oncotarget* **2016**, *7*, 16923–16935. [CrossRef]
25. Kapetanakis, N.-I.; Uzan, C.; Jimenez-Pailhes, A.-S.; Gouy, S.; Bentivegna, E.; Morice, P.; Caron, O.; Gourzones-Dmitriev, C.; Le Teuff, G.; Busson, P. Plasma miR-200b in ovarian carcinoma patients: Distinct pattern of pre/post-treatment variation compared to CA-125 and potential for prediction of progression-free survival. *Oncotarget* **2015**, *6*, 36815–36824. [CrossRef]
26. Chan, S.-H.; Wang, L.-H. Regulation of cancer metastasis by microRNAs. *J. Biomed. Sci.* **2015**, *22*, 9. [CrossRef]

27. Huang, Q.; Gumireddy, K.; Schrier, M.; Le Sage, C.; Nagel, R.; Nair, S.; Egan, D.A.; Li, A.; Huang, G.; Klein-Szanto, A.J.; et al. The microRNAs miR-373 and miR-520c promote tumour invasion and metastasis. *Nat. Cell Biol.* **2008**, *10*, 202–210. [CrossRef]
28. Li, N.; Yang, L.; Sun, Y.; Wu, X. MicroRNA-16 inhibits migration and invasion via regulation of the Wnt/β-catenin signaling pathway in ovarian cancer. *Oncol. Lett.* **2019**, *17*, 2631–2638. [CrossRef]
29. Lezina, L.; Purmessur, N.; Antonov, A.V.; Ivanova, T.; Karpova, E.; Krishan, K.; Ivan, M.; Aksenova, V.; Tentler, D.; Garabadgiu, A.V.; et al. miR-16 and miR-26a target checkpoint kinases Wee1 and Chk1 in response to p53 activation by genotoxic stress. *Cell Death Dis.* **2013**, *4*, e953. [CrossRef]
30. Choi, P.-W.; Wong, K.-K. The Functions of MicroRNA-200 Family in Ovarian Cancer: Beyond Epithelial-Mesenchymal Transition. *Int. J. Mol. Sci.* **2017**, *18*, 1207. [CrossRef]
31. Chung, V.Y.; Tan, T.Z.; Tan, M.; Wong, M.K.; Kuay, K.T.; Yang, Z.; Ye, J.; Muller, J.; Koh, C.M.; Guccione, E.; et al. GRHL2-miR-200-ZEB1 maintains the epithelial status of ovarian cancer through transcriptional regulation and histone modification. *Sci. Rep.* **2016**, *6*, 19943. [CrossRef]
32. Shi, C.; Yang, Y.; Zhang, L.; Yu, J.; Qin, S.; Xu, H.; Gao, Y. MiR-200a-3p promoted the malignant behaviors of ovarian cancer cells through regulating PCDH9. *OncoTargets Ther.* **2019**, *12*, 8329–8338. [CrossRef]
33. Brozovic, A.; Duran, G.E.; Wang, Y.C.; Francisco, E.B.; Sikic, B.I. The miR-200 family differentially regulates sensitivity to paclitaxel and carboplatin in human ovarian carcinoma OVCAR-3 and MES-OV cells. *Mol. Oncol.* **2015**, *9*, 1678–1693. [CrossRef] [PubMed]
34. Jiajie, T.; Yanzhou, Y.; Hoi-Hung, A.C.; Zi-Jiang, C.; Wai-Yee, C. Conserved miR-10 family represses proliferation and induces apoptosis in ovarian granulosa cells. *Sci. Rep.* **2017**, *7*, 41304. [CrossRef]
35. Yu, P.-N.; Yan, M.-D.; Lai, H.-C.; Huang, R.-L.; Chou, Y.-C.; Lin, W.-C.; Yeh, L.-T.; Lin, Y.-W. Downregulation ofmiR-29contributes to cisplatin resistance of ovarian cancer cells. *Int. J. Cancer* **2013**, *134*, 542–551. [CrossRef] [PubMed]
36. To, S.K.Y.; Mak, A.S.C.; Fung, Y.M.E.; Che, C.-M.; Li, S.-S.; Deng, W.; Ru, B.; Zhang, J.; Wong, A.S.T. β-catenin downregulates Dicer to promote ovarian cancer metastasis. *Oncogene* **2017**, *36*, 5927–5938. [CrossRef] [PubMed]
37. Xiao, M.; Cai, J.; Cai, L.; Jia, J.; Xie, L.; Zhu, Y.; Huang, B.; Jin, D.; Wang, Z. Let-7e sensitizes epithelial ovarian cancer to cisplatin through repressing DNA double strand break repair. *J. Ovarian Res.* **2017**, *10*, 24. [CrossRef]
38. Roush, S.; Slack, F.J. The let-7 family of microRNAs. *Trends Cell Biol.* **2008**, *18*, 505–516. [CrossRef]
39. Biamonte, F.; Santamaria, G.; Sacco, A.; Perrone, F.M.; Di Cello, A.; Battaglia, A.M.; Salatino, A.; Di Vito, A.; Aversa, I.; Venturella, R.; et al. MicroRNA let-7g acts as tumor suppressor and predictive biomarker for chemoresistance in human epithelial ovarian cancer. *Sci. Rep.* **2019**, *9*, 5668. [CrossRef]
40. Wang, X.; Cao, L.; Wang, Y.; Wang, X.; Liu, N.; You, Y. Regulation of let-7 and its target oncogenes (Review). *Oncol. Lett.* **2012**, *3*, 955–960. [CrossRef]
41. Wei, J.; Zhang, L.; Li, J.; Zhu, S.; Tai, M.; Mason, C.W.; Chapman, J.A.; Reynolds, E.A.; Weiner, C.P.; Zhou, H.H. MicroRNA-205 promotes cell invasion by repressing TCF21 in human ovarian cancer. *J. Ovarian Res.* **2017**, *10*, 33. [CrossRef]
42. Zhu, D.; Huang, X.; Liang, F.; Zhao, L. LncRNA miR503HG interacts with miR-31-5p through multiple ways to regulate cancer cell invasion and migration in ovarian cancer. *J. Ovarian Res.* **2020**, *13*, 3. [CrossRef]
43. Ibrahim, F.F.; Jamal, R.; Syafruddin, S.E.; Ab Mutalib, N.S.; Saidin, S.; MdZin, R.R.; Mollah, M.M.H.; Mokhtar, N.M. MicroRNA-200c and microRNA-31 regulate proliferation, colony formation, migration and invasion in serous ovarian cancer. *J. Ovarian Res.* **2015**, *8*, 56. [CrossRef] [PubMed]
44. Xiang, G.; Cheng, Y. MiR-126-3p inhibits ovarian cancer proliferation and invasion via targeting PLXNB2. *Reprod. Biol.* **2018**, *18*, 218–224. [CrossRef]
45. Xia, X.Y.; Yu, Y.J.; Ye, F.; Peng, G.Y.; Li, Y.J.; Zhou, X.M. MicroRNA-506-3p inhibits proliferation and promotes apoptosis in ovarian cancer cell via targeting SIRT1/AKT/FOXO3a signaling pathway. *Neoplasma* **2020**, *67*, 344–353. [CrossRef]
46. Zhou, J.; Zhang, C.; Zhou, B.; Jiang, D. miR-183 modulated cell proliferation and apoptosis in ovarian cancer through the TGF-β/Smad4 signaling pathway. *Int. J. Mol. Med.* **2019**, *43*, 1734–1746. [CrossRef]
47. Li, T.; Li, Y.; Gan, Y.; Tian, R.; Wu, Q.; Shu, G.; Yin, G. Methylation-mediated repression of MiR-424/503 cluster promotes proliferation and migration of ovarian cancer cells through targeting the hub gene KIF23. *Cell Cycle* **2019**, *18*, 1601–1618. [CrossRef]
48. Liao, Y.; Deng, Y.; Liu, J.; Ye, Z.; You, Z.; Yao, S.; He, S. MiR-760 overexpression promotes proliferation in ovarian cancer by downregulation of PHLPP2 expression. *Gynecol. Oncol.* **2016**, *143*, 655–663. [CrossRef]
49. Feng, Y.; Hang, W.; Sang, Z.; Li, S.; Xu, W.; Miao, Y.; Xi, X.; Huang, Q. Identification of exosomal and non-exosomal microRNAs associated with the drug resistance of ovarian cancer. *Mol. Med. Rep.* **2019**, *19*, 3376–3392. [CrossRef]
50. Lv, Y.L.F.; Liu, P.S. MiR-151 promotes ovarian cancer through activation of akt/mTOR signaling pathway by decreasing RhoGDIA. *Int. J. Clin. Exp. Pathol.* **2016**, *9*, 11222–11229.
51. Liu, X.; Yao, B.; Wu, Z. miRNA-199a-5p suppresses proliferation and invasion by directly targeting NF-κB1 in human ovarian cancer cells. *Oncol. Lett.* **2018**, *16*, 4543–4550. [CrossRef]
52. Meng, F.; Henson, R.; Wehbe–Janek, H.; Ghoshal, K.; Jacob, S.T.; Patel, T. MicroRNA-21 Regulates Expression of the PTEN Tumor Suppressor Gene in Human Hepatocellular Cancer. *Gastroenterology* **2007**, *133*, 647–658. [CrossRef]
53. Misso, G.; Di Martino, M.T.; De Rosa, G.; Farooqi, A.A.; Lombardi, A.; Campani, V.; Zarone, M.R.; Gullà, A.; Tagliaferri, P.; Tassone, P.; et al. Mir-34: A New Weapon Against Cancer? *Mol. Ther. Nucleic Acids* **2014**, *3*, e195. [CrossRef] [PubMed]
54. Frixa, T.; Donzelli, S.; Blandino, G. Oncogenic MicroRNAs: Key Players in Malignant Transformation. *Cancers* **2015**, *7*, 2466–2485. [CrossRef]

55. Kumar, D.; Gorain, M.; Kundu, G.; Kundu, G.C. Therapeutic implications of cellular and molecular biology of cancer stem cells in melanoma. *Mol. Cancer* **2017**, *16*, 7. [CrossRef]
56. Yeh, Y.-M.; Chuang, C.-M.; Chao, K.-C.; Wang, L.-H. MicroRNA-138 suppresses ovarian cancer cell invasion and metastasis by targeting SOX4 and HIF-1α. *Int. J. Cancer* **2013**, *133*, 867–878. [CrossRef] [PubMed]
57. Song, Q.; An, Q.; Niu, B.; Lu, X.; Zhang, N.; Cao, X. Role of miR-221/222 in Tumor Development and the Underlying Mechanism. *J. Oncol.* **2019**, *2019*, 7252013. [CrossRef]
58. Amini-Farsani, Z.; Sangtarash, M.H.; Shamsara, M.; Teimori, H. MiR-221/222 promote chemoresistance to cisplatin in ovarian cancer cells by targeting PTEN/PI3K/AKT signaling pathway. *Cytotechnology* **2018**, *70*, 203–213. [CrossRef]
59. Sun, C.; Cao, W.; Qiu, C.; Li, C.; Dongol, S.; Zhang, Z.; Dong, R.; Song, K.; Yang, X.; Zhang, Q.; et al. MiR-509-3 augments the synthetic lethality of PARPi by regulating HR repair in PDX model of HGSOC. *J. Hematol. Oncol.* **2020**, *13*, 9. [CrossRef]
60. Li, J.; Li, Q.; Huang, H.; Li, Y.; Li, L.; Hou, W.; You, Z. Overexpression of miRNA-221 promotes cell proliferation by targeting the apoptotic protease activating factor-1 and indicates a poor prognosis in ovarian cancer. *Int. J. Oncol.* **2017**, *50*, 1087–1096. [CrossRef]
61. Liu, R.; Guo, H.; Lu, S. MiR-335-5p restores cisplatin sensitivity in ovarian cancer cells through targeting BCL2L2. *Cancer Med.* **2018**, *7*, 4598–4609. [CrossRef]
62. Guan, R.; Cai, S.; Sun, M.; Xu, M. Upregulation of miR-520b promotes ovarian cancer growth. *Oncol. Lett.* **2017**, *14*, 3155–3161. [CrossRef] [PubMed]
63. Yin, M.; Lü, M.; Yao, G.; Tian, H.; Lian, J.; Liu, L.; Liang, M.; Wang, Y.; Sun, F. Transactivation of microRNA-383 by Steroidogenic Factor-1 Promotes Estradiol Release from Mouse Ovarian Granulosa Cells by Targeting RBMS1. *Mol. Endocrinol.* **2012**, *26*, 1129–1143. [CrossRef]
64. Nakayama, I.; Shibazaki, M.; Yashima-Abo, A.; Miura, F.; Sugiyama, T.; Masuda, T.; Maesawa, C. Loss of HOXD10 expression induced by upregulation of miR-10b accelerates the migration and invasion activities of ovarian cancer cells. *Int. J. Oncol.* **2013**, *43*, 63–71. [CrossRef]
65. Wang, J.; He, J.; Su, F.; Ding, N.; Hu, W.; Yao, B.; Wang, W.; Zhou, G. Repression of ATR pathway by miR-185 enhances radiation-induced apoptosis and proliferation inhibition. *Cell Death Dis.* **2013**, *4*, e699. [CrossRef]
66. Boele, J.; Persson, H.; Shin, J.W.; Ishizu, Y.; Newie, I.S.; Søkilde, R.; Hawkins, S.M.; Coarfa, C.; Ikeda, K.; Takayama, K.-I.; et al. PAPD5-mediated 3′ adenylation and subsequent degradation of miR-21 is disrupted in proliferative disease. *Proc. Natl. Acad. Sci. USA* **2014**, *111*, 11467–11472. [CrossRef]
67. Bao, J.; Yu, Y.; Chen, J.; He, Y.; Chen, X.; Ren, Z.; Xue, C.; Liu, L.; Hu, Q.; Li, J.; et al. MiR-126 negatively regulates PLK-4 to impact the development of hepatocellular carcinoma via ATR/CHEK1 pathway. *Cell Death Dis.* **2018**, *9*, 1045. [CrossRef]
68. Zou, X.; Zhu, C.; Zhang, L.; Zhang, Y.; Fu, F.; Chen, Y.; Zhou, J. MicroRNA-708 Suppresses Cell Proliferation and Enhances Chemosensitivity of Cervical Cancer Cells to cDDP by Negatively Targeting Timeless. *OncoTargets Ther.* **2020**, *13*, 225–235. [CrossRef]
69. Choi, Y.E.; Meghani, K.; Brault, M.-E.; Leclerc, L.; He, Y.; Day, T.A.; Elias, K.M.; Drapkin, R.; Weinstock, D.M.; Dao, F.; et al. Platinum and PARP Inhibitor Resistance Due to Overexpression of MicroRNA-622 in BRCA1-Mutant Ovarian Cancer. *Cell Rep.* **2016**, *14*, 429–439. [CrossRef]
70. Marchini, S.; Cavalieri, D.; Fruscio, R.; Calura, E.; Garavaglia, D.; Nerini, I.F.; Mangioni, C.; Cattoretti, G.; Clivio, L.; Beltrame, L.; et al. Association between miR-200c and the survival of patients with stage I epithelial ovarian cancer: A retrospective study of two independent tumour tissue collections. *Lancet Oncol.* **2011**, *12*, 273–285. [CrossRef]
71. Liu, J.; Zhang, X.; Huang, Y.; Zhang, Q.; Zhou, J.; Zhang, X.; Wang, X. miR-200b and miR-200c co-contribute to the cisplatin sensitivity of ovarian cancer cells by targeting DNA methyltransferases. *Oncol. Lett.* **2018**, *17*, 1453–1460. [CrossRef]
72. Cha, S.Y.; Choi, Y.H.; Hwang, S.; Jeong, J.-Y.; An, H.J. Clinical Impact of microRNAs Associated with Cancer Stem Cells as a Prognostic Factor in Ovarian Carcinoma. *J. Cancer* **2017**, *8*, 3538–3547. [CrossRef]
73. Kan, C.W.S.; Hahn, M.A.; Gard, G.B.; Maidens, J.; Huh, J.Y.; Marsh, D.J.; Howell, V.M. Elevated levels of circulating microRNA-200 family members correlate with serous epithelial ovarian cancer. *BMC Cancer* **2012**, *12*, 627. [CrossRef]
74. Iorio, M.V.; Visone, R.; Di Leva, G.; Donati, V.; Petrocca, F.; Casalini, P.; Taccioli, C.; Volinia, S.; Liu, C.-G.; Alder, H.; et al. MicroRNA Signatures in Human Ovarian Cancer. *Cancer Res.* **2007**, *67*, 8699–8707. [CrossRef]
75. Nam, E.J.; Yoon, H.; Kim, S.W.; Kim, H.; Kim, Y.T.; Kim, J.H.; Kim, S. MicroRNA Expression Profiles in Serous Ovarian Carcinoma. *Clin. Cancer Res.* **2008**, *14*, 2690–2695. [CrossRef]
76. Calura, E.; Fruscio, R.; Paracchini, L.; Bignotti, E.; Ravaggi, A.; Martini, P.; Sales, G.; Beltrame, L.; Clivio, L.; Ceppi, L.; et al. miRNA Landscape in Stage I Epithelial Ovarian Cancer Defines the Histotype Specificities. *Clin. Cancer Res.* **2013**, *19*, 4114–4123. [CrossRef] [PubMed]
77. Elgaaen, B.V.; Olstad, O.K.; Haug, K.B.F.; Brusletto, B.; Sandvik, L.; Staff, A.C.; Gautvik, K.M.; Davidson, B. Global miRNA expression analysis of serous and clear cell ovarian carcinomas identifies differentially expressed miRNAs including miR-200c-3p as a prognostic marker. *BMC Cancer* **2014**, *14*, 80. [CrossRef]
78. Yeom, K.-H.; Lee, Y.; Han, J.; Suh, M.R.; Kim, V.N. Characterization of DGCR8/Pasha, the essential cofactor for Drosha in primary miRNA processing. *Nucleic Acids Res.* **2006**, *34*, 4622–4629. [CrossRef]
79. Catalanotto, C.; Cogoni, C.; Zardo, G. MicroRNA in Control of Gene Expression: An Overview of Nuclear Functions. *Int. J. Mol. Sci.* **2016**, *17*, 1712. [CrossRef]

80. Svoronos, A.A.; Engelman, D.M.; Slack, F.J. OncomiR or Tumor Suppressor? The Duplicity of MicroRNAs in Cancer. *Cancer Res.* **2016**, *76*, 3666–3670. [CrossRef]
81. Palmero, E.; De Campos, S.G.P.; Campos, M.; De Souza, N.C.N.; Guerreiro, I.D.C.; Carvalho, A.L.; Marques, M.M.C. Mechanisms and role of microRNA deregulation in cancer onset and progression. *Genet. Mol. Biol.* **2011**, *34*, 363–370. [CrossRef]
82. Yin, F.; Liu, X.; Li, D.; Wang, Q.; Zhang, W.; Li, L. Tumor suppressor genes associated with drug resistance in ovarian cancer (Review). *Oncol. Rep.* **2013**, *30*, 3–10. [CrossRef]
83. Zhang, B.; Pan, X.; Cobb, G.; Anderson, T.A. microRNAs as oncogenes and tumor suppressors. *Dev. Biol.* **2007**, *302*, 1–12. [CrossRef]
84. Lu, L.; Schwartz, P.; Scarampi, L.; Rutherford, T.; Canuto, E.M.; Yu, H.; Katsaros, D. MicroRNA let-7a: A potential marker for selection of paclitaxel in ovarian cancer management. *Gynecol. Oncol.* **2011**, *122*, 366–371. [CrossRef]
85. Wang, Y.; Lei, X.; Gao, C.; Xue, Y.; Li, X.; Wang, H.; Feng, Y. MiR-506-3p suppresses the proliferation of ovarian cancer cells by negatively regulating the expression of MTMR6. *J. Biosci.* **2019**, *44*, 126. [CrossRef] [PubMed]
86. Bandyopadhyay, S.; Mitra, R.; Maulik, U.; Zhang, M.Q. Development of the human cancer microRNA network. *Silence* **2010**, *1*, 6. [CrossRef] [PubMed]
87. Jansson, M.D.; Lund, A.H. MicroRNA and cancer. *Mol. Oncol.* **2012**, *6*, 590–610. [CrossRef] [PubMed]
88. Berchuck, A.; Kohler, M.F.; Bast, R.C. Oncogenes in Ovarian Cancer. *Hematol. Clin. N. Am.* **1992**, *6*, 813–827. [CrossRef]
89. Vescarelli, E.; Gerini, G.; Megiorni, F.; Anastasiadou, E.; Pontecorvi, P.; Solito, L.; De Vitis, C.; Camero, S.; Marchetti, C.; Mancini, R.; et al. MiR-200c sensitizes Olaparib-resistant ovarian cancer cells by targeting Neuropilin 1. *J. Exp. Clin. Cancer Res.* **2020**, *39*, 3. [CrossRef]
90. Duran, G.E.; Wang, Y.C.; Moisan, F.; Francisco, E.B.; Sikic, B.I. Decreased levels of baseline and drug-induced tubulin polymerisation are hallmarks of resistance to taxanes in ovarian cancer cells and are associated with epithelial-to-mesenchymal transition. *Br. J. Cancer* **2017**, *116*, 1318–1328. [CrossRef]
91. Huang, R.-X.; Zhou, P.-K. DNA damage response signaling pathways and targets for radiotherapy sensitization in cancer. *Signal Transduct. Target. Ther.* **2020**, *5*, 60. [CrossRef]
92. Taylor, R.; Cullen, S.P.; Martin, S. Apoptosis: Controlled demolition at the cellular level. *Nat. Rev. Mol. Cell Biol.* **2008**, *9*, 231–241. [CrossRef]
93. Yousefi, B.; Rahmati, M.; Ahmadi, Y. The roles of p53R2 in cancer progression based on the new function of mutant p53 and cytoplasmic p21. *Life Sci.* **2014**, *99*, 14–17. [CrossRef]
94. Fukuda, T.; Yamagata, K.; Fujiyama, S.; Matsumoto, T.; Koshida, I.; Yoshimura, K.; Mihara, M.; Naitou, M.; Endoh, H.; Nakamura, T.; et al. DEAD-box RNA helicase subunits of the Drosha complex are required for processing of rRNA and a subset of microRNAs. *Nat. Cell Biol.* **2007**, *9*, 604–611. [CrossRef]
95. Gralewska, P.; Gajek, A.; Marczak, A.; Rogalska, A. Participation of the ATR/CHK1 pathway in replicative stress targeted therapy of high-grade ovarian cancer. *J. Hematol. Oncol.* **2020**, *13*, 39. [CrossRef]
96. Saravi, S.; Katsuta, E.; Jeyaneethi, J.; Amin, H.A.; Kaspar, M.; Takabe, K.; Pados, G.; Drenos, F.; Hall, M.; Karteris, E. H2A Histone Family Member X (H2AX) Is Upregulated in Ovarian Cancer and Demonstrates Utility as a Prognostic Biomarker in Terms of Overall Survival. *J. Clin. Med.* **2020**, *9*, 2844. [CrossRef] [PubMed]
97. Medina, P.P.; Slack, F.J. MicroRNAs and cancer: An overview. *Cell Cycle* **2008**, *7*, 2485–2492. [CrossRef] [PubMed]
98. Wang, Y.; Huang, J.-W.; Li, M.; Cavenee, W.K.; Mitchell, P.; Zhou, X.; Tewari, M.; Furnari, F.B.; Taniguchi, T. MicroRNA-138 Modulates DNA Damage Response by Repressing Histone H2AX Expression. *Mol. Cancer Res.* **2011**, *9*, 1100–1111. [CrossRef]
99. Georges, S.A.; Biery, M.C.; Kim, S.-Y.; Schelter, J.M.; Guo, J.; Chang, A.N.; Jackson, A.L.; Carleton, M.O.; Linsley, P.S.; Cleary, M.A.; et al. Coordinated Regulation of Cell Cycle Transcripts by p53-Inducible microRNAs, miR-192 and miR-215. *Cancer Res.* **2008**, *68*, 10105–10112. [CrossRef]
100. Agostini, A.; Brunetti, M.; Davidson, B.; Tropé, C.G.; Eriksson, A.G.Z.; Heim, S.; Panagopoulos, I.; Micci, F. The microRNA miR-192/215 family is upregulated in mucinous ovarian carcinomas. *Sci. Rep.* **2018**, *8*, 11069. [CrossRef]
101. Stucki, M.; Clapperton, J.A.; Mohammad, D.; Yaffe, M.B.; Smerdon, S.J.; Jackson, S.P. MDC1 Directly Binds Phosphorylated Histone H2AX to Regulate Cellular Responses to DNA Double-Strand Breaks. *Cell* **2005**, *123*, 1213–1226. [CrossRef]
102. Xiao, A.; Li, H.; Shechter, D.; Ahn, S.H.; Fabrizio, L.A.; Erdjument-Bromage, H.; Ishibe-Murakami, S.; Wang, B.; Tempst, P.; Hofmann, K.; et al. WSTF regulates the H2A.X DNA damage response via a novel tyrosine kinase activity. *Nat. Cell Biol.* **2008**, *457*, 57–62. [CrossRef]
103. Coster, G.; Goldberg, M. The cellular response to DNA damage: A focus on MDC1 and its interacting proteins. *Nucleus* **2010**, *1*, 166–178. [CrossRef]
104. Chapman, R.; Jackson, S.P. Phospho-dependent interactions between NBS1 and MDC1 mediate chromatin retention of the MRN complex at sites of DNA damage. *EMBO Rep.* **2008**, *9*, 795–801. [CrossRef] [PubMed]
105. Lukas, J.; Lukas, C.; Bartek, J. More than just a focus: The chromatin response to DNA damage and its role in genome integrity maintenance. *Nat. Cell Biol.* **2011**, *13*, 1161–1169. [CrossRef]
106. Nakada, S. Opposing roles of RNF8/RNF168 and deubiquitinating enzymes in ubiquitination-dependent DNA double-strand break response signaling and DNA-repair pathway choice. *J. Radiat. Res.* **2016**, *57*, i33–i40. [CrossRef] [PubMed]
107. Yu, J.; Qin, B.; Lou, Z. Ubiquitin and ubiquitin-like molecules in DNA double strand break repair. *Cell Biosci.* **2020**, *10*, 13. [CrossRef] [PubMed]

108. Price, B.D.; D'Andrea, A.D. Chromatin Remodeling at DNA Double-Strand Breaks. *Cell* **2013**, *152*, 1344–1354. [CrossRef]
109. Wang, Z.; Yin, H.; Zhang, Y.; Feng, Y.; Yan, Z.; Jiang, X.; Bukhari, I.; Iqbal, F.; Cooke, H.J.; Shi, Q. miR-214-mediated downregulation of RNF8 induces chromosomal instability in ovarian cancer cells. *Cell Cycle* **2014**, *13*, 3519–3528. [CrossRef]
110. Wan, G.; Mathur, R.; Hu, X.; Zhang, X.; Lu, X. miRNA response to DNA damage. *Trends Biochem. Sci.* **2011**, *36*, 478–484. [CrossRef]
111. Lal, A.; Pan, Y.; Navarro, F.; Dykxhoorn, D.M.; Moreau, L.; Meire, E.; Bentwich, Z.; Lieberman, J.; Chowdhury, D. miR-24–mediated downregulation of H2AX suppresses DNA repair in terminally differentiated blood cells. *Nat. Struct. Mol. Biol.* **2009**, *16*, 492–498. [CrossRef]
112. Xu, C.; Zhang, L.; Duan, L.; Lu, C. MicroRNA-3196 is inhibited by H2AX phosphorylation and attenuates lung cancer cell apoptosis by downregulating PUMA. *Oncotarget* **2016**, *7*, 77764–77776. [CrossRef]
113. Wu, J.; Sun, Y.; Zhang, P.-Y.; Qian, M.; Zhang, H.; Chen, X.; Ma, D.; Xu, Y.; Chen, X.; Tang, K.-F. The Fra-1–miR-134–SDS22 feedback loop amplifies ERK/JNK signaling and reduces chemosensitivity in ovarian cancer cells. *Cell Death Dis.* **2016**, *7*, e2384. [CrossRef]
114. Chatterjee, N.; Walker, G.C. Mechanisms of DNA damage, repair, and mutagenesis. *Environ. Mol. Mutagen.* **2017**, *58*, 235–263. [CrossRef]
115. Jackson, S.P.; Bartek, J. The DNA-damage response in human biology and disease. *Nat. Cell Biol.* **2009**, *461*, 1071–1078. [CrossRef]
116. Dietlein, F.; Thelen, L.; Reinhardt, H.C. Cancer-specific defects in DNA repair pathways as targets for personalized therapeutic approaches. *Trends Genet.* **2014**, *30*, 326–339. [CrossRef] [PubMed]
117. Alhmoud, J.F.; Woolley, J.F.; Al Moustafa, A.-E.; Malki, M.I. DNA Damage/Repair Management in Cancers. *Cancers* **2020**, *12*, 1050. [CrossRef]
118. Van Gent, D.C.; Hoeijmakers, J.H.J.; Kanaar, R. Chromosomal stability and the DNA double-stranded break connection. *Nat. Rev. Genet.* **2001**, *2*, 196–206. [CrossRef]
119. Lieber, M.R. The Mechanism of Double-Strand DNA Break Repair by the Nonhomologous DNA End-Joining Pathway. *Annu. Rev. Biochem.* **2010**, *79*, 181–211. [CrossRef]
120. Brandsma, I.; van Gent, D.C. Pathway choice in DNA double strand break repair: Observations of a balancing act. *Genome Integr.* **2012**, *3*, 9. [CrossRef]
121. Zhao, X.; Wei, C.; Li, J.; Xing, P.; Li, J.; Zheng, S.; Chen, X. Cell cycle-dependent control of homologous recombination. *Acta Biochim. Biophys. Sin.* **2017**, *49*, 655–668. [CrossRef]
122. Northall, S.J.; Ivančić-Baće, I.; Soultanas, P.; Bolt, E.L. Remodeling and Control of Homologous Recombination by DNA Helicases and Translocases that Target Recombinases and Synapsis. *Genes* **2016**, *7*, 52. [CrossRef] [PubMed]
123. Kolinjivadi, A.M.; Sannino, V.; De Antoni, A.; Técher, H.; Baldi, G.; Costanzo, V. Moonlighting at replication forks—A new life for homologous recombination proteins BRCA1, BRCA2 and RAD51. *FEBS Lett.* **2017**, *591*, 1083–1100. [CrossRef] [PubMed]
124. Moskwa, P.; Buffa, F.M.; Pan, Y.; Panchakshari, R.; Gottipati, P.; Muschel, R.J.; Beech, J.; Kulshrestha, R.; Abdelmohsen, K.; Weinstock, D.M.; et al. miR-182-Mediated Downregulation of BRCA1 Impacts DNA Repair and Sensitivity to PARP Inhibitors. *Mol. Cell* **2011**, *41*, 210–220. [CrossRef]
125. Pannunzio, N.R.; Watanabe, G.; Lieber, M.R. Nonhomologous DNA end-joining for repair of DNA double-strand breaks. *J. Biol. Chem.* **2018**, *293*, 10512–10523. [CrossRef] [PubMed]
126. Chiruvella, K.K.; Liang, Z.; Wilson, T.E. Repair of Double-Strand Breaks by End Joining. *Cold Spring Harb. Perspect. Biol.* **2013**, *5*, a012757. [CrossRef]
127. Yan, D.; Ng, W.L.; Zhang, X.; Wang, P.; Zhang, Z.; Mo, Y.-Y.; Mao, H.; Hao, C.; Olson, J.J.; Curran, W.J.; et al. Targeting DNA-PKcs and ATM with miR-101 Sensitizes Tumors to Radiation. *PLoS ONE* **2010**, *5*, e11397. [CrossRef]
128. Chakraborty, A.; Uechi, T.; Kenmochi, N. Guarding the 'translation apparatus': Defective ribosome biogenesis and the p53 signaling pathway. *Wiley Interdiscip. Rev. RNA* **2011**, *2*, 507–522. [CrossRef]
129. Dancy, B.M.; Cole, P.A. Protein Lysine Acetylation by p300/CBP. *Chem. Rev.* **2015**, *115*, 2419–2452. [CrossRef]
130. Hasty, P.; Christy, B.A. p53 as an intervention target for cancer and aging. *Pathobiol. Aging Age-Relat. Dis.* **2013**, *3*. [CrossRef]
131. Chen, X.; Che, T. Roles of MicroRNA in DNA Damage and Repair. In *DNA Repair*; Kruman, I., Ed.; InTech; ISBN 978-953-307-697-3. Available online: http://www.intechopen.com/books/dnarepair/roles-of-microrna-in-dna-damage-and-repair (accessed on 13 April 2021).
132. Jones, M.F.; Lal, A. MicroRNAs, wild-type and mutant p53: More questions than answers. *RNA Biol.* **2012**, *9*, 781–791. [CrossRef]
133. Yamakuchi, M.; Ferlito, M.; Lowenstein, C.J. miR-34a repression of SIRT1 regulates apoptosis. *Proc. Natl. Acad. Sci USA* **2008**, *105*, 13421–13426. [CrossRef]
134. Park, S.-Y.; Lee, J.H.; Ha, M.; Nam, J.-W.; Kim, V.N. miR-29 miRNAs activate p53 by targeting p85α and CDC42. *Nat. Struct. Mol. Biol.* **2008**, *16*, 23–29. [CrossRef]
135. Fornari, F.; Gramantieri, L.; Ferracin, M.; Veronese, A.; Sabbioni, S.; A Calin, G.; Grazi, G.L.; Giovannini, C.; Croce, C.M.; Bolondi, L.; et al. MiR-221 controls CDKN1C/p57 and CDKN1B/p27 expression in human hepatocellular carcinoma. *Oncogene* **2008**, *27*, 5651–5661. [CrossRef] [PubMed]
136. Fornari, F.; Gramantieri, L.; Giovannini, C.; Veronese, A.; Ferracin, M.; Sabbioni, S.; Calin, G.A.; Grazi, G.L.; Croce, C.M.; Tavolari, S.; et al. MiR-122/Cyclin G1 Interaction Modulates p53 Activity and Affects Doxorubicin Sensitivity of Human Hepatocarcinoma Cells. *Cancer Res.* **2009**, *69*, 5761–5767. [CrossRef]
137. Corney, D.C.; Flesken-Nikitin, A.; Godwin, A.K.; Wang, W.; Nikitin, A.Y. MicroRNA-34b and MicroRNA-34c Are Targets of p53 and Cooperate in Control of Cell Proliferation and Adhesion-Independent Growth. *Cancer Res.* **2007**, *67*, 8433–8438. [CrossRef]

138. Raver-Shapira, N.; Marciano, E.; Meiri, E.; Spector, Y.; Rosenfeld, N.; Moskovits, N.; Bentwich, Z.; Oren, M. Transcriptional Activation of miR-34a Contributes to p53-Mediated Apoptosis. *Mol. Cell* **2007**, *26*, 731–743. [CrossRef] [PubMed]
139. Chang, T.-C.; Wentzel, E.A.; Kent, O.A.; Ramachandran, K.; Mullendore, M.; Lee, K.H.; Feldmann, G.; Yamakuchi, M.; Ferlito, M.; Lowenstein, C.J.; et al. Transactivation of miR-34a by p53 Broadly Influences Gene Expression and Promotes Apoptosis. *Mol. Cell* **2007**, *26*, 745–752. [CrossRef]
140. Asslaber, D.; Piñón, J.D.; Seyfried, I.; Desch, P.; Stöcher, M.; Tinhofer, I.; Egle, A.; Merkel, O.; Greil, R. microRNA-34a expression correlates with MDM2 SNP309 polymorphism and treatment-free survival in chronic lymphocytic leukemia. *Blood* **2010**, *115*, 4191–4197. [CrossRef]
141. Bryant, H.E.; Schultz, N.; Thomas, H.D.; Parker, K.M.; Flower, D.; Lopez, E.; Kyle, S.; Meuth, M.; Curtin, N.J.; Helleday, T. Specific killing of BRCA2-deficient tumours with inhibitors of poly(ADP-ribose) polymerase. *Nat. Cell Biol.* **2005**, *434*, 913–917. [CrossRef]
142. Meghani, K.; Fuchs, W.; Detappe, A.; Drané, P.; Gogola, E.; Rottenberg, S.; Jonkers, J.; Matulonis, U.; Swisher, E.M.; Konstantinopoulos, P.A.; et al. Multifaceted Impact of MicroRNA 493-5p on Genome-Stabilizing Pathways Induces Platinum and PARP Inhibitor Resistance in BRCA2-Mutated Carcinomas. *Cell Rep.* **2018**, *23*, 100–111. [CrossRef]
143. Neijenhuis, S.; Bajrami, I.; Miller, R.; Lord, C.J.; Ashworth, A. Identification of miRNA modulators to PARP inhibitor response. *DNA Repair* **2013**, *12*, 394–402. [CrossRef] [PubMed]
144. Wang, Y.; Huang, J.-W.; Calses, P.; Kemp, C.J.; Taniguchi, T. MiR-96 Downregulates REV1 and RAD51 to Promote Cellular Sensitivity to Cisplatin and PARP Inhibition. *Cancer Res.* **2012**, *72*, 4037–4046. [CrossRef]
145. Liu, Y.; Niu, Z.; Lin, X.; Tian, Y. MiR-216b increases cisplatin sensitivity in ovarian cancer cells by targeting PARP1. *Cancer Gene Ther.* **2017**, *24*, 208–214. [CrossRef] [PubMed]
146. Yang, N.; Kaur, S.; Volinia, S.; Greshock, J.; Lassus, H.; Hasegawa, K.; Liang, S.; Leminen, A.; Deng, S.; Smith, L.; et al. MicroRNA Microarray Identifies Let-7i as a Novel Biomarker and Therapeutic Target in Human Epithelial Ovarian Cancer. *Cancer Res.* **2008**, *68*, 10307–10314. [CrossRef] [PubMed]
147. Cortez, D.; Guntuku, S.; Qin, J.; Elledge, S.J. ATR and ATRIP: Partners in Checkpoint Signaling. *Science* **2001**, *294*, 1713–1716. [CrossRef]
148. Cimprich, K.A.; Cortez, D. ATR: An essential regulator of genome integrity. *Nat. Rev. Mol. Cell Biol.* **2008**, *9*, 616–627. [CrossRef]
149. Feng, Y.-J.; Tian, R.; You, A.-G.; Wang, J.; Wu, Y.-J.; Wang, W.; Zhou, A.-Y.; Wei, X.-L.; He, Q.-D.; Feng, X.; et al. Expression and significance of DNA methyltransferase in sera of patients with lung cancer. *Zhonghua Yi Xue Za Zhi* **2013**, *93*, 3822–3825.
150. Bartek, J.; Lukas, J. Chk1 and Chk2 kinases in checkpoint control and cancer. *Cancer Cell* **2003**, *3*, 421–429. [CrossRef]
151. Buisson, R.; Boisvert, J.L.; Benes, C.H.; Zou, L. Distinct but Concerted Roles of ATR, DNA-PK, and Chk1 in Countering Replication Stress during S Phase. *Mol. Cell* **2015**, *59*, 1011–1024. [CrossRef]
152. Xu, J.; Li, Y.; Wang, F.; Wang, X.; Cheng, B.; Ye, F.; Xie, X.; Zhou, C.; Lu, W. Suppressed miR-424 expression via upregulation of target gene Chk1 contributes to the progression of cervical cancer. *Oncogene* **2012**, *32*, 976–987. [CrossRef] [PubMed]
153. Li, Y.; Yao, L.; Liu, F.; Hong, J.; Chen, L.; Zhang, B.; Zhang, W. Characterization of microRNA expression in serous ovarian carcinoma. *Int. J. Mol. Med.* **2014**, *34*, 491–498. [CrossRef] [PubMed]
154. Liu, B.; Qu, J.; Xu, F.; Guo, Y.; Wang, Y.; Yu, H.; Qian, B. MiR-195 suppresses non-small cell lung cancer by targeting CHEK1. *Oncotarget* **2015**, *6*, 9445–9456. [CrossRef]
155. Su, B.-B.; Zhou, S.-W.; Gan, C.-B.; Zhang, X.-N. MiR-330-5p regulates tyrosinase and PDIA3 expression and suppresses cell proliferation and invasion in cutaneous malignant melanoma. *J. Surg. Res.* **2016**, *203*, 434–440. [CrossRef]
156. Lee, K.-H.; Chen, Y.-L.; Yeh, S.-D.; Hsiao, M.; Lin, J.-T.; Goan, Y.-G.; Lu, P.-J. MicroRNA-330 acts as tumor suppressor and induces apoptosis of prostate cancer cells through E2F1-mediated suppression of Akt phosphorylation. *Oncogene* **2009**, *28*, 3360–3370. [CrossRef]
157. Bibby, B.A.S.; Reynolds, J.V.; Maher, S.G. MicroRNA-330-5p as a Putative Modulator of Neoadjuvant Chemoradiotherapy Sensitivity in Oesophageal Adenocarcinoma. *PLoS ONE* **2015**, *10*, e0134180. [CrossRef]
158. Kong, R.; Liu, W.; Guo, Y.; Feng, J.; Cheng, C.; Zhang, X.; Ma, Y.; Li, S.; Jiang, J.; Zhang, J.; et al. Inhibition of NOB1 by microRNA-330-5p overexpression represses cell growth of non-small cell lung cancer. *Oncol. Rep.* **2017**, *38*, 2572–2580. [CrossRef]
159. Shao, S.; Tian, J.; Zhang, H.; Wang, S. LncRNA myocardial infarction-associated transcript promotes cell proliferation and inhibits cell apoptosis by targeting miR-330-5p in epithelial ovarian cancer cells. *Arch. Med. Sci.* **2018**, *14*, 1263–1270. [CrossRef] [PubMed]
160. Gong, D.; Feng, P.-C.; Ke, X.-F.; Kuang, H.-L.; Pan, L.-L.; Ye, Q.; Wu, J.-B. Silencing Long Non-coding RNA LINC01224 Inhibits Hepatocellular Carcinoma Progression via MicroRNA-330-5p-Induced Inhibition of CHEK1. *Mol. Ther. Nucleic Acids* **2020**, *19*, 482–497. [CrossRef]
161. Xie, Y.; Wei, R.-R.; Huang, G.-L.; Zhang, M.-Y.; Yuan, Y.-F.; Wang, H.-Y. Checkpoint kinase 1 is negatively regulated by miR-497 in hepatocellular carcinoma. *Med. Oncol.* **2014**, *31*, 844. [CrossRef]
162. Wei, L.; He, Y.; Bi, S.; Li, X.; Zhang, J.; Zhang, S. miRNA-199b-3p suppresses growth and progression of ovarian cancer via the CHK1/E-cadherin/EMT signaling pathway by targeting ZEB1. *Oncol. Rep.* **2020**, *45*, 569–581. [CrossRef] [PubMed]
163. Kim, M.K.; James, J.; Annunziata, C.M. Topotecan synergizes with CHEK1 (CHK1) inhibitor to induce apoptosis in ovarian cancer cells. *BMC Cancer* **2015**, *15*, 196. [CrossRef] [PubMed]

164. Wang, T.; Wang, G.; Zhang, X.; Wu, D.; Yang, L.; Wang, G.; Hao, D. The expression of miRNAs is associated with tumour genome instability and predicts the outcome of ovarian cancer patients treated with platinum agents. *Sci. Rep.* **2017**, *7*, 14736. [CrossRef] [PubMed]
165. Ferreira, P.; Roela, R.A.; Lopez, R.V.M.; Estevez-Diz, M.D.P. The prognostic role of microRNA in epithelial ovarian cancer: A systematic review of literature with an overall survival meta-analysis. *Oncotarget* **2020**, *11*, 1085–1095. [CrossRef] [PubMed]

Article

Circulating miRNAs as Novel Non-Invasive Biomarkers to Aid the Early Diagnosis of Suspicious Breast Lesions for Which Biopsy Is Recommended

Marta Giussani [1,†], Chiara Maura Ciniselli [2,†], Loris De Cecco [3], Mara Lecchi [2], Matteo Dugo [4], Chiara Gargiuli [3], Andrea Mariancini [5], Elisa Mancinelli [6], Giulia Cosentino [7], Silvia Veneroni [8], Biagio Paolini [9], Rosaria Orlandi [7], Massimiliano Gennaro [10], Marilena Valeria Iorio [7], Catherine Depretto [11], Claudio Ferranti [11], Gabriella Sozzi [12], Marialuisa Sensi [3], Mario Paolo Colombo [13], Gianfranco Scaperrotta [11], Elda Tagliabue [7,*,†] and Paolo Verderio [2,†]

1 Laboratory Medicine Unit, Diagnostic Pathology and Laboratory Department, Fondazione IRCCS Istituto Nazionale dei Tumori, 20133 Milan, Italy; marta.giussani@istitutotumori.mi.it
2 Bioinformatics and Biostatistics Unit, Department of Applied Research and Technological Development, Fondazione IRCCS Istituto Nazionale dei Tumori, 20133 Milan, Italy; chiara.ciniselli@istitutotumori.mi.it (C.M.C.); mara.lecchi@istitutotumori.mi.it (M.L.); paolo.verderio@istitutotumori.mi.it (P.V.)
3 Platform of Integrated Biology, Department of Applied Research and Technological Development, Fondazione IRCCS Istituto Nazionale dei Tumori, 20133 Milan, Italy; loris.dececco@istitutotumori.mi.it (L.D.C.); chiara.gargiuli@istitutotumori.mi.it (C.G.); marialiusa.sensi@istitutotumori.mi.it (M.S.)
4 Department of Medical Oncology, IRCCS Ospedale San Raffaele, 20132 Milan, Italy; dugo.matteo@hsr.it
5 Department of Biomedical Sciences, Humanitas University, 20090 Pieve Emanuele, Italy; andrea.mariancini@humanitasresearch.it
6 Chemical-Clinical and Microbiological Analyses, ASST Grande Ospedale Metropolitano Niguarda, 20162 Milan, Italy; elisa.mancinelli@ospedaleniguarda.it
7 Molecular Targeting Unit, Department of Research, Fondazione IRCCS Istituto Nazionale dei Tumori, 20133 Milan, Italy; giulia.cosentino@istitutotumori.mi.it (G.C.); rosaria.orlandi@istitutotumori.mi.it (R.O.); marilena.iorio@istitutotumori.mi.it (M.V.I.)
8 Tissue Biobank, Department of Applied Research and Technological Development, Fondazione IRCCS Istituto Nazionale dei Tumori, 20133 Milan, Italy; silvia.veneroni@istitutotumori.mi.it
9 Anatomic Pathology A Unit, Department of Diagnostic Pathology and Laboratory, Fondazione IRCCS Istituto Nazionale dei Tumori, 20133 Milan, Italy; biagio.paolini@istitutotumori.mi.it
10 Breast Surgery Unit, Fondazione IRCCS Istituto Nazionale dei Tumori, 20133 Milan, Italy; massimiliano.gennaro@istitutotumori.mi.it
11 Breast Unit, Imaging, Fondazione IRCCS Istituto Nazionale dei Tumori, 20133 Milan, Italy; Catherine.Depretto@istitutotumori.mi.it (C.D.); claudio.ferrantii@istitutotumori.mi.it (C.F.); Gianfranco.scaperrotta@istitutotumori.mi.it (G.S.)
12 Tumor Genomics Unit, Department of Research, Fondazione IRCCS Istituto Nazionale dei Tumori, 20133 Milan, Italy; gabriella.sozzi@istitutotumori.mi.it
13 Molecular Immunology Unit, Department of Research, Fondazione IRCCS Istituto Nazionale dei Tumori, 20133 Milan, Italy; mario.colombo@istitutotumori.mi.it
* Correspondence: elda.tagliabue@istitutotumori.mi.it; Tel.: +39-02-2390-3013
† These authors contributed equally to this work.

Citation: Giussani, M.; Ciniselli, C.M.; De Cecco, L.; Lecchi, M.; Dugo, M.; Gargiuli, C.; Mariancini, A.; Mancinelli, E.; Cosentino, G.; Veneroni, S.; et al. Circulating miRNAs as Novel Non-Invasive Biomarkers to Aid the Early Diagnosis of Suspicious Breast Lesions for Which Biopsy Is Recommended. *Cancers* **2021**, *13*, 4028. https://doi.org/10.3390/cancers13164028

Academic Editor: Paola Tucci

Received: 16 July 2021
Accepted: 6 August 2021
Published: 10 August 2021

Simple Summary: In population-based screens, tissue biopsy remains the standard practice for women with imaging that suggests breast cancer. We examined circulating microRNAs as minimally invasive diagnostic biomarkers to discriminate malignant from benign breast lesions. A retrospective cohort of plasma samples divided into training and testing sets and a prospective cohort of women with suspicious imaging findings who underwent tissue biopsy were investigated through a global microRNA profile by OpenArray. Seven signatures, involving 5 specific miRNAs (miR-625, miR-423-5p, miR-370-3p, miR-181c, and miR-301b), were identified and validated in the testing set. Among the 7 signatures, the discriminatory performances of 5 of them were confirmed in the prospective cohort.

Abstract: In population-based screens, tissue biopsy remains the standard practice for women with imaging that suggests breast cancer. We examined circulating microRNAs as minimally invasive diagnostic biomarkers to discriminate malignant from benign breast lesions. miRNAs were analyzed by OpenArray in a retrospective cohort of plasma samples including 100 patients with malignant (T), 89 benign disease (B), and 99 healthy donors (HD) divided into training and testing sets and a prospective cohort (BABE) of 289 women with suspicious imaging findings who underwent tissue biopsy. miRNAs associated with disease status were identified by univariate analysis and then combined into signatures by multivariate logistic regression models. By combining 16 miRNAs differentially expressed in the T vs. HD comparison, 26 signatures were also able to significantly discriminate T from B disease. Seven of them, involving 5 specific miRNAs (miR-625, miR-423-5p, miR-370-3p, miR-181c, and miR-301b), were statistically validated in the testing set. Among the 7 signatures, the discriminatory performances of 5 were confirmed in the prospective BABE Cohort. This study identified 5 circulating miRNAs that, properly combined, distinguish malignant from benign breast disease in women with a high likelihood of malignancy.

Keywords: breast cancer; diagnosis; circulating biomarkers; microRNAs

1. Introduction

Breast cancer (BC) is the most frequent cancer among women, and despite screens for its early diagnosis, it remains a leading cause of cancer-related mortality worldwide [1]. The Breast Imaging Report and Data System (BI-RADS) lexicon was introduced by the American College of Radiology to score the risk of suspected BC in imaging studies [2,3] and determine the need for image-guided biopsy. BI-RADS categories 4 and 5 classify suspicious lesions for which biopsy is recommended. However, while BI-RADS 5 findings are greatly suggestive of BC, BI-RADS 4 lesions are highly variable in the outcome group, having a probability of malignancy ranging from 3% to 95%. Thus, some patients have benign lesions but undergo unnecessary biopsies or, in some cases, surgery. Biopsy remains required to prove that suspicious imaging findings are malignant or benign in 7% to 10% of women who undergo breast cancer screens, as reported by the National Centre for screening monitoring (https://www.osservatorionazionalescreening.it/, accessed on 9 August 2021).

Tissue biopsy is an invasive procedure that represents a cost for the health system. Thus, a simple and minimally invasive test to overcome these drawbacks remains an unmet clinical need. One such option is to monitor circulating molecular markers in blood that distinguish benign from malignant breast disease. Over the last 20 years, the advent of "omics" strategies has led to novel approaches in the search for noninvasive biomarkers for diagnosing BC. Circulating carcinoma antigens, tumor cells, cell-free tumor DNA and RNA, and extracellular vesicles in the peripheral blood have appeared as potential biomarkers that supplement current clinical tools [4].

MiRNAs are a class of short noncoding, single-stranded RNAs that regulate gene expression at the post-transcriptional level by binding to target mRNAs. miRNAs are commonly dysregulated in various human cancers, becoming oncogenes or tumor suppressor genes and regulating several steps in neoplastic transformation (reviewed in [5,6]). Differences in miRNA expression in various malignancies have been examined primarily as biomarkers for the diagnosis, prognosis, and response to treatment in cancer. miRNAs can be secreted by several cell types into the extracellular space and then shuttled to peripheral blood in a form that is resistant to digestion by RNases through their encapsulation by extracellular vesicles or binding to lipoproteins. Because miRNAs are stable in routinely collected clinical liquid samples, in contrast to mRNA, these molecules constitute a class of reliable, minimally invasive cancer biomarkers that merit interest in the detection of early-onset disease (reviewed in [7]).

Circulating miRNAs are indicative of BC [8–19], and the combination of certain circulating miRNAs distinguishes BC from normal and healthy controls [20–24]. However, benign breast lesions that may yield diagnostic images that indicate BC have rarely been included in these studies, and the number of samples that have been considered has been limited [25–28]. Thus, the development of an accurate and reliable panel of circulating miRNAs for the early diagnosis of BC in women with suspicious diagnostic images remains a challenge.

In this study, we attempted to discriminate malignant from benign breast disease by analyzing circulating miRNAs in a training set and a testing set of retrospectively collected plasma samples from BC patients, women with breast benign disease, and healthy donors and performing a prospective clinical study of women with suspicious imaging findings (BI-RADS 4–5) who underwent biopsy to obtain a correct histopathological diagnosis of malignant or benign disease.

2. Materials and Methods

2.1. Plasma Samples

Two independent cohorts of plasma samples were retrospectively (Retrospective Cohort) and prospectively (BABE—BreAst Blood Early diagnosis) collected at Fondazione IRCCS Istituto Nazionale dei Tumori di Milano (INT) between 2013 and 2017 (Figure 1).

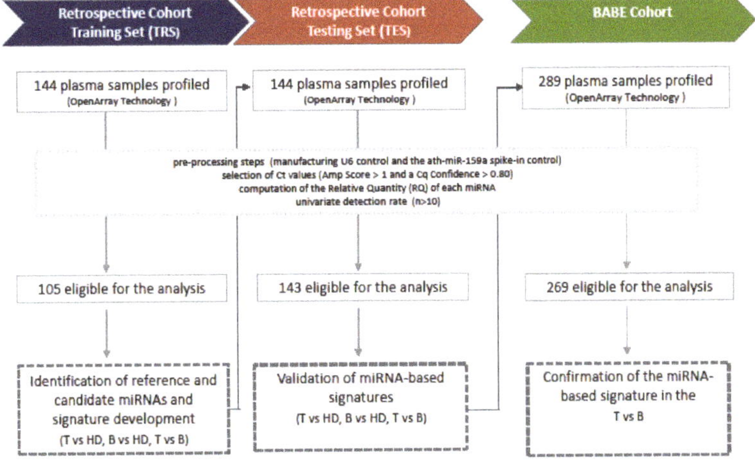

Figure 1. Study workflow. Graphical representation of plasma samples profiled and analyzed by OpenArray technology for each cohort.

For the Retrospective Cohort, a total of 288 plasma samples were collected between 2013 and 2015, stored in the Biobank of INT, and randomly split into a training (TRS) and testing (TES) set by annotated disease status. Overall, Retrospective Cohort consisted of 99 healthy donor women (HD, 50 in the TRS and 49 in the TES), 100 patients with a breast tumor (T, 50 in the TRS and 50 in the TES), and 89 patients with a benign breast lesion (B, 44 in the TRS and 45 in the TES). Two-hundred eighty-nine plasma samples from the BABE study were prospectively collected between 2015 and 2017 from women with no previous diagnosis of cancer. These women underwent a biopsy to determine whether abnormal areas, identified by breast ultrasonography [maximum diameter of 20 mm (BI-RADS 4-5)], were malignant or benign lesions. Plasma samples were also collected at 12 ± 3 months of follow-up from 29 women at INT after being diagnosed with a malignant lesion (BABE-FU Cohort). Institutional approval from our independent ethics committee was obtained for this study (approval numbers INT111-13, INT144-14, and INT66-15). Patients gave informed consent to the use of their samples. All procedures were conducted per the Declaration of Helsinki.

Table 1 summarizes the clinicopathological characteristics of the tumors in the cohorts by WHO classification [29]. The median age was 50 years (interquantile range, IQR: 42–56), 47 (IQR: 41–53), and 59 (IQR: 49–72) for HD, B, and T in the Retrospective Cohort, respectively; similarly, in the BABE Cohort, the median age was 46 (IQR: 41–52) and 55 (IQR: 48–70) for B and T, respectively. In the BABE-FU Cohort, the median age at surgery was 56 (IQR: 50–72). With regard to histology, benign breast disease was represented primarily by fibroadenoma (26%) and benign epithelial proliferations (47%) in the Retrospective Cohort and by fibroadenoma (35%) and benign epithelial proliferations (51%) in the BABE Cohort. All 29 BC patients of the BABE-FU Cohort received radiotherapy, 19 patients received hormone therapy alone and 5 patients received chemotherapy in addition to hormone therapy.

2.2. *Blood Collection, Plasma Separation, and RNA Extraction*

Blood was withdrawn before surgery from patients with T or B (Retrospective Cohort) in collaboration with the INT Biobank and before a core biopsy from women with imaging that was suggestive of breast cancer (BABE Cohort). For BABE patients in follow-up, blood was taken before surgery (T0) and 12 $\pm$ 3 months after surgery (T1) during a scheduled clinical evaluation. Blood was obtained from HDs at the time of blood donation in the Immunohematology and Transfusion Medicine Service of INT. Whole blood was collected in commercially available EDTA-treated tubes and centrifuged at 2200$\times$ g for 20 min at 4 °C to remove cells, and the recovered plasma was frozen immediately at −80 °C. Total RNA was extracted from 200 µL of plasma using the mirVana PARIS Kit, catalog number AM 1556, (Thermo Fisher Scientific, Waltham, MA, USA) according to the manufacturer's protocol and eluted in 50 µL of buffer.

To determine the influence of hemolysis on miRNA expression, an ad hoc forced hemolysis experiment was implemented. Hemolysis was artificially introduced into the plasma sample from an HD of the TRS by adding serial 1:4 dilutions of red blood cells (0.004–0.25% v/v) and uncontaminated plasma. Each sample was analyzed in triplicate over the entire absorbance spectrum on a NanoDrop™ 1000 (Thermo Fisher Scientific) before RNA extraction to check that hemolysis had occurred. The samples were then profiled for miRNA expression using OpenArray and analyzed per the simultaneous confidence intervals approach [30].

2.3. *miRNA Profile*

The miRNA in each sample was profiled by qRT-PCR using the OpenArray Human microRNA panel (OA), catalog number 4,470,189 (ThermoFisher Scientific), a fixed-content panel that contains 754 validated human TaqMan miR assays that were designed in miRBase, RRID:SCR_003152 v14.0. Briefly, the miRNAs in each sample were amplified with the manufacturer's replicates of internal controls, including U6 and ath-miR-159a spike-in. Reverse-transcription and preamplification were performed using Megaplex RT Primers Human Pools A (v2.1) and B (v3.0), catalog number 4,444,750, per the manufacturer's instructions. The samples, master mix, and Taqman reactions were arranged in a 384-well plate and transferred automatically to OpenArray plates using a QuantStudio OpenArray AccuFill System. The loaded OpenArray plate was sealed immediately, filled with OpenArray Immersion Fluid, and sealed by inserting the OpenArray Plug into the loading port. qRT-pCR was performed on a QuantStudio 12K Flex (Thermo Fisher). Primary data were retrieved using QuantStudio 12K Flex, v1.2.3.

Table 1. Clinicopathological characteristics of breast cancer patients.

		Retrospective Cohort (n = 100)	BABE Cohort (n = 125)	BABE FU Cohort (n = 29)
		n (%)	n (%)	n (%)
Histology	IDC [a]	74 (74)	87 (69)	17 (59)
	ILC [b]	10 (10)	18 (15)	5 (18)
	IDC + ILC	5 (5)	1 (1)	1 (3)
	In situ [c]	3 (3)	8 (6)	4 (14)
	IDC mixed [d]	3 (3)	4 (3)	1 (3)
	Special Types [e]	5 (1)	6 (5)	1 (3)
	Normal Tissue	-	1 (1)	-
IHC Histotype [f]	Luminal A	17 (17)	34 (27)	12 (41)
	Luminal B	45 (45)	60 (48)	12 (41)
	Luminal HER2	11 (11)	5 (4)	-
	HER2	5 (5)	7 (6)	1 (3)
	Triple-Negative	19 (19)	5 (4)	-
	In situ	3 (3)	8 (6)	4 (15)
	Not determined	-	6 (5)	-
Grade	I	8 (8)	14 (11)	3 (11)
	II	43 (43)	72 (58)	21 (72)
	III	49 (49)	38 (30)	5 (17)
	Not determined	-	1 (1)	-
Tumor Size [g]	T1	68 (68)	101 (81)	27 (93)
	T2	32 (32)	21 (17)	2 (7)
	Not determined	-	3 (2)	-
Lymph node	Negative	62 (62)	83 (66)	22 (76)
	Positive	38 (38)	24 (19)	7 (24)
	Not determined	-	18 (15)	-
ER [h]	Positive	74 (74)	105 (84)	26 (90)
	Negative	26 (26)	14 (11)	3 (10)
	Not determined	-	6 (5)	-
PgR [h]	Positive	63 (63)	95 (76)	19 (66)
	Negative	36 (36)	24 (19)	10 (34)
	Not determined	1 (1)	6 (5)	-
HER2 [i]	Positive	17 (17)	14 (11)	2 (7)
	Negative	83 (83)	105 (84)	27 (93)
	Not determined	-	6 (5)	-
Ki-67 [l]	Positive	79 (79)	79 (63)	13 (45)
	Negative	19 (19)	36 (29)	16 (55)
	Not determined	2 (2)	10 (8)	-
Age	Median (interquartile range)	59 (49–72)	55 (48–70)	56 (50–72)

[a] IDC infiltrating ductal carcinoma; [b] ILC infiltrating lobular carcinoma; [c] Ductal in situ and intracystic tumor; [d] IDC plus mucinous or iperplasia or in situ; [e] Other invasive tumors: Apocrine, Tubular, Mucinous, Metaplastic and Papillary. [f] IHC Subtype: Luminal A: ER^+, PgR^{+or-}, Ki-67^-, Luminal B: ER^+, PgR^{+or-}, Ki-67^+, Luminal HER2: ER^+, PgR^{+or-}, $HER2^+$, HER2: ER^-, PgR^-, $HER2^+$, Triple-Negative: ER^-, PgR^-, $HER2^-$; [g] T2 when size > 2 cm; [h] ER- and PgR-positive > 10% cell positivity by IHC; [i] HER2 positive scored 3+ by IHC or 2+/FISH-positive; [l] Ki-67-positive > 14% cell positivity by IHC.

2.4. Statistical Analysis

2.4.1. Preprocessing Step

For all cohorts, quality control of the data was performed to identify critical samples. The number of wells with a low ROX signal (ROX < 1000) and the number of detected miRNAs (Amp Score > 1 and Cq Confidence > 0.80) were evaluated for each sample. Outliers were flagged using the Hampel filter (values outside of the interval between the median of the distribution $\pm 3 \times$ the median absolute deviation were considered outliers). A hierarchical clustering of the correlation of expression profiles for all possible pairs of samples was also performed to assess the homogeneity of the data. Samples with detectable U6 manufacturing control and ath-miR-159a spike-in were included in the subsequent statistical analysis workflow. Ct values were analyzed in terms of Amp Score and Cq Confidence: only Ct values with Amp Score > 1 and Cq Confidence > 0.80 were considered in the subsequent statistical analysis.

TRS preprocessed data were analyzed to identify a subset of reference miRNAs and a set of candidate miRNAs that were to be combined into signatures. Reference miRNAs were identified by running an updated version of the NqA R-function [31,32]. The relative quantity (RQ) of each miRNA, expressed on a logarithmic scale ($\log_2 RQ = -dCt$), was then considered to be the pivotal variable for the subsequent statistical analysis. The same normalization was then used to analyze the TES and BABE data.

2.4.2. Retrospective Cohort Analysis

Two disease-specific comparisons were first considered for Retrospective Cohort: patients with breast tumor vs. healthy donors (T vs. HD) and patients with benign breast lesion vs. healthy donors (B vs. HD). In this step for the TRS data, only miRNAs that were detected in at least 10 subjects/disease were considered for the univariate analysis [33]. Differentially expressed miRNAs were identified in the univariate analysis by a nonparametric Kruskal–Wallis test. Candidate hemolysis-free miRNAs (according to the forced hemolysis experiment) that showed specific-disease statistical significance in the T vs. HD or B vs. HD comparison, but not both, were selected for the multivariate analysis. According to the required number of events per variable (EPV) [34], a standard method or the Penalized Maximum Likelihood Estimation (PMLE) approach [35,36] was used to combine significant miRNAs by multivariate analysis (i.e., all subset analyses). For each fitted model, the area under the receiver operating characteristic (ROC) curve (AUC) and its corresponding 95% confidence interval (95% CI) were calculated. Signatures that showed significant performance on the TRS, in terms of AUC values (i.e., lower 95% CI > 0.50), in the T vs. HD comparison but not in B vs. HD and vice versa were then evaluated on the TES data. Signatures that retained their significance on the TES were examined further in between T vs. B) by applying the same regression coefficients as in the TRS [33], to mimic the application to the subsequent BABE Cohort.

2.4.3. BABE Cohort Analysis

The most promising signatures in the Retrospective Cohort with regard to the T vs. B comparison were assessed in the BABE plasma samples alone or as extended models that included the CA15-3 epitope of the large transmembrane glycoprotein MUC1, that was tested in heparin plasma samples on an automatic electrochemiluminescence immunoassay system, catalog number 03045838122 (Cobas 6000 e601, Roche Diagnostics, Germany). The expression profiles of BABE-FU samples before surgery (T0) and 12 $\pm$ 3 months after surgery (T1) were compared by Wilcoxon signed rank (WSR) test for paired data.

All statistical analyses were carried out with SAS (Statistical Analysis System, RRID:SCR_008567, version 9.4.; SAS Institute, Inc., Cary, NC, USA) and R, adopting an α level of 5%.

3. Results

3.1. Retrospective Cohort Analysis

Of the 144 TRS plasma samples that were profiled on the OpenArray plates, 105 (46 HD, 31 T, and 28 B) passed the preprocessing steps, and 255 miRNAs were considered in subsequent statistical analyses (Figure 1). By NqA 31, 4 miRNAs (hsa-miR-143-002249, hsa-miR-152-000475, hsa-miR-185-002271, hsa-miR-139-5p-002289) were identified for data normalization. Hemolysis-free miRNAs that were associated with disease status (T vs. HD or B vs. HD) were identified by univariate analysis by Kruskal–Wallis test. Specifically,

16 miRNAs (10 upregulated and 6 downregulated) were differentially expressed only in T vs. HD comparison, versus 14 (3 upregulated and 11 downregulated) only in the B vs. HD comparison (Table 2, Supplementary Figure S1).

Table 2. List of differentially expressed miRNAs in the two disease-specific comparisons in the TRS.

T vs. HD	miRNA	#T	#HD	KW-*p* Value	Direction
	hsa-miR-423-5p-002340	31	45	0.0003	up
	hsa-miR-21-000397	29	46	0.0006	up
	hsa-miR-148a-000470	30	46	0.0011	up
	hsa-miR-218-000521	31	42	0.0037	up
	dme-miR-7-000268	24	37	0.0046	up
	hsa-miR-324-3p-002161	31	45	0.0067	up
	hsa-miR-502-3p-002083	30	46	0.0067	up
	hsa-miR-625-002431	27	45	0.0081	down
	hsa-miR-18a-002422	31	46	0.0120	up
	hsa-miR-142-5p-002248	31	46	0.0127	down
	hsa-miR-301b-002392	21	43	0.0148	down
	hsa-miR-186-002285	31	46	0.0153	down
	hsa-miR-370-002275	16	43	0.0155	up
	hsa-miR-548c-5p-002429	20	35	0.0182	up
	hsa-miR-181c-000482	30	44	0.0190	down
	mmu-miR-134-001186	18	43	0.0237	down
B vs. HD	**miRNA**	**#B**	**#HD**	**KW-*p* Value**	**Direction**
	hsa-miR-128a-002216	26	45	0.0008	down
	hsa-miR-24-000402	27	46	0.0009	down
	hsa-miR-598-001988	26	45	0.0013	down
	hsa-miR-27a-000408	28	46	0.0027	down
	hsa-miR-133a-002246	27	46	0.0028	down
	hsa-miR-30c-000419	28	46	0.0048	down
	hsa-miR-320-002277	28	46	0.0051	up
	hsa-miR-148b-000471	27	46	0.0068	down
	hsa-miR-204-000508	27	45	0.0107	up
	hsa-miR-376a-000565	28	45	0.0126	down
	hsa-miR-331-000545	28	46	0.0133	down
	hsa-miR-324-5p-000539	27	46	0.0140	down
	hsa-miR-330-000544	24	42	0.0142	down
	hsa-miR-502-001109	15	27	0.0216	up

T: breast tumor, B: benign breast lesion, HD: healthy donor women, KW: Kruskal–Wallis Test.

According to each comparison (T vs. HD or B vs. HD), candidate miRNAs were combined in multivariate manner (i.e., signatures) using the TRS data. In the T vs. HD scenario, 52 signatures retained their significant performance in the 143 samples that passed the preprocessing steps in the TES (i.e., T-promising signatures). No signatures showed significant performance in discriminating B vs. HD within the TES data. Among the 52 T-promising signatures, 26 had significant discriminatory performance in the T vs. B comparison for the TES. Supplementary Table S1 reports several descriptive statistics of the AUC values of these 26 signatures (i.e., TB-promising signatures) in the TRS and TES data.

Among these 26 TB-promising signatures, 7 (M1–M7, top signatures) retained their significant performance in the TES even by applying the same regression coefficients that were obtained in the TRS (Table 3).

These top signatures were specific combinations of 5 miRNAs, from a maximum of 4 to a minimum of 2 miRNAs. Figure 2 reports the ROC curves of the top 7 signatures in the TRS and TES, with AUC values ranging from 0.680 to 0.769 and 0.632 to 0.708, respectively.

Table 3. Performance of the 7 validated signatures (M1-M7) in the breast tumor vs. benign breast lesion.

Model	TRS Data AUC (95% CI)	TES Data AUC (95% CI)	n. miRNAs Included	miRNAs Included
M1	0.726 (0.556; 0.897)	0.708 (0.580; 0.837)	4	hsa-miR-423-5p-002340; hsa-miR-181c-000482; hsa-miR-625-002431; hsa-miR-301b-002392
M2	0.769 (0.562; 0.976)	0.683 (0.546; 0.820)	4	hsa-miR-423-5p-002340; hsa-miR-181c-000482; hsa-miR-301b-002392; hsa-miR-370-002275
M3	0.712 (0.527; 0.897)	0.696 (0.564; 0.828)	3	hsa-miR-181c-000482; hsa-miR-625-002431; hsa-miR-301b-002392
M4	0.753 (0.559; 0.946)	0.675 (0.539; 0.812)	3	hsa-miR-423-5p-002340; hsa-miR-625-002431; hsa-miR-370-002275
M5	0.688 (0.515; 0.861)	0.657 (0.522; 0.791)	3	hsa-miR-423-5p-002340; hsa-miR-625-002431; hsa-miR-301b-002392
M6	0.763 (0.557; 0.970)	0.660 (0.522; 0.799)	3	hsa-miR-181c-000482; hsa-miR-301b-002392; hsa-miR-370-002275
M7	0.680 (0.511; 0.849)	0.632 (0.507; 0.758)	2	hsa-miR-181c-000482; hsa-miR-301b-002392

TRS: training set; TES: testing set, AUC: area under the ROC Curve; CI: confidence interval.

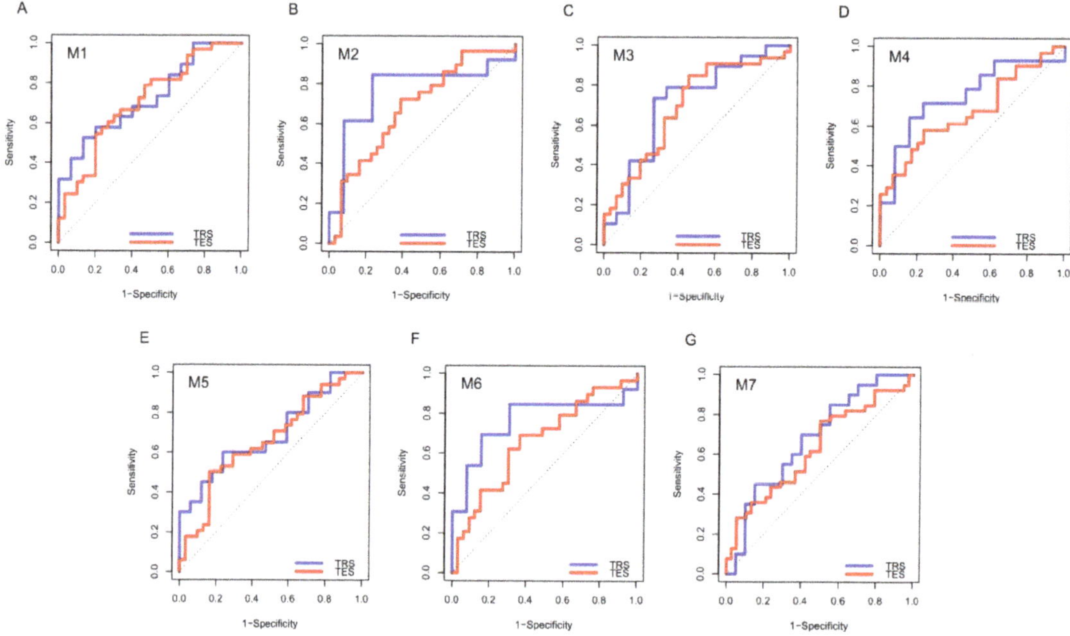

Figure 2. ROC curves of 7 miRNA-based signatures (M1–M7) in the TRS and TES sets. Each panel, (**A**): M1 signature; (**B**): M2 signature; (**C**): M3 signature; (**D**): M4 signature; (**E**): M5 signature; (**F**): M6 signature; (**G**): M7 signature, depicts the ROC curves of the 7 miRNA-based signatures in the training (TRS, in blue) and testing (TES, in red) sets of the Retrospective Cohort.

3.2. BABE Cohort Analysis

To confirm the relevance of all 7 signatures in a clinical setting, 289 EDTA-recovered plasma samples from the prospective BABE study and their technical controls were allocated and profiled in OpenArray plates, as performed for the TRS and TES data of the Retrospective Cohort. After the preprocessing steps, 269 samples (115 T and 154 B) were considered in subsequent statistical analyses. By fitting the M1-M7 signatures to the BABE data, the discriminatory performance of 5 signatures was confirmed (in terms of AUC value) but borderline significant for the remaining 2 (Figure 3).

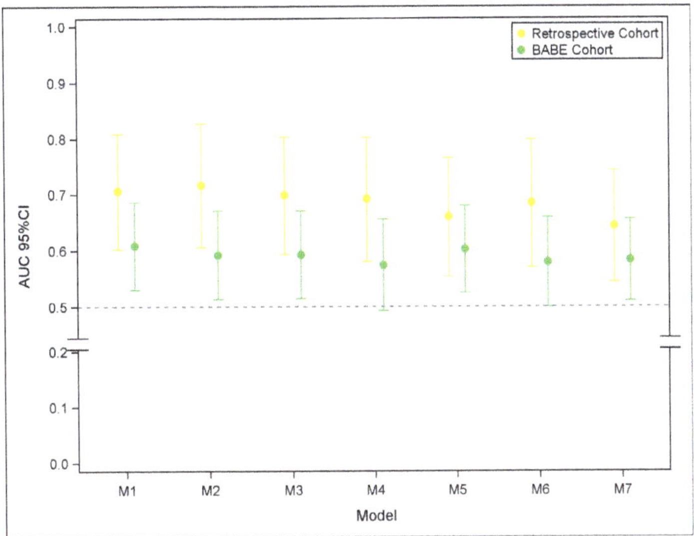

Figure 3. Performance of the 7-miRNA-based signatures (M1–M7) in the Retrospective and BABE Cohorts. Vertical bars representing 95% confidence interval of AUC values.

A significant association (KW *p*-value: 0.036) between CA15-3 level and disease status was noted in the BABE Cohort, with higher levels in breast tumor samples (Supplementary Figure S2). However, when the analysis extended the 7 miRNA-based signatures (M1-M7) with CA15-3 levels, no significant increase in AUC values was observed (Supplementary Table S2), suggesting that an objective assessment of these candidate molecules could mitigate the limited diagnostic performance of currently available soluble markers.

3.3. *BABE-FU Cohort Analysis*

To examine the evolution of candidate miRNAs after surgical removal of the tumor, the circulating levels of the 5 miRNAs in signatures M1-M7 were measured in 29 BABE patients with a histological diagnosis of a tumor (Table 1), using matched plasma samples that were collected before surgery (T0) and 12 $\pm$ 3 months after surgery (T1). Their relative expression levels, according to the overall mean approach [37], were compared at the 2 time points. All patients were disease-free at T1. A significant increase in the $\log_2$(RQ) of miR-625 was observed after surgery (WSR test *p*-value: 0.044), but the 4 remaining miRNAs did not differ significantly at the 2 time points (Figure 4), suggesting that the origin of these 5 miRNAs was not from cancer cells.

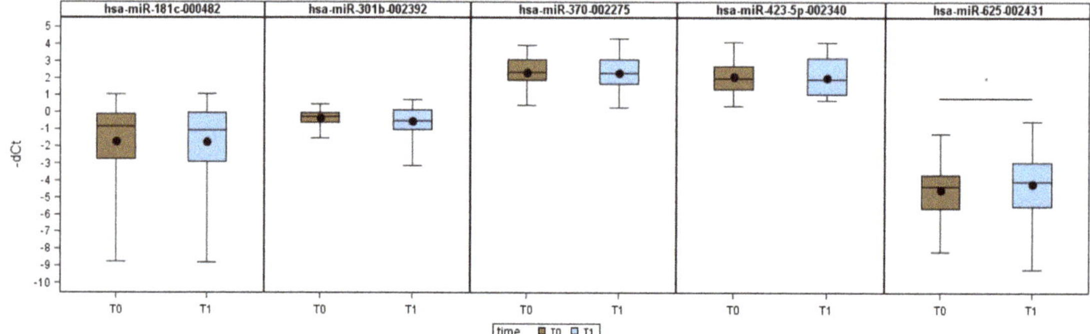

Figure 4. Evaluation of signature miRNAs at 2 time points. Distribution of each of the 5 miRNAs in signatures M1-M7 before surgery (T0) and 12 ± 3 months after surgery (T1). Each box indicates the 25th and 75th percentiles. The horizontal line in the box indicates the median, and the whiskers indicate the extremes. The black dot indicates the mean value. * if p-value < 0.05.

4. Discussion

Although mammography remains the pillar diagnostic method in the early diagnosis of BC, current image-based approaches are associated with an increased frequency of biopsies to determine the malignant or benign nature of abnormal areas. Thus, reliable minimally invasive blood-based tests are long cherished to increase the compliance, while reducing cost, of population-based screens for BC.

In this study, we analyzed circulating microRNAs in search of diagnostic biomarkers able to discriminate the benign and malignant nature of abnormal breast areas with imaging suggestive of BC (BI-RADS 4-5). We have identified 5 miRNAs that, when properly combined to form 7 miRNA-based signatures, can be applied to fluid biopsies to support diagnostic imaging. These results were obtained examining a retrospective cohort and confirmed in a prospective clinical cohort consecutively enrolled during the study. Using high-throughput OpenArray technology, over 700 microRNAs were analyzed in a retrospective cohort of plasma samples from age matched HDs and T or B patients, split into training and testing sets. Although distinct microRNAs emerged from the T vs. HD and B vs. HD comparisons in the TRS set, only signatures that discriminated between T and HD—not B and HD—were confirmed in the TES, highlighting the challenges of identifying circulating molecules that reflect the presence of benign breast disease. Nevertheless, out of 52 signatures distinguishing between T and HD, 26 significantly discriminated T from B lesions in the TES and, notably, 7 miRNA-based signatures comprising ad hoc combinations of 5 miRNAs retained significant performance even when the same regression coefficients obtained in the TRS was applied. Although the differences in blood samples from B patients were minimal with respect to HD, these results argue in favor of dissimilarities between malignant and benign breast blood samples that could be exploited in making a differential diagnosis.

Consistently, 5 of the 7 combinations of 5 miRNAs maintained their ability to discriminate malignant from benign disease in our large BABE prospective cohort. These signatures were applicable to 93% of women with uncertain tumor or benign disease, indicating that the 5 constituent miRNAs are readily detected in plasma samples and that their absence (9 tumors, 11 benign lesions) is independent of disease status. Thus, it is conceivable that a small tumor, as in the early screening of the BABE Cohort, harbors circulating miRNAs that are sufficiently differentially expressed compared with benign breast disease. This is consistent with the detection of various circulating miRNAs, in a spontaneous model of mammary carcinogenesis, that are differently expressed from the non-transgenic siblings and that are maintained or differently represented along the stage of transformation [38].

Notably, one microRNA has a human homolog (has-miR-370) which belongs to the 7 signatures discriminating T from B patients. The miR-423-5p, detected at higher levels in T than B plasma samples, has been shown highly expressed in plasma and blood exosomes of breast cancer patients in comparison with healthy controls and significantly associated with clinical stage and Ki-67 levels [39]. The miR-625, which we found decreased in T versus B plasma samples, has been reported at lower levels in ductal lavage from patients with unilateral breast cancer versus ductal lavage of the contralateral normal breast [40]. Remarkably, the level of miR-625-5p increased after surgical removal of the tumor indicating its properness in combination with the other four miRNAs to form the diagnostic models for the presence of malignancies. In addition, this finding suggests that the source of miR-625-5p is not from neoplastic cells that rather negatively regulate its expression. Several datasets of microRNA expression in normal cell populations, show miR-625-5p expressed at high level in T lymphocytes (Supplementary Figure S3) and, therefore tumor cells might find benefit from lowering its expression in the attempt to escape immune surveillance. Although the presence of a minimal residual disease not detectable by conventional detection strategies cannot be excluded, the lack of change in the levels of miR-423-5p, miR-370-3p, miR-181c, and miR-301b after surgery also indicates their origin from cells other than the primary tumor. Accordingly, datasets of microRNA expression in normal cell populations [41–43] showed inflammatory cells, endothelial cells, fibroblasts and adipocytes as possible source of the 4 miRs. Specifically, the miR-423-5p (up in T vs. B plasma samples) was found enriched in immune populations, particularly B lymphocytes, compared with epithelium and endothelium (Supplementary Figure S3). Moreover, based on miRNA databases, in human plasma (PRJNA296772) and plasma-derived exosomes (PRJNA196121), miR-423-5p was among the top 20 most abundant circulating miRNAs [44]. Regarding the other 3 miRNAs, miR-301b-3p was higher in monocytes and endothelial cells; miR-181c-5p was expressed in T lymphocytes, along with miR-625-5p, and expressed in neutrophils and mast cells. Finally, miR-370-3p, was enriched in mesenchymal stem cells and mesenchymal-derived lineages, such as fibroblasts and adipocytes (including preadipocytes). Thus, no changes in the levels of miR-423-5p, miR-370-3p, miR-181c, and miR-301b after surgery could be indicative that inflammation and breast healing still occur at 1-year follow-up likely due to the adjuvant therapy and radiotherapy the patients received at T1 blood withdrawal.

Defining the biological meaning of circulating miRNAs proves to be particularly challenging due to their unknown origin and cell/tissue specific mechanism of action. Based on lists of predicted targets of the five miRNAs of interest, the most represented pathways are related to the control of cell cycle and senescence, but there is also the so called "Human T-cell leukemia virus 1 infection" KEGG pathway, which is implicated in chronic inflammatory diseases [45]. The literature provides data on validated targets, such as KRAS for miR-181c-5p [46], or SOX2 for miR-625-5p [47], which are important oncogenes in breast cancer.

Although the analysis of a defined set of miRNAs by qRT-PCR on the OpenArray platform might have overlooked additional or better-performing candidates in detecting BC, our signatures, obtained by properly combining miR-625, miR-423-5p, miR-370-3p, miR-181c, and miR-301b, significantly discriminated a tumor from a benign nodule, even in a prospectively recruited cohort of women with a high likelihood of malignancy, such as in the BABE Cohort. The collection of plasma samples from patients enrolled in the BABE-FU is still ongoing to evaluate possible time trends in the change of miRNA levels as well as to evaluate their potential prognostic values in predicting patients' outcome.

5. Conclusions

In this study, we identified and confirmed on a prospective clinical study five miRNAs-based signatures able to discriminate malignant from benign breast disease. Even though our signatures are unlikely to be used alone to make accurate BC predictions, our work supports the use of circulating miRNAs in distinguishing malignant from benign breast disease to complement imaging-based screens for the early diagnosis of BC and perhaps to spare unnecessary biopsies to a large fraction of women. Further studies are needed to confirm the analytically performance of our signature and to fully assess their clinical utility. To this end, an easy-to-use assay with the discovered miRNA signatures should be firstly developed and evaluated in the current clinical setting program on patients from the same target population.

Supplementary Materials: The following are available online at https://www.mdpi.com/article/10.3390/cancers13164028/s1, Figure S1: Distribution of expression levels of de-regulated miRNAs in the TRS. Panel A and B report the miRNAs distribution tumors vs. healthy donors comparison and in the benign lesions vs. healthy donors comparison, respectively. Each box indicates the 25th and 75th percentiles. The horizontal line inside the box indicates the median, and whiskers indicate the extreme measured values; Figure S2: Association between CA15.3 expression levels and disease status. Distribution of CA15.3 expression levels according to the disease status (malignant or benign lesions) of the BABE Cohort. Each box indicates the 25th and 75th percentiles. The horizontal line inside the box indicates the median, and the whiskers indicate the extreme measured values. Figure S3: In silico miRNA expression analysis in human primary cells. miR-423-5p, miR-181c-5p, miR-301b-3p, miR-625-5p and miR-370-3p expression levels in 26 human primary cells, comprising 19 blood cells, 6 stromal cells and mammary epithelial cells. Data are expressed as average reads per million miRNA reads (RPM) (± SD) and are not otherwise normalized. Sample number (*n*) for each cell type is indicated below expression bars; Table S1: Descriptive statistics (in terms of AUC) of the 26 TB-promising signatures; Table S2: AUC values and their corresponding 95% CI for each signature alone or with CA15.3 in the model.

Author Contributions: Conceptualization, M.G. (Marta Giussani), C.M.C., G.S. (Gabriella Sozzi), M.P.C., E.T. and P.V.; Data curation, M.G. (Marta Giussani), C.M.C., L.D.C., M.D., E.M., G.C., S.V., B.P., R.O., M.G. (Massimiliano Gennaro), M.V.I., C.D., C.F., M.S. and G.S. (Gianfranco Scaperrotta); Formal analysis, C.M.C., M.L., M.D., C.G., A.M., E.M., E.T. and P.V.; Investigation, M.G. (Marta Giussani) and L.D.C.; Methodology, C.M.C., L.D.C., M.L., E.T. and P.V.; Resources, G.S. (Gabriella Sozzi), M.S., M.P.C. and E.T.; Supervision, E.T. and P.V.; Writing—original draft, M.G. (Marta Giussani), C.M.C., L.D.C., M.L., M.D., G.C., C.F., G.S. (Gianfranco Scaperrotta), E.T. and P.V.; Writing—review & editing, M.V.I., G.S. (Gabriella Sozzi), M.P.C., G.S. (Gianfranco Scaperrotta), E.T. and P.V. All authors have read and agreed to the published version of the manuscript.

Funding: This work was supported by funding from Fondazione Associazione Italiana per la Ricerca sul Cancro (Fondazione AIRC) to E.T., M.S., M.P.C. (GLs) and G.S. (Gabriella Sozzi) (PI) (Special Program—5 per mille 2011 No 12162).

Institutional Review Board Statement: Institutional approval from our independent ethics committee was obtained for this study (approval numbers INT111-13, INT144-14, and INT66-15). Patients gave informed consent to the use of their samples. All procedures were conducted in accordance with the Declaration of Helsinki.

Informed Consent Statement: Informed consent was obtained from all subjects involved in the study.

Data Availability Statement: All data generated and analyzed during the current study are available from the corresponding author on reasonable request.

Acknowledgments: The authors would like to thank all the women who agreed to the use of their own plasma samples for this study; Mirella Ferruccio and Chiara Ghidoli for performing the blood withdrawal and plasma collection; and the INT Tissue and Blood Biobank for storing plasma samples. We thank Laura Mameli for secretarial assistance.

Conflicts of Interest: The authors declare no competing interest.

References

1. Siegel, R.L.; Miller, K.D.; Jemal, A. Cancer statistics, 2019. *CA Cancer J. Clin.* **2019**, *69*, 7–34. [CrossRef]
2. Mendelson, E.B.; Baum, J.; Berg, W.; Merritt, C.; Rubin, E. ACR BIRADS Ultrasound. In *ACR BI-RADS Atlas, Breast Imaging Reporting and Data System*, 4th ed.; ACR BI-RADS Atlas, Breast Imaging Reporting and Data System; American College of Radiology: Reston, VA, USA, 2003.

3. Mendelson, E.B. ACR BIRADS Ultrasound. In *ACR BI-RADS Atlas, Breast Imaging Reporting and Data System*, 5th ed.; ACR BI-RADS Atlas, Breast Imaging Reporting and Data System; American College of Radiology: Reston, VA, USA, 2013.
4. Li, J.; Guan, X.; Fan, Z.; Ching, L.M.; Li, Y.; Wang, X.; Cao, W.M.; Liu, D.X. Non-Invasive Biomarkers for Early Detection of Breast Cancer. *Cancers* **2020**, *12*, 2767. [CrossRef]
5. Iorio, M.V.; Croce, C.M. microRNAs in cancer: Small molecules with a huge impact. *J. Clin. Oncol.* **2009**, *27*, 5848–5856. [CrossRef] [PubMed]
6. Iorio, M.V.; Croce, C.M. MicroRNA dysregulation in cancer: Diagnostics, monitoring and therapeutics. A comprehensive review. *EMBO Mol. Med.* **2017**, *9*, 852. [CrossRef] [PubMed]
7. Nassar, F.J.; Talhouk, R.; Zgheib, N.K.; Tfayli, A.; El Sabban, M.; El Saghir, N.S.; Boulos, F.; Jabbour, M.N.; Chalala, C.; Boustany, R.M.; et al. microRNA Expression in Ethnic Specific Early Stage Breast Cancer: An Integration and Comparative Analysis. *Sci. Rep.* **2017**, *7*, 16829–16978. [CrossRef] [PubMed]
8. Zhu, W.; Qin, W.; Atasoy, U.; Sauter, E.R. Circulating microRNAs in breast cancer and healthy subjects. *BMC Res. Notes* **2009**, *2*, 89. [CrossRef]
9. Heneghan, H.M.; Miller, N.; Kelly, R.; Newell, J.; Kerin, M.J. Systemic miRNA-195 differentiates breast cancer from other malignancies and is a potential biomarker for detecting noninvasive and early stage disease. *Oncologist* **2010**, *15*, 673–682. [CrossRef]
10. Asaga, S.; Kuo, C.; Nguyen, T.; Terpenning, M.; Giuliano, A.E.; Hoon, D.S. Direct serum assay for microRNA-21 concentrations in early and advanced breast cancer. *Clin. Chem.* **2011**, *57*, 84–91. [CrossRef]
11. Roth, C.; Rack, B.; Muller, V.; Janni, W.; Pantel, K.; Schwarzenbach, H. Circulating microRNAs as blood-based markers for patients with primary and metastatic breast cancer. *Breast Cancer Res.* **2010**, *12*, R90. [CrossRef]
12. Jung, E.J.; Santarpia, L.; Kim, J.; Esteva, F.J.; Moretti, E.; Buzdar, A.U.; Di Leo, A.; Le, X.F.; Bast, R.C., Jr.; Park, S.T.; et al. Plasma microRNA 210 levels correlate with sensitivity to trastuzumab and tumor presence in breast cancer patients. *Cancer* **2012**, *118*, 2603–2614. [CrossRef]
13. Cuk, K.; Zucknick, M.; Heil, J.; Madhavan, D.; Schott, S.; Turchinovich, A.; Arlt, D.; Rath, M.; Sohn, C.; Benner, A.; et al. Circulating microRNAs in plasma as early detection markers for breast cancer. *Int. J. Cancer* **2013**, *132*, 1602–1612. [CrossRef] [PubMed]
14. Si, H.; Sun, X.; Chen, Y.; Cao, Y.; Chen, S.; Wang, H.; Hu, C. Circulating microRNA-92a and microRNA-21 as novel minimally invasive biomarkers for primary breast cancer. *J. Cancer Res. Clin. Oncol.* **2013**, *139*, 223–229. [CrossRef] [PubMed]
15. Sun, Y.; Wang, M.; Lin, G.; Sun, S.; Li, X.; Qi, J.; Li, J. Serum microRNA-155 as a potential biomarker to track disease in breast cancer. *PLoS ONE* **2012**, *7*, e47003. [CrossRef]
16. Chan, M.; Liaw, C.S.; Ji, S.M.; Tan, H.H.; Wong, C.Y.; Thike, A.A.; Tan, P.H.; Ho, G.H.; Lee, A.S. Identification of circulating microRNA signatures for breast cancer detection. *Clin. Cancer Res.* **2013**, *19*, 4477–4487. [CrossRef] [PubMed]
17. Cuk, K.; Zucknick, M.; Madhavan, D.; Schott, S.; Golatta, M.; Heil, J.; Marmé, F.; Turchinovich, A.; Sinn, P.; Sohn, C.; et al. Plasma microRNA panel for minimally invasive detection of breast cancer. *PLoS ONE* **2013**, *8*, e76729. [CrossRef]
18. Zearo, S.; Kim, E.; Zhu, Y.; Zhao, J.T.; Sidhu, S.B.; Robinson, B.G.; Soon, P.S. MicroRNA-484 is more highly expressed in serum of early breast cancer patients compared to healthy volunteers. *BMC Cancer* **2014**, *14*, 200–214. [CrossRef] [PubMed]
19. Sochor, M.; Basova, P.; Pesta, M.; Dusilkova, N.; Bartos, J.; Burda, P.; Pospisil, V.; Stopka, T. Oncogenic microRNAs: miR-155, miR-19a, miR-181b, and miR-24 enable monitoring of early breast cancer in serum. *BMC Cancer* **2014**, *14*, 448. [CrossRef]
20. Ng, E.K.; Li, R.; Shin, V.Y.; Jin, H.C.; Leung, C.P.; Ma, E.S.; Pang, R.; Chua, D.; Chu, K.M.; Law, W.L.; et al. Circulating microRNAs as specific biomarkers for breast cancer detection. *PLoS ONE* **2013**, *8*, e53141. [CrossRef]
21. Mar-Aguilar, F.; Mendoza-Ramirez, J.A.; Malagon-Santiago, I.; Espino-Silva, P.K.; Santuario-Facio, S.K.; Ruiz-Flores, P.; Rodriguez-Padilla, C.; Reséndez-Pérez, D. Serum circulating microRNA profiling for identification of potential breast cancer biomarkers. *Dis. Markers* **2013**, *34*, 163–169. [CrossRef]
22. Liu, J.; Mao, Q.; Liu, Y.; Hao, X.; Zhang, S.; Zhang, J. Analysis of miR-205 and miR-155 expression in the blood of breast cancer patients. *Chin. J. Cancer Res.* **2013**, *25*, 46–54. [CrossRef] [PubMed]
23. Kodahl, A.R.; Lyng, M.B.; Binder, H.; Cold, S.; Gravgaard, K.; Knoop, A.S.; Ditzel, H.J. Novel circulating microRNA signature as a potential non-invasive multi-marker test in ER-positive early-stage breast cancer: A case control study. *Mol. Oncol.* **2014**, *8*, 874–883. [CrossRef]
24. Hamam, R.; Hamam, D.; Alsaleh, K.A.; Kassem, M.; Zaher, W.; Alfayez, M.; Aldahmash, A.; Alajez, N.M. Circulating microRNAs in breast cancer: Novel diagnostic and prognostic biomarkers. *Cell Death Dis.* **2017**, *8*, e3045. [CrossRef]
25. Shimomura, A.; Shiino, S.; Kawauchi, J.; Takizawa, S.; Sakamoto, H.; Matsuzaki, J.; Ono, M.; Takeshita, F.; Niida, S.; Shimizu, C.; et al. Novel combination of serum microRNA for detecting breast cancer in the early stage. *Cancer Sci.* **2016**, *107*, 326–334. [CrossRef] [PubMed]
26. Joosse, S.A.; Muller, V.; Steinbach, B.; Pantel, K.; Schwarzenbach, H. Circulating cell-free cancer-testis MAGE-A RNA, BORIS RNA, let-7b and miR-202 in the blood of patients with breast cancer and benign breast diseases. *Br. J. Cancer* **2014**, *111*, 909–917. [CrossRef] [PubMed]
27. Schwarzenbach, H.; Milde-Langosch, K.; Steinbach, B.; Muller, V.; Pantel, K. Diagnostic potential of PTEN-targeting miR-214 in the blood of breast cancer patients. *Breast Cancer Res. Treat.* **2012**, *134*, 933–941. [CrossRef]

28. Zou, R.; Loke, S.Y.; Tan, V.K.; Quek, S.T.; Jagmohan, P.; Tang, Y.C.; Madhukumar, P.; Tan, B.K.; Yong, W.S.; Sim, Y.; et al. Development of a microRNA Panel for Classification of Abnormal Mammograms for Breast Cancer. *Cancers* **2021**, *13*, 2130. [CrossRef]
29. Lakhani, S.R.; Ellis, I.O.; Schnitt, S.J.; Tan, P.H.; Van de Vijver, M.J. WHO Classification of Tumours. In *WHO Classification of Tumours of the Breast*, 4th ed.; WHO Classification of Tumours of the Breast; WHO Press: Geneva, Switzerland, 2016; ISBN 978-92-832-2433-4.
30. Pizzamiglio, S.; Zanutto, S.; Ciniselli, C.M.; Belfiore, A.; Bottelli, S.; Gariboldi, M.; Verderio, P. A methodological procedure for evaluating the impact of hemolysis on circulating microRNAs. *Oncol. Lett.* **2017**, *13*, 315–320. [CrossRef] [PubMed]
31. Verderio, P.; Bottelli, S.; Ciniselli, C.M.; Pierotti, M.A.; Gariboldi, M.; Pizzamiglio, S. NqA: An R-based algorithm for the normalization and analysis of microRNA quantitative real-time polymerase chain reaction data. *Anal. Biochem.* **2014**, *461*, 7–9. [CrossRef]
32. Zanutto, S.; Ciniselli, C.M.; Belfiore, A.; Lecchi, M.; Masci, E.; Delconte, G.; Primignani, M.; Tosetti, G.; Dal, F.M.; Fazzini, L.; et al. Plasma miRNA-based signatures in CRC screening programs. *Int. J. Cancer* **2020**, *146*, 1164–1173. [CrossRef] [PubMed]
33. Di Cosimo, S.; Appierto, V.; Pizzamiglio, S.; Tiberio, P.; Iorio, M.V.; Hilbers, F.; De Azambuja, E.; de la Pena, L.; Izquierdo, M.A.; Huober, J.; et al. Plasma microRNA levels for predicting therapeutic response to neoadjuvant treatment in HER2-positive breast cancer: Results from the NeoALTTO trial. *Clin. Cancer Res.* **2019**, *25*, 3887–3895. [CrossRef]
34. Verderio, P.; Bottelli, S.; Pizzamiglio, S.; Ciniselli, C.M. Developing miRNA signatures: A multivariate prospective. *Br. J. Cancer* **2016**, *115*, 1–4. [CrossRef] [PubMed]
35. Harrell, F.E., Jr. *Regression Modeling Strategies with Applications to Linear Models, Logistic Regression, and Survival Analysis*, 1st ed.; Springer: New York, NY, USA, 2001; ISBN 0-387-95232-2.
36. Moons, K.G.; Donders, A.R.; Steyerberg, E.W.; Harrell, F.E. Penalized maximum likelihood estimation to directly adjust diagnostic and prognostic prediction models for overoptimism: A clinical example. *J. Clin. Epidemiol.* **2004**, *57*, 1262–1270. [CrossRef]
37. Mestdagh, P.; Van Vlierberghe, P.; De Weer, A.; Muth, D.; Westermann, F.; Speleman, F.; Vandesompele, J. A novel and universal method for microRNA RT-qPCR data normalization. *Genome Biol.* **2009**, *10*, R64. [CrossRef] [PubMed]
38. Chiodoni, C.; Cancila, V.; Renzi, T.A.; Perrone, M.; Tomirotti, A.M.; Sangaletti, S.; Botti, L.; Dugo, M.; Milani, M.; Bongiovanni, L.; et al. Transcriptional Profiles and Stromal Changes Reveal Bone Marrow Adaptation to Early Breast Cancer in Association with Deregulated Circulating microRNAs. *Cancer Res.* **2020**, *80*, 484–498. [CrossRef]
39. Liu, D.; Li, B.; Shi, X.; Zhang, J.; Chen, A.M.; Xu, J.; Wang, W.; Huang, K.; Gao, J.; Zheng, Z.; et al. Cross-platform genomic identification and clinical validation of breast cancer diagnostic biomarkers. *Aging* **2021**, *13*, 4258–4273. [CrossRef]
40. Do Canto, L.M.; Marian, C.; Willey, S.; Sidawy, M.; Da Cunha, P.A.; Rone, J.D.; Li, X.; Gusev, Y.; Haddad, B.R. MicroRNA analysis of breast ductal fluid in breast cancer patients. *Int. J. Oncol.* **2016**, *48*, 2071–2078. [CrossRef]
41. McCall, M.N., Kim, M.S.; Adil, M.; Patil, A.H.; Lu, Y.; Mitchell, C.J.; Leal-Rojas, P.; Xu, J.; Kumar, M.; Dawson, V.L.; et al. Toward the human cellular microRNAome. *Genome Res.* **2017**, *27*, 1769–1781. [CrossRef] [PubMed]
42. De Rie, D.; Abugessaisa, I.; Alam, T.; Arner, E.; Arner, P.; Ashoor, H.; Astrom, G.; Babina, M.; Bertin, N.; Burroughs, A.M.; et al. An integrated expression atlas of miRNAs and their promoters in human and mouse. *Nat. Biotechnol.* **2017**, *35*, 872–878. [CrossRef] [PubMed]
43. Juzenas, S.; Venkatesh, G.; Hubenthal, M.; Hoeppner, M.P.; Du, Z.G.; Paulsen, M.; Rosenstiel, P.; Senger, P.; Hofmann-Apitius, M.; Keller, A.; et al. A comprehensive, cell specific microRNA catalogue of human peripheral blood. *Nucleic Acids Res.* **2017**, *45*, 9290–9301. [CrossRef]
44. Aparicio-Puerta, E.; Jaspez, D.; Lebron, R.; Koppers-Lalic, D.; Marchal, J.A.; Hackenberg, M. liqDB: A small-RNAseq knowledge discovery database for liquid biopsy studies. *Nucleic Acids Res.* **2019**, *47*, D113–D120. [CrossRef]
45. Raudvere, U.; Kolberg, L.; Kuzmin, I.; Arak, T.; Adler, P.; Peterson, H.; Vilo, J. g:Profiler: A web server for functional enrichment analysis and conversions of gene lists (2019 update). *Nucleic Acids Res.* **2019**, *47*, W191–W198. [CrossRef] [PubMed]
46. Hashimoto, Y.; Akiyama, Y.; Otsubo, T.; Shimada, S.; Yuasa, Y. Involvement of epigenetically silenced microRNA-181c in gastric carcinogenesis. *Carcinogenesis* **2010**, *31*, 777–784. [CrossRef] [PubMed]
47. Wang, Z.; Qiao, Q.; Chen, M.; Li, X.; Wang, Z.; Liu, C.; Xie, Z. miR-625 down-regulation promotes proliferation and invasion in esophageal cancer by targeting Sox2. *FEBS Lett.* **2014**, *588*, 915–921. [CrossRef] [PubMed]

Article

Circulating Exosomal MicroRNA-1307-5p as a Predictor for Metastasis in Patients with Hepatocellular Carcinoma

Jung Woo Eun [1], Chul Won Seo [1,2], Geum Ok Baek [1], Moon Gyeong Yoon [1], Hye Ri Ahn [1,2], Ju A. Son [1,2], Suna Sung [1,2], Do Wan Kim [3], Soon Sun Kim [1], Hyo Jung Cho [1,*] and Jae Youn Cheong [1,*]

1 Department of Gastroenterology, Ajou University School of Medicine, Suwon 16499, Korea; jetaimebin@gmail.com (J.W.E.); countpas@naver.com (C.W.S.); ptok99@hanmail.net (G.O.B.); ymk8028@hanmail.net (M.G.Y.); rhkwp37@naver.com (H.R.A.); gracia1429@ajou.ac.kr (J.A.S.); tjdtnsk1203@naver.com (S.S.); soonsunkim@aumc.ac.kr (S.S.K.)

2 Department of Biomedical Sciences, Ajou University Graduate School of Medicine, Suwon 16499, Korea

3 Ajou Translational Omics Center, Ajou University Medical Center, Suwon 16499, Korea; kdowan@ajou.ac.kr

* Correspondence: pilgrim8107@hanmail.net (H.J.C.); jaeyoun620@gmail.com (J.Y.C.); Tel.: +82-31-219-7824 (H.J.C.); +82-31-219-5119 (J.Y.C.)

Received: 26 November 2020; Accepted: 15 December 2020; Published: 18 December 2020

Simple Summary: Exosomal microRNAs (exo-miRs) significantly contribute to cancer metastasis. However, few studies have investigated the role of exosomes as metastasis mediators in hepatocellular carcinoma (HCC) despite recent advancements in liquid biopsy. We aimed to identify pro-metastatic circulating exo-miRs potentially predicting metastasis onset in HCC through comprehensive and systematic integrative analyses of plasma exo-miR sequencing data and publicly available RNA expression datasets, and accordingly propose a potential mechanism of action of pro-metastatic miRs, including promoting epithelial–mesenchymal transition (EMT). We found that circulating exo-miR-1307-5p is a predictive marker for metastasis in patients with HCC, and EMT promotion through SEC14L2 and ENG downregulation could be the potential downstream pathway of miR-1307-5p. We believe that our study makes a significant contribution to the literature because our findings provide novel insights into the role of circulating exo-miRs in the pathogenesis and progression of HCC and suggest that exo-miRs are a potential treatment target in HCC.

Abstract: Exosomal microRNAs (exo-miRs) contribute to cancer metastasis. To identify pro-metastatic circulating exo-miRs in hepatocellular carcinoma (HCC), next-generation sequencing-based plasma exo-miR profiles of 14 patients with HCC (eight non-metastatic and six with metastasis within 1 year of follow-up) were analyzed. Sixty-one miRs were significantly overexpressed among patients with metastatic HCC. Candidate miRs were selected through integrative analyses of two different public expression datasets, GSE67140 and The Cancer Genome Atlas liver hepatocellular carcinoma (TCGA_LIHC). Integrative analyses revealed 3 of 61 miRs (miR-106b-5p, miR-1307-5p, and miR-340-5p) commonly overexpressed both in metastasis and vascular invasion groups, with prognostic implications. Validation was performed using stored blood samples of 150 patients with HCC. Validation analysis showed that circulating exo-miR-1307-5p was significantly overexpressed in the metastasis group ($p = 0.04$), as well as in the vascular invasion and tumor recurrence groups. Circulating exo-miR-1307-5p expression was significantly correlated with tumor stage progression ($p < 0.0001$). Downstream signaling pathways of miR-1307 were predicted using TargetScan and Ingenuity Pathway Analysis. On comprehensive bioinformatics analysis, the downstream pathway of miR-1307-5p, promoting epithelial–mesenchymal transition (EMT), showed SEC14L2 and ENG downregulation. Our results show that circulating exo-miR-1307-5p promotes metastasis and helps predict metastasis in HCC, and SEC14L2 and ENG are target tumor suppressor genes of miR-1307 that promote EMT.

Keywords: hepatocellular carcinoma; metastasis; exosome; microRNA; bioinformatics analysis

---

1. Introduction

Hepatocellular carcinoma (HCC) is the sixth most common malignancy and the third leading cause of cancer-related mortality worldwide [1]. In the past several decades, the prognosis of HCC has significantly improved owing to advancements in diagnostic and treatment approaches for HCC [2]. However, the prognosis of patients with advanced-stage HCC remains poor with a median survival of 4–6 months [3]. Metastasis is a key determinant of the treatment strategy among patients with HCC because locoregional therapies are no longer effective to control extrahepatic metastasis [4]. Therefore, determination of the metastatic status during the initial staging process is essential for generating an appropriate treatment strategy directly associated with survival. In most cases, extrahepatic metastasis is detected through conventional imaging modalities including computed tomography and bone scintigraphy; however, they require considerable effort and are costly, and sometimes they cannot detect small metastatic lesions. The identification of metastasis driver molecules in blood before diagnosis through conventional imaging modalities would help classify patients in accordance with the stratified risk of metastasis, thus, potentially facilitating the implementation of precision medicine approaches.

Liquid biopsy is performed to detect tumor-derived genetic factors in body fluids including blood, urine, and saliva [5–7]. Liquid biopsy can target various classes of circulating tumor molecules including cell-free DNA, circulating tumor cells, and tumor cell-derived extracellular vesicles [8]. Exosomes (approximately 30–100 nm diameter) are extracellular vesicles delivering genetic factors from the donor cell to the recipient cell [9,10]. Exosomal content has been extensively assessed as major targets of liquid biopsy [11]. Among the exosomal genetic factors, microRNAs (miRs) received increasing attention because the loading of specific miRs into exosomes is suggested to result from active selection in accordance with the properties of donor cells [12,13]. Recent studies have shown that exosomes promote the generation of a metastatic niche by transferring functional molecules activating epithelial–mesenchymal transition (EMT) at different sites and promoting downstream signaling in recipient cells [14,15]. In particular, exosomal miRs (exo-miRs) significantly contribute to cancer metastasis. However, few studies have investigated the role of exosomes as metastasis mediators in HCC despite recent advancements in liquid biopsy.

In this study, we aimed to identify pro-metastatic circulating exo-miRs that potentially predict metastasis onset in HCC through comprehensive and systematic integrative analyses of plasma exo-miR sequencing data and publicly available RNA expression datasets. Accordingly, we propose a potential mechanism of action of pro-metastatic miRs, including promoting EMT.

2. Results

2.1. Confirmation of Isolated Circulating Exosomes and Identification of Overexpressed Exo-miRs in Metastatic HCC Patients

Figure 1 shows a schematic representation of systematic integrative analyses performed herein to identify circulating exo-miRs potentially promoting metastasis. First, circulating exosomes were isolated and their baseline characteristics were evaluated. Transmission electron microscopy (TEM) revealed that the samples contained spherical vesicles of 30–100 nm in diameter (Figure 2a). The concentration and size distribution of these vesicles were determined through Nanoparticle tracking analysis (NTA) (Figure 2b). NTA revealed that particles of 30–100 nm diameter were strongly enriched in the samples. The results of western blotting showed that the isolated vesicles were positive for exosomal markers

(CD63, CD81, and CD9) and negative for an endoplasmic reticulum marker, Grp78 (Figure 2c and Figure S1). These results indicate that the circulating exosomes were well isolated, and exosomal RNA was appropriately extracted. Small RNA sequencing libraries were successfully generated from 14 of the 52 exosomal RNA preparation samples from the Plasma-HCC cohort. These 14 samples were used herein, of which, eight were obtained from the metastasis-free group and six from the metastasis group. Sequencing data of plasma exo-miRs were analyzed, and 61 exo-miRs predominantly overexpressed in the metastasis group were identified (>2-fold, $p < 0.05$). The heatmap reveals the overexpression of the 61 exo-miRs in the metastasis group (Figure 2d).

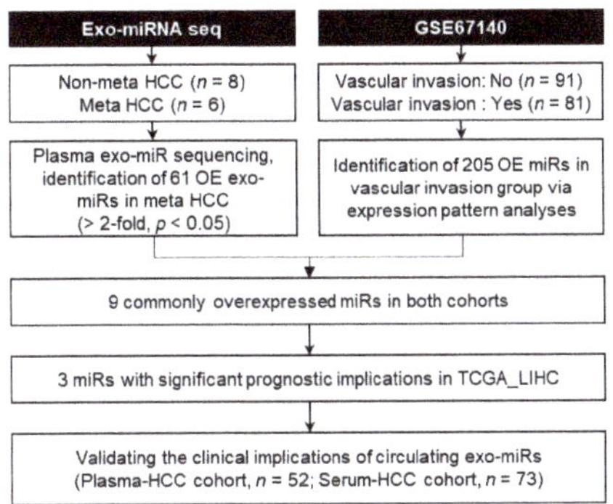

Figure 1. Schematic representation of the systematic integrative analyses performed herein to identify metastasis-stimulating circulating exosomal microRNAs. Exo-miRNA seq, exosomal microRNA sequencing; HCC, hepatocellular carcinoma; Meta, metastasis; exo-miR, exosomal microRNA; OE, overexpressed.

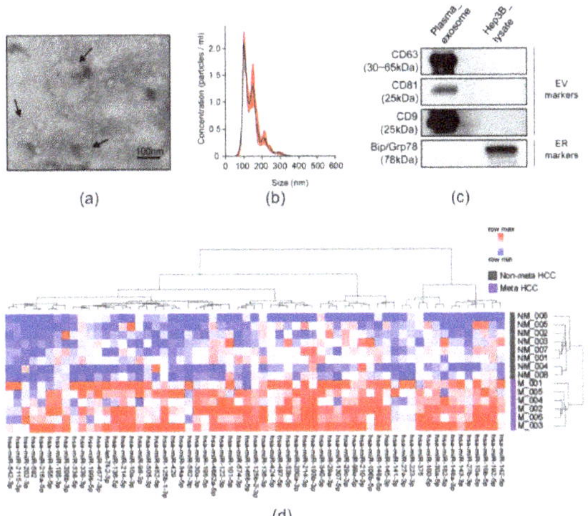

Figure 2. Confirmation of isolated circulating exosomes and identification of overexpressed exo-miRs in the metastasis group. (**a**) Transmission electron microscopic observation of separated circulating exosomes obtained from patients with hepatocellular carcinoma. Arrows indicate exosomes. Scale bar = 100 nm. (**b**) Nanoparticle tracking analysis (NTA) size distribution and concentration of exosomes. (**c**) Western blot analysis. Plasma exosomes were positive for exosome markers (CD63, CD81, and CD9) and negative for Grp78. Hep3B lysate was used as a negative control. (**d**) Heatmap of the differentially expressed 61 exo-miRs in accordance with the metastasis status. NM, non-metastasis; M, metastasis.

2.2. Integrative Analyses of Public Gene Expression Datasets to Select Potential Candidate Pro-Metastatic miRs

To further select pro-metastatic miRs, systematic integrative analyses were performed in two public gene expression datasets, GSE67140 and TCGA_LIHC [16]. The GSE67140 cohort comprised 91 patients with HCC with vascular invasion and 81 without vascular invasion. In the GSE67140 dataset, pattern analysis was performed in accordance with vascular invasion using CLICK algorithm [17].

As a result, miRs could be categorized into three clusters by the expression pattern according to vascular invasion status (Figure 3a). Figure 3b illustrates heatmaps of the miR expression in each cluster according to vascular invasion status. Among the three clusters, 185 miRs of cluster 1 and 20 miRs of cluster 3 showed significantly increased values in the vascular invasion group than in the non-vascular invasion group. Thus, the 205 miRs in cluster 1 or 3 were analyzed using a Venn diagram, which revealed that nine miRs—miR-106b-5p, miR-1307-5p, miR-193b-3p, miR-202-3p, miR-33b-5p, miR-340-5p, miR-455-3p, miR-542-3p, and miR-574-3p—were commonly overexpressed in both the metastasis group of the Plasma-HCC group and vascular invasion group of GSE67140 (Figure 3c).

The expression of the nine miRs in the 14 plasma exosomal small RNA sequencing (RNA-Seq) dataset was visualized in accordance with the metastasis status in Figure 3d. All nine miRs were significantly overexpressed in the metastasis group. Figure 3e shows the expression of the candidate miRs in accordance with their vascular invasion status in the GSE67140 cohort. All nine miRs were significantly overexpressed in the vascular invasion group. Using TCGA_LIHC datasets, the Overall survival (OS) was analyzed in accordance with the expression of the nine miRs (Figure 3f). As a result, overexpression of miR-106b-5p or miR-1307-5p was significantly associated with a poor OS; however, the expression levels of other miRs did not influence the OS.

2.3. Validation of the Clinical Implications of Candidate Exo-miRs in the Validation Cohort

Expression levels of plasma exo-miR-106b-5p and miR-1307-5p were determined in the Plasma-HCC cohort ($n = 52$) to validate the clinical implications of the selected miRs as metastasis predictors. Figure 4a shows the expression of the two plasma exo-miRs in the metastasis-free ($n = 27$) and metastasis groups ($n = 25$). Plasma miR-1307-5p expression levels were significantly higher in the metastasis group than in the non-metastasis group ($p = 0.04$), while that of exo-miR-106b-5p did not significantly differ between the two groups.

Further validation analysis was performed to investigate the clinical implication of the selected circulating exo-miRs in the Serum-HCC cohort owing to its limited size and the lack of several clinical data regarding the Plasma-HCC cohort. Figure 4b shows the expression of exo-miR-106b-5p and miR-1307-5p in accordance with the vascular invasion status in the Serum-HCC cohort. Patients with vascular invasion had significantly higher serum exo-miR-106b-5p and miR-1307-5p expression levels. Figure 4c shows the expression of serum exo-miRs in accordance with the tumor recurrence status. Both of serum exo-miRs were significantly overexpressed in the tumor recurrence group. Figure 4d shows the expression of the three exo-miRs in accordance with the modified Union for International Cancer Control (mUICC) stages. Expression of the two serum exo-miRs were gradually upregulated according to tumor stage progression significantly.

Taken together, both of circulating exo-miR-1307-5p and exo-miR-106-5p were identified as potential biomarker for predicting vascular invasion, tumor recurrence, and advanced tumor stage in patients with HCC. However, in the aspect of metastasis, only exo-miR-1307-5p showed significant association with extrahepatic metastasis. Further, we tried to identify if circulating exo-miR-1307-5p could be used as a potential biomarker in the detection of HCC. Expression of serum exo-miR-1307-5p was compared between the HCC patients and normal healthy control (Figure S2). Serum exo-miR-1307-5p was markedly overexpressed in the HCC group compared to normal control ($p < 0.0001$). The AUC of exo-miR-1307-5p for detecting HCC was calculated as 0.958.

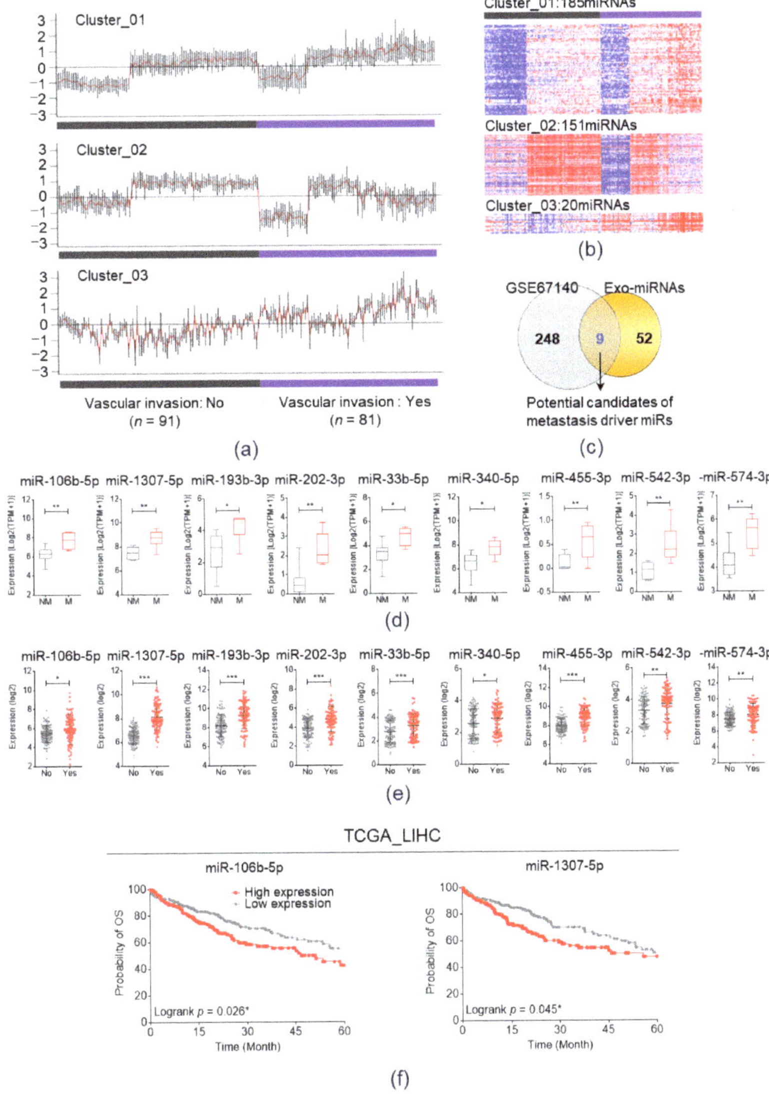

Figure 3. Integrative analyses of two different public RNA expression datasets (**a**) On Cluster analysis, miRs could be categorized into three clusters on the basis of their expression patterns according to vascular invasion status in GSE67140. (**b**) Heatmaps of miR expression in each cluster based on the vascular invasion status. (**c**) Venn diagram analysis to select potential candidate metastasis driver miRs. (**d**) Comparison of the expression of nine miRs between metastasis (M) and non-metastasis (NM) groups based on the expression of 14 plasma exosomal small RNA sequencing data. * $p < 0.05$; ** $p < 0.01$. (**e**) Comparison of the expression levels of nine miRs based on the vascular invasion status in the GSE67140 cohort. * $p < 0.05$; ** $p < 0.01$; *** $p < 0.001$. (**f**) Kaplan–Meier plot of overall survival based on the expression of miR-106b-5p and miR-1307-5p in TCGA_LIHC. * $p < 0.05$. NM, non-metastasis; Meta, metastasis; TCGA_LIHC, The Cancer Genomic Atlas Liver Hepatocellular Carcinoma.

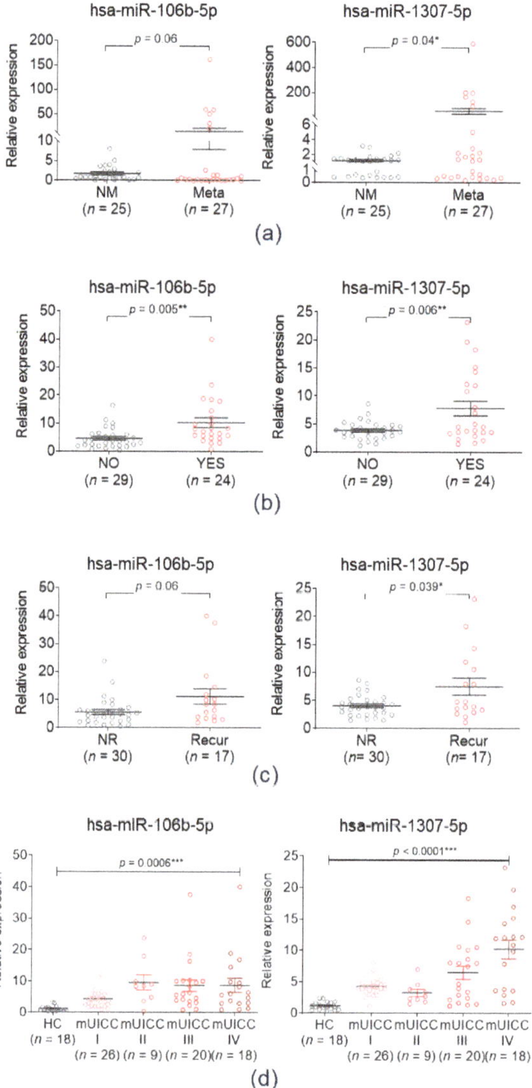

Figure 4. Validation of the clinical implications of candidate exo-miRs in the validation cohort. (**a**) Expression of two plasma exo-miRs in the metastasis-free group (NM, n = 27) and the metastasis group (Meta, n = 25). * $p < 0.05$. (**b**) Expression of the serum exo-miRs according to the vascular invasion status. * $p < 0.05$; ** $p < 0.01$. (**c**) Expression of serum exo-miRs according to the recurrence status. (**d**) Expression of the serum exo-miRs according to the modified UICC stage. *** $p < 0.001$. NM, non-metastasis; Meta, metastasis; UICC, Union for International Cancer Control.

2.4. In Silico Prediction of the Target Genes of miR-1307-5p

We attempted to identify the downstream target genes of miR-1307-5p that promote extrahepatic metastasis. Target gene prediction using TargetScan 7.2 identified 120 candidates as miR-1307-5p target genes (Figure 5a). Thereafter, the expression levels of 120 genes were evaluated in the TCGA_LIHC cohort. As miRs negatively regulate their target genes, we attempted to identify genes downregulated in the HCC tissues. In the TCGA_LIHC cohort, 16 of 120 genes were downregulated in HCC tissue rather than in adjacent non-tumor tissue, and 9 of 16 genes, namely, *ALDH8A1*, *C11orf96*, *CLYBL*, *EFNB3*, *ENG*, *NPC1L1*, *PIM3*, *SEC14L2*, and *SLC8A1*, displayed a significant difference ($p < 0.05$) (Figure 5a and Figure S3).

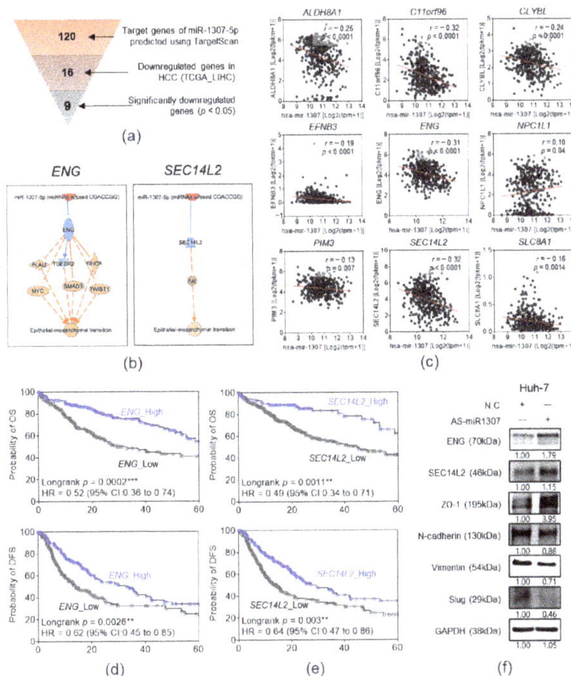

Figure 5. Prediction of target genes of miR-1307-5p in hepatocellular carcinoma through bioinformatics analysis. (**a**) Selection of potential target genes of miR-1307-5p via TargetScan 7.2 and expression data in TCGA_LIHC. (**b**) Pearson's correlation analysis using the expression data in TCGA_LIHC to identify inversely correlated genes. (**c**) Pathway analysis with functional annotation of the epithelial–mesenchymal transition using the IPA software on miR-1307-5p and the target gene candidates, ENG and SEC14L2. (**d**) Kaplan–Meier plot of overall survival and disease-free survival based on ENG expression in TCGA_LIHC. ** $p < 0.01$; *** $p < 0.001$. (**e**) Kaplan–Meier plot of overall survival and disease-free survival based on SEC14L2 expression in TCGA_LIHC. ** $p < 0.01$. (**f**) Western blot analysis of ENG, SEC14L2, and EMT markers after miR-1307-5p inhibition by AS-miR-1307-5p in human HCC Huh-7 cells.

To identify genes inversely associated with miR-1307-5p, Pearson's correlation analysis was performed using the expression data in the TCGA_LIHC database. The expression levels of five of nine genes, namely, *ALDH8A1*, *C11orf96*, *CLYBL*, *ENG*, and *SEC14L2*, displayed a significant inverse correlation with miR-1307-5p expression ($r \leq -0.3$ and $p < 0.05$) (Figure 5b). We performed pathway analysis with functional annotation of the EMT, using the Ingenuity Pathway Analysis (IPA) software on miR-1307-5p and the five target candidate genes (Figure 5c and Figure S4). Consequently, miR-1307-5p/SEC14L2/Akt and miR-1307-5p/ENG signaling pathways were associated with the EMT. Survival analyses based on the expression of *ENG* and *SEC14L2* was performed using expression data from the TCGA_LIHC database. Figure 5d,e display Kaplan–Meier plots of OS and disease-free survival (DFS) based on the expression of ENG and SEC14L2, respectively. Compared to the high expression group, the low ENG expression group had a significantly poor OS ($p = 0.0002$) and poor

DFS survival ($p = 0.0026$). Furthermore, compared to the high expression group, the low *SEC14L2* group had a poor OS ($p = 0.011$) and DFS ($p = 0.003$).

To validate the downstream pathway of miR-1307-5p proposed by TargetScan and IPA, we treated antisense (AS)-miR-1307-5p, an inhibitor of miR-1307-5p, to immortalized HCC cell line- Huh-7 cell. Then, protein expression level of ENG and SEC14L2, which were proposed target molecules of miR-1307-5p, and expression of EMT markers (ZO-1, N-cadherin, vimentin, and slug) were evaluated by western blotting in Huh-7 (Figure 5f). As a result, expression level of ENG and SEC14L2 was increased by inhibition of miR-1307-5p, and the expressions of the EMT markers were altered as a direction of EMT promotion after treatment of AS-miR-1307-5p. Taken together, we could confirm the downstream pathway of miR-1307-5p in HCC, which down-regulates ENG/SEC14L2 and promotes EMT process.

3. Discussion

Increasing evidence indicates that exosomes deliver pro-metastatic molecules to recipient cells, resulting in a pre-metastatic niche [14,15]. This study was based on the assumption that the expression of specific exo-miRs potentially increased during systemic circulation prior to extrahepatic metastasis, thus, promoting metastasis in patients with HCC.

To confirm this assumption, next generation sequencing-based circulating exo-miR profiles were analyzed, and differentially expressed circulating exo-miRs were identified between the metastasis-free group and the metastasis group during the follow-up period. Among the 61 predominantly overexpressed exo-miRs in the metastasis group, candidate miRs were further selected through systematic integrative analyses of publicly available RNA expression datasets. Consequently, exo-miR-1307-5p was identified as potential candidate pro-metastatic molecule. In the validation study, circulating exo-miR-1307-5p was significantly overexpressed in the metastasis group. Furthermore, SEC14L2 and ENG downregulation and the promotion of the EMT were considered potential downstream pathways of miR-1307-5p upon comprehensive bioinformatics analyses. We validated it by demonstrating up-regulation of ENG/SEC14L2 and EMT marker expression alteration after AS-miR-1307-5p treatment.

Exosomes contain unique cargo from donor cells, and exosomal cargo is considered a promising cancer biomarker. Several recent studies have shown circulating exo-miRs as potential diagnostic biomarkers for early-stage HCC [18,19]. Moreover, aberrantly regulated exo-miRs can promote HCC progression and metastasis by altering the genetic network [20]. An in vitro study by Lin et al., exosome-mediated miR delivery was shown to promote the EMT and metastasis in HCC [21]. However, few studies have investigated circulating exo-miR profiles as metastasis predictors or promoters in HCC. Herein, circulating exo-miR-1307-5p was considered a potential candidate metastasis predictor and metastasis driver in patients with HCC. Although miR-1307 is known as an onco-miR in diverse cancers as well as HCC, the clinical implication of the circulating exo-miR-1307-5p as cancer biomarker was only evaluated in ovarian cancer [22–25]. We identified exo-miR-1307-5p as a potential candidate metastasis driver and predictor in HCC. To our knowledge, this is the first study to identify circulating exo-miR-1307-5p as a novel metastasis promoter and predictor in patients with HCC.

Vascular invasion is considered as a pathognomonic hallmark of HCC invasiveness and poor prognosis [26]. Furthermore, it has been reported as a principal predictive marker for tumor recurrence and extrahepatic metastasis in HCC [27]. Herein, circulating exo-miR-1307-5p was significantly overexpressed in patients with vascular invasion as well as metastasis and tumor recurrence. Considering that vascular invasion is closely associated with subsequent extrahepatic metastasis and tumor recurrence in patients with HCC, exo-miR-1307-5p may potentially serve as a prognostic biomarker in patients with HCC.

Epithelial cells lose their epithelial phenotype and display a mesenchymal phenotype during the EMT [28]. EMT markedly promotes tumor invasiveness and metastasis by obliterating cell-cell adhesion [29]. Herein, we proposed SEC14L2/Akt and ENG-related signaling pathways as downstream pathways of miR-1307-5p for promoting the EMT in patients with HCC. Pathway analysis using IPA revealed that SEC14L2 is downregulated by miR-1307, in turn activating the Akt pathway, thus, promoting the EMT. SEC14L2 is a potent tumor suppressor gene in various malignancies [30]. Li et al. reported that SEC14L2, a novel master regulator gene, exerts an anti-proliferative effect in HCC cells and strongly suppresses tumor growth in a mouse model [31]. ENG (CD105), a transmembrane glycoprotein, is a transforming growth factor-β co-receptor [32]. ENG is involved in angiogenesis in solid tumors including HCC [33]. Several studies have shown that ENG downregulation in HCC tissue and its serum levels potentially serve as a poor prognostic marker in patients with HCC [32,34,35]. However, the mechanisms of action of ENG in HCC progression remain unclear. The present study demonstrated that miR-1307-5p down-regulates ENG and that downregulation of ENG is associated with the EMT promotion.

This study has two limitations. First, the patient cohort sizes were small. Herein, the metastasis group included patients with extrahepatic metastasis after initial blood sampling. As patients with HCC with available blood samples prior to the occurrence of metastasis were rare, we could not enroll enough patients to obtain a high statistical power. Furthermore, owing to the shortage of blood samples from patients with metastatic HCC, we assume that the present results provide valuable and potentially useful information regarding pro-metastatic exo-miRs. Further validation studies with larger cohorts are needed to verify the present results. Second, we could not determine the mechanism underlying the promotion of extrahepatic metastasis by exo-miR-1307-5p. To overcome this limitation, we implemented an in silico analysis strategy. Downstream pathways of miR-1307-5p promoting the EMT were predicted through in silico analysis. Hence, further studies are required to confirm the underlying signaling pathway.

4. Materials and Methods

4.1. Patients and Sample Collection

To identify candidate circulating exo-miRs with pro-metastatic potential, the medical records of patients with HCC with available plasma samples during diagnosis were reviewed and patients were included in accordance with the inclusion and exclusion criteria. The inclusion criteria were as follows: (1) patients newly diagnosed with HCC in accordance with the American Association for the Study of Liver Diseases, criteria [36]; (2) patients without extrahepatic metastasis during diagnosis; (3) patients treated with local or systemic therapy in accordance with the tumor burden or location [37]; (4) Child-Pugh class A or B; and (5) the availability of follow-up imaging data for evaluating the tumor burden and metastasis status every 3 months for >1 year. Patients lost to follow-up 1 year before metastasis onset were excluded. Extrahepatic metastasis occurred in 25 of 52 patients meeting the inclusion criteria (metastasis group), and the remaining 27 patients were metastasis-free during follow-up evaluation (metastasis-free group). This cohort was called the Plasma-HCC cohort.

Owing to the low strength and lack of several clinical data in the Plasma-HCC cohort, patients with HCC with available pre-treatment serum samples during diagnosis were included to validate the clinical implications of selected circulating exo-miRs. This validation cohort was called the Serum-HCC cohort, comprising 91 serum samples from 73 patients with HCC and 28 healthy controls. A healthy control was defined as an individual without any medical history, who visited the Ajou Health Promotion Center for a regular health check-up. Data on the vascular invasion status, metastasis status, and tumor stage based on the modified Union for International Cancer Control (mUICC) staging system were obtained. The baseline characteristics of patients in the Plasma-HCC and Serum-HCC cohorts are elucidated in Table 1.

Table 1. Baseline characteristics of the Plasma-HCC cohort and the Serum-HCC cohrort.

Variables	Plasma-HCC Cohort (n = 52)	Serum-HCC Cohort (n = 91)	
	HCC (n = 52)	Healthy Control (n = 28)	HCC (n = 73)
Age (years), mean ± SD	54.22 ± 9.98	34.96 ± 8.18	54.62 ± 9.06
Male sex, n (%)	47 (85.5)	3 (10.7)	57 (78.1)
AST, IU/mL	72.73 ± 77.36	16.82 ± 4.06	73.81 ± 92.28
ALT, IU/mL	46.11 ± 32.85	14.14 ± 7.77	49.48 ± 63.53
Platelet, x109/L	161.59 ± 82.04	314± 63.33	169.85 ± 85.41
AFP (ng/mL), mean ± SD	8000.26 ± 17,325.08	1.65 ± 0.58	4394.77 ± 14,600.73
Etiology, n HBV/HCV/alcohol/others	–		65/4/3/1
Albumin (g/L), mean ± SD	4.07 ± 0.60		4.23 ± 0.55
Bilirubin (mg/dL), mean ± SD	1.22 ± 1.81		1.49 ± 3.91
INR, mean ± SD	1.16 ± 0.14		1.18 ± 0.19
Modified UICC stage, n (%) I/II/III/IVa/IVb	–		26 (36)/9 (12)/20 (27)/11 (15.4)/7 (9.6)
Metastasis, n (%) Yes/No	25 (48.1)/27 (51.9)		7 (9.6)/66 (90.4)
Vascular invasion, n (%) Yes/No	–		29 (54.7)/24 (45.3)

HCC, hepatocellular carcinoma; AST, aspartate transaminase; ALT, alanine transaminase; AFP, alpha-feto protein; HBV, hepatitis B virus; HCV, hepatitis C virus; INR, international normalized ratio; UICC, Union for International Cancer Control; –Dashes denote lack of reliable data.

The study protocol was approved by the Institutional Review Board of the Ajou University Hospital, Suwon, South Korea (AJRIB-BMR-OBS-16-344). Anonymous blood samples and clinical data were provided by the Ajou Human Bio-Resource Bank. Informed consent was waived.

4.2. *Analysis of Gene Expression Omnibus (GEO) Database and TCGA_LIHC*

To estimate miR expression levels in HCC, public genomic data were obtained from TCGA_LIHC (https://cancergenome.nih.gov) and the GEO database (GSE67140) [16] of the National Center for Biotechnology Information.

4.3. *Cell Culture and AS-miR-1307-5p Transfection*

Hep3B and Huh-7 cells (ATCC, Manassas, VA, USA) were cultured in EMEM or DMEM mudium (GenDEPOT, Barker, TX, USA) containing 10% FBS (Invitrogen, Waltham, MA, USA) and 100 units/mL penicillin-streptomycin (GenDEPOT), at 37 °C in a humidified incubator with 5% CO_2. Antisense inhibitor miR-1307-5p or scrambled control antisense inhibitor (Bioneer, Daejeon, Korea) was transfected into Huh7 cells using Lipofectamine 2000 (Invitrogen) according to the manufacturer's protocol.

4.4. *Blood Exosome Isolation and Total Exosomal RNA Extraction*

Blood samples were obtained from the Biobank of Ajou University Hospital, a member of the Korea Biobank Network. Five milliliters of blood were collected from each individual directly into EDTA-containing tubes (for plasma) or serum-separating tubes (for serum) and centrifuged at 2000× g for 5 min at 4 °C and the resultant plasma or sera were aliquoted into 1.5 mL tubes and stored at −80 °C until use. To isolate exosomes from blood, the ExoQuick reagent-cat# EXOQ5A-1 (System Biosciences, Mountain View, CA, USA) was used for serum exosome isolation and the ExoQuick Plasma Prep and Exosome Precipitation Kit (Cat# EXOQ5TMA-1) was used for plasma exosome isolation. Briefly, plasma samples were ultra-centrifuged for 15 min (13,000 rpm) to remove partial cells and their debris. Then, to remove fibrinogen and fibrin in plasma, 5uL SBI Thrombin Reagent was added to the supernatant to convert fibrinogen into fibrin and centrifuged at 10,000× g for 5 min to make fibrin pellet. Then, the supernatant was transferred to a new microfuge tube and the fibrin pellet was discarded. Exosomes were isolated from the supernatant using the ExoQuick Exosome Precipitation Solution in accordance with the manufacturer's protocol. Finally, the exosome pellet was re-suspended in 100 uL of PBS and stored at −80 °C for subsequent extraction of RNAs and proteins. RNA from blood-derived exosomes was extracted using the SeraMir Exosome RNA Amplification kit (System Biosciences). Thereafter, the total RNA concentration and purity were assessed using a NanoDrop 2000 spectrophotometer (Thermo Fisher Scientific, Waltham, MA, USA).

4.5. Small RNA Sequencing

Small RNA libraries were constructed from total RNA using the Illumina HiSeq 2000 system (Illumina Inc., San Diego, CA, USA). After small-RNA sequencing, reads were trimmed by cutadapt program for removing adapter and low-quality sequences, 18~26 bp in length considering the length of mature miRNA. Then, the trimmed reads were collapsed to remove duplicates and estimate abundance for the same sequence, and annotated using blast with miRBase. For comparison between samples, count of each sample was normalized in units of Transcripts Per Million (TPM).

4.6. TEM

TEM was performed to confirm the presence and sizes of exosomes. Samples were fixed in 2% glutaraldehyde and 4% paraformaldehyde for 2 h at room temperature and embedded responded with 0.13% methylcellulose and 0.4% uranyl acetate. Exosomes were then observed using a Hitachi H-7600 TEM (Hitachi High-Tech, Tokyo, Japan).

4.7. NTA

NTA was used to measure the size distribution and concentration of exosomes on the basis of light scattering and Brownian motion. NanoSight NS300 (Malvern Panalytical Ltd., Malvern, UK) equipped with a 405 nm laser with a frame rate of 30 frames/s was used for recording particle movement, and the data were evaluated using the NTA software (version 3.0, Malvern Panalytical Ltd.).

4.8. Western Blotting

Proteins were extracted from exosome and cell lysates, using radio immunoprecipitation (RIPA) buffer containing Halt Protease Inhibitor Cocktail (Thermo Fisher Scientific). Total proteins were separated by sodium dodecyl sulfate polyacrylamide gel electrophoresis and transferred to a polyvinylidene difluoride membrane (Merck Millipore, Burlington, MA, USA). The membranes were blocked with 5% non-fat milk in Tris-buffer saline and 0.1% Tween-20 and probed with the following primary antibodies: mouse anti-CD63 (1:1000, Abcam, Cambridge, MA, USA), rabbit anti-CD9 (1:2000, Abcam), mouse anti-CD81 (1:250, Invitrogen), and mouse anti-Bip/Grp78 (1:1000, BD Biosciences, San Jose, CA, USA), rabbit anti-ENG (1:1000, Abcam), rabbit anti-SEC14L2 (1:1000, Abcam), mouse anti-ZO-1 (1:1000, Thermo Fisher Scientific), mouse anti-N-cadherin (1:2000, BD bioscience), rabbit anti-Vimentin (1:5000, GeneTex, Alton, CA, USA), rabbit anti-Slug (1:1000, Cell Signaling Technology, Danvers, MA, USA), and mouse anti-GAPDH (1:1000, Santa Cruz Biotechnology, Santa Cruz, CA, USA). Chemiluminescence signals were detected using Clarity™ Western ECL Substrate and ChemiDoc (both from Bio-Rad Laboratories, Hercules, CA, USA).

4.9. Quantitative Reverse-Transcription Polymerase Chain Reaction (qRT-PCR) Analysis

Circulating exo-miR expression levels were quantified using qRT-PCR. Each miR sequence was obtained from the miRBase database. [38] Primer sequences used herein are listed in Table S1. cDNA was synthesized from exosomal RNA using the miScript RT II kit (Qiagen, Hilden, Germany) in accordance with the manufacturer's instructions, amplified using the Amfisure qGreen qPCR Master Mix (GenDEPOT), and monitored in real time using CFX Connect™ Real-Time PCR Detection System (Bio-rad Laboratories). The cycling conditions were as follows: 2 min at 95 °C, 40 cycles of 15 s at

95 °C, 34 s at 58 °C or 60 °C, and 30 s at 72 °C. Individual miR expression levels were determined from triplicate reactions and normalized with that of hsa-miR-1228-3p. The relative standard curve method (2−ΔΔCT) was used to determine relative expression levels.

4.10. Prediction of miR Targets

miR-1307-5p targets were predicted in silico using TargetScan 7.2 [39].

4.11. IPA

Signaling pathways downstream of miR-1307-5p and its target genes were subjected to functional annotation of EMT via IPA (Qiagen Inc., Redwood City, CA, USA) [40].

4.12. Statistical Analysis

All experiments were performed at least three times and all samples were analyzed in triplicate. Between-group differences were analyzed using a paired *t*-test or unpaired Welch's *t*-test with GraphPad prism version 5.0 software (GraphPad Software Inc., San Diego, CA, USA). OS and DFS were plotted using the Kaplan–Meier method, and the significant differences were analyzed using the log-rank test. Differences were considered statistically significant when p was < 0.05.

5. Conclusions

In conclusion, this study shows that circulating exo-miR-1307-5p is a novel metastasis promoter and predictive marker for metastasis in patients with HCC, through systematic integrative analyses. EMT promotion through SEC14L2 and ENG downregulation could be the potential downstream pathway of miR-1307-5p, as revealed through comprehensive bioinformatics analyses.

Supplementary Materials: The following are available online at http://www.mdpi.com/2072-6694/12/12/3819/s1, Figure S1: The whole blot (uncropped blots) showing all the bands with all molecular weight markers, Figure S2: Serum exo-miR-1307-5p expression and ROC curves in normal healthy controls and the patients with HCC, Figure S3: Expression of the nine target gene candidates of miR-1307-5p in The Cancer Genomic Atlas Liver Hepatocellular Carcinoma (TCGA_LIHC), Figure S4: Pathway analysis with a functional annotation of the epithelial–mesenchymal transition (EMT) using the Ingenuity Pathway Analysis (IPA) software on miR-1307-5p and the target gene candidates, *ALDH8A1*, *C11orf96*, and *CLYBL*, Table S1: miRNA sequence used for qRT-PCR.

Author Contributions: Conceptualization, J.W.E., J.Y.C. and H.J.C.; methodology, C.W.S.; validation, G.O.B., M.G.Y. and H.R.A.; formal analysis, J.W.E.; investigation, J.A.S. and S.S.; software, D.W.K.; resources, S.S.K.; data curation, J.W.E.; writing—original draft preparation, H.J.C.; writing—review and editing, J.Y.C.; visualization, J.W.E.; supervision, S.S.K.; funding acquisition, J.W.E., J.Y.C. and H.J.C. All authors have read and agreed to the published version of the manuscript.

Funding: This research was funded by National Research Foundation (NRF) of Korea (NRF-2017M3A9B6061509, NRF-2019R1C1C1007366, and NRF-2018M3A9E8023861) and the intramural research fund of Ajou University medical center.

Acknowledgments: The biospecimens and data used for this study were provided by the Biobank of Ajou University Hospital, a member of Korea Biobank Network.

Conflicts of Interest: The authors declare no conflict of interest.

References

1. Rowe, J.H.; Ghouri, Y.A.; Mian, I. Review of hepatocellular carcinoma: Epidemiology, etiology, and carcinogenesis. *J. Carcinog.* **2017**, *16*, 1. [CrossRef] [PubMed]
2. Wong, R.; Frenette, C. Updates in the Management of Hepatocellular Carcinoma. *Gastroenterol. Hepatol.* **2011**, *7*, 16–24.
3. Crissien, A.M.; Frenette, C. Current Management of Hepatocellular Carcinoma. *Gastroenterol. Hepatol.* **2014**, *10*, 153–161.
4. Heimbach, J.K.; Kulik, L.M.; Finn, R.S.; Sirlin, C.B.; Abecassis, M.M.; Roberts, L.R.; Zhu, A.X.; Murad, M.H.; Marrero, J.A. AASLD guidelines for the treatment of hepatocellular carcinoma. *Hepatology* **2018**, *67*, 358–380. [CrossRef] [PubMed]

5. Cheung, A.H.-K.; Chow, C.; To, K. Latest development of liquid biopsy. *J. Thorac. Dis.* **2018**, *10*, S1645–S1651. [CrossRef]
6. Diaz, L.A., Jr.; Bardelli, A. Liquid biopsies: Genotyping circulating tumor DNA. *J. Clin. Oncol.* **2014**, *32*, 579. [CrossRef]
7. Buder, A.; Tomuta, C.; Filipits, M. The potential of liquid biopsies. *Curr. Opin. Oncol.* **2016**, *28*, 130–134. [CrossRef]
8. Gold, B.; Cankovic, M.; Furtado, L.V.; Meier, F.; Gocke, C.D. Do circulating tumor cells, exosomes, and circulating tumor nucleic acids have clinical utility? A report of the association for molecular pathology. *J. Mol. Diagn.* **2015**, *17*, 209–224. [CrossRef]
9. Peterson, M.F.; Otoc, N.; Sethi, J.K.; Gupta, A.; Antes, T.J. Integrated systems for exosome investigation. *Methods* **2015**, *87*, 31–45. [CrossRef]
10. Bebelman, M.P.; Smit, M.J.; Pegtel, D.M.; Baglio, S.R. Biogenesis and function of extracellular vesicles in cancer. *Pharmacol. Ther.* **2018**, *188*, 1–11. [CrossRef]
11. Contreras-Naranjo, J.C.; Wu, H.-J.; Ugaz, V.M. Microfluidics for exosome isolation and analysis: Enabling liquid biopsy for personalized medicine. *Lab Chip* **2017**, *17*, 3558–3577. [CrossRef] [PubMed]
12. Zhang, J.; Li, S.; Li, L.; Li, M.; Guo, C.; Yao, J.; Mi, S. Exosome and Exosomal MicroRNA: Trafficking, Sorting, and Function. *Genom. Proteom. Bioinform.* **2015**, *13*, 17–24. [CrossRef] [PubMed]
13. Safdar, A.; Saleem, A.; Tarnopolsky, M.A. The potential of endurance exercise-derived exosomes to treat metabolic diseases. *Nat. Rev. Endocrinol.* **2016**, *12*, 504–517. [CrossRef] [PubMed]
14. Syn, N.; Wang, L.; Sethi, G.; Thiery, J.P.; Goh, B.-C. Exosome-Mediated Metastasis: From Epithelial–Mesenchymal Transition to Escape from Immunosurveillance. *Trends Pharmacol. Sci.* **2016**, *37*, 606–617. [CrossRef]
15. Greening, D.W.; Gopal, S.K.; Mathias, R.A.; Liu, L.; Sheng, J.; Zhu, H.-J.; Simpson, R.J. Emerging roles of exosomes during epithelial–mesenchymal transition and cancer progression. *Semin. Cell Dev. Biol.* **2015**, *40*, 60–71. [CrossRef]
16. Kim, T.-H.; Kim, S.Y.; Tang, A.; Yoon, J.H. Comparison of international guidelines for noninvasive diagnosis of hepatocellular carcinoma: 2018 update. *Clin. Mol. Hepatol.* **2019**, *25*, 245–263. [CrossRef]
17. Korean Liver Cancer Association (KLCA). 2018 Korean Liver Cancer Association–National Cancer Center Korea Practice Guidelines for the Management of Hepatocellular Carcinoma. *Korean J. Radiol.* **2019**, *20*, 1042–1113. [CrossRef]
18. Xie, Q.Y.; Almudevar, A.; Whitney-Miller, C.L.; Barry, C.T.; McCall, M.N. A microRNA biomarker of hepatocellular carcinoma recurrence following liver transplantation accounting for within-patient heterogeneity. *BMC Med. Genom.* **2016**, *9*, 18. [CrossRef]
19. Sharan, R.; Maron-Katz, A.; Shamir, R. CLICK and EXPANDER: A system for clustering and visualizing gene expression data. *Bioinformatics* **2003**, *19*, 1787–1799. [CrossRef]
20. Cho, H.J.; Eun, J.W.; Baek, G.O.; Seo, C.W.; Ahn, H.R.; Kim, S.S.; Cho, S.W.; Cheong, J.Y. Serum exosomal microRNA, miR-10b-5p, as a potential diagnostic biomarker for early-stage hepatocellular carcinoma. *J. Clin. Med.* **2020**, *9*, 281. [CrossRef]
21. Cho, H.J.; Baek, G.O.; Seo, C.W.; Ahn, H.R.; Sung, S.; Son, J.A.; Kim, S.S.; Cho, S.W.; Jang, J.W.; Nam, S.W.; et al. Exosomal microRNA-4661-5p-based serum panel as a potential diagnostic biomarker for early-stage hepatocellular carcinoma. *Cancer Med.* **2020**. [CrossRef] [PubMed]
22. Qu, Z.; Wu, J.; Wu, J.; Ji, A.; Qiang, G.; Jiang, Y.; Jiang, C.; Ding, Y. Exosomal miR-665 as a novel minimally invasive biomarker for hepatocellular carcinoma diagnosis and prognosis. *Oncotarget* **2017**, *8*, 80666–80678. [CrossRef] [PubMed]
23. Lin, Q.; Zhou, C.R.; Bai, M.J.; Zhu, D.; Chen, J.W.; Wang, H.F.; Li, M.A.; Wu, C.; Li, Z.R.; Huang, M.S. Exosome-mediated miRNA delivery promotes liver cancer EMT and metastasis. *Am. J. Transl. Res.* **2020**, *12*, 1080–1095. [PubMed]
24. Tang, R.; Qi, Q.; Wu, R.; Zhou, X.; Wu, D.; Zhou, H.; Mao, Y.; Li, R.; Liu, C.; Wang, L. The polymorphic terminal-loop of pre-miR-1307 binding with MBNL1 contributes to colorectal carcinogenesis via interference with Dicer1 recruitment. *Carcinogenesis* **2015**, *36*, 867–875. [CrossRef] [PubMed]
25. Han, S.; Zou, H.; Lee, J.-W.; Han, J.; Kim, H.C.; Cheol, J.J.; Kim, L.-S.; Kim, H. miR-1307-3p Stimulates Breast Cancer Development and Progression by Targeting SMYD4. *J. Cancer* **2019**, *10*, 441–448. [CrossRef]
26. Su, Y.Y.; Sun, L.; Guo, Z.R.; Li, J.C.; Bai, T.T.; Cai, X.X.; Li, W.H.; Zhu, Y.F. Upregulated expression of serum exosomal miR-375 and miR-1307 enhance the diagnostic power of CA125 for ovarian cancer. *J. Ovarian Res.* **2019**, *12*, 6. [CrossRef]

27. Chen, S.; Wang, L.; Yao, B.; Liu, Q.; Guo, C. miR-1307-3p promotes tumor growth and metastasis of hepatocellular carcinoma by repressing DAB2 interacting protein. *Biomed. Pharmacother.* **2019**, *117*, 109055. [CrossRef]
28. European Association for the Study of the Liver. European Organisation for Research and Treatment of Cancer EASL–EORTC Clinical Practice Guidelines: Management of hepatocellular carcinoma. *J. Hepatol.* **2012**, *56*, 908–943. [CrossRef]
29. Hashimoto, M.; Kobayashi, T.; Ishiyama, K.; Ide, K.; Ohira, M.; Tahara, H.; Kuroda, S.; Hamaoka, M.; Iwako, H.; Okimoto, S.; et al. Predictive Independent Factors for Extrahepatic Metastasis of Hepatocellular Carcinoma Following Curative Hepatectomy. *Anticancer Res.* **2017**, *37*, 2625–2631. [CrossRef]
30. Blackwell, R.H.; Foreman, K.E.; Gupta, G. The Role of Cancer-Derived Exosomes in Tumorigenicity & Epithelial-to-Mesenchymal Transition. *Cancers* **2017**, *9*, 105. [CrossRef]
31. Gotzmann, J.; Mikula, M.; Eger, A.; Schulte-Hermann, R.; Foisner, R.; Beug, H.; Mikulits, W. Molecular aspects of epithelial cell plasticity: Implications for local tumor invasion and metastasis. *Mutat. Res.* **2004**, *566*, 9–20. [CrossRef]
32. Wang, X.; Ni, J.; Hsu, C.-L.; Johnykutty, S.; Tang, P.; Ho, Y.-S.; Lee, C.-H.; Yeh, S. Reduced Expression of Tocopherol-Associated Protein (TAP/Sec14L2) in Human Breast Cancer. *Cancer Investig.* **2009**, *27*, 971–977. [CrossRef] [PubMed]
33. Li, Z.; Lou, Y.; Tian, G.; Wu, J.; Lu, A.; Chen, J.; Xu, B.; Shi, J.; Yang, J. Discovering master regulators in hepatocellular carcinoma: One novel MR, SEC14L2 inhibits cancer cells. *Aging* **2019**, *11*, 12375–12411. [CrossRef]
34. Kasprzak, A.; Adamek, A. Role of Endoglin (CD105) in the Progression of Hepatocellular Carcinoma and Anti-Angiogenic Therapy. *Int. J. Mol. Sci.* **2018**, *19*, 3887. [CrossRef] [PubMed]
35. Ho, J.W.; Poon, R.T.; Sun, C.K.; Xue, W.-C.; Fan, S.-T. Clinicopathological and prognostic implications of endoglin (CD105) expression in hepatocellular carcinoma and its adjacent non-tumorous liver. *World J. Gastroenterol.* **2005**, *11*, 176–181. [CrossRef]
36. Yagmur, E.; Rizk, M.; Stanzel, S.; Hellerbrand, C.; Lammert, F.; Trautwein, C.; Wasmuth, H.E.; Gressner, A.M. Elevation of endoglin (CD105) concentrations in serum of patients with liver cirrhosis and carcinoma. *Eur. J. Gastroenterol. Hepatol.* **2007**, *19*, 755–761. [CrossRef] [PubMed]
37. Teama, S.; Fawzy, A.; Teama, S.; Helal, A.; Drwish, A.D.; Elbaz, T.; Desouky, E. Increased Serum Endoglin and Transforming Growth Factor β1 mRNA Expression and Risk of Hepatocellular Carcinoma in Cirrhotic Egyptian Patients. *Asian Pac. J. Cancer Prev.* **2016**, *17*, 2429–2434.
38. miRBase. Available online: http://www.mirbase.org (accessed on 17 May 2020).
39. Agarwal, V.; Bell, G.W.; Nam, J.W.; Bartel, D.P. TargetScanHuman: Prediction of microRNA targets. Available online: http://www.targetscan.org/vert_72 (accessed on 27 May 2020).
40. QIAGEN Ingenuity Oathway Analysis (QIAGEN IPA). Available online: https://digitalinsights.qiagen.com/products-overview/discovery-insights-portfolio/analysis-and-visualization/qiagen-ipa (accessed on 3 June 2020).

Article

A Pilot Study Comparing the Efficacy of Lactate Dehydrogenase Levels Versus Circulating Cell-Free microRNAs in Monitoring Responses to Checkpoint Inhibitor Immunotherapy in Metastatic Melanoma Patients

Matias A. Bustos [1,*], Rebecca Gross [1], Negin Rahimzadeh [1], Hunter Cole [2], Linh T. Tran [3], Kevin D. Tran [3], Ling Takeshima [1], Stacey L. Stern [4], Steven O'Day [2] and Dave S. B. Hoon [1,3]

1. Department of Translational Molecular Medicine, John Wayne Cancer Institute (JWCI), Providence Saint John's Health Center (SJHC), Santa Monica, CA 90404, USA; Rebecca.Gentry@providence.org (R.G.); Negin.Rahimzadeh@providence.org (N.R.); TakeshimaL@jwci.org (L.T.); HoonD@jwci.org (D.S.B.H.)
2. Department of Immuno-Oncology and Clinical Research, JWCI, Providence SJHC, Santa Monica, CA 90404, USA; ColeH@jwci.org (H.C.); O'DayS@jwci.org (S.O.)
3. Department of Genomic Sequencing Center, JWCI, Providence SJHC, Santa Monica, CA 90404, USA; Linh.Tran3@providence.org (L.T.T.); Kevin.Tran@providence.org (K.D.T.)
4. Department of Biostatistics, JWCI, Providence SJHC, Santa Monica, CA 90404, USA; SternS@jwci.org

* Correspondence: BustosM@jwci.org

Received: 9 September 2020; Accepted: 9 November 2020; Published: 13 November 2020

Simple Summary: Improvement in melanoma patients with metastatic disease is needed to better assess immunotherapies. Lactate dehydrogenase (LDH) is currently an accepted biomarker for stage IV, but it has limited utility for stage III melanoma patients. Thus, finding biomarkers for metastatic melanoma is important not only to identify progressive melanoma tumors, but also to monitor patients under checkpoint inhibitor immunotherapy (CII). The aim of this pilot study was to demonstrate the utility of circulating cell-free microRNAs (cfmiRs) as potential blood biomarkers for stage III and IV melanoma patients compared to LDH. To accomplish this aim, we profiled for cfmiR the plasma of metastatic melanoma patients before and during CII treatment, and compared them to normal healthy donors' samples. The cfmiR profiling was performed using an NGS-based miRNA assay, which requires no extraction and a small volume input. We found specific cfmiR signatures in stage III and IV metastatic melanoma patients. As a proof of concept, our results showed that certain cfmiRs are associated with CII outcomes.

Abstract: Serum lactate dehydrogenase (LDH) is a standard prognostic biomarker for stage IV melanoma patients. Often, LDH levels do not provide real-time information about the metastatic melanoma patients' disease status and treatment response. Therefore, there is a need to find reliable blood biomarkers for improved monitoring of metastatic melanoma patients who are undergoing checkpoint inhibitor immunotherapy (CII). The objective in this prospective pilot study was to discover circulating cell-free microRNA (cfmiR) signatures in the plasma that could assess melanoma patients' responses during CII. The cfmiRs were evaluated by the next-generation sequencing (NGS) HTG EdgeSeq microRNA (miR) Whole Transcriptome Assay (WTA; 2083 miRs) in 158 plasma samples obtained before and during the course of CII from 47 AJCC stage III/IV melanoma patients' and 73 normal donors' plasma samples. Initially, cfmiR profiles for pre- and post-treatment plasma samples of stage IV non-responder melanoma patients were compared to normal donors' plasma samples. Using machine learning, we identified a 9 cfmiR signature that was associated with stage IV melanoma patients being non-responsive to CII. These cfmiRs were compared in pre- and post-treatment plasma samples from stage IV melanoma patients that showed good responses. Circulating miR-4649-3p, miR-615-3p, and miR-1234-3p demonstrated potential prognostic utility in assessing CII responses. Compared to LDH levels during CII, circulating miR-615-3p levels were consistently more efficient in detecting melanoma patients undergoing CII who developed progressive disease. By combining stage III/IV patients, 92 and 17 differentially expressed cfmiRs were identified in pre-treatment plasma samples from responder and non-responder patients, respectively. In conclusion, this pilot study demonstrated cfmiRs that identified treatment responses and could allow for real-time monitoring of patients receiving CII.

Keywords: serum LDH; blood biomarker; miRNA; circulating microRNA; plasma; immunotherapy; immune checkpoint inhibitors; metastatic melanoma

---

1. Introduction

Over the past decade, checkpoint inhibitor immunotherapy (CII) has significantly improved the outcomes of metastatic melanoma patients [1]. The CII monoclonal antibodies approved to treat metastatic melanoma patients include ipilimumab (targeting cytotoxic T lymphocyte-associated antigen 4, CTLA-4) [2], nivolumab and pembrolizumab (targeting programmed cell death protein-1, PD-1) [3]. Ipilimumab, nivolumab, and pembrolizumab represent the standard of care and are the most commonly utilized CII for treating metastatic melanoma patients [4]. One of the advantages of specific CII regimens is the durable response observed in melanoma patients even after treatment discontinuation, which can vary depending on the individual or combinatory CII implemented. Unfortunately, the complete response (CR) rate in melanoma patients is about 12–15% [5,6]. Major limitations for CIIs are primary and acquired CII resistance. Another limitation is the development of severe immune-related adverse events (IRAE), which forces the oncologist to discontinue the patient's treatment [7]. Different tumor responses, tumor microenvironment changes, and host systemic immune responses play interactive roles in CII resistance, and IRAE [7,8]. Unfortunately, no key consistent findings and biomarkers have been found to identify these induced CII events earlier on patients.

Lactate dehydrogenase (LDH) is an enzyme involved in glucose metabolism that is highly expressed in rapidly growing tumors [9,10]. Due to the high energy demand from the tumor cells, glycolysis shifts from aerobic to anaerobic in a process called the Warburg effect [9]. Consequently, LDH expression increases in the cytosol of tumor cells, but in general will only reach the blood stream when the damaged cells release LDH [9]. Several prognostic blood biomarkers have been proposed for melanoma, but only serum LDH has been accepted as a prognostic biomarker for stage IV metastatic melanoma by the American Joint Committee on Cancer (AJCC) [11]. Therefore, the prognostic value of LDH has been assessed in metastatic melanoma patients receiving CII. In a prospective study, LDH and S100B have both been shown to be indicators of disease progression, although S100B was shown to be a better predictor of the development of distant metastasis [12]. Nevertheless, both failed at identifying high-risk patients with loco-regional metastasis and low tumor burden [12]. Elevated baseline LDH is an independent prognostic factor for overall survival (OS) in melanoma patients receiving ipilimumab [13], pembrolizumab, or ipilimumab and nivolumab combined [14]. Moreover, among different prognostic factors (LDH, tumor size, tumor PD-L1 status, ECOG performance status, *BRAF* mutation status, prior *BRAF* inhibitor targeted therapy, prior line of therapies, size of metastasis, and albumin levels), only low LDH baseline levels were associated with a CR to pembrolizumab [5]. Additionally, elevated LDH baseline levels were reduced at the first scan in melanoma patients receiving nivolumab or pembrolizumab, who had a better objective rate response when compared to patients with progressive disease (PD) [15]. To summarize, baseline LDH is a strong prognostic blood biomarker for stage IV melanoma patients, but has limitations. However, serum LDH assessment does not have informative utility for assessing stage III melanoma patients receiving CII.

Blood biomarkers are necessary for real-time monitoring of metastatic melanoma patients during treatment to allow for more effective decision making on treatment strategies. In the past several years, our group and others have shown that circulating cell-free nucleic acids (cfNA) have utility in monitoring metastatic melanoma patients undergoing treatment, particularly using circulating cell-free DNA (cfDNA) and circulating tumor cells (CTCs) [16–23]. The limitations of studying cfDNA in melanoma blood samples are the poor extraction efficacy from plasma, large volume of plasma required for assays, and the limited frequency of genomic aberrations in specific genes that are detectable [24,25]. The limitations in monitoring CTCs are the robustness of the isolation method used and the heterogeneity of the CTCs that can limit the interpretation of the findings. To find robust blood molecular biomarkers, our group has also focused on finding microRNAs (miRs) in melanoma patients' blood [26] and tumor tissues [27,28]. MiRs are short sequence nucleic acids of 18–22 base pairs length that have a longer half-life and degrade minimally compared to cfDNA [29,30]. MiRs play significant roles in controlling and regulating mRNA expression, and thus lead to the activation/deactivation of specific molecular pathways [29,30]. In most of cancers, including melanoma, miRs are aberrantly expressed which affects molecular pathways controlling different cellular processes. These miRs can also be referred to as oncomiRs as they promote tumor development and progression. In melanoma several miRs have been proposed as tumor biomarkers to determine disease progression [29,31]. Also, significant efforts have been made in determining circulating cell-free miRs (cfmiRs) and exosomal miRs [29,31]. Recently, by using HTG EdgeSeq miR WTA, we found cfmiR signatures in plasma samples from patients with melanoma brain metastasis (MBM) [32]. Furthermore, we unraveled common cfmiR signatures in pre-operative plasma samples taken from stage III and IV melanoma patients receiving CIIs [32]. The advantage of using HTG EdgeSeq miR WTA to study cfmiRs is that we can directly profile and quantify >2000 miRs found in plasma samples by next-generation sequencing (NGS) to identify signature patterns [32]. Moreover, compared to other cfNA assays, the assay requires a minimal amount of plasma and no tedious extraction procedures.

Our hypothesis is that specific cfmiR signatures found in metastatic melanoma patients' plasma samples allows us to perform multiple assessments and provides the clinician with the opportunity to monitor CII response in real-time. This is important in metastatic melanoma patients' treatment management, as resistance to CII followed by rapid disease progression requires immediate decisions in order to prolong survival. In this study, we compared cfmiR expression to the standard blood protein biomarker LDH in stage IV melanoma patients. To carry this out, we screened for specific cfmiRs that were indicative of metastatic melanoma disease in pre- and post-treatment plasma samples from stage IV melanoma patients compared to normal donors' plasma samples. By using machine learning we identified cfmiR signatures that were associated with CII response in stage IV responder and non-responder patients. Then, we compared the utility of these cfmiRs in predicting CII response in comparison to LDH levels at baseline and throughout the patients' follow-ups. CfmiRs produced consistent results in predicting CII responses compared to elevated LDH levels at baseline and in longitudinal clinical assessment in stage IV melanoma patients. Finally, we identified cfmiRs that have potential in determining CII responses in both stage III and IV melanoma patients.

2. Results

2.1. LDH Levels at Treatment Baseline as a Predictive Factor for CII Response in Metastatic Melanoma Patients

In order to identify cfmiRs associated with metastatic melanoma, we assessed plasma from a cohort of 47 melanoma patients (AJCC 8th edition stage III (n = 24) and IV (n = 23)) seen at the JWCI/SJHC clinic (Table 1). For each patient a range of 3–6 blood samples were collected and the samples were categorized as pre- or post-treatment according to the CII start date. Only plasma samples were included in the study and from this point on all the samples will be referred to as plasma. The samples were all analyzed using the HTG EdgeSeq miR WTA [32]. All of the patients analyzed had a median follow-up of 9.7 months and received CII (ipilimumab, nivolumab, pembrolizumab, or the combination of ipilimumab and nivolumab) as first line treatment. The 47 patients were divided into four different cohorts based on stage (III and IV) and CII response (responders and non-responders), which were analyzed by Response Evaluation Criteria In Solid Tumors (RECIST) 1.1 (Figure 1A–D). The four groups were as follows: stage III responder (group A); stage III non-responder (group B); stage IV responder (group C); and stage IV non-responder (group D). All of the patients had an LDH assessment taken at baseline and on longitudinal LDH assessments (average of 11 samples per patient) during CII (Figure 1A–D). LDH was considered elevated if patients had values taken >1 the upper limit normal (ULN) [23] (Table 1). Since LDH values at baseline were shown to be predictive of CII response in previous clinical studies [12–15], we initially compared the LDH levels at baseline for stage III and IV responder and non-responder patients. Although the sample size for this analysis is limited, the results showed a significantly higher expression of LDH levels at baseline in the stage IV non-responder group D when compared to the stage IV responder patients group C (Figure 1F). As expected, no differences were observed in responder and non-responder stage III patients (Figure 1E). Similar results were observed when the LDH values were assessed at 3 months after CII in both groups C and D (Figure 1G,H). These results are in agreement with previous observations showing that the LDH baseline levels predicts response in stage IV patients undergoing CII [15]. However, when assessing individual patients, the LDH levels were not of prognostic utility, since only ~54% of stage IV patients (7 of 13 patients) showed a correlation between high LDH levels and positive CII response. Importantly, the LDH values did not offer any advantage for stage III melanoma patients in relation to their response to CII.

Table 1. Clinical pathological information for metastatic melanoma patients receiving CII [1] analyzed for cfmiRs [2] in plasma samples.

	Melanoma Patients (n = 47)
Variables	**n (%)**
Age at diagnosis, mean (SD [4])	62.0 (13.9)
Age at treatment, mean (SD [4])	65.9 (13.5)
<60	14 (29.8)
≥60	33 (70.2)
Gender	
Male	30 (63.8)
Female	17 (36.2)
Treatment regimen	
Anti-PD-1	31 (65.95)
Anti-PD-1/anti-CTLA-4	16 (34.05)
AJCC 8th ed. Stages [3]	
III b/c	24 (51.1)
IV a/b/c/d	23 (48.9)
BRAF mutation	
Positive	25 (53.2)
Negative	22 (46.8)
CII-response based on RECIST [5] 1.1	
Responders	22 (46.8)
Non-responders	25 (53.2)
Number of metastasis	
1	24 (51.06)
≥2	16 (34.04)
unknown	7 (14.90)
LDH [6] level at baseline	
≤1X [7] ULN	35 (74.5)
>1X ULN	12 (23.5)

[1] CII = checkpoint immune inhibitor. [2] Cell-free microRNAs = cfmiRs. [3] AJCC 8th stage = American Joint Committee on Cancer 8th edition determined at the start date of CII. [4] SD = standard deviation. [5] RECIST = Respond Evaluation Criteria In Solid Tumors. [6] LDH = lactate dehydrogenase. [7] ULN = upper limit normal.

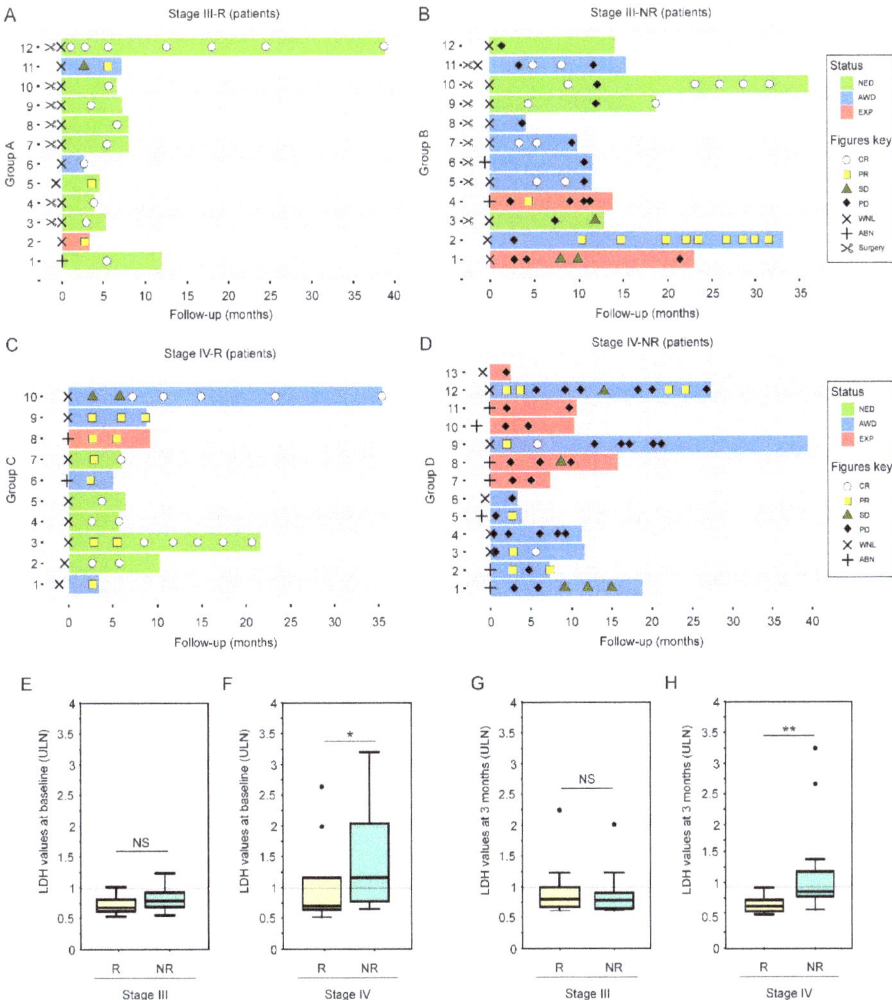

Figure 1. Melanoma patients and LDH assessment. (**A**,**B**) Swimmer plots showing disease status (NED, no evidence of disease; AWD, alive with disease; EXP, expired), RECIST 1.1 criteria (CR, complete response; PR, partial response; SD, stable disease; PD, progressive disease), surgery, and LDH levels (WNL, within normal; ABN, above normal) in stage III responder (Group A) (**A**) and non-responder (Group B) (**B**) melanoma patients. (**C**,**D**) Swimmer plots showing (NED, no evidence of disease; AWD, alive with disease; EXP, expired), RECIST 1.1 criteria (CR, complete response; PR, partial response; SD, stable disease; PD, progressive disease), surgery, and LDH levels (WNL, within normal; ABN, above normal) in stage IV responder (Group C) (**C**) and non-responder (Group D) (**D**) melanoma patients. (**E**) Boxplot showing the LDH values (ULN, upper limit normal) at baseline in stage III responder and non-responder melanoma patients (NS, non-significant). (**F**) Boxplot showing the LDH values (ULN) at baseline in stage IV responder and non-responder melanoma patients (* $p < 0.05$). (**G**) Boxplot showing the LDH values (ULN) at three months follow-up in stage III responder and non-responder melanoma patients (NS, non-significant). (**H**) Boxplot showing the LDH values (ULN) at three months follow-up in stage IV responder and non-responder melanoma patients (** $p < 0.01$). Dots represent outliers in each condition.

2.2. Identification of cfmiRs in Pre- and Post-Treatment Samples from Patients Non-Responsive to CII

In evaluating the utility of cfmiRs, it is important to find cfmiRs that have applicability in real-time monitoring of melanoma patient's disease status before and during CII(s) to evaluate response. Recently, we have shown that specific cfmiR patterns found in MBM patients' plasma may have utility in monitoring melanoma patients undergoing treatment [32]. Our hypothesis is that specific cfmiRs have a better utility compared to serum LDH levels in the assessment of melanoma patients undergoing CII. To address this hypothesis, pre-treatment samples (n = 13) from 13 stage IV melanoma patients who progressed (group D) were compared to normal donors' samples (n = 73). A total of 162 differentially expressed (DE) cfmiRs were observed in the melanoma samples, of which 89 were upregulated and 73 were downregulated. To determine which cfmiRs classify metastatic melanoma patients from normal donors' samples, we implemented a Random Forest algorithm to the 162 DE cfmiRs identified. The analysis generated a cfmiR classifier signature consisting of 12 cfmiRs (Figure 2A, Figure S1, and Table 2). To identify DE cfmiRs associated with disease progression during CII, 26 post-treatment samples collected from 13 stage IV non-responder (group D) melanoma patients were compared to normal donors' samples. In each analysis 215 and 202 DE cfmiRs were found. Random Forest algorithm was applied to the 215 and 202 DE cfmiRs identified (Figure 2A, Figures S2 and S3). The top and commonly identified nine cfmiRs were selected as potential cfmiR biomarkers to monitor disease progression on melanoma patients undergoing CII (Table 2 and Figure 2B). Then, the levels of those nine cfmiRs were compared in pre-treatment, post-treatment, and normal donors' samples. Of the nine cfmiRs identified, eight (miR-1234-3p, miR-3175, miR-4271, miR-4649-3p, miR-4745-3p, miR-615-3p, miR-6511-3p, and miR-6794-5p) were further evaluated since they showed significant changes in pre- and post-treatment samples from stage IV non-responders (Figure S4A–I). To summarize, using 13 paired blood samples (13 pre- and 26 post-treatment samples) nine cfmiRs were found as a potential biomarker for stage IV non-responder (group D) melanoma patients. Only eight of the nine cfmiRs were significantly DE in melanoma patients' compared to normal donors' samples.

2.3. MiR-615-3p Correlates with Melanoma Response to CII

To determine whether the cfmiRs identified in stage IV non-responder patients had clinical utility to monitor patients' treatment, we selected nine pre- and post-treatment samples from stage IV patients (group C) that responded to CII-treatment (achieved objective rate response, PR or CR). Of those nine patients, four reached CR (Figure 3A) and five patients had a partial response (PR) (Figure 3B). All of the samples were analyzed to determine the levels of the eight cfmiRs in the pre- and post-treatment samples. MiR-4649-3p, miR-1234-3p, and miR-615-3p levels significantly decreased in the post-treatment samples of the stage IV responder patients who had a CR (Figure 3C–E), but the levels did not change significantly for miR-3175, miR-4271, miR-4745-3p, miR-6511-3p, and miR-6794-5p (Figure S5A–E). On the contrary, in patients who had a PR no significant differences were observed in pre- and post-treatment samples for any of the eight cfmiRs assessed (Figure 3F–H and Figure S5F–J). To validate our observation, we assessed the expression levels of miR-615-3p in plasma samples from two stage IV responder and non-responder patients. Stage IV non-responder patients who progressed during CII had a significant increase in the expression of miR-615-3p (Figure 3I,J). In both cases LDH levels were unable to detect melanoma disease progression (Figure 3I,J). On the contrary, responder patients showed a decrease in miR-615-3p levels in post-treatment samples (Figure 3K,L). Similarly, LDH levels were also unable to detect CII response (Figure 3K,L). Then, we analyzed the detection levels of miR-615-3p for its ability to monitor stage III patients. To do that we compared pre- and post-treatment samples. The post-treatment samples were selected based on the patients' RECIST 1.1 criteria. All of the patients had PD at some point during treatment, but at the time point of blood collection, only 8 patients had PD. MiR-651-3p was significantly increased in melanoma patients with PD compared to pre-treatment samples (Figure S6A). More importantly, miR-615-3p was able to monitor stage III non-responder melanoma patients during CII (Figure S6B). Finally, we compared the expression of miR-615-3p in pre- and post-treatment samples from stage III responders. For the

12 post-treatment samples selected, the patients achieved CR at the time point of blood collection. No significant differences were observed for miR-615-3p in stage III responder patients (Figure S6C). Similar analysis were performed for miR-4649-3p and the results were consistent with those observed for miR-615-3p (Figure S7A–E). To summarize, the cfmiR signature was successful in identifying stage IV responders during CII-treatment. MiR-4649-3p, miR-1234-3p, and miR-615-3p levels were associated with a CR in stage IV patients undergoing CIIs and were useful in monitoring responses of stage IV melanoma patients undergoing CII. Also, the results demonstrated differences for miR-615-3p in detecting stage III patients with PD, but failed to identify stage III patients with CR.

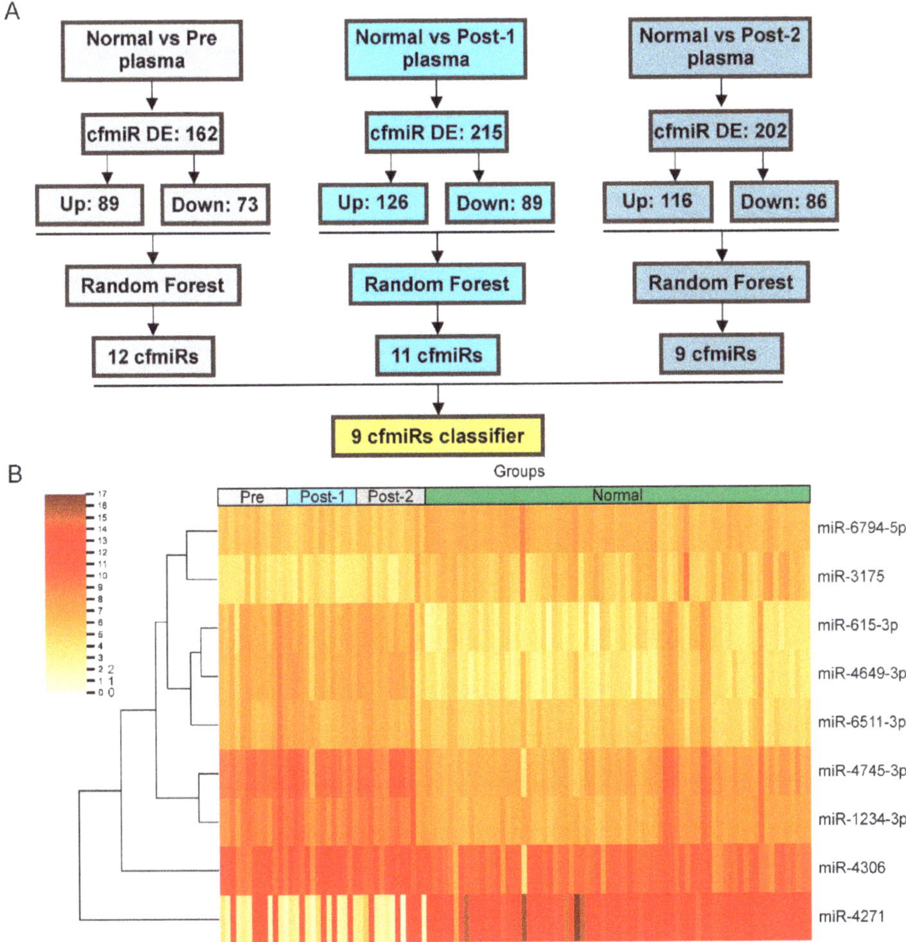

Figure 2. Identification of DE cfmiRs in normal donors' and melanoma patients' plasma samples (**A**) Shown are the DE cfmiRs in normal plasma samples versus pre- and post-treatment (Post-1 and Post-2) plasma samples. DE cfmiRs in each comparison were analyzed by the Random Forest algorithm. Specific classifiers were obtained for each analysis. A nine cfmiR classifier was commonly identified in all groups. (**B**) Heatmap showing the nine DE cfmiRs that were commonly identified in pre- and post-treatment plasma samples from stage IV non-responder melanoma patients. Scale bar showing the Log2 of normalized counts (ncounts).

Table 2. CfmiR [1] classifiers commonly identified by Random Forest in stage IV patients that had PD [2].

Probe	Pre-Treatment FIS [3]	Post-Treatment-1 FIS [3]	Post-Treatment-2 FIS [3]
miR-4271	0.06	0.05	0.07
miR-3175	0.05	0.05	0.04
miR-4745-3p	0.04	0.01	0.02
miR-6862-3p	0.03	N/A	N/A
miR-4649-3p	0.03	0.04	0.02
miR-6510-3p	0.02	0.01	N/A
miR-4306	0.01	0.02	0.02
miR-1234-3p	0.01	0.05	0.01
miR-6511a-3p	0.01	0.04	0.01
miR-615-3p	0.01	0.02	0.02
miR-6794-5p	0.01	0.02	0.03
miR-1301-5p	0.01	0.02	N/A

[1] CfmiR = cell-free miRNA. [2] PD = progressive disease. [3] FIS = Feature Importance Scores.

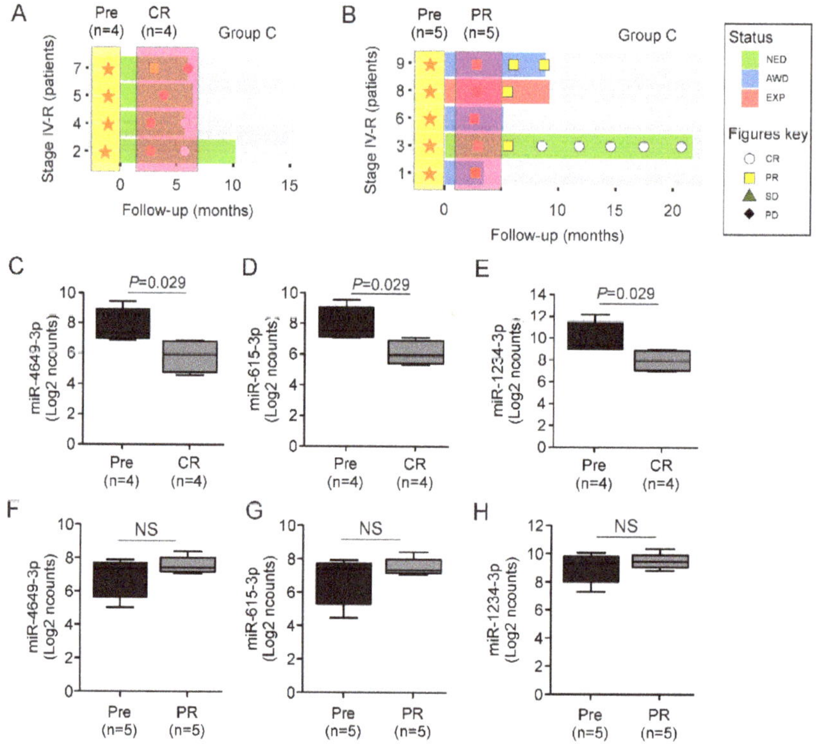

Figure 3. *Cont.*

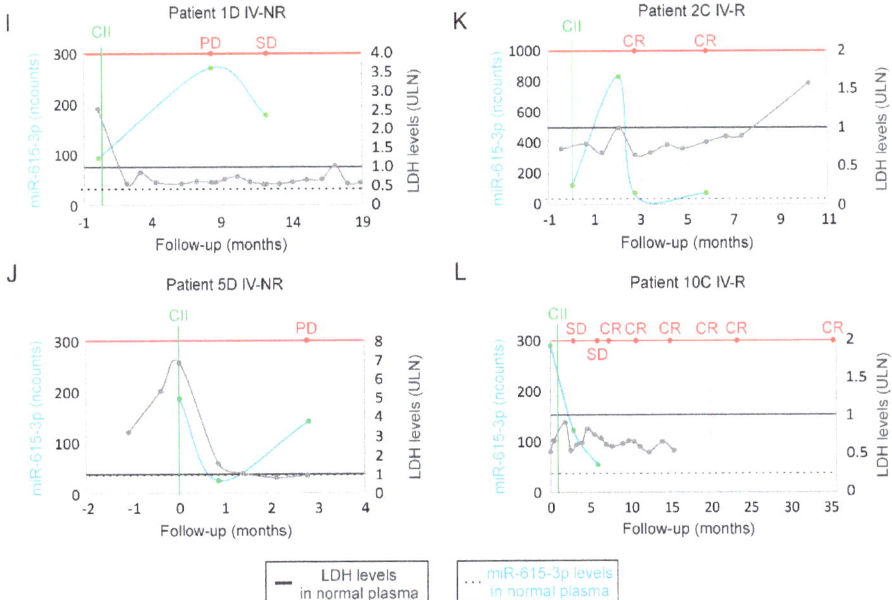

Figure 3. Validation of cfmiRs identified in assessing patient response to CIIs. (**A**,**B**) Disease status in stage IV patients who had a complete response (**A**, CR) or partial response (**B**, PR). Orange stars indicate pre-treatment (Pre) plasma samples. Red circles indicate blood collected in patients at CR. Red squares indicate blood collected in patients at PR. (**C**–**E**) Boxplots showing the changes in miR-4649-3p (**C**), miR-615-3p (**D**), and miR-1234-3p (**E**) levels in patients who achieved a CR (p values are indicated on top) compared to pre-treatment samples. (**F**–**H**) Boxplots showing the changes in miR-4649-3p (**F**) and miR-615-3p (**G**), and miR-1234-3p (**H**) levels in patients who achieved PR compared to pre-treatment samples (NS, non-significant). (**I**–**L**) Graph showing four melanoma patients: stage IV non-responder (IV-NR) patient 1D (**I**) or patient 5D (**J**); stage IV responders (IV-R) patient 2C (**K**) or patient 10C7 (**L**). Shown is the follow-up in months, LDH levels (labeled as light gray), and miR-615-3p levels (labeled as light blue; normalized counts, ncounts) at the indicated time points. Red line points to RECIST 1.1. Gray solid line indicates the upper limit normal (ULN) for LDH. Black dotted line indicates the average level of miR-615-3p detected in normal healthy donors' plasma samples. Green solid line indicates the start of CII.

2.4. A cfmiR Signature to Assess CII Responses in Stage III Melanoma Patients

To find specific cfmiRs associated with stage III and CII response, 24 stage III patients (22 stage IIIC and 2 stage IIIB) undergoing CII were examined, of which 12 were responders (group A; Figure 1A) and 12 were non-responders (group B; Figure 1B). Initially, we compared the cfmiR expression in pre-treatment samples taken from stage III responders (group A) versus non-responders (group B). Surprisingly, miR-3197 was the only significantly DE cfmiR in the comparison (Figure 4A,B). MiR-3197 differentiated stage III responders from non-responders in pre-treatment samples (Figure 4B). Additionally, miR-3197 showed significant differences when comparing pre-treatment samples from responders versus normal donors' samples (Figure 4B). This suggested that the cfmiRs detected in stage III patients are not significantly changing compared to normal donors' samples. This is likely related to low tumor burden and low doubling time of stage III tumors being treated with CII.

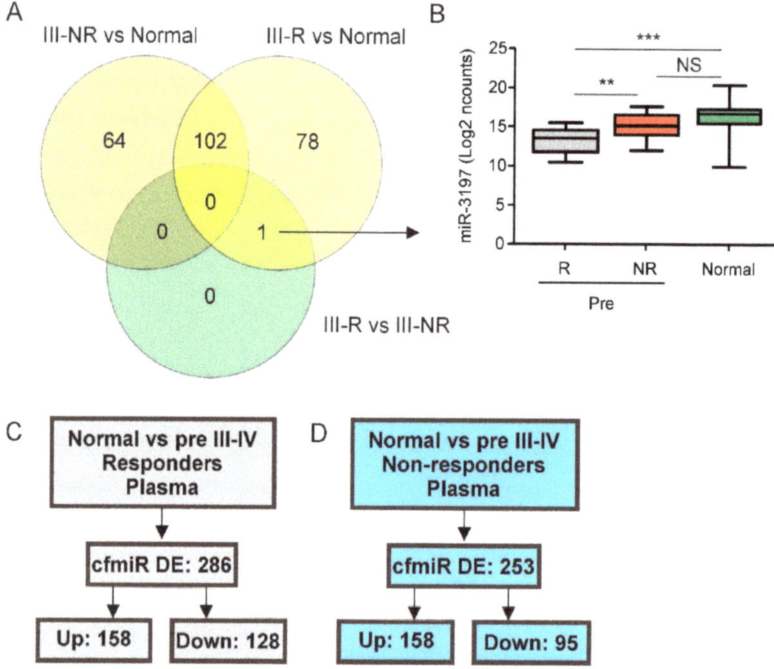

Figure 4. Characterization of cfmiRs in stage III melanoma patients. (**A**) Venn diagram showing the number of cfmiRs found in each comparison: stage III non-responders versus normal donors (III-NR vs. N); stage III responders versus normal donors (III-R vs. N); stage III responders versus non-responders (III-R vs. III-NR). (**B**) Boxplot showing the levels of miR-3197 in stage III responders (R), non-responders (NR), and normal donors' plasma samples. (** $p < 0.01$, *** $p < 0.001$, NS, non-significant). (**C**) Shown are the DE cfmiRs in normal donors' versus pre-treatment plasma samples from stage III/IV responder melanoma patients. (**D**) Shown are the DE cfmiRs in normal donors versus pre-treatment plasma samples in stage III/IV non-responders melanoma patients.

In order to find biomarkers to monitor metastatic melanoma patients and increase our sample size, we combined stage III/IV melanoma patients and grouped them as non-responders and responders to CII. Then, non-responder and responder samples were compared to normal donors' samples, respectively. A total of 286 DE cfmiRs (158 upregulated and 128 downregulated) were found in CII pre-treatment samples from the responder group compared to normal samples (Figure 4C). We then compared the pre-treatment samples from the non-responder patients versus normal donors' samples. In the analysis, 253 DE cfmiRs (158 upregulated and 95 downregulated) were observed in non-responder patients compared to normal donors' samples (Figure 4C). It is important to find cfmiRs that are useful for the monitoring of CII responses and to help distinguish metastatic melanoma responders from non-responders. Therefore, we focused on the detection of DE cfmiRs that were observed associated with non-response or response to CII. Therefore, we calculated the ratio of the FCs obtained in responders versus normal donors' samples and in non-responders versus normal donors' samples. Only cfmiRs with a ratio FC <0.75 were included. A total of 92 cfmiRs were DE in responders' compared to normal donors' and non-responders' samples (Table S1).

We proposed that specific cfmiRs have the potential to identify patients who will respond to CII. By applying the same strategy but considering a ratio FC >1.25, 17 DE cfmiRs were found in stage III/IV non-responders' compared to normal donors' and responders' samples (Table S2). The cfmiRs identified may represent potential biomarkers to determine patients who will likely develop PD to CII.

MiR-1273e, miR-584-5p, and miR-1290 were found increased in non-responders stage III/IV melanoma patients. Surprisingly, the same cfmiRs were also found elevated in pre-operative MBM plasma samples as previously described by our group [32]. To summarize, 92 cfmiRs found in pre-treatment samples distinguished stage III/IV responders' from non-responders' and normal donors' samples. On the contrary, 17 cfmiRs differentiated stage III/IV non-responders from responders and normal donors' samples.

3. Discussion

Notwithstanding the large number of clinical and translational research studies, there is still a dire need for more reliable and informative blood biomarkers to better evaluate CII responses in real-time in melanoma patients. Metastatic melanoma progression can be rapid once tumors develop resistance to CIIs and bypass the host systemic immune control. Better blood biomarkers that can identify real-time changes in the patient's disease status and allow for active monitoring could translate into earlier treatment decision making. There is evidence showing that higher baseline LDH values are associated with CII responses [13–15,33] and can allow for monitoring CII [15], but often the levels of LDH do not correlate with disease progression in patients receiving CII. Thus, it is difficult to rely on longitudinal LDH level assessment to make early clinical decisions in patients who are undergoing unsuccessful CII. Our study provides a detailed profiling of cfmiRs that potentially allow for the monitoring of stage III and IV melanoma patients during CII.

Despite the significant advances in improving progression-free survival (PFS) and OS, a high percentage of patients will still develop resistance and experience recurrence within the first year of starting CII [34]. Several studies have been conducted to identify miR biomarkers in melanoma tissues and/or plasma/serum that could predict melanoma progression [35,36]. However, most of the proposed cfmiRs are not validated or they represent single cfmiRs with limited reproducibility, and non-specific overlapping with benign diseases or normal healthy donor levels. In identifying biomarkers for CII response, some groups have focused on specific deregulated miRs in the tumor that can modulate the immune response against melanoma tumors, and thus control CII response. For example, miR-30b is upregulated in melanoma patients' tissues and correlates to different clinical variables such as stage, metastatic potential, and shorter OS. MiR-30b promotes immunosuppression by targeting *GALNT7* (N-Acetylgalactosaminyltransferase 7) and increasing IL-10 production [37]. We observed an increased level of circulating cfmiR-30b in both responder and non-responder melanoma stage III/IV patients compared to normal donors' samples. Also, an increase in cfmiR-30b levels was observed in responder compared to non-responder patients. In another study, miR-210 was shown to be upregulated in hypoxic areas of the tumor controlling cytotoxic T lymphocytes meditated lysis [38]. To summarize, miR-210 mediates its effects by targeting *PTPN1*, *HOXA1*, and *TP53I11* [38]. These studies support the role of elevated miRs in promoting melanoma progression in response to CII. However, the translational value of these findings into clinical biomarkers would require an assessment of the miRs in longitudinal biopsies of the tumor, which is not always feasible.

Blood biomarkers represent the most logistical and promising way to actively monitor patients in real-time during CII. Other studies have shown an eight cfmiR signature (miR-146a, miR-155, miR-125b, miR-100, let-7e, miR-125a, miR-146b, and miR-99b) found in extracellular vesicles released by metastatic melanoma tumors which were found to be associated with an increase in myeloid-derived suppressor cells and resistance to ipilimumab and nivolumab therapy [39]; however, not all of the cfmiRs identified were DE in melanoma patients when compared to normal donors' samples. Our findings revealed that cfmiRs (miR-1234-3p, miR-3175, miR-4271, miR-4649-3p, miR-4745-3p, miR-615-3p, miR-6511-3p, and miR-6794-5p) are detected in pre-treatment plasma samples. Only miR-1234-3p, miR-4649-3p and miR-615-3p were significantly enhanced in post-treatment samples from stage IV non-responder patients. Accordantly, miR-4649-3p, miR-1234-3p, and miR-615-3p decreased in post-treatment samples of stage IV responder patients who had a CR during CII. Whereas, no significant differences were observed in stage IV responder patients who had a PR in comparison to pre-treatment samples.

On longitudinal blood assessment, miR-615-3p and miR-4649-3p showed promising clinical utility in monitoring CII response in stage IV responder and non-responder patients. MiR-615-3p was previously detected and listed as a potential cfmiR for metastatic melanoma [40], but its function in melanoma is unknown [41]. To our knowledge, there is not report of the miR-4649-3p function in melanoma; however, it was previously reported that miR-4649-3p inhibits cell proliferation by targeting protein tyrosine phosphatase SHP-1 in nasopharyngeal carcinoma cells [42].

Identifying informative cfmiR biomarkers for stage IIIB-D melanoma is challenging, as the tumor size is variable ranging from multinodal micrometastasis to macrometastasis, with often variable growth rates. In this study, plasma samples derived from 24 melanoma patients undergoing CII (22 stage IIIC and 2 stage IIIB) were examined. As shown in previous studies [13–15,33] and in the present study, LDH baseline level assessment was successful in identifying most of stage IV patients, but it was not a reliable prognostic factor for stage III patients. When comparing stage III responders versus non-responders, only miR-3197 was found DE. Factors influencing the detection of cfmiRs changing could be associated with individual cfmiR variability, tumor burden, and tumor heterogeneity. To address this problem and to identify cfmiRs able to monitor metastatic melanoma, stage III and IV melanoma responders and non-responders were compared to normal donor samples. We found 92 cfmiRs associated with CII response. Whether any of these cfmiRs can be used as a robust biomarker will require further investigation. Also, 17 cfmiRs have shown potential applicability to determine stage III/IV melanoma patients who will not respond to CII. Relevant to this, we observed that miR-1273e, miR-584-5p, and miR-1290 have also been detected in MBM patients' plasma. These cfmiRs may be indicators of III/IV melanoma patients who will eventually develop MBM. Recently, Walbrecq et al. identified miR-1290 as a novel hypoxia-associated miR, which is highly abundant in hypoxic extracellular vesicles released by melanoma cells [43].

Similar to CII resistance, different mechanisms have been associated with BRAF and MEK inhibitors resistance in metastatic melanoma. Often, these mechanisms over-activate the mitogen-activated protein kinase (MAPK) pathway and overcome BRAF and MEK inhibitors effects [44–46]. It has been proposed that similar mechanisms may drive CII resistance, as MAPK pathway can be over-active in CII-treated melanoma tumors [45]. Thus, it is also important to determine whether elevated miRs regulate MAPK pathway. In a previous study, high levels of miR-125b-5p were shown to be associated with Vemurafenib (BRAF inhibitor) resistance [47,48]. Accordingly, we observed that miR-125b was elevated in CII non-responder patients, but decreased in responding patients to CII (Table S1). Thus, miR-125b-5p may represent an example of overlapping roles for miRs in promoting cross-resistance to both BRAF and MEK inhibitors and CII.

We understand the limitations of our study in analyzing melanoma patients receiving different types of CII. Therefore, these findings may represent cfmiRs associated with responses to different CII. Future analyses are required to confirm and validate whether the cfmiRs have the ability to determine treatment response in well-defined cohort of patients receiving specific CII. To the best of our knowledge, this is the first report showing the potential ability of cfmiRs to distinguish patients who are non-responsive to CII from normal donors' plasma samples. Further studies are needed to validate our observations in prospective clinical trials on larger sample sizes of metastatic melanoma patients undergoing CII(s).

4. Materials and Methods

4.1. Consent to Participate and Patient Specimen Accrual

This single-center study followed the principles found in the Declaration of Helsinki. All human samples and clinical information for this study were obtained according to the protocol guidelines approved by Saint John's Health Center (SJHC)/John Wayne Cancer Institute (JWCI) Joint Institutional Review Board (IRB): JWCI Universal Consent (Providence Health and Services Portland IRB: JWCI-18-0401) and Western IRB: MORD-RTPCR-0995. Informed consent was obtained from all participants. The study was a prospective study designed to assess cfNA in CII-treated patients seen at JWCI/SJHC. All specimens were de-identified and entered into a restricted access database by a database operator.

4.2. Blood Sample Collection

Blood samples of healthy donors and melanoma patients were prospectively collected at SJHC/JWCI. Briefly, all blood samples were collected from 2016–2020 in Streck blood collection tubes (Streck, La Vista, NE, USA). Blood samples were accrued and processed to obtain plasma. Plasma was centrifuged, filtered, aliquoted, barcoded, and cryopreserved at −80°C as previously described [16]. Aliquots of plasma were thawed only once, mixed, and centrifuged before being analyzed by HTG EdgeSeq miR WTA.

For HTG EdgeSeq miR WTA analysis plasma samples (n = 73) were collected from normal healthy donors ranging in age from 21–65 years old, of which 41 were females and 32 were males. Pre-treatment samples (n = 47) were collected from AJCC 8th stage III and IV melanoma patients who received CII (Table 1). From the same CII- treated patients 2–5 blood samples (n = 111) that were collected post-treatment (after the first dose and during CII). All plasma samples were analyzed by HTG EdgeSeq miR WTA analysis. Overall, 158 melanoma plasma samples were analyzed from 47 patients. The melanoma patients analyzed had detailed clinical follow-up information and treatment response assessments as described in Section 4.3 below. The clinical demographics information for the 47 melanoma patients analyzed is summarized in Table 1.

4.3. CII Response

Every patient had a follow-up evaluation at the JWCI/SJHC cancer clinic as recommended in the current standard of care. The median follow-up was 9.7 months for the 47 patients analyzed. Each patient included in the study received at least four doses of the approved CII drugs (ipilimumab, nivolumab, pembrolizumab, or the combination of ipilimumab and nivolumab [16]) and were assessed for the RECIST 1.1. Briefly, CII responses were assessed using computerized tomography/magnetic resonance imaging every three months according to RECIST 1.1 criteria, denoting PD, SD (stable disease), PR, and CR. Based on RECIST 1.1 criteria the patients were stratified into responders (PR/CR) and non-responders (PD). Stage III patients who received surgery before receiving CII were considered NED until evaluated for RECIST 1.1. Stage III patients (3A, 4A, 7A, 8A, 9A, 10A, and 12A) from group A and patients (3B, 5B, 6B, 7B, 8B, 9B, 10B, and 11B) from group B received surgery and adjuvant treatment. This prospective study was performed in accordance with the REMARK guidelines [49,50].

LDH levels were evaluated using Dimension Vista LDH (LDI) Flex reagent cartridge (cat# K2054) an in vitro diagnostic test for the quantitative measurement of LDH in human serum on the Dimension Vista System analyzer (Siemens Medical Solutions Inc., PA, USA) at the SJHC Clinical Chemistry Department. LDH baseline and subsequent values were obtained for each patient. At least 3 LDH values were collected at different time points for all of the patients. Elevated LDH levels were considered in patients with >1X ULN (>240 U/L).

4.4. Sample Processing for HTG WTA

Melanoma patients' and normal donors' plasma samples were computer coded and de-identified during processing and assessing. The melanoma patients' and normal donors' plasma samples processing and NGS library preparation, quality control, normalized, and pooled was performed as described previously [32]. The pool library was sequenced on MiSeq or NextSeq 550 instruments following the respective Illumina instrument sequencing protocols. FASTQ files were generated from raw sequencing data using Illumina BaseSpace BCL to FASTQ software version 2.2.0 and Illumina Local Run Manager Software version 2.0.0. FASTQ files were analyzed with HTG EdgeSeq Parser software version v5.1.724.4793 to generate raw counts for 2083 miRs per sample. An .xls file containing the final counts for 2083 miRs per sample was generated for downstream data analysis. Data normalization was performed as discussed in Biostatistical analysis. Each HTG EdgeSeq miR WTA included negative (CTRL_ANT1, CTRL_ANT2, CTRL_ANT3, CTRL_ANT4, CTRL_ANT5) and positive (CTRL_miR_POS) miR controls. In all runs, Human Brain Total RNA (Ambion, Inc., Austin, TX, USA) was used as a control for library preparation, but they were not sequenced. All the samples that did not pass the quality control set by the HTG REVEAL software version 2.0.1 (Tuscon, AR, USA) were excluded from the analysis.

4.5. Biostatistical Analysis

The DESeq2 data normalization, analyses, and statistical comparisons for the melanoma (pre- and post-treatment) and normal donors' plasma samples were all performed using the HTG REVEAL software version 2.0.1. In all of the comparison only cfmiRs with a $\log_2$ fold-change (Log_2FC) >1.2 or <−1.2, a false discovery rate (FDR) > 0.05, and normalized counts greater than 30 were only considered. Ratio of the FCs was calculated by dividing the FC in non-responder versus normal to the FC of responder versus normal. Data normal distribution was evaluated by Shapiro-Wilk normality test. According to data normal distribution Kruskal-Wallis (non-normal distribution) tests were performed to determine differences among three or more groups. Mann-Whitney U-test (non-normal distribution) analysis was performed to compare differences between two groups. Box plots were performed with GraphPad Prism 5 (GraphPad Software Inc., La Jolla, CA, USA). To visualize the sequence and duration of treatments, patient response, and LDH levels, swimmer plots were employed using ggplot2 package version 3.3.2.9000 [51,52]. The swimmer plots were carried out using R version 4.0.0 (R Core Team) [52]. Data processing and Random Forest algorithm were performed using Python 3.7.7 using Scikit-learn and other packages as previously described in [32]. A two-sided p-value < 0.05 was considered statistically significant: * $p < 0.05$; ** $p < 0.01$; *** $p < 0.001$, and a p-value > 0.05 was considered non-significant (NS). The figures were processed using CorelDraw graphics suite 8X (Corel Corporation, Ottawa, Canada).

4.6. Data Deposit

The data generated and discussed in this study has been deposited in the NCBI's Gene Expression Omnibus (GEO) and is accessible through the GEO Series accession number GSE157370.

5. Conclusions

In this prospective study, specific cfmiR signatures were found in plasma samples from metastatic melanoma patients. Three cfmiRs that were elevated in pre- and post-treatment plasma samples of stage IV non-responder patients were found to be downregulated in post-treatment plasma samples from patients who responded to CII and vice versa (see the Graphical Abstract). In addition, we proposed cfmiRs that may have the potential prognostic value to assess stage III/IV melanoma patients who will progress during CII. The present pilot study revealed specific cfmiRs that can help in monitoring CII response. MiR-615-3p and miR-4649-3p demonstrated a higher efficiency than LDH at baseline or during CII to monitor stage IV patients undergoing CII.

Supplementary Materials: The following are available online at http://www.mdpi.com/2072-6694/12/11/3361/s1, Figure S1: Random Forest algorithm in pre-treatment samples from stage IV non-responder patients, Figure S2: Random Forest algorithm in post-treatment-1 samples from stage IV non-responder patients, Figure S3: Random Forest algorithm in post-treatment-2 samples from stage IV non-responder patients, Figure S4: CfmiRs detection levels in pre- and post-treatment-1 and 2 samples from stage IV non-responder patients, Figure S5: CfmiR levels in pre- and post-CII-treated melanoma patients, Figure S6: MiR-615-3p levels in longitudinal bloods from stage III non-responder patients, Figure S7: MiR-4649-3p levels in longitudinal bloods from stage IV non-responder patients, Table S1: CfmiR identified in pre-treatment samples of stage III/IV responders, Table S2: CfmiR identified in pre-treatment samples of stage III/IV non-responders.

Author Contributions: M.A.B.: conceptualization and design, methodology, data analysis and interpretation, writing original draft preparation, reviewing, and editing; R.G.: specimen and clinical data organization and procurement, reviewing and editing; N.R.: bioinformatics analysis and visualization, reviewing, and editing; H.C.: data acquisition, clinical data, and blood specimens consenting and procurement, and reviewing; L.T.T.: HTG assays data acquisition, reviewing, and editing; K.D.T.: HTG assays, data acquisition, reviewing, and editing; L.T.: blood sample processing, plasma isolation, and reviewing; S.L.S.: clinical database mining and organization, and reviewing; S.O.: patient treatment, clinical data, and evaluating patients clinical status, and reviewing; D.S.B.H.: conceptualization and design, supervision of the study, funding, writing, reviewing, and editing. All authors have read and agreed to the published version of the manuscript.

Funding: This research was funded by Miriam and Sheldon G. Adelson Medical Research Foundation and Gonda Foundation award to D.S.B.H.; the Fashion Footwear Association of New York (FFANY) award to M.A.B. and D.S.B.H.; the Borstein Family Foundation to S.O.

Acknowledgments: The authors thank the Department of Translational Molecular Medicine staff at JWCI and the cancer clinic and pathology staff at SJHC for their kind advisory and technical assistance.

Conflicts of Interest: S.O. is an Advisory Board: Biothera, BMS, BionTech, Exicure, Immunsys, Merck; Consultant: Agenus, Biothera, Immunsys; Research: Agenus, Amgen, Biothera, BMS, Exicure, Genocea, Incyte Merck, Ultimovacs, Viralytics; and Speaker's Bureau: BMS. All other authors declare no competing interests.

Abbreviations

AJCC	American Joint Committee on Cancer
CfmiR	Cell-free miR
CfDNA	Cell-free DNA
CfNA	Cell-free Nucleic Acids
CII	Checkpoint Inhibitor Immunotherapy
CR	Complete Response
CTLA-4	Cytotoxic T Lymphocyte-Associated Antigen 4
IRAE	Immune-Related Adverse Events
DE	Differentially Expressed
FC	Fold-Change
FDR	False Discovery Rate
GEO	Gene Expression Omnibus
MBM	Melanoma Brain Metastasis
miR	MicroRNA
NED	No Evidence of Disease
NGS	Next-Generation Sequencing
PD-1	Programmed Cell Death Protein-1
PD-L1	Programmed Cell Death Protein-1 Ligand
PD	Progressive Disease
PFS	Progressive-Free Survival
PR	Partial Response
NS	Non-Significant
RECIST	Response Evaluation Criteria In Solid Tumors 1.1
REMARK	Reporting Recommendations For Tumour Marker
SD	Stable Disease
LDH	Lactate Dehydrogenase
ULN	Upper Limit Normal
WTA	Whole Transcriptome Assay

References

1. Berk-Krauss, J.; Stein, J.A.; Weber, J.; Polsky, D.; Geller, A.C. New Systematic Therapies and Trends in Cutaneous Melanoma Deaths among US Whites, 1986–2016. *Am. J. Public Health* **2020**, *110*, 731–733. [CrossRef]
2. Robert, C.; Thomas, L.; Bondarenko, I.; O'Day, S.; Weber, J.; Garbe, C.; Lebbe, C.; Baurain, J.-F.; Testori, A.; Grob, J.-J.; et al. Ipilimumab plus Dacarbazine for Previously Untreated Metastatic Melanoma. *N. Engl. J. Med.* **2011**, *364*, 2517–2526. [CrossRef] [PubMed]

3. Schachter, J.; Ribas, A.; Long, G.V.; Arance, A.; Grob, J.J.; Mortier, L.; Daud, A.; Carlino, M.S.; McNeil, C.; Lotem, M.; et al. Pembrolizumab versus ipilimumab for advanced melanoma: Final overall survival results of a multicentre, randomised, open-label phase 3 study (KEYNOTE-006). *Lancet* **2017**, *390*, 1853–1862. [CrossRef]
4. Ribas, A.; Lawrence, D.; Atkinson, V.; Agarwal, S.; Miller, W.H., Jr.; Carlino, M.S.; Fisher, R.; Long, G.V.; Hodi, F.S.; Tsoi, J.; et al. Combined BRAF and MEK inhibition with PD-1 blockade immunotherapy in BRAF-mutant melanoma. *Nat. Med.* **2019**, *25*, 936–940. [CrossRef] [PubMed]
5. Robert, C.; Ribas, A.; Hamid, O.; Daud, A.; Wolchok, J.; Joshua, A.M.; Hwu, W.-J.; Weber, J.S.; Gangadhar, T.C.; Joseph, R.W.; et al. Durable Complete Response After Discontinuation of Pembrolizumab in Patients With Metastatic Melanoma. *J. Clin. Oncol.* **2018**, *36*, 1668–1674. [CrossRef]
6. Haslam, A.; Prasad, V. Estimation of the Percentage of US Patients With Cancer Who Are Eligible for and Respond to Checkpoint Inhibitor Immunotherapy Drugs. *JAMA Netw. Open* **2019**, *2*, e192535. [CrossRef]
7. Mckean, W.B.; Moser, J.C.; Rimm, D.; Hu-Lieskovan, S. Biomarkers in Precision Cancer Immunotherapy: Promise and Challenges. *Am. Soc. Clin. Oncol. Educ. Book* **2020**, *40*, e275–e291. [CrossRef]
8. Gide, T.N.; Wilmott, J.S.; Scolyer, R.A.; Long, G.V. Primary and Acquired Resistance to Immune Checkpoint Inhibitors in Metastatic Melanoma. *Clin. Cancer Res.* **2017**, *24*, 1260–1270. [CrossRef]
9. Van Wilpe, S.; Koornstra, R.; Brok, M.D.; De Groot, J.W.; Blank, C.; De Vries, J.; Gerritsen, W.; Mehra, N. Lactate dehydrogenase: A marker of diminished antitumor immunity. *OncoImmunology* **2020**, *9*, 1731942. [CrossRef]
10. Mishra, D.; Banerjee, D. Lactate Dehydrogenases as Metabolic Links between Tumor and Stroma in the Tumor Microenvironment. *Cancers* **2019**, *11*, 750. [CrossRef]
11. Amin, M.B.; Greene, F.L.; Edge, S.B.; Compton, C.C.; Gershenwald, J.E.; Brookland, R.K.; Meyer, L.; Gress, D.M.; Byrd, D.R.; Winchester, D.P. The Eighth Edition AJCC Cancer Staging Manual: Continuing to build a bridge from a population-based to a more "personalized" approach to cancer staging. *CA Cancer J. Clin.* **2017**, *67*, 93–99. [CrossRef] [PubMed]
12. Egberts, F.; Hitschler, W.N.; Weichenthal, M.; Hauschild, A. Prospective monitoring of adjuvant treatment in high-risk melanoma patients: Lactate dehydrogenase and protein S-100B as indicators of relapse. *Melanoma Res.* **2009**, *19*, 31–35. [CrossRef] [PubMed]
13. Kelderman, S.; Heemskerk, B.; Van Tinteren, H.; Brom, R.R.H.V.D.; Hospers, G.A.P.; Eertwegh, A.J.M.V.D.; Kapiteijn, E.W.; De Groot, J.W.B.; Soetekouw, P.; Jansen, R.L.; et al. Lactate dehydrogenase as a selection criterion for ipilimumab treatment in metastatic melanoma. *Cancer Immunol. Immunother.* **2014**, *63*, 449–458. [CrossRef] [PubMed]
14. Wagner, N.B.; Forschner, A.; Leiter, U.; Garbe, C.; Eigentler, T. S100B and LDH as early prognostic markers for response and overall survival in melanoma patients treated with anti-PD-1 or combined anti-PD-1 plus anti-CTLA-4 antibodies. *Br. J. Cancer* **2018**, *119*, 339–346. [CrossRef] [PubMed]
15. Diem, S.; Kasenda, B.; Spain, L.; Martin-Liberal, J.; Marconcini, R.; Gore, M.; Larkin, J. Serum lactate dehydrogenase as an early marker for outcome in patients treated with anti-PD-1 therapy in metastatic melanoma. *Br. J. Cancer* **2016**, *114*, 256–261. [CrossRef]
16. Lin, S.Y.; Huang, S.K.; Huynh, K.T.; Salomon, M.P.; Chang, S.-C.; Marzese, D.M.; Lanman, R.B.; Talasaz, A.; Hoon, D.S. Multiplex Gene Profiling of Cell-Free DNA in Patients With Metastatic Melanoma for Monitoring Disease. *JCO Precis. Oncol.* **2018**, *2*, 1–30. [CrossRef]
17. Goh, J.Y.; Feng, M.; Wang, W.; Oguz, G.; Yatim, S.M.J.M.; Lee, P.L.; Bao, Y.; Lim, T.H.; Wang, P.; Tam, W.L.; et al. Chromosome 1q21.3 amplification is a trackable biomarker and actionable target for breast cancer recurrence. *Nat. Med.* **2017**, *23*, 1319–1330. [CrossRef]
18. Leung, F.; Kulasingam, V.; Diamandis, E.P.; Hoon, D.S.; Kinzler, K.; Pantel, K.; Alix-Panabières, C. Circulating Tumor DNA as a Cancer Biomarker: Fact or Fiction? *Clin. Chem.* **2016**, *62*, 1054–1060. [CrossRef]
19. Huynh, K.; Hoon, D.S. Liquid Biopsies for Assessing Metastatic Melanoma Progression. *Crit. Rev. Oncog.* **2016**, *21*, 141–154. [CrossRef]
20. Huang, S.K.; Hoon, D.S. Liquid biopsy utility for the surveillance of cutaneous malignant melanoma patients. *Mol. Oncol.* **2016**, *10*, 450–463. [CrossRef]

21. Hoshimoto, S.; Faries, M.B.; Morton, D.L.; Shingai, T.; Kuo, C.; Wang, H.-J.; Elashoff, R.; Mozzillo, N.; Kelley, M.C.; Thompson, J.F.; et al. Assessment of Prognostic Circulating Tumor Cells in a Phase III Trial of Adjuvant Immunotherapy After Complete Resection of Stage IV Melanoma. *Ann. Surg.* **2012**, *255*, 357–362. [CrossRef] [PubMed]
22. Hoshimoto, S.; Shingai, T.; Morton, D.L.; Kuo, C.; Faries, M.B.; Chong, K.; Elashoff, D.; Wang, H.-J.; Elashoff, R.M.; Hoon, D.S. Association Between Circulating Tumor Cells and Prognosis in Patients with Stage III Melanoma with Sentinel Lymph Node Metastasis in a Phase III International Multicenter Trial. *J. Clin. Oncol.* **2012**, *30*, 3819–3826. [CrossRef] [PubMed]
23. Lin, S.Y.; Chang, S.-C.; Lam, S.; Ramos, R.I.; Tran, K.; Ohe, S.; Salomon, M.P.; Bhagat, A.A.S.; Lim, C.T.; Fischer, T.D.; et al. Prospective Molecular Profiling of Circulating Tumor Cells from Patients with Melanoma Receiving Combinatorial Immunotherapy. *Clin. Chem.* **2019**, *66*, 169–177. [CrossRef] [PubMed]
24. Fleischhacker, M.; Schmidt, B. Circulating nucleic acids (CNAs) and cancer—A survey. *Biochim. Biophys. Acta (BBA)-Bioenerg.* **2007**, *1775*, 181–232. [CrossRef]
25. Schwarzenbach, H.; Hoon, D.S.B.; Pantel, K. Cell-free nucleic acids as biomarkers in cancer patients. *Nat. Rev. Cancer* **2011**, *11*, 426–437. [CrossRef]
26. Ono, S.; Oyama, T.; Lam, S.; Chong, K.; Foshag, L.J.; Hoon, D.S. A direct plasma assay of circulating microRNA-210 of hypoxia can identify early systemic metastasis recurrence in melanoma patients. *Oncotarget* **2015**, *6*, 7053–7064. [CrossRef]
27. Iida, Y.; Ciechanover, A.; Marzese, D.M.; Hata, K.; Bustos, M.; Ono, S.; Wang, J.; Salomon, M.P.; Tran, K.; Lam, S.; et al. Epigenetic Regulation of KPC1 Ubiquitin Ligase Affects the NF-κB Pathway in Melanoma. *Clin. Cancer Res.* **2017**, *23*, 4831–4842. [CrossRef]
28. Bustos, M.A.; Ono, S.; Marzese, D.M.; Oyama, T.; Iida, Y.; Cheung, G.; Nelson, N.; Hsu, S.C.; Yu, Q.; Hoon, D.S.B. MiR-200a Regulates CDK4/6 Inhibitor Effect by Targeting CDK6 in Metastatic Melanoma. *J. Investig. Dermatol.* **2017**, *137*, 1955–1964. [CrossRef]
29. Gajos-Michniewicz, A.; Czyz, M. Role of miRNAs in Melanoma Metastasis. *Cancers* **2019**, *11*, 326. [CrossRef]
30. Mumford, S.L.; Towler, B.P.; Pashler, A.L.; Gilleard, O.; Martin, Y.; Newbury, S.F. Circulating MicroRNA Biomarkers in Melanoma: Tools and Challenges in Personalised Medicine. *Biomolecules* **2018**, *8*, 21. [CrossRef]
31. Thyagarajan, A.; Tsai, K.Y.; Sahu, R.P. MicroRNA heterogeneity in melanoma progression. *Semin. Cancer Biol.* **2019**, *59*, 208–220. [CrossRef] [PubMed]
32. Bustos, M.A.; Tran, K.D.; Rahimzadeh, N.; Gross, R.; Lin, S.Y.; Shoji, Y.; Murakami, T.; Boley, C.L.; Tran, L.T.; Cole, H.; et al. Integrated Assessment of Circulating Cell-Free MicroRNA Signatures in Plasma of Patients with Melanoma Brain Metastasis. *Cancers* **2020**, *12*, 1692. [CrossRef] [PubMed]
33. Deichmann, M.; Benner, A.; Bock, M.; Jäckel, A.; Uhl, K.; Waldmann, V.; Näher, H. S100-Beta, Melanoma-Inhibiting Activity, and Lactate Dehydrogenase Discriminate Progressive From Nonprogressive American Joint Committee on Cancer Stage IV Melanoma. *J. Clin. Oncol.* **1999**, *17*, 1891. [CrossRef] [PubMed]
34. Yokota, K.; Uchi, H.; Uhara, H.; Yoshikawa, S.; Takenouchi, T.; Inozume, T.; Ozawa, K.; Ihn, H.; Fujisawa, Y.; Qureshi, A.; et al. Adjuvant therapy with nivolumab versus ipilimumab after complete resection of stage III / IV melanoma: Japanese subgroup analysis from the phase 3 CheckMate 238 study. *J. Dermatol.* **2019**, *46*, 1197–1201. [CrossRef] [PubMed]
35. Hanniford, D.; Zhong, J.; Koetz, L.; Gaziel-Sovran, A.; Lackaye, D.J.; Shang, S.; Pavlick, A.C.; Shapiro, R.; Berman, R.S.; Darvishian, F.; et al. A miRNA-Based Signature Detected in Primary Melanoma Tissue Predicts Development of Brain Metastasis. *Clin. Cancer Res.* **2015**, *21*, 4903–4912. [CrossRef] [PubMed]
36. Ba, N.H.F.; Zhong, J.; Silva, I.C.; De Miera, E.V.-S.; Brady, B.; Han, S.W.; Hanniford, D.; Wang, J.; Shapiro, R.L.; Hernando, E.; et al. Serum-based miRNAs in the prediction and detection of recurrence in melanoma patients. *Cancer* **2014**, *121*, 51–59. [CrossRef]
37. Gaziel-Sovran, A.; Segura, M.F.; Di Micco, R.; Collins, M.K.; Hanniford, D.; De Miera, E.V.-S.; Rakus, J.F.; Dankert, J.F.; Shang, S.; Kerbel, R.S.; et al. miR-30b/30d Regulation of GalNAc Transferases Enhances Invasion and Immunosuppression during Metastasis. *Cancer Cell* **2011**, *20*, 104–118. [CrossRef]
38. Noman, M.Z.; Buart, S.; Romero, P.; Ketari, S.; Janji, B.; Mari, B.; Mami-Chouaib, F.; Chouaib, S. Hypoxia-Inducible miR-210 Regulates the Susceptibility of Tumor Cells to Lysis by Cytotoxic T Cells. *Cancer Res.* **2012**, *72*, 4629–4641. [CrossRef]

39. Huber, V.; Vallacchi, V.; Fleming, V.; Hu, X.; Cova, A.; Dugo, M.; Shahaj, E.; Sulsenti, R.; Vergani, E.; Filipazzi, P.; et al. Tumor-derived microRNAs induce myeloid suppressor cells and predict immunotherapy resistance in melanoma. *J. Clin. Investig.* **2018**, *128*, 5505–5516. [CrossRef]
40. Carpi, S.; Polini, B.; Fogli, S.; Nieri, P.; Romanini, A. Circulating MicroRNAs in Cutaneous Melanoma Diagnosis and Prognosis. Management Malignant Melanoma. 2016. Available online: http://www.smgebooks.com/management-of-malignant-melanoma/chapters/MMM-16-02.pdf (accessed on 8 September 2020).
41. Godínez-Rubí, M.; Ortuño-Sahagún, D. miR-615 Fine-Tunes Growth and Development and Has a Role in Cancer and in Neural Repair. *Cells* **2020**, *9*, 1566. [CrossRef]
42. Pan, X.; Peng, G.; Liu, S.; Sun, Z.; Zou, Z.-W.; Wu, G. MicroRNA-4649-3p inhibits cell proliferation by targeting protein tyrosine phosphatase SHP-1 in nasopharyngeal carcinoma cells. *Int. J. Mol. Med.* **2015**, *36*, 559–564. [CrossRef] [PubMed]
43. Walbrecq, G.; Lecha, O.; Gaigneaux, A.; Fougeras, M.R.; Philippidou, D.; Margue, C.; Nomigni, M.T.; Bernardin, F.; Dittmar, G.; Behrmann, I.; et al. Hypoxia-Induced Adaptations of miRNomes and Proteomes in Melanoma Cells and Their Secreted Extracellular Vesicles. *Cancers* **2020**, *12*, 692. [CrossRef] [PubMed]
44. Da-Via, M.; Solimando, A.G.; Garitano-Trojaola, A.; Barrio, S.; Munawar, U.; Strifler, S.; Haertle, L.; Rhodes, N.; Teufel, E.; Vogt, C.; et al. CIC Mutation as a Molecular Mechanism of Acquired Resistance to Combined BRAF-MEK Inhibition in Extramedullary Multiple Myeloma with Central Nervous System Involvement. *Oncology* **2019**, *25*, 112–118. [CrossRef]
45. Motti, M.L.; Minopoli, M.; Di Carluccio, G.; Ascierto, P.A.; Carriero, M.V. MicroRNAs as Key Players in Melanoma Cell Resistance to MAPK and Immune Checkpoint Inhibitors. *Int. J. Mol. Sci.* **2020**, *21*, 4544. [CrossRef] [PubMed]
46. Wang, B.; Krall, E.B.; Aguirre, A.J.; Kim, M.; Widlund, H.R.; Doshi, M.B.; Sicinska, E.; Sulahian, R.; Goodale, A.; Cowley, G.S.; et al. ATXN1L, CIC, and ETS Transcription Factors Modulate Sensitivity to MAPK Pathway Inhibition. *Cell Rep.* **2017**, *18*, 1543–1557. [CrossRef]
47. Vergani, E.; Di Guardo, L.; Dugo, M.; Rigoletto, S.; Tragni, G.; Ruggeri, R.; Perrone, F.; Tamborini, E.; Gloghini, A.; Arienti, F.; et al. Overcoming melanoma resistance to vemurafenib by targeting CCL2-induced miR-34a, miR-100 and miR-125b. *Oncotarget* **2015**, *7*, 4428–4441. [CrossRef]
48. Varrone, F.; Caputo, E. The miRNAs Role in Melanoma and in Its Resistance to Therapy. *Int. J. Mol. Sci.* **2020**, *21*, 878. [CrossRef]
49. McShane, L.M.; for the Statistics Subcommittee of the NCI-EORTC Working Group on Cancer Diagnostics; Altman, D.G.; Sauerbrei, W.; Taube, S.E.; Gion, M.; Clark, G.M. REporting recommendations for tumour MARKer prognostic studies (REMARK). *Br. J. Cancer* **2005**, *93*, 387–391. [CrossRef]
50. Sauerbrei, W.; Taube, S.E.; McShane, L.M.; Cavenagh, M.M.; Altman, D.G. Reporting Recommendations for Tumor Marker Prognostic Studies (REMARK): An Abridged Explanation and Elaboration. *J. Natl. Cancer Inst.* **2018**, *110*, 803–811. [CrossRef]
51. Chia, P.L.; Gedye, C.; Boutros, P.C.; Wheatley-Price, P.; John, T. Current and Evolving Methods to Visualize Biological Data in Cancer Research. *J. Natl. Cancer Inst.* **2016**, *108*, djw031. [CrossRef]
52. *R: A Language and Environment for Statistical Computing*; R Foundation for Statistical Computing: Vienna, Austria, 2020. Available online: https://www.gbif.org/tool/81287/r-a-language-and-environment-for-statistical-computing (accessed on 18 August 2020).

Publisher's Note: MDPI stays neutral with regard to jurisdictional claims in published maps and institutional affiliations.

MDPI
St. Alban-Anlage 66
4052 Basel
Switzerland
Tel. +41 61 683 77 34
Fax +41 61 302 89 18
www.mdpi.com

Cancers Editorial Office
E-mail: cancers@mdpi.com
www.mdpi.com/journal/cancers

www.ingramcontent.com/pod-product-compliance
Lightning Source LLC
LaVergne TN
LVHW071523170726
843515LV00008B/2044

* 9 7 8 3 0 3 6 5 4 4 1 6 8 *